Ant-Plant Interactions

Ants are probably the most dominant insect family on earth, and flowering plants have been the dominant plant group on land for more than 100 million years. In recent decades, human activities have degraded natural environments with unparalleled speed and scale, making it increasingly apparent that interspecific interactions vary not only under different ecological conditions and across habitats, but also according to anthropogenic global change. This is the first volume entirely devoted to the anthropogenic effects on the interactions between these two major components of terrestrial ecosystems. A first-rate team of contributors report their research from a variety of temperate and tropical ecosystems worldwide, including South, Central, and North America, Africa, Japan, Polynesia, Indonesia, and Australia.

The volume provides an in-depth summary of the current understanding for researchers already acquainted with insect-plant interactions, yet is written at a level that offers a window into the ecology of ant-plant interactions for the mostly uninitiated international scientific community.

Paulo S. Oliveira is Professor of Ecology at the Universidade Estadual de Campinas, São Paulo, Brazil. His research is on the natural history, ecology, and behavior of tropical ants. He is past President and Honorary Fellow of the Association for Tropical Biology and Conservation. He has coauthored and coedited two books.

Suzanne Koptur is Professor of Biological Sciences at Florida International University, USA, teaching courses in ecology and botany to undergraduate and graduate students. She has published extensively on plant-animal interactions, especially antiherbivore defense of plants and pollination ecology.

Ant-Plant Interactions

Impacts of Humans on Terrestrial Ecosystems

EDITED BY

PAULO S. OLIVEIRA
Universidade Estadual de Campinas

SUZANNE KOPTUR
Florida International University

CAMBRIDGE
UNIVERSITY PRESS

University Printing House, Cambridge CB2 8BS, United Kingdom

One Liberty Plaza, 20th Floor, New York, NY 10006, USA

477 Williamstown Road, Port Melbourne, VIC 3207, Australia

4843/24, 2nd Floor, Ansari Road, Daryaganj, Delhi – 110002, India

79 Anson Road, #06-04/06, Singapore 079906

Cambridge University Press is part of the University of Cambridge.

It furthers the University's mission by disseminating knowledge in the pursuit of education, learning, and research at the highest international levels of excellence.

www.cambridge.org
Information on this title: www.cambridge.org/9781107159754
DOI: 10.1017/9781316671825

© Cambridge University Press 2017

First published 2017

Printed in the United Kingdom by TJ International Ltd. Padstow Cornwall in July 2017

A catalogue record for this publication is available from the British Library.

Library of Congress Cataloging-in-Publication Data
Names: Oliveira, Paulo S., editor. | Koptur, Suzanne, editor.
Title: Ant-plant interactions: impacts of humans on terrestrial ecosystems /
edited by Paulo S. Oliveira, Universidade Estadual de Campinas, São Paulo,
Brazil, Suzanne Koptur, Florida International University.
Description: Cambridge: Cambridge University Press, 2017. |
Includes bibliographical references and index.
Identifiers: LCCN 2017009647 | ISBN 9781107159754 (hardback : alk. paper)
Subjects: LCSH: Ants – Ecology. | Plant ecology. | Insect-plant relationships. |
Nature – Effect of human beings on.
Classification: LCC QL568.F7 A557 2017 | DDC 595.79/6–dc23
LC record available at https://lccn.loc.gov/2017009647

ISBN 978-1-107-15975-4 Hardback

Contents

The color plates section is found between pages 332 and 333.

Contributors

Alan N. Andersen
Tropical Ecosystems Research Centre, CSIRO Land and Water, Winnellie, NT, Australia

Gabriela B. Arcoverde
Graduate Program in Plant Biology, Federal University of Pernambuco, Recife, Brazil

Inge Armbrecht
Department of Biology, University of Valle, Cali, Colombia

Julia Backé
Plant Ecology and Systematics, University of Kaiserslautern, Kaiserslautern, Germany

Andrew J. Beattie
Department of Biological Sciences, Macquarie University, Sydney, NSW, Australia

Ana G. D. Bieber
State University of Southwestern Bahia, Itapetinga, Bahia, Brazil

Rumsaïs Blatrix
Center of Functional and Evolutionary Ecology, UMR 5175, CNRS, Montpellier, France

Nico Blüthgen
Department of Biology, Darmstadt Technical University, Darmstadt, Germany

Mark A. Bradford
Yale School of Forestry and Environmental Studies, Yale University, New Haven, CT, USA

Kaitlin U. Campbell
Department of Biology, University of North Carolina, Pembroke, NC, USA

Nathalia Chavarro-Rodríguez
Department of Biology and Chemistry, Los Llanos University, Villavicencio, Colombia

Lacy D. Chick
Department of Biology, Case Western Reserve University, Cleveland, OH, USA

Alexander V. Christianini
Department of Environmental Sciences, Federal University of São Carlos, Sorocaba, Brazil

Yann Clough
Centre for Environmental and Climate Research, Lund University, Lund, Sweden

Thomas O. Crist
Department of Biology, Miami University, Oxford, OH, USA

Cecilia Díaz-Castelazo
Multitrophic Interactions Network, Institute of Ecology, AC, Xalapa, México

Alejandro G. Farji-Brener
Regional University Center of Bariloche, National University of Comahue, Bariloche, Argentina

Tom M. Fayle
Institute of Entomology, Biology Centre of the Czech Academy of Sciences and Faculty of Science, University of South Bohemia, Ceske Budejovice, Czechia; Forest Ecology and Conservation Group, Imperial College London, Ascot, UK

Marcia González-Teuber
Department of Biology, La Serena University, La Serena, Chile

Martin Heil
Department of Genetic Engineering, Cinvestav, Irapuato, Mexico

David A. Holway
Division of Biological Sciences, University of California, San Diego, CA, USA

Ian M. Jones
Department of Biological Sciences and International Center for Tropical Botany, Florida International University, Miami, FL, USA

Robert R. Junker
Department of Ecology and Evolution, University of Salzburg, Salzburg, Austria

Christopher Kaiser-Bunbury
Department of Biology, Darmstadt Technical University, Darmstadt, Germany

Joshua R. King
Biology Department, University of Central Florida, Orlando, FL, USA

Petr Klimes
Institute of Entomology, Biology Centre of the Czech Academy of Sciences, Ceske Budejovice, Czech Republic

Suzanne Koptur
Department of Biological Sciences and International Center for Tropical Botany, Florida International University, Miami, FL, USA

Lori Lach
ARC Research Fellow, James Cook University, Cairns, QLD, Australia

Inara R. Leal
Department of Botany, Federal University of Pernambuco, Recife, Brazil

Laura C. Leal
Departament of Biological Sciences, Federal University of São Paulo, Diadema, Brazil

María N. Lescano
Regional University Center of Bariloche, National University of Comahue, Bariloche, Argentina

Hong Liu
Department of Earth and Environment and International Center for Tropical Botany, Florida International University, Miami, FL, USA

Doyle McKey
Center of Functional and Evolutionary Ecology, UMR 5175, CNRS, Montpellier, France

Joshua H. Ness
Biology Department and Environmental Studies Program, Skidmore College, New York, USA

Fernanda M. P. Oliveira
Graduate Program in Plant Biology, Federal University of Pernambuco, Recife, Brazil

Paulo S. Oliveira
Department of Animal Biology, State University of Campinas, São Paulo, Brazil

Todd M. Palmer
Department of Zoology, University of Florida, Gainesville, FL, USA

Ivette Perfecto
School of Natural Resources and Environment, University of Michigan, Ann Arbor, MI, USA

Stacy Philpott
Environmental Studies Department, University of California, Santa Cruz, CA, USA

Marco A. Pizo
Department of Zoology, Rio Claro, São Paulo State University, São Paulo, Brazil

Victor Rico-Gray
Institute of Neuroethology, Veracruzana University, Xalapa, México

Felipe F. S. Siqueira
Graduate Program in Plant Biology, Federal University of Pernambuco, Recife, Brazil

Marcelo Tabarelli
Department of Botany, Federal University of Pernambuco, Recife, Brazil

Mariana Tadey
Regional University Center of Bariloche, National University of Comahue, Bariloche, Argentina

Teja Tscharntke
Department of Crop Sciences, Georg-August-University of Göttingen, Göttingen, Germany

Edgar C. Turner
Insect Ecology Group, University Museum of Zoology, Cambridge, UK

Chua Wanji
Institute for Tropical Biology and Conservation, University of Malaysia Sabah, Kota Kinabalu, Malaysia

Robert J. Warren II
Department of Biology, SUNY Buffalo State, Buffalo, NY, USA

Rainer Wirth
Plant Ecology and Systematics, University of Kaiserslautern, Kaiserslautern, Germany

Akira Yamawo
Faculty of Agriculture and Life Science, Hirosaki University, Hirosaki, Japan

Truman P. Young
Department of Plant Sciences, University of California, Davis, CA, USA

Kalsum M. Yusah
Institute for Tropical Biology and Conservation, University of Malaysia Sabah, Kota Kinabalu, Malaysia

Preface
Ants and Plants: A Prominent Interaction in a Changing World

The chief enemy of an ant is another ant.
This is the measure of the ants' success.
It is true of only one other living creature – man.
Derek W. Morley 1953

In the end, I suspect it will all come down to a decision of ethics – how we value the natural worlds in which we evolved and now, increasingly, how we regard our status as individuals.
Edward O. Wilson 1988

Ants and flowering plants are dominant in most terrestrial ecosystems, and their evolutionary histories have been crossing paths for at least 100 million years (Wilson & Hölldobler 2005). The study of ant-plant interactions has increased markedly over the past century with the monographs by Bequart (1922) and Wheeler (1942), but most especially in the past 50 years since Daniel Janzen's pioneering experimental work on the Central American ant-inhabited acacias (Janzen 1966). During this time, different aspects of the natural history and evolutionary ecology of ant-plant systems have been highlighted in a number of books (Buckley 1982; Beattie 1985; Huxley & Cutler 1991; Gorb & Gorb 2003; Rico-Gray & Oliveira 2007; Hölldobler & Wilson 2010).

Research on ant-plant interactions comprises a wide range of topics, including plant defense, pollination, seed dispersal, damage by leaf-cutting ants, seed predation, ant-fed plants, the association of ants with trophobionts (exudate-producing insect herbivores), applied ant ecology (as related to agriculture and restoration), and all forms of combinations among these subjects. Ant-plant interactions are geographically widespread and have already been studied in many types of terrestrial communities. Indeed, they offer an excellent opportunity to investigate the effects of both historical and ecological factors (including global change) on the evolution of mutualistic as well as antagonistic ant-plant systems (Heil & McKey 2003). Most interspecific interactions, however, are highly facultative, and the diversity of species involved in ant-plant associations can vary considerably even over short geographic ranges, suggesting that a landscape approach should be employed when investigating such interaction systems (Bronstein 2015). Our understanding of the ecology and evolution of ant-plant interactions and their effects on community organization requires an assessment of how the diversity of ants and their use of plants varies across regions, landscapes, habitat gradients, and, most importantly, under the effect of different patterns of global change (Kiers et al. 2010).

The Problem

In the past few decades human activities have altered and degraded natural environments at an unparalleled speed and scale. With the continual loss of global biodiversity, current ecological research has focused on assessing the factors that cause species loss as well as on the ways through which ecosystem functioning can be assured (Ellis 2015). Given that current biodiversity is the product of diversification of both species and the interactions among them, conservation programs need to put increased emphasis on the preservation not only of species but also of their interactions (Price 2002). Thus the knowledge that a species is present or absent is not sufficient to predict its impact on communities and ecosystems. Not only are species unevenly distributed, but their interactions also vary along environmental gradients and under different disturbance regimes. In the past few decades it has become increasingly apparent that the outcome of interspecific interactions may vary not only under different ecological conditions and across habitats, but also according to human impact and global change (Tylianakis et al. 2008).

Habitat disturbance and change are significant threats to the functioning of communities and ecosystems since they may not only alter the outcome of interspecific interactions, including ant-plant mutualisms, but also may make them vulnerable to colonization by opportunistic invasive species (Ness & Bronstein 2004). Accumulated evidence from recent decades has demonstrated that human activities can affect the habitats where ants (and plants) live in diverse ways, including deforestation and/or fragmentation, urbanization, agriculture, pasture, soil, fire, land management, and climate change (Andersen & Majer 2004). The effects of anthropogenic disturbance on ant communities may include loss of diversity, change in species composition, change in interspecific interactions, change in trophic interactions with plants and insect herbivores, as well as alterations in ant-derived ecosystem services such as seed dispersal, plant protection, and soil productivity (Lach et al. 2010). Although ant-plant mutualisms have evolved in spatially and temporally variable environments, little is known about the degree of resilience of these associations in the face of the current global changes and man-induced disturbances.

Scope of This Book

This volume is about man-induced effects on the interactions between two major components of terrestrial ecosystems: ants and plants. The book is broad in scope and is aimed at raising relevant ecological questions from an array of topics within ant-plant interactions in altered habitats around the world. Chapters provide an in-depth summary of current understanding for researchers already acquainted with general insect-plant interactions, and a window into the ecology of ant-plant interactions for the mostly uninitiated international community. We hope that this volume will serve as an important resource for undergraduate and graduate students in conservation biology and its applications, as well as working conservationists and natural resource managers around the globe. The research-level text is suitable to readers from a wide range of backgrounds, including basic and applied entomologists, botanists, behavioral, evolutionary, and applied ecologists, agronomists, conservationists, and landscape managers. This book brings together research from a variety of regions and environments worldwide, including South, Central, and North America, Africa, Japan, Polynesia, Indonesia, and Australia. Agricultural systems from regions apart are also treated in the volume since they have extensively reshaped natural landscapes, modifying the number and types of interspecific interactions compared to those present in natural ecosystems. A number of chapters included in this book were presented in the Symposium on "Ant-Plant Interactions in a Changing World" that took place at the XXV International Congress of Entomology, held in Orlando (Florida, USA) in September 2016.

We aimed for a worldwide assessment of man-induced effects on ant-plant interactions, incorporating a diversity of approaches that range from landscape levels to finer scales of interactions and processes. Although we have organized the book into six parts in accordance with the scale and/or type of interactions, the reader will often make connections between and among parts given the complexity and interdependence of ecological processes at various levels. Part I deals with the major effects of habitat heterogeneity (vegetation mosaics), fragmentation, and plantations on arboreal and ground ant communities, ant functional groups, and on the multiple impacts of leaf-cutting ants on vegetation. Part II examines man-induced effects (e.g., climate change, grazing, fire, fragmentation, human settlements) on primary and secondary seed dispersal by ants in temperate forests and neotropical ecosystems. Part III addresses how facultative and obligate ant-plant protective mutualisms are affected by habitat heterogeneity and human disturbance in temperate and tropical habitats. Topics include species producing ant-attractants (and their variable quality), plasticity in defense strategies against herbivory (mediated or not by visiting ants), complex obligate ant-plant systems affected by vertebrate defaunation, and evolutionary responses of ant-plant systems to global change. Part IV examines the effect of invasive ant species on interaction systems involving native ants, insect trophobionts, associated herbivores, flower visitors, and plants. The topic is addressed in relation to ant-plant systems in continental and insular habitats. Part V applies the accumulated knowledge on ant-plant-herbivore interactions from natural ecosystems in insect pest management programs of agricultural systems. Potential ecosystem services (and disservices) provided by ants in the recovery of human-modified landscapes are also treated. Part VI provides an overview of the major challenges created by global changes on interspecific interactions, particularly ant-plant mutualisms (obligate and facultative), pointing out evolutionary consequences, promising avenues of investigation embracing both fundamental and applied aspects of ant-plant interactions, and possible strategies and approaches to maintain viable conserved communities.

Acknowledgments

This book was conceived during a three-month visit by Paulo Oliveira to Suzanne Koptur's laboratory at Florida International University, sponsored by the São Paulo Research Foundation (FAPESP) and the National Council for Scientific and Technological Development (CNPq). In Miami we shaped the scope of the volume, established the main research areas to be covered, invited prospective authors, adjusted chapter contents, and wrote the book proposal.

Each chapter was improved with comments and suggestions from external reviewers. We are indebted to the following colleagues for their constructive feedback: Nigel Andrew, Nico Blüthgen, Guillaume Chomicki, Jaeson Clayborn, Evan Economo, Alejandro Farji-Brener, Tom Fayle, Heike Feldhaar, Silvia Gallegos, Heather Gamper, Crisanto Gomez, Aaron Gove, Ben Hoffmann, Jerome Howard, Ian Jones, Mattias Jonsson, Sabine Kasel, Lori Lach, Edward LeBrun, Maurice Leponce, Deborah Letourneau, Hong Liu, Bette Loiselle, Jonathan Majer, Sebastian Meyer, Antonio Manzaneda, Terry McGlynn, James Montoya-Lerma, Anselmo Nogueira, Joachim Offenberg, Todd Palmer, Elizabeth Pringle, Kirsten Prior, Victor Rico-Gray, Flavio Roces, Gustavo Romero, Sebastian Sendoya, Andy Suarez, James Trager, and Robert Warren II. Marianne Azevedo-Silva, Hélio Soares Jr., and André Freitas provided invaluable help with the final formatting of the figures.

Finally, but far from least, we thank life sciences editor Dominic Lewis, editorial assistant Timothy Hyland, content managers Sarah Payne and Lindsey Tate, and all of Cambridge University Press for encouragement and advice on the development of this project. We are especially grateful to the two anonymous reviewers of the book proposal for Cambridge University Press for their constructive suggestions on the initial book project.

Maristela Oliveira and John Palenchar helped with encouragement, support, and patience.

Paulo S. Oliveira and Suzanne Koptur

References

Andersen, A. N. & Majer, J. D. (2004). Ants show the way down under: invertebrates as bioindicators in land management. *Frontiers in Ecology and the Environment*, 2, 291–8.

Beattie, A. J. (1985). *The Evolutionary Ecology of Ant-Plant Mutualisms*. Cambridge University Press, Cambridge.

Bequaert, J. (1922). Ants in their diverse relations to the plant world. *Bulletin of the American Museum of Natural History*, 45, 333–621.

Bronstein, J. L. (editor). (2015). *Mutualism*. Oxford University Press, Oxford.

Buckley, R. C. (1982). *Ant-Plant Interactions in Australia*. Dr. W. Junk, The Hague.

Ellis, E. C. (2015). Ecology in an anthropogenic biosphere. *Ecological Monographs*, 85, 287–331.

Gorb, E. V. & Gorb, S. N. (2003). *Seed Dispersal by Ants in a Deciduous Forest Ecosystem: Mechanisms, Strategies, Adaptations*. Kluwer, The Hague.

Heil, M. & McKey, D. (2003). Protective ant-plant interactions as model systems in ecological and evolutionary research. *Annual Review of Ecology, Evolution, and Systematics*, 34, 425–53.

Hölldobler, B. & Wilson, E. O. (2010). *The Leafcutter Ants: Civilization by Instinct*. W. W. Norton & Company, New York.

Huxley, C. R. & Cutler, D. F. (editors). (1991). *Ant-Plant Interactions*. Oxford University Press, Oxford.

Janzen, D. H. (1966). Coevolution of mutualism between ants and acacias in Central America. *Evolution*, 20, 249–75.

Kiers, E. T., Palmer, T. M., Ives, A. R., Bruno, J. F., & Bronstein, J. L. (2010). Mutualisms in a changing world: an evolutionary perspective. *Ecology Letters*, 13, 1459–74.

Lach, L., Parr, C. L., & Abbott, K. L. (editors). (2010). *Ant Ecology*. Oxford University Press, Oxford.

Morley, D. W. (1953). *The Ant World: Their Evolutionary History, Their Behavior, Their Many Varieties and the Organization of Their Society*. Penguin Books Ltd., Harmondsworth, Middlesex, UK.

Ness, J. H. & Bronstein, J. L. (2004). The effects of invasive ants on prospective ant mutualists. *Biological Invasions*, 6, 445–61.

Price, P. W. (2002). Species interactions and the evolution of biodiversity. In: C. M. Herrera & O. Pellmyr (editors). *Plant-Animal Interactions: An Evolutionary Approach*. Blackwell Science, Oxford. pp. 3–25.

Rico-Gray, V. & Oliveira, P. S. (2007). *The Ecology and Evolution of Ant-Plant Interactions*. University of Chicago Press, Chicago.

Tylianakis, J. M., Didham, R. K., Bascompte, J., & Wardle, D. A. (2008). Global change and species interactions in terrestrial ecosystems. *Ecology Letters*, 11, 1351–63.

Wheeler, W. M. (1942). Studies of neotropical ant-plants and their ants. *Bulletin of the Museum of Comparative Zoology*, 90, 3–262.

Wilson, E. O. (1988). The current state of biological diversity. In: E. O. Wilson (editor). *Biodiversity*. National Academy Press, Washington, DC. pp. 3–18.

Wilson, E. O. & Hölldobler, B. (2005). The rise of the ants: a phylogenetical and ecological explanation. *Proceedings of the National Academy of Sciences of the United States of America*, 102, 7411–14.

Part I

Landscape Mosaics, Habitat Fragmentation, and Edge Effects

1 Ant Biodiversity and Functional Roles in Fragmented Forest and Grassland Ecosystems of the Agricultural Midwest, North America

Thomas O. Crist and Kaitlin U. Campbell[*]

Regional Ant Biodiversity of the Midwest

The native grassland and forest ecosystems of the north-central United States – a region that is historically and culturally called "the Midwest" or "Midwestern United States" – are highly modified by agricultural and urban development (Figure 1.1). Most of the native grasslands in the western Great Plains were replaced by cultivation or are now managed as grazing lands (Samson & Knopf, 1994). In the Allegheny Plateau to the east, over a century of logging and mining in mixed oak and mesophytic forests continues until present despite gradual increases in second-growth forest (Strittholt & Boerner, 1995). The native beech-maple forests of the northern Till Plain are now dominated by intensive agriculture – largely corn and soybeans – with smaller areas of riparian forest, scattered woodlots, remnant prairies, or reconstructed grasslands (Medley et al., 1995; Renwick et al., 2008). Exurban expansion around the industrial cities of the Midwest continues to modify landscapes along urban corridors and the urban-rural fringe (Grimm et al., 2008). The resulting landscape mosaics of agricultural, natural, and urban ecosystems have sharp habitat gradients in vegetation structure and soil disturbance, resulting in ant species assemblages that are admixtures of habitat specialists and cosmopolitan generalists.

[*] We thank Paulo Oliveira and Suzanne Koptur for inviting us to participate in this edited volume and for their comments on an earlier version of the manuscript. Two anonymous reviewers also improved the manuscript. Several people helped in field data collection and laboratory analyses for our forest and grassland ant studies, including Mike Mahon, Mike Cunningham-Minnick, Valerie Peters, Anita Schaefer, Jason Nelson, Garrett Dienno, Mayrolin García, Sam Stephenson, Tia Loyke, Kelsey Seaman, Aaron Coleman, Carol Ramos, Natalie Konig, Amanda McDonald, Heather Brewster, and Becca King. Research was supported by the Nature Conservancy, the National Geographic Society, Ohio Biological Survey, Prairie Biotic Research Inc., and Miami University.

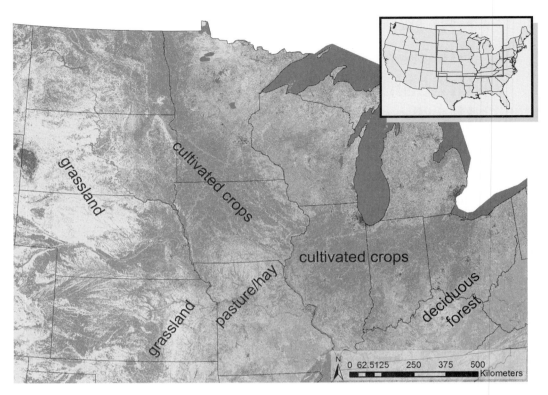

Figure 1.1.   Land use and land cover map of the Midwest US region of North America. The major
land cover types in this region include cultivated crops (brown), deciduous forest (green),
pasture/hay (yellow), grassland (tan), low-intensity developed (pink), and high-intensity
developed (red) (Homer et al., 2015, reproduced with permission.) (A black-and-white
version of this figure will appear in some formats. For the color version, please refer to the
plate section.)

The ants of the Midwest are well-described and comprise about 200 species, a
relatively small fauna compared to those in the highly diverse tropics. Two valuable
keys to the ants of Ohio (Coovert, 2005) and New England (Ellison et al., 2012)
and a checklist of the Michigan ant fauna (Wheeler et al., 1994) provide estimates
for species richness of the Midwest and neighboring regions. Ellison et al. (2012)
recorded 132 species (32 genera) and estimates 153 species for the New England
states. Similarly, Coovert (2005) documented 118 species (36 genera) in Ohio, with
an estimated 178 species. Wheeler et al. (1994) noted 113 species (27 genera), 87 of
which could be found in one large nature reserve (Talbot, 1975). The majority of the
ant species of the Midwest fall into two subfamilies, Myrmicinae and Formicinae,
with only a few species in subfamilies Dolichoderinae, Ponerinae, Amblyoponinae,
Proceratiinae, and Dorylinae. Although we focus on the Midwest region, we draw
upon studies from surrounding regions that have overlapping ant faunas.

Habitat Associations of Ants

Forests and Grasslands

Many ant species specialize in forests or open habitats. For example, Hill et al. (2008) found distinct compositional and functional group differences between ants in open (prairie and pasture) and forest habitats in Mississippi but more subtle differences in species composition between prairie and pasture. Ant communities may also differ among patches of similar habitats, depending on soils, disturbance, or surrounding land uses (Menke et al., 2015; Campbell & Crist, 2017).

Ants specialized in open, grassland habitats tend to be ground-dwelling ants including Dolichoderinae, *Lasius* and *Formica* species (Formicinae), and small heat-tolerant Myrmicinae (e.g., *Monomorium, Pheidole, Solenopsis*) (Hill & Brown, 2010; Moranz et al., 2013; Campbell & Crist, 2017). In contrast, forests harbor diverse assemblages of cryptic litter-dwelling ants (e.g., *Strumigenys, Proceratium, Stenamma*) and are dominated by opportunists such as *Aphaenogaster, Nylanderia,* and *Myrmica* and several species of carpenter ants (*Camponotus*) (Lessard et al., 2007; Ivanov et al., 2009; Mahon et al., 2017). Species richness in regional studies of these two habitats range from 14–53 species in restored and remnant grasslands (Phipps, 2006; Friedrich, 2010; Hill & Brown, 2010; Campbell & Crist, 2017) and 17–40 species in forest fragments (Talbot, 1934; Wang et al., 2001; Ivanov et al., 2009; Mahon et al., 2017) to 53 species in large preserves of the Great Smoky Mountains to the southeast (Lessard et al., 2007) (Table 1.1).

Native prairies may have hundreds of plant species, but restored and constructed grasslands usually have much lower plant species richness (Nemec, 2014; Peters et al., 2016). The Conservation Reserve Program (CRP) and other programs provide incentives for farmers to retire crop fields on highly erodible land and seed them with native plants (USDA, 2016). Although constructed grasslands have lower plant diversity than native prairies, plant species richness is often unrelated to ant species richness in restored and constructed grasslands (Dahms et al., 2005; Peters et al., 2016). Instead, the duration since disturbance, soil characteristics, or surrounding land cover types are more important determinants of the diversity and composition of grassland ant communities (Phipps, 2006; Menke et al., 2015; Campbell & Crist, 2017). As the site recovers from disturbance, species composition and functional group dominance shift toward habitat specialists and subterranean foragers (Dauber & Wolters, 2005; Campbell & Crist, 2017) (Figure 1.2).

Urban and Agriculture

Relatively few studies have quantified ant biodiversity of urban and agricultural environments in the Midwest, despite the importance of these environments in the region and around the globe. Many ant species thrive in urban environments because of the diversity of habitat types and the heterogeneity of environmental

Table 1.1 Ant Diversity and Typical Ant Species in Multiple Habitat Types Based on Records from 24 Ant Biodiversity Studies in the Midwest and Surrounding Regions

Habitat Type	State	No. Sites	Mean Richness	Total Richness	Dominant Species	Reference
Northern hardwood forest	NY	1	–	33	*Aphaenogaster rudis, Myrmica punctiventris, Formica neogagates*	Ellison et al., 2007
Urban forest fragments	OH	9	–	31	*Aphaenogaster picea, Lasius alienus, Myrmica punctiventris*	Ivanov et al., 2009
Mixed hardwood forest	TN	22	–	53	*Aphaenogaster rudis*	Lessard et al., 2007
Hardwood and pine forest	TN	4	14.7	23	*Aphaenogaster rudis, Ponera pennsylvanica, Nylanderia faisonensis, Crematogaster cerasi*	Martelli et al., 2004
Mixed oak forest	VA, WV	18	–	31	*Camponotus pennsylvanicus, Aphaenogaster rudis, Formica neogagates*	Wang et al., 2001
Mixed oak forest	NC	6	–	17	*Aphaenogaster rudis, Prenolepis imparis, Myrmica punctiventris*	Lessard et al., 2009
Mixed hardwood forest	OH, IN	5	15.3	25	*Camponotus, Aphaenogaster*	Campbell & Crist, in preparation
Mixed hardwood forest	OH	5	20	34	*Aphaenogaster rudis, Lasius alienus, Tapinoma sessile, Temnothorax curvispinosus, Ponera pennsylvanica*	Mahon et al., 2017
Oak-hickory forest	MS	6	25	–	–	Hill et al., 2008
Urban forest fragments	OH	8	–	26	*Tetramorium caespitum, Lasius neoniger*	Uno et al., 2010
Mixed hardwood and pine forest	NC	10	9.8	31	*Aphaenogaster carolinensis*	Menke et al., 2011
Mixed hardwood and pine forest	AR	14	20.6	40	–	General & Thompson, 2008
Remnant prairies	NE	65	–	–	*Lasius, Myrmica* (genus level only)	Jurzenski et al., 2012
Restored and remnant savannah	IL	21	13.7	37	*Lasius alienus, Aphaenogaster picea, Camponotus pennsylvanicus*	Menke et al., 2015

Restored and remnant prairies	IL	47	10.5	42	*Lasius neoniger, Myrmica fracticornis, Solenopsis molesta*	Menke et al., 2015
Prairie remnants	AL, MS	23	12.4	53	*Forelius mccooki, Monomorium minimum, Solenopsis molesta, Pheidole tysoni*	Hill & Brown, 2010
Restored and remnant Prairies	IA, MO	12	–	14	*Formica montana, Temnothorax ambiguus*	Moranz et al., 2013
Restored grassland	IL	1	–	11	*Lasius alienus*	Peterson et al., 1998
Constructed grasslands	OH	23	10.6	31	*Lasius neoniger, Monomorium minimum, Myrmica americana*	Campbell & Crist, 2017
CRP grasslands	MO	12	13.5	28	*Tapinoma sessile, Lasius neoniger, Temnothorax ambiguus*	Phipps, 2006
Remnant prairies	OH	17	7.3	32	*Aphaenogaster treatae, Formica dolosa, Nylanderia parvula, Lasius neoniger*	Friedrich, 2010
C3 grassland pastures	MS	3	5.2	–	–	Hill et al., 2008
Remnant prairies	MS	3	17.1	–	–	Hill et al., 2008
Grasslands	OK	1	–	7	*Crematogaster punctulata, Lasius neoniger, Monomorium minimum, Pheidole dentata*	Albrecht & Gotelli, 2001
Park – prairie	AR	2	13	17	–	General & Thompson, 2008
Business	NC	10	8.5	27	*Pheidole dentata, Solenopsis invicta*	Menke et al., 2011
Greenway	NC	9	6.6	25	*Solenopsis molesta, Lasius alienus, Aphaenogaster carolinensis*	Menke et al., 2011
Park	NC	10	7.5	32	*Solenopsis molesta, Monomorium minimum, Aphaenogaster carolinensis*	Menke et al., 2011
Industrial	NC	10	5.8	24	*Solenopsis invicta, Nylanderia vividula, Monomorium minimum*	Menke et al., 2011

(continued)

Table 1.1 (cont.)

Habitat Type	State	No. Sites	Mean Richness	Total Richness	Dominant Species	Reference
Residential	NC	30	7.2	32	*Pheidole dentata, Solenopsis molesta, Tapinoma sessile*	Menke et al., 2011
Traffic medians	NY	44	–	13	*Tetramorium caespitum, Lasius neoniger*	Pecarevic et al., 2010
Vacant lots	OH	8	–	20	*Tetramorium caespitum, Lasius neoniger*	Uno et al., 2010
Gardens	OH	8	–	14	*Tetramorium caespitum, Lasius neoniger*	Uno et al., 2010
Park – mowed	AR	4	10.8	19	–	General & Thompson, 2008
Soybeans and corn	OH	5	15.6	26	*Lasius neoniger, Prenolepis imparis, Tetramorium caespitum*	Cunningham-Minnick and Campbell, unpublished data
Corn	NC, VA	17	2*	19	*Lasius alienus, Forelius pruinosus, Monomorium minimum*	Peck et al., 1998
Soybeans	NC, VA	10	1.5*	8	*Lasius alienus, neoniger, Pheidole bicarinata*	Peck et al., 1998
Hay	NC, VA	17	4*	23	*Solenopsis texana, Lasius alienus*	Peck et al., 1998
Wheat	NC, VA	6	3*	18	*Lasius alienus, neoniger, Pheidole bicarinata*	Peck et al., 1998
Agriculture (various)	NC	10	9	27	*Lasius neoniger, Monomorium minimum, Pheidole bicarinata*	Menke et al., 2011

* Denotes median species richness rather than mean.

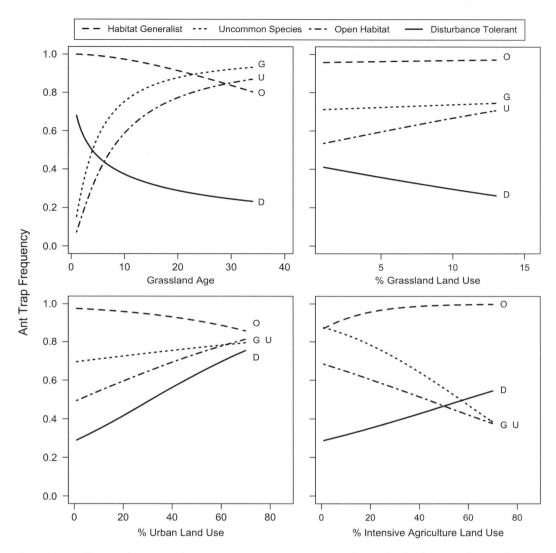

Figure 1.2. Changes in ant species occurrences among constructed grasslands along gradients of grassland age (time since planting) and different amounts of surrounding land use/land cover types in agricultural landscapes of southwest Ohio. Ant trap frequency represents the proportional occurrence of ant species among pitfall traps at each site, with ant species grouped according to their habitat affinities (groupings from Campbell & Crist, 2017).

conditions. Unlike most vertebrates, which require extensive habitat areas, ants can inhabit small patches, and several species occur in isolated groups of trees, small plantings, or vacant lots (Uno et al., 2010; Guénard et al., 2015). A citizen science project in urban North Carolina has recorded over 140 species, including both rare and invasive ants (Lucky et al., 2014). Another study found 79 ant species, ranging from the expected exotic species, *Solenopsis invicta* and *Tetramorium* sp. E, to rarely recorded hypogaeic ant genera such as *Discothyrea* and *Proceratium* (Guénard et al., 2015). At smaller scales, the urban street medians of New York City host at least 13 species of ants, with patterns of species occurrence and richness that depend on median area (Pećaravić et al., 2010). Some less-disturbed sites such as the urban forests, gardens, and vacant lots of Detroit, Michigan, and Toledo, Ohio, have 33 and 28 ant species, respectively, with significant differences in ant species composition among habitats due to the heterogeneity and isolation of urban habitat islands (Uno et al., 2010).

The opposite is true for the agroecosystems in the Midwest, characterized by crop monocultures (primarily corn-soybean rotations), high chemical input, and frequent soil disturbance. The agricultural biodiversity of ants in the Midwest, and the United States in general, has not been quantified beyond species records and anecdotal observations. Peck et al. (1998) found 41 species of ants in 90 fields (10 crop varieties) across North Carolina and Virginia, with 18 ant species occurring across 27 corn and soybean fields. Cunningham-Minnick and Campbell (unpublished data) recorded 26 ant species across 5 corn and soybean fields in Ohio, the most abundant of which were *Lasius neoniger*, *Prenolepis imparis*, and *Tetramorium caespitum*. Ants respond negatively to disturbance by pesticides and intensive tillage. House (1989) found that fields with conservation tillage have higher structural complexity, less soil disturbance, and increased food resources, such as soil arthropods. Together, these factors result in higher ant richness, biomass, and predaceous activities in fields with conservation compared to conventional tillage (House & Stinner, 1983; Peck et al., 1998; Haddad et al., 2011).

Invasive Ant Species

Interestingly, the Midwest has few exotic species and none of the dominant invasive species, such as the Red Imported Fire Ant (*Solenopsis invicta*) and the Argentine Ant (*Linepithema humile*), which are common in southern and western United States. Several exotic species (*Cardiocondyla obscurior*, *Tapinoma melanocephalum*, *Hypoponera punctatissima*, *Monomorium pharaonis*, *Monomorium floricola*, *Paratrechina longicornis*, and *Pheidole flavens*) are found in greenhouses and other heated buildings, but have not become established beyond these structures. The most abundant introduced species is *Tetramorium* sp. E (formerly *Tetramorium caespitum*), the pavement ant, first introduced in the 1800s from Europe. It has become established across most of the United States where it dominates disturbed habitats, especially urban spaces (Uno et al., 2010; Penick et al., 2015). *Tetramorium* sp. E also displaces other ants in sites that are already disturbed (Steiner et al., 2008; Uno et al., 2010), and may be an agricultural pest (Merickel & Clark, 1994). Because *Tetramorium* sp. E has been naturalized for over a century, impacts are limited primarily to highly disturbed habitats. A more aggressive *Tetramorium* species (*T. tsushimae*) has recently been discovered as an invasive species in Missouri and Illinois, and could potentially spread throughout the Midwest with greater impacts on native ant communities (Steiner et al., 2006). *Nylanderia flavipes*, introduced in the early 1900s from Asia, has been detected in the Midwest and Great Lakes regions but seems to blend into the community with little recorded impact beyond displacing the ecologically similar *Nylanderia faisonensis* (Wetterer, 2011). *Myrmica rubra* and *M. scabrinodis*, invasive in New England (Wetterer & Radchenko, 2011; Ellison et al., 2012), have not been detected in the Midwest but may also spread to this region with time.

Landscape Variation in Ant Diversity

Landscape-level variation in topography, soils, and vegetation structure act as strong environmental filters of the regional species pool to limit ant species richness in local ecosystems (Bestelmeyer & Wiens, 2001; Spiesman & Cumming, 2008). Land use, management practices, and invasive species are superimposed on these natural drivers of ant species diversity, and may have strong effects on ant communities (Holway et al., 2002; Morrison, 2002; Crist, 2009; Chapter 5). The ant species richness and composition found in local remnant ecosystems, such as a grassland or forest, are therefore shaped by historical factors as well as more recent land-use changes (Dunn et al., 2010; Philpott et al., 2010). In more mesic areas of the Midwest, forests now occur primarily as riparian buffers or small woodlots (Medley et al., 1995). Likewise, grasslands established through conservation incentives are often on marginal agricultural lands that are too wet or sloping for cultivation. Most ground-nesting ant species prefer well-drained soils, so that topography and soil texture may act as a strong filter for establishment (Campbell & Crist, 2017). The surrounding land uses also differ in their suitability to ant species, which may further limit the ant species that are available to colonize isolated forest or grass-land habitats.

Variation in structure of natural habitats and surrounding land use may limit the number of ant species found within habitats (alpha diversity) and cause large turnover in species composition among habitats (beta diversity) (see Anderson et al., 2011, for a review of these concepts). Crist (2009) found that fragmented habitats averaged 75 percent turnover in ant species composition (as measured by an additive measure of beta diversity), a pattern that was consistent across habitat type, spatial extent, and region. In studies from the Midwest and eastern United States (Table 1.1), turnover in ant species composition averages 49 percent of the total species richness in forest habitats, and 70 percent of the total richness in grassland and urban habitats. Crist (2009) focused on studies of fragmented habitats that are predicted to have greater levels of beta diversity, whereas those in Table 1.1 include both large, continuous habitats as well as fragmented patches. The lower turnover in species composition among forests than in grassland patches may also be due to a greater size or age of forest compared to grassland patches. Land uses in the surrounding landscape may also influence local ant diversity and turnover in species composition among habitats. Our studies in constructed grasslands showed that groups of ant species had contrasting responses to surrounding land uses in ways that reflected their known habitat affinities (Figure 1.2). In other regions, the surrounding landscape configuration and composition explained significant variation in ant species composition (Spiesman & Cumming, 2008). Studies of reconstructed or restored habitats show that habitat age may also be important to developing a diverse local ant community (Phipps, 2006; Menke et al., 2015; Campbell & Crist, 2017). This suggests that dispersal of alates from other suitable habitats in the landscape is a limiting factor to local ant diversity, and may contribute to high levels of species turnover among isolated habitats (Crist, 2009).

Ant Functional Roles in Ecosystems

Forests

Forest ant communities can be divided into three layers of vertical stratification: (1) ants that forage within and beneath the leaf litter (hypogaeic); (2) ants that forage above the leaf litter (epigaeic); and (3) ants that forage on tree trunks and into the canopy. Subterranean ants include root-aphid tenders in the genus *Lasius* (subgenera *Cautolasius*, *Chthonolasius*, and *Acanthomyops*) and predatory *Ponera* and *Stigmatomma* ants. Inhabiting the leaf litter layer are small predatory and cryptic species, especially *Temnothorax curvispinosus*, which often makes its nests in acorns, and tiny trap jaw ants in the genus *Strumigenys* that are specialist predators of springtails. The epigaeic community is dominated by several *Aphaenogaster* species, *Lasius alienus*, and *Camponotus*, especially large species in the subgenus *Camponotus*. *Aphaenogaster* species are important seed dispersers for elaiosome-bearing understory herbs, including bloodroot, violets, and wild ginger, although several other medium- and large-bodied ants also contribute to myrmecochore dispersal (Beattie & Culver, 1981).

Ant communities in the forest canopy are poorly known. In a regional study of southern Ohio, we found 25 species in 96 trees and 24 forest stands using canopy fogging techniques (Campbell & Crist, unpublished data). The most numerically dominant members of the canopy community were *Camponotus* (*Myrmentoma*). Several species including *C. caryae* and *Colobopsis* spp., and *Aphaenogaster mariae* are exclusively found in the canopy or on tree trunks. Vertical stratification of tropical ant communities is well documented (Longino & Nadkarni, 1990); our work suggests that temperate regions may also have distinct litter, ground, and canopy specialists with a greater degree of overlap among strata (Figure 1.3).

Grasslands

Ants may affect the diversity and productivity of plants indirectly as ecosystem engineers and herbivore mutualists. Many ground-dwelling ants create hot spots

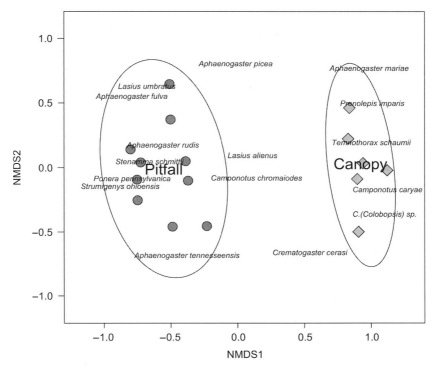

Figure 1.3. Non-metric multidimensional scaling of ant species composition using Bray-Curtis dissimilarities of log-transformed ant abundances in pitfall traps (60 traps from 10 plots in 5 forest stands) and canopy fogging samples (24 trees from 6 forest stands) in deciduous forests of SW Ohio. Sample scores are derived from ant species abundances in six pitfall traps in 20 × 20 m plots and fogging four tree species in each forest stand. Species scores are shown for the most common ant species in each stratum. Ellipses are 95 percent confidence intervals among sample scores.

of soil nutrient deposition, mineralization, and aeration by tunneling and storing resources in and around their nests. The most significant soil engineers in grasslands are those that create large mounds including multiple *Formica* species and *Lasius* (*Acanthomyops*) *claviger* (Lane & BassiriRad, 2005). *Formica* mounds, including those of *F. montana*, a known prairie associate, can structure plant community composition on and around nest mounds, enrich soil nutrients, and alter soil bulk density and moisture (Henderson & Jeanne, 1992; Lane & BassiriRad, 2005). Over 18 ant species tend and protect native aphids in grasslands, including *Lasius neoniger*, which will interfere with parasitism by braconid wasps (Wyckhuys et al., 2007), and *F. montana*, which has higher aphid infestations on plants near nests (Henderson & Jeanne, 1992). *Formica* also are facultative hosts of butterfly caterpillars (Lycaenidae) (Fiedler, 2001; Coovert, 2005).

Some grassland ants can directly affect the diversity of plants through seed predation and dispersal. The midwestern seed predators are primarily in the genus

Pheidole. Despite the hyperdiversity of this ant genus worldwide, midwestern *Pheidole* are not speciose, dominant, or conspicuous members of the ant community. *Pheidole* are often restricted to sandier soils and are more abundant in grasslands than in other habitats because of this preference for well-drained, dry habitats (Wilson, 2003; Coovert, 2005; Nemec, 2014; Campbell & Crist, 2017). The most common *Pheidole* species make use of alternative food sources including sugars from aphids and floral sources and insect carrion (Wilson, 2003; Coovert, 2005).

Agroecosystems

Ants in agricultural fields can affect both above- and belowground communities by interacting with multiple trophic levels including parasites, herbivores, pollinators, predators and soil organisms (Chapters 16, 17). Through these interactions, ants may be beneficial or detrimental to crop yield, and their net effects have been tested in only a few studies. Cacao plantations in Indonesia with diverse ant communities had significantly higher yields than those where the ant community was dominated by an invasive ant that reduced native ant species richness and evenness (Wielgoss et al., 2014). In the southern United States, cotton and soybean crops are often dominated by the invasive red imported fire ant (*Solenopsis invicta*). Eubanks (2001) found that, although *S. invicta* eliminated other biocontrol agents through intraguild predation, there was a net reduction of all herbivorous insects, including some of the most severe insect pests.

Ants tend, disperse, and protect a diversity of aphid species and other sap-sucking hemipteran insects in return for honeydew secretions, and can indirectly affect plant growth and crop yield through these mutualistic relationships. Aphid tending by ants increases aphid population sizes by shifting aphid growth rates from density dependent to density independent growth (Johnson, 2008). The soybean aphid (*Aphis glycines*) emerged as a serious economic pest for the $27 billion USD soybean industry, since its detection in the agricultural Midwest in 2000 (Ragsdale et al. 2011). Many ant species will tend soybean aphids including those commonly found in agricultural fields: *Monomorium minimum, Lasius neoniger, Prenolepis imparis*, and *Formica subsericea* (Cunningham-Minnick & Campbell, unpublished data). *M. minimum* also harasses or kills beneficial insects including predators and parasitoids such as lady beetles, pirate bugs, and braconid wasps (Herbert & Horn, 2008).

Ant modification of the agricultural soils can alter soil characteristics and belowground communities. *Lasius neoniger*, one of the most abundant ants in agricultural fields, can have up to 30,000 nest openings ha^{-1}, and dig 70 cm deep (Wang et al., 1995). *L. neoniger* nests represent localized regions of decreased soil bulk density and increased organic matter, phosphorus, and potassium levels (Wang et al., 1995). Ant engineering of soils alters water movement, nutrient availability, mineralization, density, and turnover rates (Lobry de Bruyn, 1999; Chapter 18).

Urban Ecosystems

A handful of ant species are significant structural pests in our region such as *Camponotus* species, *Tapinoma sessile*, and *Tetramorium* cf. *caespitum* (Klotz et al., 2008). These ants are higher in abundance in anthropogenic than in native habitats and can cause damage to houses by living in wood or entering houses for food and water. Some ant species, for example *T.* cf. *caespitum*, have become so integrated into human habitats that their use of human food can be detected through stable isotope enrichments of δC^{13} and δN^{15} (Penick et al., 2015). Ants like these are the most familiar to humans because they are encountered on a daily basis; however, they give ants a bad reputation. Most of the ant species in urban environments do not directly interact with humans and may provide ecosystem services, such as seed dispersal in forest patches and parks (Thompson & McLachlan, 2007), soil aeration and nutrient deposition (Sanford et al., 2009), and suppression of herbivore pests (Klotz et al., 2008).

North Temperate Ants in a Changing World

Climate Change

As the anthropogenic drivers of climate change, land use, and species invasions continue to alter regional ant faunas and local community structure, significant changes are predicted in ant biodiversity and functional roles in natural and managed ecosystems. Fitzpatrick et al. (2011) used species distribution models (SDM) and generalized dissimilarity modeling (GDM) to predict changes in richness and composition of 66 genera of North American ants. The greatest changes in the composition of ant genera were predicted to occur in the southern United States, with 0.30–0.55 changes in Bray-Curtis dissimilarity compared to present-day distributions in those regions. The Midwest was predicted to show a 0.20–0.30 change in dissimilarity compared to current distributions of ant genera. Ant genera that included social parasites or litter and soil predators were expected to decrease in the Midwest suggesting ecosystem changes in the importance of these functional groups (Fitzpatrick et al., 2011). A more focused study on range shifts in ants of eastern deciduous forests using more finely resolved ant functional groups predicted that ant species with roles as wood decomposers, seed dispersers, and leaf-litter regulators would undergo range contractions as well as northward range expansions (Del Toro et al., 2015).

The Midwest US climate is projected to become warmer and wetter during winter, and warmer with more highly variable precipitation in summer, resulting in shifts in tree species composition in forests (Pryor et al., 2014). Range shifts in ant species are also likely, but may depend strongly on dispersal limitation between suitable habitats (Fitzpatrick et al., 2011) and the degree to which the surrounding land use types may act as barriers to dispersal between isolated forest or grassland ecosystems, factors which are poorly known for most ant species.

At finer scales, laboratory and field experiments are beginning to elucidate temperature responses of individual ant species under climate change scenarios. Pelini et al. (2012) conducted laboratory experiments with *Aphaenogaster rudis* and *Temnothorax curvispinosus* colonies collected along a latitudinal gradient. Ants in these common garden experiments were exposed to summer temperature regimes from their source locations as well as those predicted in the future by climate change models. *A. rudis* colonies had very low survival at elevated temperatures; however, the response of *T. curvispinosus* depended on source location, with those from cooler source locations showing decreased survival compared to those from warmer locations. These laboratory results largely confirmed initial results from a long-term field-warming experiment on ant communities demonstrating declines in the activities of seed-dispersing ants, like *A. rudis* (Pelini et al., 2011).

Together, the macroscale species-distribution models and community-level experiments in eastern forests suggest that climate change will result in substantial changes to ant species diversity and their roles in ecosystem functioning over the next 50 years. Ant communities and tree composition of eastern Atlantic forests are similar to those of the Midwest United States, but the climate-change predictions for summer precipitation are more variable and extreme for the continental Midwest (Horton et al., 2014; Pryor et al., 2014). Species distribution models also predict differences in the responses of ant functional groups between the two areas of eastern deciduous forests (Fitzpatrick et al., 2011). Dispersal limitation of forest ants and other species may play a greater role in the range shifts in the Midwest, where forests are more isolated by agriculture.

Agricultural Intensification

Agriculture is the predominant land use in the Midwest (Figure 1.1) but most changes over the past 50 years have been intensification of agricultural practices and reduced crop diversity rather than extensification in the amount of land in cultivation (Lubowski et al., 2006). There have even been slight declines in area under cultivation over the past 20 years, mostly through conservation incentives to convert cropland to CRP grasslands, riparian forests, or pasture and hayfields (Lubowski et al., 2006; Renwick et al., 2008). During the same time, however, intensification of practices has increased considerably, with larger field sizes, increased use of fertilizers and pesticides, fewer crop types, and greater use of genetically modified corn and soybeans (Renwick et al., 2008). Despite increased intensification, the use of no-till or conservation tillage practices have increased significantly in the past 20 years (Renwick et al., 2008). Reduced tillage has positive effects on ants and beneficial arthropods, such as spiders and carabid beetles (e.g., Menalled et al., 2007; Haddad et al., 2011), although these benefits may be partially offset by the increased use of herbicides with conservation tillage (House, 1989). In addition to the roles of ants in soil modification and food-web interactions, no-till crop management may support a more diverse and abundant ant community, and facilitate dispersal between isolated natural habitats.

Urbanization

The large cities of the Midwest were major industrial centers in the twentieth century, and in recent decades have continued to expand through exurban development (Theobald, 2002; Grimm et al., 2008). The changing boundaries of the urban-rural fringe in the Midwest usually involve the conversion of farmland into urban development (Theobald, 2002). Exurban development is characterized by large residential parcels that form mosaics of low-intensity housing, pasture and forest, and farmland. These landscapes differ significantly from the habitats of urban cores comprising industrial areas, high-density housing, parks, boulevards, and vacant lots. We would therefore predict diverse and heterogeneous ant assemblages in Midwest exurban landscapes with urban, disturbance-tolerant, grassland, and forest species present in different landscape elements. Where invasive ant species are more important, however, a different set of ant community dynamics may occur. For example, studies of ant communities in remnant scrub habitat of California revealed important influences of surrounding urbanization on local ant communities, particularly in how they were influenced by the invasive Argentine ant, *Linepithema humile* (Holway et al., 2002; Bolger, 2007).

Invasive Species

Few exotic ant species in the Midwest exhibit the dominant effects on native ant communities found in warmer climates, such as *S. invicta* and *L. humile*. Future climate change, however, is likely to result in significant range expansions of invasive ant species. For example, Morrison et al. (2005) predicted *S. invicta* to spread into the lower Midwest region under future climate conditions. The ecological impacts of invasive fire ants are well documented in southeastern ecosystems, where they range from strong negative to positive effects on native ant communities, depending on habitat type, disturbance, and productivity (Morrison & Porter 2003; Chapters 13–15).

Ant communities in Midwest forests are also likely to change due to direct and indirect effects of tree mortality from invasive forest insect pests. The emerald ash borer (*Agrilus planipennis*) has caused almost complete mortality of native ash trees (*Fraxinus* spp., Oleaceae) in the Midwest as it spreads from the Great Lakes southward (Herms & McCullough, 2014). The treefall gaps created by ash mortality are anticipated to have multiple ecological impacts in forests, including alteration of microclimate, increased coarse woody debris, and facilitation of invasive understory plants (Herms & McCullough, 2014). The consequences for ant communities in maple-ash forests could be substantial and are likely to cause shifts in ant species composition that depend on the thermal tolerances and nesting habitats of different ant species. In addition, canopy-dwelling ants, such as some *Camponotus* spp., may be directly affected by the loss of host trees.

Conclusions and Future Directions

The ant fauna of Midwest North America is well described and we know a great deal about the ant species diversity and composition in natural and managed ecosystems of the region. Nonetheless, our knowledge of the regional ant fauna and patterns of species diversity belies our understanding of the roles of ants in ecosystems and their community responses to disturbance and management practices. On the basis of known ant species ranges and habitat associations, efforts are underway to predict how species distributions will shift with climate change, land use, or invasive species. Given the ubiquitous nature of ants, these changes in ant species distribution and abundance are likely to have important consequences for human-dominated ecosystems.

We identify three areas of inquiry for future study: (1) ant communities and functional roles in crop systems, especially given the potential for increasing importance of ants in conservation tillage and no-till systems; (2) ant functional roles and species interactions in multiple strata in forest ecosystems (soil, leaf litter, canopy), and how these are modified by tree mortality from invasive forest insect pests; and (3) evaluation of ants as bioindicators in constructed or restored grassland and forest habitats.

Our knowledge of ant biodiversity and functional roles in agricultural production areas is limited to a handful of papers in regions with dissimilar ant faunas, climate, and crop systems (i.e. South America, southeastern United States, Europe). Most studies to date consider specific roles in predation, plant protection, or aphid interactions, but the combined or net effects of ant communities in crop systems remains unexplored (but see Eubanks, 2001; Wielgoss, 2014). Additional research is needed to determine the effects of agricultural practices (cover crops, pesticides, and decreased tillage regimes, use of genetically modified crops) on ant biodiversity and ecosystem services (especially crop yield). In particular, we need a clearer understanding of the net effects of species interactions by various ant species involved in plant protection from herbivores and the possible disservices of ants that occur from aphid protection.

Ground-foraging ants in temperate forests are well studied, and their significance as seed dispersers links them to understory herbaceous plant communities. There are substantial gaps in our knowledge, however, in how different ant species interact with herbivores and predators in the forest canopy, and how ants interact with belowground detrital food webs. Future work could explore several questions on the roles of canopy ants (Chapters 2 and 3). How do ants, such as *Camponotus* and *Formica* species, mediate plant-herbivore and plant-homopteran interactions? What is the role of host trees in determining the species diversity and functional roles of canopy ants? How will changes in tree composition and forest structure due to climate change and forest invaders (e.g. emerald ash borer, Asian longhorn beetle, invasive shrubs) alter ant biodiversity and functional roles in forests? Similarly, several interesting questions remain about the roles of ants in leaf litter and soils.

What are the effects of subterranean ant species on litter and detrital food webs, or rates of litter decomposition and nutrient cycling? Controlled removals or meso-cosm experiments may help determine the extent to which cryptic ant species influence the soil and litter food webs, the microbial community, and ecosystem fluxes such as decomposition, respiration, and mineralization.

Ants are increasingly used as bioindicators of environmental change, restoration success, or the diversity of other taxa. The majority of work has focused on ant functional groups in Australia and South America (Andersen, 1995; Bestelmeyer & Wiens, 1996; Andersen & Majer, 2004; Underwood & Fisher, 2006). In these studies, ant functional groups are determined at the genus level, based on thermal tolerance and interspecific behavioral interactions of ants. The goals of this approach are to characterize and predict ant community responses to environmental gradients, anthropogenic disturbance (mining or farming), or management practices (grazing or logging). These efforts have been successful in using ants as bioindicators of environmental change (Andersen & Major, 2004). Ant functional groups based on genera have also been applied in some regions of North America (Andersen, 1997; Ellison, 2012; Moranz et al., 2013). In constructed grasslands of the Midwest, we found that ant indicators based on habitat fidelity and life history traits were better predictors of grassland habitat quality as measured by the biodiversity of several other animal taxa. In addition, ants and plants were complementary indicators, each predicting the biodiversity of different taxa and species that were habitat specialists (Peters et al., 2016).

Ant functional groups based on genera, thermal tolerance, and behavior differ from functional groups defined by species traits that relate to specific ecosystem functions, such as predation, seed dispersal, or plant protection (Crist, 2009). For example, the ant species that disperse seeds in temperate forests fall into Opportunists, Generalized Myrmicines, Subordinate Camponotini, and Cold-Climate Specialists functional groups, making it difficult to observe how shifts in any one of these functional groups might alter ecosystem function. One might use either approach to functional group classification depending on the study objectives, but ecological studies of most taxa use functional groups to represent species with similar patterns of resource use or similar roles in ecosystem processes. Plant functional groups, for example, are based on growth form, growth rate, reproductive strategies, nitrogen fixation, and life history traits (Voigt et al., 2007). Bee functional groups are determined by plant specificity, tongue length, body size, or sociality (Hoehn et al., 2008). Ground beetles are used in Europe to assess restoration and environmental quality with functional groups based on species traits including body size, dispersal abilities, and reproduction (Homburg et al., 2013). Because ant species are well known in temperate systems, it should be possible to build on our knowledge of species and ecological traits that reflect resource use and functional roles in ecosystems, and use them to understand how ant functional diversity changes in response to climate change or land use (e.g. Fitzpatrick et al., 2011; Del Torro et al., 2015).

Climate change, agricultural intensification, and urbanization pose significant challenges in predicting shifts in species ranges because climatic niche models typically assume that suitable habitat is available as species ranges shift over time. Ants exhibit a wide range of habitat affinities and thermal tolerances, making them good models for understanding the interactions between climate, land use, and invasive species. The great numerical abundance of ants also means that we can be assured they will have important roles in the novel ecosystems of the future.

References

Albrecht, M. and Gotelli, N. J. (2001). Spatial and temporal niche partitioning in grassland ants. *Oecologia*, 126, 134–141.

Andersen, A. N. (1995). A classification of Australian ant communities, based on functional groups which parallel plant life-forms in relation to stress and disturbance. *Journal of Biogeography*, 22, 15–29.

   (1997). Functional groups and patterns of organization in North American ant communities: a comparison with Australia. *Journal of Biogeography*, 24, 433–460.

Andersen, A. N. and Majer, J. D. (2004). Ants show the way down under: Invertebrates as bioindicators in land management. *Frontiers in Ecology and the Environment*, 2, 291–298.

Anderson, M. J., Crist, T. O., Chase, J. M. et al. (2011). Navigating the multiple meanings of β diversity: a road map for the practicing ecologist. *Ecology Letters*, 14, 19–28.

Beattie, A. J. and Culver, D. C. (1981). The guild of myrmecochores in the herbaceous flora of West Virginia forests. *Ecology*, 62, 107–115.

Bestelmeyer, B. T. and Wiens, J. A. (1996). The effects of land use on the structure of ground-foraging ant communities in the Argentine Chaco. *Ecological Applications*, 6, 1225–1240.

   (2001). Ant biodiversity in semiarid landscape mosaics: the consequences of grazing vs. natural heterogeneity. *Ecological Applications*, 11, 1123–1140.

Bolger, D. T. (2007). Spatial and temporal variation in the Argentine ant edge effect: implications for the mechanism of edge limitation. *Biological Conservation*, 136, 295–305.

Campbell, K. U. and Crist, T. O. (2017). Ant species assembly in constructed grasslands is structured at the patch and landscape levels. *Insect Conservation and Diversity,* 10, 180–191.

Coovert, G. A. (2005). *The Ants of Ohio*. Columbus, OH: Ohio Biological Survey, Inc.

Crist, T. O. (2009). Biodiversity, species interactions, and functional roles of ants (Hymenoptera, Formicidae) in fragmented landscapes: a review. *Myrmecological News*, 12, 3–13.

Dahms, H., Wellstein, C., Wolters, V., and Dauber, J. (2005). Effects of management practices on ant species richness and community composition in grasslands (Hymenoptera: Formicidae). *Myrmecologische Nachrichten*, 7, 9–16.

Dauber, J. and Wolters, V. (2005). Colonization of temperate grassland by ants. *Basic and Applied Ecology*, 6, 83–91.

Del Toro, I., Silva, R. R., and Ellison, A. M. (2015). Predicted impacts of climate change on ant functional diversity and distributions in eastern North American forests. *Diversity and Distributions*, 21, 781–791.

Dunn, R. R., Guénard, B., Weiser, M. D., and Sanders, N. J. (2010). Geographic Gradients. In: L. Lach, C. L. Parr, and K. L. Abbott (editors). *Ant Ecology*. Oxford: Oxford University Press, pp. 38–58.

Ellison, A. M. (2012). Out of Oz: opportunities and challenges for using ants (Hymenoptera: Formicidae) as biological indicators in north-temperate cold biomes. *Myrmecological News*, 17, 105–119.

Ellison, A. M., Gotelli, N. J., Farnsworth, E. J., and Alpert, G. D. (2012). *A Field Guide to the Ants of New England*. New Haven, CT: Yale University Press.

Ellison, A. M., Record, S., Arguello, A., and Gotelli, N. J. (2007). Rapid inventory of the ant assemblage in a temperate hardwood forest: species composition and assessment of sampling methods. *Environmental Entomology*, 36, 766–775.

Eubanks, M. D. (2001). Estimates of the direct and indirect effects of red imported fire ants on biological control in field crops. *Biological Control*, 21(1), 35–43.

Fiedler, K. (2001). Ants that associate with Lycaeninae butterfly larvae: diversity, ecology and biogeography. *Diversity and Distributions*, 7, 45–60.

Fitzpatrick, M. C., Sanders, N. J., Ferrier, S., Longino, J. T., Weiser, M. D., and Dunn, R. (2011). Forecasting the future of biodiversity: a test of single- and multi-species models for ants in North America. *Ecography*, 34, 836–847.

Friedrich, R. L. (2010). *The Short Term Impacts of Burning and Mowing on Prairie Ant Communities of the Oak Openings Region*. Toledo: University of Toledo.

General, D. and Thompson, L. (2008). Ants of Arkansas Post National Memorial: how and where collected. *Journal of the Arkansas Academy of Science*, 62, 52–60.

Grimm, N. B., Foster, D., Groffman, P. et al. (2008). The changing landscape: ecosystem responses to urbanization and pollution across climatic and societal gradients. *Frontiers in Ecology and the Environment*, 6, 264–272.

Guénard, B., Cardinal-De Casas, A., and Dunn, R. R. (2015). High diversity in an urban habitat: are some animal assemblages resilient to long-term anthropogenic change? *Urban Ecosystems*, 18, 449–463.

Haddad, G. Q., Cividanes, F. J., and Martins, I. C. F. (2011). Species diversity of myrmeco-fauna and araneofauna associated with agroecosystem and forest fragments and their interaction with Carabidae and Staphylinidae (Coleoptera). *Florida Entomologist*, 94, 500–509.

Henderson, G. and Jeanne, R. L. (1992). Population biology and foraging ecology of prairie ants in Southern Wisconsin (Hymenoptera: Formicidae). *Journal of the Kansas Entomological Society*, 65, 16–29.

Herbert, J. J. and Horn, J. D. (2008). Effect of ant attendance by *Monomorium minimum* (Buckley) (Hymenoptera: Formicidae) on predation and parasitism of the soybean aphid *Aphis glycines* Matsumura (Hemiptera: Aphididae). *Environmental Entomology*, 37, 1258–1263.

Herms, D. A. and McCullough, D. G. (2014). Emerald ash borer invasion of North America: history, biology, ecology, impacts, and management. *Annual Review of Entomology*, 59, 13–30.

Hill, J. G. and Brown, R. L. (2010). The ant (Hymenoptera: Formicidae) fauna of Black Belt Prairie remnants in Alabama and Mississippi. *Southeastern Naturalist*, 9, 73–84.

Hill, J. G., Summerville, K. S., and Brown, R. L. (2008). Habitat associations of ant species (Hymenoptera: Formicidae) in a heterogeneous Mississippi landscape. *Environmental Entomology*, 37, 453–463.

Hoehn, P., Tscharntke, T., Tylianakis, J. M., and Steffan-Dewenter, I. (2008). Functional group diversity of bee pollinators increases crop yield. *Proceedings of the Royal Society B*, 275, 2283–2291.

Holway, D. A., Lach, L., Suarez, A. V., Tsutsui, N. D., and Case, T. J. (2002). The causes and consequences of ant invasions. *Annual Review of Ecology and Systematics*, 33, 181–233.

Homburg, K., Homburg, N., Schäfer, F., Schuldt, A., and Assmann, T. (2013). Carabids. org – a dynamic online database of ground beetle species traits (Coleoptera, Carabidae). *Insect Conservation and Diversity*, 7, 195–205.

Homer, C. G., Dewitz, J. A., Yang, L. et al. (2015). Completion of the 2011 National Land Cover Database for the conterminous United States – representing a decade of land cover change information. *Photogrammetric Engineering and Remote Sensing*, 81, 345–354.

Horton, R., Yohe, G., Easterling, W. et al. (2014). Chapter 16: Northeast. *Climate Change Impacts in the United States: The Third National Climate Assessment*. In: J. M. Melillo, T. C. Richmond, and G. W. Yohe, editors. US Global Change Research Program, pp. 371–395.

House, G. J. (1989). Soil arthropods from weed and crop roots of an agroecosystem in a wheat-soybean-corn rotation: impact of tillage and herbicides. *Agriculture, Ecosystems and Environment*, 25, 233–244.

House, G. J. and Stinner, B. R. (1983). Arthropods in no-tillage soybean agroecosystems: community composition and ecosystem interactions. *Environmental Management*, 7, 23–28.

Ivanov, K., Milligan, J., and Keiper, J. (2009). Efficiency of the winkler method for extracting ants (Hymenoptera: Formicidae) from temperate-forest litter. *Myrmecological News*, 13, 73–79.

Johnson, M. T. J. (2008) Bottom-up effects of plant genotype on aphids, ants, and predators. *Ecology*, 89(1), 141–154.

Jurzenski, J., Albrecht, M., and Hoback, W. W. (2012). Distribution and diversity of ant genera from selected ecoregions across Nebraska. *The Prairie Naturalist* 44, 17–29.

Klotz, J., Hansen, L., Pospischil, R. and Rust, M. (2008). *Urban Ants of North America and Europe: Identification, Biology, and Management*. Ithaca, NY: Cornell University Press.

Lane, D. R. and BassiriRad, H. (2005). Diminishing effects of ant mounds on soil heterogeneity across a chronosequence of prairie restoration sites. *Pedobiologia*, 49, 359–366.

Lessard, J., Dunn, R. R., Parker, C. R., and Sanders, N. J. (2007). Rarity and diversity in forest ant assemblages of Great Smoky Mountains National Park. *Southeastern Naturalist*, 6, 215–228.

Lessard, J. P., Dunn, R. R., and Sanders, N. J. (2009). Temperature-mediated coexistence in temperate forest ant communities. *Insectes Sociaux*, 56, 149–156.

Lobry de Bruyn, L. A. (1999). Ants as bioindicators of soil function in rural environments. *Agriculture, Ecosystems and Environment*, 74, 425–441.

Longino, J. T. and Nadkarni, N. M. (1990). A comparison of ground and canopy leaf litter ants (Hymenoptera: Formicidae) in a neotropical montane forest. *Psyche*, 97, 81–93.

Lubowski, R. N., Bucholtz, S., Claassen, R. et al. (2006). Environmental effects of agricultural land-use change: the role of economics and policy. *US Department of Agriculture, Economic Research Report*, 25, pp. 1–75.

Lucky, A., Savage, A. M. Nichols, L. M. et al. (2014). Ecologists, educators, and writers collaborate with the public to assess backyard diversity in The School of Ants Project. *Ecosphere*, 5, 1–23.

Mahon, M. B., Campbell, K. U. and Crist, T. O. (2017). Effectiveness of Winkler litter extraction and pitfall traps in sampling ant communities and functional groups in a temperate forest. Environmental Entomology, 46, 470–479.

Martelli, M. G., Ward, M. M., and Fraser, A. M. (2004). Ant diversity sampling on the southern Cumberland Plateau: a comparison of litter sifting and pitfall trapping. *Southeastern Naturalist*, 3, 113–126.

Medley, K. E., Okey, B. W., Barrett, G. W., Lucas, M. F. and Renwick, W. H. (1995). Landscape change with agricultural intensification in a rural watershed, southwestern Ohio, U.S.A. *Landscape Ecology*, 10, 161–176.

Menalled, F. D., Smith, R. G., Dauer, J. T., and Fox, T. B. (2007). Impact of agricultural management on carabid communities and weed seed predation. *Agriculture, Ecosystems, and Environment*, 118, 49–54.

Menke, S. B., Gaulke, E., Hamel, A., and Vachter, N. (2015). The effects of restoration age and prescribed burns on grassland ant community structure. *Environmental Entomology*, 44(5), 1336–1347.

Menke, S. B., Guénard, B., Sexton, J. O. et al. (2011). Urban areas may serve as habitat and corridors for dry-adapted, heat tolerant species; an example from ants. *Urban Ecosystems*, 14, 135–163.

Merickel, F. W. and Clark, W. H. (1994). *Tetramorium caespitum* (Linnaeus) and *Liometopum luctuosum* WM Wheeler (Hymenoptera: Formicidae): New state records for Idaho and Oregon, with notes on their natural history. *Pan-Pacific Entomologist*, 70, 148–158.

Moranz, R. A., Debinski, D. M., Winkler, L. et al. (2013). Effects of grassland management practices on ant functional groups in central North America. *Journal of Insect Conservation*, 17, 699–713.

Morrison, L. W. (2002). Long-term impacts of an arthropod-community invasion by the imported fire ant, *Solenopsis invicta*. *Ecology*, 83, 2337–2345.

Morrison, L. W. and Porter, S. D. (2003). Positive association between densities of the Red Imported Fire Ant, Solenopsis invicta (Hymenoptera: Formicidae) and generalized ant and arthropod diversity. *Environmental Entomology*, 32, 548–554.

Morrison, L. W., Korzukhin, D., and Porter, S. D. (2005). Predicted range expansion of the invasive fire ant, Solenopsis invicta, in the eastern United States based on the VEMAP global warming scenario. *Diversity and Distributions*, 11, 199–204.

Nemec, K. T. (2014). Tallgrass prairie ants: their species composition, ecological roles, and response to management. *Journal of Insect Conservation*, 18, 509–521.

Pećaravić, M., Danoff-Burg, J., and Dunn, R. R. (2010). Biodiversity on broadway – enigmatic diversity of the societies of ants (Formicidae) on the streets of New York City. *Public Library of Science Biology*, 5, 1–8.

Peck, S. L., McQuaid, B. and Campbell, C. L. (1998). Using ant species (Hymenoptera: Formicidae) as a biological indicator of agroecosystem condition. *Environmental Entomology*, 27, 1102–1110.

Pelini, S. L., Boudreau, M. McCoy, N. et al. (2011). Effects of short-term warming on low and high latitude forest ant communities. *Ecosphere*, 2, art 62.

Pelini, S. L., Diamond, S. E., MacLean, H. et al. (2012). Common garden experiments reveal uncommon responses across temperatures, locations, and species of ants. *Ecology and Evolution*, 2, 3009–3015.

Penick, C. A, Savage, A. M. and Dunn, R. R. (2015). Stable isotopes reveal links between human food inputs and urban ant diets. *Proceedings of the Royal Society B*, 282(1806), 1–8.

Peters, V. E., Campbell, K. U., Dienno, G. et al. (2016). Ants and plants as indicators of biodiversity, ecosystem services and conservation value in constructed grasslands. *Biodiversity and Conservation,* 25, 1481–1501.

Peterson, D. E., Zwolfer, K., and Frandkin, J. (1998). Ant fauna of reconstructed tallgrass prairie in Northeastern Illinois. *Transactions of the Illinois State Academy of Science,* 91, 85–90.

Philpott, S. M., Perfecto, I., Armbrecht, I., and Parr, C. L. (2010). Ant diversity and function in disturbed and changing habitats. In L. Lach, C. L. Parr, and K. L. Abbott, editors. *Ant Ecology.* Oxford: Oxford University Press, pp. 137–156.

Phipps, S. J. (2006). *Biodiversity of Ants (Hymenoptera: Formicidae) in Restored Grasslands of Different Ages.* University of Missouri, Columbia.

Pryor, S. C., Scavia, D. Downer, C. et al. (2014). Chapter 18: Midwest. In: J. M. Melillo, T. C. Richmond, and G. W. Yohe, editors. *Climate Change Impacts in the United States: The Third National Climate Assessment.* US Global Change Research Program, pp. 418–440.

Ragsdale, D. W., Landis, D. A., Brodeur, J., Heimpel, G. E., and Desneux, N. (2011). Ecology and management of the soybean aphid in North America. *Annual Review of Entomology,* 56, 375–399.

Renwick, W. H., Vanni, M. J., Zhang, Q., and Patton, J. (2008). Water quality trends and changing agricultural practices in a Midwest U.S. watershed, 1994–2006. *Journal of Environmental Quality,* 37, 1862–1874.

Samson, F. and Knopf, F. (1994). Prairie conservation in North America. *BioScience,* 44(6), 418–421.

Sanford, M. P., Manley, P. N. and Murphy, D. D. (2009). Effects of urban development on ant communities: Implications for ecosystem services and management. *Conservation Biology,* 23, 131–141.

Spiesman, B. J. and Cumming, G. S. (2008). Communities in context: the influences of multiscale environmental variation on local ant community structure. *Landscape Ecology,* 23, 313–325.

Steiner, F. M., Schlick-Steiner, B. C., Trager, J. C. et al. (2006). *Tetramorium tsushimae,* a new invasive ant in North America. *Biological Invasions,* 8, 117–123.

Steiner, F. M., Schlick-Steiner, B. C., Vanderwal, J. et al. (2008). Combined modelling of distribution and niche in invasion biology: a case study of two invasive *Tetramorium* ant species. *Diversity and Distributions,* 14, 538–545.

Strittholt, J. R. and Boerner, R. E. J. (1995). Applying biodiversity gap analysis in a regional nature reserve design for the Edge-of-Appalachia, Ohio (U.S.A.). *Conservation Biology,* 9, 1492–1505.

Talbot, M. (1934). Distribution of ant species in the Chicago region with reference to ecological factors and physiological toleration. *Ecology,* 15, 416–439.

   (1975). A list of the ants of the Edwin S. George Reserve. *The Great Lakes Entomologist,* 8, 245–246.

Theobald, D. M. (2002). Land-use dynamics beyond the American urban fringe. *The Geographical Review,* 91, 544–564.

Thompson, B. and McLachlan, S. (2007). The effects of urbanization on ant communities and myrmecochory in Manitoba, Canada. *Urban Ecosystems,* 10, 43–52.

Underwood, E. C. and Fisher, B. L. (2006). The role of ants in conservation monitoring: if, when, and how. *Biological Conservation,* 132, 166–182.

Uno, S., Cotton, J. and Philpott, S. M. (2010). Diversity, abundance, and species composition of ants in urban green spaces. *Urban Ecosystems*, 13, 425–441.

US Department of Agriculture (2016). Farm Service Agency, Conservation Reserve Program. From: www.fsa.usda.gov/programs-and-services/conservation-programs (Accessed September 2016).

Voigt, W., Perner, J., and Hefin Jones, T. (2007). Using functional groups to investigate community response to environmental changes: two grassland case studies. *Global Change Biology*, 13, 1710–1721.

Wang, C., Strazanac, J. S., and Butler, L. (2001). Association between ants (Hymenoptera: Formicidae) and habitat characteristics in oak-dominated mixed forests. *Environmental Entomology*, 30, 842–848.

Wang, D., McSweeney, K., Lowery, B., and Norman, J. M. (1995). Nest structure of ant *Lasius neoniger* Emery and its implications to soil modification. *Geoderma*, 66, 259–272.

Wetterer, J. K. (2011). Worldwide spread of the yellow-footed ant, *Nylanderia flavipes* (Hymenoptera: Formicidae). *Florida Entomologist*, 94, 582–587.

Wetterer, J. K. and Radchenko, A. G. (2011). Worldwide spread of the ruby ant, *Myrmica rubra* (Hymenoptera: Formicidae). *Myrmecological News*, 14, 87–96.

Wheeler, G. C., Wheeler, J. N., and Kannowski, P. B. (1994). Checklist of the ants of Michigan (Hymenoptera: Formicidae). *The Great Lakes Entomologist*, 26, 297–310.

Wielgoss, A., Tscharntke, T., Rumede, A. et al. (2014). Interaction complexity matters: disentangling services and disservices of ant communities driving yield in tropical agroecosystems. *Proceedings of the Royal Society B*, 281, 1–10.

Wilson, E. O. (2003). *Pheidole in the New World: A Dominant, Hyperdiverse Ant Genus.* Cambridge, MA: Harvard University Press.

Wyckhuys, K. A. G., Koch, R. L. and Heimpel, G. E. (2007). Physical and ant-mediated refuges from parasitism: implications for non-target effects in biological control. *Biological Control*, 40, 306–313.

2 Diversity and Specificity of Ant-Plant Interactions in Canopy Communities: Insights from Primary and Secondary Tropical Forests in New Guinea

Petr Klimes[*]

Introduction

Tropical equatorial forests represent one of the most diverse ecosystems on Earth and harbour the highest species diversity of trees and ants (Lach et al., 2010; Basset et al., 2012). These ecosystems are under increasing threat from human-mediated environmental changes and disturbances. In particular, a large proportion of forests have been altered to become structurally simpler and fragmented habitats like secondary forests and plantations. Ants are extremely abundant and relatively species-rich in the canopies of lowland tropical rainforests, where they play various ecological roles in plant-insect food webs. However, their species diversity and composition is significantly affected by habitat disturbance, which may have serious consequences for the ant-plant communities. Although a wide range of interactions between ants and their host plants have been studied in detail in the tropics using particular myrmecophytic plant species and their mutualistic ant partners as models (e.g. Janzen, 1966; Rico-Gray & Oliveira, 2007), a comparative analysis of

[*] This work builds on the ideas from my doctoral thesis. I am thankful to the editors of the book for inviting me to write this chapter. I am obliged to Papua New Guinean customary landowners of Wanang for allowing me to work in their forests and for their logistical support during the project. Special thanks are to Cliffson Idigel and Maling Rimandai from New Guinea Binatang Research Center, who helped with data collecting and management. Vojtech Novotny, Milan Janda, Jan Hrcek and George D. Weiblen are greatly acknowledged for advice on data collection and support for the project. Philip T. Butterill helped with data-sorting and database maintenance, and Petr Smilauer advised me on multivariate analyses, Martin Libra helped with sorting ant samples. Archie J. McArthur, Bonnie Blaimer, Rudolf J. Kohout, Eli M. Sarnat, Steven O. Shattuck, Robert W. Taylor and Philip S. Ward are acknowledged for assistance with ant species identifications. Kipiro Damas, Kenneth Molem, Brus Isua and Timothy J. S. Whitfeld provided tree species identifications. Carl Wardhaugh, Vojtech Novotny, Tom M. Fayle, Maurice Leponce and an anonymous reviewer made valuable and constructive comments on the manuscript. This study was supported by the Czech Science Foundation Center for Tropical Biology (14-36098G) and European Research Council Advanced Grant (GA 669609).

diversity and specificity of all ant-tree interactions, including non-symbiotic ones, at the whole forest level is lacking. This is because to map all ant and tree partners within such species-rich ecosystem is a challenging task, even at small spatial scales. Consequently, changes in the overall diversity and specificity of interactions between host trees and their ant communities, including trends through different forest successional stages, remain poorly explored. This knowledge is particularly important as it allows us to predict how ant communities will react to human-mediated disturbance of forests in the future.

In this chapter I review current knowledge of changes due to tropical forest disturbance and fragmentation in ant communities in the canopies of primary (old growth) and disturbed (secondary and logged) forest ecosystems. Furthermore, I examine arboreal ant communities and their host trees in primary and secondary lowland forest in New Guinea, where a plot-based census of all tree-dwelling ants and their nests was conducted. I show how the diversity and host-specificity of the interactions between ants, their host tree species and nest microhabitats within host trees change between the two forest stages, and discuss potential consequences for the conservation of diverse canopy communities.

Ecological Importance of Ant Communities in Tropical Forest Canopies and Changes due to Forest Disturbance and Fragmentation

Tree canopies have been estimated to host approximately half of all ant species living in the tropical forests, according to sampling of trees in Bornean (Floren et al., 2014) and Amazonian forests (Ryder Wilkie et al., 2010). Ants are also very abundant in forest canopies, sometimes reaching up to two-thirds of the total biomass of arboreal invertebrates (Davidson et al., 2003). Some species forage between as well as within trees using lianas and branches connecting individual trees, or climb to the canopy from the ground (Yanoviak et al., 2012; Klimes et al., 2015). Others belong to tree-dwelling species that build their nests directly in trees, inhabiting various readily-available nesting resources, such as leaves, hollow twigs, or epiphytes, or even build their own nests from various plant materials (Figure 2.1; Chapters 10, 11). The forest canopy is an important source of energy-rich resources, offered by plants to ants both directly, for example plant nectaries, food bodies, and indirectly, for example via honeydew from hemipteran symbionts and/or attracting insect prey (Blüthgen et al., 2000; Rico-Gray & Oliveira, 2007). In return, it is assumed that most trees benefit from hosting ants, as their workers protect the plant hosts against natural enemies. This has been experimentally confirmed in multiple ant-plant systems (Rico-Gray & Oliveira, 2007). However, interactions between ants, herbivores and host plants are very complex, and range from negative through neutral to positive depending on the insect species involved and their feeding strategy (Oliveira et al., 2002; Frederickson & Gordon, 2007).

Disturbance, clearance and fragmentation of tropical forests by humans are currently the biggest threat to biodiversity in these species-rich ecosystems (Butler &

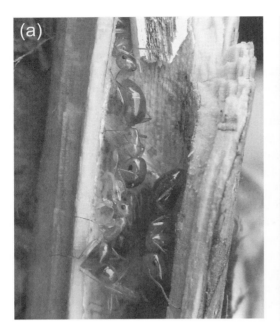 

Figure 2.1.   Arboreal ant nests in tree canopies of New Guinea lowland forests. (a) Nest of *Colobopsis* cf. *macrocephala* in a dead twig – example of internal nest site in host tree. (b) Carton nest of *Polyrhachis luteogaster* – example of external nest site on a leaf of host tree (photographs by P. Klimes). See also Table 2.1. (A black-and-white version of this figure will appear in some formats. For the color version, please refer to the plate section.)

Laurance, 2008; Shearman & Bryan, 2011). Tree size distribution, stem density, plant taxonomic diversity and canopy-connectivity are vegetation parameters that usually change considerably along successional and disturbance gradients (Chazdon, 2014). These structural changes have important consequences for tree-dwelling ant communities. When forests are severely disturbed, i.e. temporarily cleared for shifting agriculture, logged or converted to new agro-ecosystems as plantations, the species richness of arboreal ant communities declines considerably and species composition is fundamentally changed (Floren et al., 2001; Fayle et al., 2010; Klimes et al., 2015; Chapter 3). During secondary regeneration, ant communities begin to recover to pre-disturbance levels of species richness and composition, particularly if young successional forests are embedded in a landscape consisting of a matrix of primary forest (Vasconcelos & Bruna, 2012). However, it may take decades or longer until they completely recover (Bihn et al., 2008). For arboreal ants, even selective logging and moderate degradation can have a relatively strong effect on diversity and/or taxonomic composition (Widodo et al., 2004). Conversely, moderate disturbances may be less significant for ground-nesting ant species (Woodcock et al., 2011; Vasconcelos & Bruna, 2012). The changes to ant communities due to forest disturbance and fragmentation have important consequences for all forest strata and the whole ecosystem, because degraded forests and plantations are more vulnerable to invasion by non-native species that alter ant interspecific interactions (Lach et al., 2010; Fayle et al., 2013; Chapters 12–15). Furthermore, the structure of food-webs involving ants can shift with forest disturbance (Blüthgen et al., 2003; Woodcock et al., 2013). Similarly, forest fragmentation also affects ant communities in primary forest fragments, depending on fragment size and edge effects (Vasconcelos & Bruna, 2012). Nevertheless, there is increasing evidence that secondary and logged forests harbour a comparable overall diversity of epigeic and leaf-litter ants as pristine forests if considered over the whole region in a mosaic landscape, due to higher species turnover among forests with disturbance history (Woodcock et al., 2011; de Castro Solar et al., 2015). While ground-dwelling species have received some attention, the landscape consequences of forest changes for canopy biodiversity and function with respect to ants are still poorly known. However, it is well documented that invertebrates, including ants, occupy vital functional roles in the tropical forests (e.g. predation, redistribution of nutrients) (Dejean et al., 2007; Ewers et al., 2015).

Methods to Access the Rainforest Canopy and Sample the Tree-Dwelling Ant Fauna

Despite the ecological importance of arboreal ants, they are relatively understudied at the community level compared to their epigeic counterparts due to relatively difficult access to tree canopies. Consequently, most of the protocols recommended for standardised sampling of ants focus on the ground level (Agosti et al., 2000). A common technique to sample canopy ants and other arboreal arthropods is *canopy-fogging*, where a tree crown is fumigated with an insecticide and dead insects (ants) are then collected on plastic sheets spread beneath the tree (Adis et al., 1984; Floren & Linsenmair, 1997). Such a remote and destructive sampling, however, cannot provide details about the ecology of ant species, whether they were nesting in the tree or not, or which food resources they utilised (Floren, 2005; Klimes et al., 2015). Hence, a number of techniques have been developed to access tree canopies, such as canopy cranes, balloons, canopy-bridges and walkways, as well as a range of tree-climbing techniques (Basset et al., 2012; Lowman et al., 2012). These methods allow researchers to observe ants foraging at experimental baits, search for nests and explore interactions between ants, host plants, and other insect species, as well as conduct experiments (e.g. Blüthgen & Fiedler, 2002; Tanaka et al., 2010). Although combinations of some of these methods have been tested in recent years (e.g. Dejean et al., 2010; Yusah et al., 2012), they are usually unable to provide accurate estimates of the overall diversity of ants in trees, or record more complex information on interactions with host tree and its nest sites. More time-consuming protocols such as destructive sampling of nests are typically needed to gather such information (Tanaka et al., 2010). This is because a significant number of arboreal ant species nest cryptically inside tree structures or epiphytes (Klimes & McArthur, 2014; Fayle et al., 2015), while other species forage across multiple adjacent trees

(Yanoviak et al., 2012; Klimes et al., 2015). One solution to this problem is to fell trees and collect ants directly from the cryptic microhabitats in which they choose to nest. This is a feasible and suitable method, however, only when taking advantage of the existing deforestation, such as during logging operations or slash-and-burn agriculture.

A Case Study of Complete Ant Communities in the Canopies of Primary and Secondary Rainforest in New Guinea

In New Guinea, the arboreal ant fauna was surveyed by destructively sampling nests in felled trees within one ha plot of primary and one ha plot of secondary lowland forest (Klimes et al., 2012). New Guinean rainforests have one of the highest levels of diversity of ant and tree species, with a single hectare typically supporting over 200 tree species and perhaps a similar number of ant species (Whitfeld et al., 2012; Klimes et al., 2015). New Guinean forests are an ideal model ecosystem to explore the extraordinarily species-rich interactions between trees and ants. The island itself represents one of the last three regions on Earth (together with Amazonia and the Congo) where large areas of pristine tropical rainforest remain (Novotny, 2010; Shearman & Bryan, 2011).

Sampling was conducted with the cooperation of indigenous landowners of the forests (Wanang, Madang Province, Papua New Guinea), who practise slash-and-burn agriculture in small gardens up to approximately one ha in area. Hence, arboreal ant nests could be exhaustively censused on all vegetation by sampling trees that were gradually felled for the establishment of food gardens, without contributing to further deforestation. On the contrary, forest conservation as an alternative to logging has been successfully supported at the site because this research activity brought employment to the local community linked to rainforest conservation (Novotny, 2010). The primary forest plot was in an old-growth rainforest that had not been disturbed for at least 50 years, while the secondary forest plot represented 10 years of forest regeneration from complete clearance. The diversity and composition of ant communities and their nest sites were studied in all trees with a DBH (diameter at breast height) above 5 cm. The forest canopy was defined as all above-ground biomass, including trunks. More details about sampling methodologies, forest site characteristics and ant species abundances can be found in Klimes et al. (2015).

Previous studies from this project focused on the diversity and composition of the ant communities in these felled forest plots (Klimes et al., 2012; Klimes & McArthur, 2014; Klimes et al., 2015). They revealed that secondary forest supports only half the number of ant species found in primary forest. This difference was attributed entirely to turnover in ant species between individual trees, as the secondary and primary plots hosted a similar number of species per tree. Null models based on nest-site presences using the secondary forest as a template dataset, and the primary forest as a pool of host trees, showed that 50 percent

of the decrease in total ant species richness can be explained by the overall differences in tree density, size and taxonomic composition between the two forests
(Klimes et al., 2012). In particular, the effects of tree taxa on ants was rather low
as taxonomic dissimilarities between individual trees and between their respective
ant communities were weakly correlated in both forest types, but the correlation
was stronger at the species level than at higher taxonomic levels (family, genus).
The results suggested that the reduction in tree species diversity is not the predominant driver of ant species loss in secondary forests compared to primary
forests. However, species-level ant-plant interactions have not been assessed, and
the level of specialisation to particular microhabitats (nest sites) and tree species
in primary and secondary forests remain unknown. As the secondary forest not
only hosts a less diverse community of ants compared to primary forest, but also
a very different ant fauna dominated by invasive species (Klimes et al., 2015),
one can expect changes in the strength and frequency of interactions between ant
and tree species due to shifts in the identity of partners (Rico-Gray & Oliveira,
2007). Indeed, it is likely that host plant identity also plays some role in shaping
ant-plant interactions in this system since the composition of ant assemblages
was more similar between conspecific than allospecific trees in both forest types
(Klimes et al., 2012). However, to what extent this may change along a forest disturbance gradient, and how it is correlated with the range of nest microhabitats
provided by trees is unknown.

Analytical Approach

Previously published work using the dataset from felled trees in New Guinea was
restricted to single 0.3-ha sections (80 × 40 m) in each of the two one ha plots
(Klimes et al., 2015). In this chapter, I examine the diversity and distribution of ant
nests in trees in 0.4-ha areas (i.e. extension to 100 × 40 m in each of the two forest
types). In total, data from 326 trees and 278 trees with ant nests in primary and secondary forest respectively are studied here. Ants were also sampled in the remaining
area of the plots, but this was restricted to collecting common foraging species and
is therefore not relevant to the current study. Active nest sites were assigned to 12
categories, based on the microhabitats in the host trees utilised by ants (described
in detail in Table 2.1). These categories do not distinguish myrmecophytic and non-
myrmecophytic hosts as the focus here is on all tree species within the two forests,
with most trees hosting ants in living tissues and other microhabitats. Furthermore,
distinguishing myrmecophytic and non-myrmecophytic tree species is challenging
because the literature on ant-plant symbioses in New Guinea is scarce. Instead,
ants were recorded as inhabiting either living tissues (twigs, branches) or dead
parts of the host tree when dissecting the trees in the field. The only exception
was for myrmecophytic epiphytes, i.e. nests inside obligate ant plants of the genera
Myrmecodia and *Hydnophytum* (Rubiaceae), which grow attached to the canopy
trees and develop special tuber structures for hosting ants (Huxley, 1978).

Table 2.1 Definition of Nest Site Categories Used for Characterisation of Nesting Habits of Ant Species in Trees in Papua New Guinea Lowland Rainforests

Ant nest site	Location	Definition
Dead twig	Internal	Nest in cavity of a small, dry, hollow branch (≤ 5 cm diameter)
Dead branch	Internal	Nest in cavity of a large, dry, hollow branch (> 5 cm diameter)
Live twig	Internal	Nest in cavity (gallery) of a small live branch (≤ 5 cm diameter)
Live branch	Internal	Nest in cavity (gallery) of a large live branch (> 5 cm diameter)
In trunk	Internal	Nest inside cavity (gallery) in the trunk of host tree
Under bark	Internal	Nest in space under the bark of the host tree
Epiphytic roots	External	Nest in aerial soil and leaf litter at the base of epiphytes (i.e. under roots of orchids, ferns, mosses)
Liana	External	Nest in cavity in (or under) the stem of a liana (climber) attached to the host tree
On bark	External	Carton nest on the bark of the host tree (made by ants or termites from wood or soil material)
On leaves	External	Carton or silk nest on a leaf (or on several leaves rolled together)
Myrmecophytic epiphyte	External	Nest inside of myrmecophytic epiphyte, i.e. plants of genera *Myrmecodia* and *Hydnophytum*
Trunk base	External	Soil nest at the base of the host tree or on its stalk roots

Location describes whether the nest site was inside the host tree (internal), or outside the host tree on its surface and/or in other plants attached to it (external). Note that the last nest sites (in italics) are present in < 3 tree individuals and are not part of the analysed data (Table 2.2). See also Figure 2.1.

Each 0.4-ha plot was divided into 10 sub-plots (20 × 20m, 400 m²) to allow for an analysis of the spatial variation in ant, nest site and tree species density and diversity in order to ascertain the variability within and between the two forest types in these measures at this smaller scale (Figure 2.2). All trees, including unoccupied ones, and all taxa were included. Subsequent analyses were restricted to the occupied trees in each plot, as the focus was on the host-specificity of realised interactions between ants and host tree species (Figures 2.3–2.6). Furthermore, the data were limited to the ant species and nest site categories that were replicated over at least three tree individuals in each forest plot, when also excluding simultaneously all singleton and doubleton tree species. This was necessary to minimise possible false positives in specific ant-plant and ant-nest site relationships, for example, cases where multiple nests of a single ant species were found in only one or two individuals of rare tree species. These limitations reduce the dataset to approximately half of all nest records and less than half of ant and tree species found in each forest plot (see Table 2.2 for further details).

Nesting preferences within ant communities from each forest plot were assessed using multivariate analyses with variation partitioning to test for separate effects of the nest sites and tree species on ants within each forest community. The analysed dataset represented a matrix of tree individuals (as samples) and nest occurrences for each ant species. As the gradient in the data was long and matrices contained many zeros (i.e. relatively few ant species were present in each tree from the species pool), canonical correspondence analyses (CCA) with forward step-wise regression was considered the most suitable multivariate model (Šmilauer & Lepš, 2014). The overall variance explained by CCA was then partitioned and compared between the groups of variables. The explanatory power of this analysis is expressed as the percentage of the variability in ant communities that is significantly explained by each of the two sets of variables separately and combined for: (1) *environmental nest predictors*: tree size (DBH in cm), height of the nest above ground (m), and nest site composition in each tree; (2) *tree species*, and (3) *their shared effect*. As host trees and not individual nests were considered as samples, in the cases where multiple nests were found in a tree the mean height of nests in the tree and the proportions of nest occurrences per nest-site category were used in analyses. Analyses were conducted in CANOCO 5 using the conservative approach of comparing the adjusted variation and p-value (999 random permutations, $p_{adj.} < 0.05$) as recommended for datasets with lot of species and variables (Šmilauer & Lepš, 2014).

The level of specialisation of individual ant species to particular nest sites and tree species was assessed using two host-specificity measures, *Llloyd's* index and HS_k, since both have previously been used for analogous analyses of the interactions between herbivorous insects and their host trees (Novotny et al., 2004; Wardhaugh et al., 2013). Both indices are frequency-based measures that depend on the distribution of individual ant species across the respective categories (nest sites or tree species), where higher values indicate a higher specificity of the ant species to the tree host species or nest site type. Although these specificity indices are often correlated, no single measurement is considered superior, and they can vary depending on the number of categories and the distribution of records among them. I therefore used both Xindices in the analyses. Lloyd's index is defined for each ant species as:

$$L = 1 + (S_x^2 - \overline{X}) / \overline{X}^2$$

where S_x^2 is the variance and $\overline{X}$ is the mean number of nests per nest site category. The resulting value is a number > 0. In contrast, HS_k index (i.e. host specificity of an ant species to a nest category (k)) is calculated as:

$HS_k = (\text{Number of nests on the preferred nest site}) / (\text{Total number of nests}),$

Table 2.2 Data on Ant-Occupied Trees and Their Ant Nests in the Primary and Secondary Lowland Forest Plots (0.4 ha Each) in Papua New Guinea

	Ant species		No. of tree		No. of nest		No. of nest sites (ordered as in Table 2.1)	
Code	Name		Individuals	Species	Individuals	Site types	Dead twig	Dead branch
Primary forest		–	–	–	–	–	–	–
ANON001	*Anonychomyrma* cf. *scrutator*		22	13	33	3	–	–
CAMP001*	*Colobopsis vitrea*		54	24	60	8	5	**18**
CAMP005	*Colobopsis* aff. *conithorax*		4	4	4	4	1	1
CAMP006	*Colobopsis conithorax*		8	6	8	5	1	1
CAMP007	*Camponotus* sp.7 aff. *trajanus*		8	6	11	4	–	**4**
CAMP008	*Colobopsis* aff. *sanguinifrons*		17	3	26	4	1	–
CAMP010*	*Colobopsis* cf. *macrocephala*		3	3	4	2	**3**	–
CAMP016	*Camponotus dorycus confusus*		5	3	5	4	–	1
CREM002	*Crematogaster elysii*		13	10	17	6	**6**	1
CREM003	*Crematogaster polita*		66	22	106	9	2	13
CREM007*	*Crematogaster* sp.7 aff. *fritzi*		26	15	27	9	2	2
DIAC001	*Diacamma rugosum*		11	10	11	2	2	–
PARA001	*Paraparatrechina pallida*		23	16	25	6	–	3
PARA003	*Paraparatrechina minutula*		20	12	22	8	1	3
PARA005	*Nylanderia* sp.5 aff. *vaga*		5	4	5	3	–	**2**
PHEI004	*Pheidole hospes*		7	6	7	3	–	–
PHEI007*	*Pheidole* sp.7 aff. *gambogia*		3	3	5	3	–	–
PHEI014	*Pheidole* sp.14 aff. *gambogia*		4	3	4	3	–	–
PHEI024	*Pheidole* sp.24 aff. *amber*		13	11	13	7	–	2
PHIL001	*Philidris* cf. *cordata*		3	3	4	2	–	–
PODO002	*Podomyrma* sp.2 aff. *basalis*		4	3	4	3	–	1
PODO003	*Podom.* sp.3 aff. *laevifrons*		4	3	8	3	–	–
POLY001	**Polyrhachis esuriens**		8	7	9	1	–	–
POLY004	**Polyrhachis debilis**		7	6	10	1	–	–
POLY008	**Polyrhachis alphea**		32	14	43	1	–	–
POLY010	**Polyrhachis luteogaster**		20	10	27	1	–	–
POLY011*	**Polyrhachis queenslandica**		5	5	5	1	–	–
POLY015	*Polyrhachis waigeuensis*		6	5	6	3	–	–
POLY020	**Polyrhachis dolomedes**		4	3	5	1	–	–
ROGE001	**Rogeria stigmatica**		3	3	3	1	–	–
SOLE004	*Solenopsis papuana*		13	7	15	6	1	1
STRU002	*Strumigenys szalayi*		3	3	3	3	–	1
TECH003*	*Technomyrmex difficilis*		3	3	4	3	1	–
TETP001*	*Tetraponera laeviceps*		7	5	10	2	–	–
TETR006	*Tetramorium* aff. *bicarinatum*		6	5	6	3	–	**3**
Tested data	35 ant species		221	30	555	10	–	–
All data	85 ant species		326	103	871	12	–	–

No. of nest sites (ordered as in Table 2.1)								Lloyd index		HS_k index	
Live twig	Live branch	In trunk	Under bark	Epiph. roots	Liana	On bark	On leaves	Nest sites	Tree spp.	Nest sites	Tree spp.
5	**16**	12	–	–	–	–	–	3.92	1.94	0.48	0.18
–	11	2	1	4	16	3	–	2.06	1.33	0.30	0.11
–	1	–	–	–	1	–	–	0.17	0.22	0.25	0.25
–	**3**	1	–	–	2	–	–	1.42	3.00	0.38	0.38
–	**4**	–	1	–	2	–	–	2.38	3.00	0.36	0.38
11	**13**	–	–	–	–	1	–	4.30	19.79	0.50	0.82
–	–	–	–	1	–	–	–	4.33	0.31	0.75	0.33
–	–	1	–	1	–	**2**	–	1.00	5.14	0.40	0.40
–	2	3	4	–	1	–	–	1.88	1.51	0.35	0.23
4	4	2	2	–	4	**74**	1	5.44	2.16	0.70	0.15
1	**6**	3	4	**6**	2	1	–	1.21	1.57	0.22	0.19
–	–	–	–	**9**	–	–	–	6.79	0.57	0.82	0.18
–	7	3	2	**9**	1	–	–	2.21	1.07	0.36	0.13
–	1	1	3	**9**	3	1	–	2.01	2.03	0.41	0.20
–	–	–	1	**2**	–	–	–	1.89	2.66	0.40	0.40
–	–	1	–	**5**	1	–	–	4.58	1.38	0.71	0.29
–	1	–	1	**3**	–	–	–	2.78	0.31	0.60	0.33
–	–	1	**2**	1	–	–	–	1.56	4.10	0.50	0.50
1	1	3	1	**4**	1	–	–	1.29	0.78	0.31	0.15
–	–	–	–	2	–	**2**	–	2.94	0.31	0.50	0.33
–	**2**	–	–	1	–	–	–	1.56	4.10	0.50	0.50
–	**4**	3	–	1	–	–	–	3.15	4.10	0.50	0.50
–	–	–	–	–	–	–	**9**	**9.89**	1.06	**1.00**	0.25
–	–	–	–	–	–	–	**10**	**10.00**	1.38	**1.00**	0.29
–	–	–	–	–	–	–	**43**	**10.77**	2.00	**1.00**	0.19
–	–	–	–	–	–	–	**27**	**10.63**	2.50	**1.00**	0.20
–	–	–	–	–	–	–	**5**	**9.00**	0.17	**1.00**	0.20
–	–	1	–	–	–	1	**4**	3.78	1.86	0.67	0.33
–	–	–	–	–	–	–	**5**	9.00	4.10	**1.00**	0.50
–	–	–	–	**3**	–	–	–	**7.67**	0.31	**1.00**	0.33
–	–	1	1	**10**	1	–	–	4.41	2.98	0.67	0.23
1	–	1	–	–	–	–	–	0.26	0.31	0.33	0.33
–	–	–	**2**	1	–	–	–	1.56	0.31	0.50	0.33
7	3	–	–	–	–	–	–	5.33	3.91	0.70	0.43
–	–	2	–	1	–	–	–	2.54	1.86	0.50	0.33
–	–	–	–	–	–	–	–	~4.11	~2.40	~0.59	~0.31
–	–	–	–	–	–	–	–	–	–	–	–

(continued)

Table 2.2 *(cont.)*

Code	Ant species Name	No. of tree Individuals	Species	No. of nest Individuals	Site types	No. of nest sites (ordered as in Table 2.1) Dead twig	Dead branch
Secondary Forest							
ANOP001	*Anoplolepis gracilipes*	5	5	6	6	–	1
CAMP001*	*Colobopsis vitrea*	7	4	7	4	1	2
CAMP004	*Colobopsis aruensis*	18	6	19	3	6	5
CAMP010*	*Colobopsis* cf. *macrocephala*	89	13	113	5	<u>**54**</u>	19
CAMP011	*Camponotus* aff. *pictostriatus*	4	2	7	2	2	–
CAMP012	*Camponotus* cf. *chloroticus*	14	9	20	6	<u>6</u>	2
CREM005	*Crematogaster flavitarsis*	38	8	45	4	13	12
CREM007*	*Crematogaster* sp.7 aff. *fritzi*	33	13	37	9	4	<u>13</u>
MONO001	*Monomorium floricola*	29	10	34	6	9	<u>11</u>
MONO002	*Monomorium intrudens*	14	7	15	3	5	1
PARA010	*Paraparatrechina* sp.10	5	3	8	5	–	1
PHEI007*	*Pheidole* sp.7 aff. *gambogia*	17	10	24	7	2	<u>10</u>
POLY009	***Polyrhachis* aff. *neptunus***	45	11	83	2	1	–
POLY011*	***Polyrhachis queenslandica***	7	6	9	1	–	–
TECH001	*Technomyrmex brunneus*	58	11	86	7	16	14
TECH003*	*Technomyrmex difficilis*	3	2	3	3	1	1
TETP001*	*Tetraponera laeviceps*	5	3	5	2	<u>**4**</u>	1
TETR002	*Tetramorium kydelphon*	4	3	4	3	1	<u>**2**</u>
Tested data	18 ant species	219	15	525	9		
All data	50 ant species	278	49	674	11		

Only ant species and nest site categories present in > 2 individual trees in each forest plot are included; rare tree species are removed (only species with > 2 individuals are retained). Specialists (HS_k > 0.9) to the particular nest site type or tree species are given in bold letters, the most used nest site types by individual ant species are underlined.

* Species common for both forest types; ~ mean index per ant species.

where the preferred nest site is the category with the highest number of nests. HS_k is therefore a simple proportion of the records in the most commonly used site, and is maximised ($HS_k = 1$), if all nests for that ant species occur only in one category. Both indices were also used to calculate the specificity of each ant species to tree species by using the number of occupied tree individuals instead of the numbers of nest records. The mean specificity of ant species to nest sites and tree species (i.e. mean L and HS_k across each ant community) and its variance was then compared between primary and secondary forests. Furthermore, nest-site specificity was compared for each forest type separately between ants that nest internally on trees (inside of trunk, twigs, branches and under the bark) and external nesters (the other nest sites) (Figure 2.1, Table 2.1). Following studies on herbivores (e.g. Wardhaugh et al., 2013), ant species where HS_k > 0.9 on either particular nest sites or host tree

	No. of nest sites (ordered as in Table 2.1)							Lloyd index		HS_k index	
Live twig	Live branch	In trunk	Under bark	Epiph. roots	Liana	On bark	On leaves	Nest sites	Tree spp.	Nest sites	Tree spp.
–	1	1	1	–	1	–	1	0.06	0.14	0.17	0.20
–	–	–	–	–	**3**	1	–	1.69	2.05	0.43	0.29
–	–	–	–	–	**8**	–	–	2.91	2.96	0.42	0.28
1	3	–	–	–	36	–	–	3.43	3.13	0.48	0.36
–	–	–	–	–	**5**	–	–	4.58	6.22	0.71	0.75
1	2	3	–	–	**6**	–	–	1.70	1.48	0.30	0.29
–	1	–	–	–	**19**	–	–	3.05	4.41	0.42	0.47
1	3	6	2	–	4	2	2	1.55	1.70	0.35	0.24
–	–	2	3	–	7	–	2	1.96	1.69	0.32	0.21
–	–	–	–	–	**9**	–	–	4.09	1.32	0.60	0.21
–	1	–	1	–	1	**4**	–	1.91	4.00	0.50	0.60
–	1	2	6	–	2	1	–	2.14	1.22	0.42	0.24
–	–	–	–	–	–	–	**82**	**9.65**	3.43	**0.99**	0.40
–	–	–	–	–	–	–	**9**	**9.00**	0.74	**1.00**	0.29
–	–	3	11	–	**24**	8	10	1.58	2.57	0.28	0.26
–	–	–	–	–	1	–	–	0.25	3.86	0.33	0.67
–	–	–	–	–	–	–	–	4.96	4.00	0.80	0.60
–	–	–	–	–	1	–	–	1.42	2.21	0.50	0.50
								~3.11	~2.62	~0.50	~0.38
								–	–	–	–

species were considered specialists (Table 2.2). The normality and homoscedasticity of the data were checked visually and by Shapiro-Wilk and Barlett tests whenever comparing the specificity of two groups (e.g. forest types, microhabitats) and one-way ANOVA or non-parametric tests were then used accordingly. Host and nest site specificity analyses were conducted in R (R Core Team, 2015) using the *stats* and *sciplot* packages.

I expected that ant species in primary forest would be more host-specific due to a lack of habitat disturbance and a higher occurrence of natural interactions predicted among host trees and ants. I also predicted that ants nesting inside trees would be more host-specific than externally nesting species due to interspecific variation in the availability of appropriate nest sites and the closer physical relationship of these ants with their hosts.

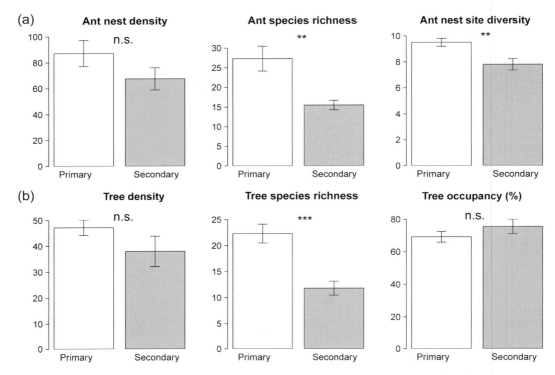

Figure 2.2.   Successional trends in the characteristics of tree-dwelling ant communities (a) and their host trees (b) in primary and secondary lowland tropical forests in New Guinea. The bars show mean value ± S.E. for 10 subplots (20 x 20 m each) of a 0.4-ha forest plot surveyed for each forest type. n.s. not significant difference among the forest types ($p > 0.05$), *** ($p < 0.001$), ** ($p < 0.01$) in one-way ANOVA (d.f. = 1).

Abundance, Diversity and Variability of the Ant and Tree Communities

Lower nest abundance is expected in secondary forests, as they host typically smaller trees and fewer high-canopy trees compared to primary forest. Indeed, abundances of ants were lower in the secondary forest plot, although the densities of nests and trees did not differ significantly between primary and secondary forest at small spatial scale (20 × 20m; Figure 2.2). There was, however, a relatively large variation in nest and tree densities in both forests. Similarly, occupancy of trees by ants was comparable in both forest types. Differences in species richness were much more pronounced, with primary forest supporting approximately twice as many species of ants and trees as secondary forest. For instance, in 400 m² of primary forest one can expect to find about 30 tree-dwelling ant species, while in the same area of secondary forest only about 15 species could be found. Similar patterns were also found at larger spatial scales, as 85 ant and 103 tree species were recorded in 0.4 ha of primary forest, but only 50 ant and 49 tree species occupied a comparable area of secondary forest (Table 2.2). Interestingly, these differences in

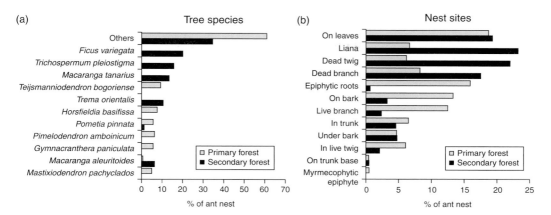

Figure 2.3. Ant nest distribution among the 11 most common tree species (a) and the nest site categories (b) in primary and secondary lowland tropical forest in New Guinea.

species diversity were not reflected in differences in nest site diversity, since most of the 12 nest categories were present in both forest types, even at small spatial scales (Figure 2.2). One exception was the absence of myrmecophytic epiphytes in secondary forest (Figure 2.3). However, when the overall composition of tree species and nest site diversity is considered, each forest type supports very different tree taxa and offers relatively different micro-environments for ant nests. For example, the most common nest sites in primary forest were within epiphytic roots, live twigs and live branches, while dead twigs and lianas were the most occupied habitats in secondary forest (Figure 2.3). Nevertheless, nests constructed on the leaves in the canopy were similarly distributed in both forest types suggesting that larger foliage biomass provided by primary forest trees (Whitfeld et al., 2012) does not necessary translate to higher establishment of nests on leaves.

Multivariate Analyses of the Effects of Host Tree Species and Microhabitat Types on Ant Communities

Partitioning of variation in ant community composition showed that all the significant environmental predictors together explain 11.2 percent of the total variability in primary forest and 12 percent in secondary forest. Similar proportions of the variability were explained by *environmental nest predictors* themselves (9.9 percent in primary and 9.4 percent in secondary forest), while the remaining variance linked to *tree species* was very low (~1 percent). The overall percentage of variation in community composition significantly explained by variables was relatively high for both datasets as most of the variance was linked to the first two CCA axes (efficiency, i.e. proportion of unconstrained variance explained in each axis by the model was between 43 percent and 71 percent, Figure 2.4). Only one tree species of 30 in primary forest and only 2 tree species of 15 species in secondary

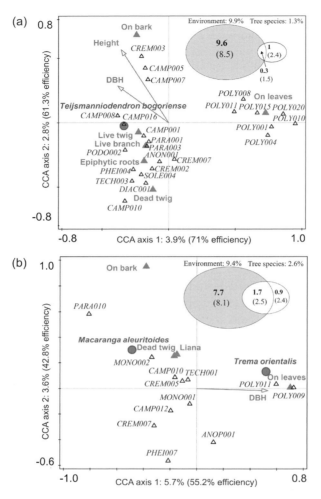

Figure 2.4. CCA ordination and Venn diagrams of the effects of explanatory variables on tree-dwelling ant communities in primary (a) and secondary (b) lowland tropical forest in New Guinea. Variation of ant community composition (nest records of ant species in trees) in ordination diagrams is related to the first two ordination axes and two sets of the explanatory variables: *Environmental predictors* (nest site categories as dark triangles; nest height and DBH as arrows) and *Tree species* (as circles). Only variance explained by individual variables that were significant in forward selection is considered ($p_{adj.} < 0.05$; see main text for more). Empty triangles with shortcuts represent the ant species most affected by the gradient (see Table 2.2 for the full species names and used data). Venn diagrams visualise area proportional to the percentage of explained variance in CCA analysis and partitioned between the two sets of variables; the proportion of the variance significantly explained is noted in bold letters and the overall variance explained by all tree species and all measured variables in the brackets.

forest significantly influenced ant species composition (Figure 2.4). In contrast, six of ten nest site types significantly affected ant communities in primary forest and four of nine in secondary forest. Moreover, the effect of the significant tree species was inter-correlated with tree size (DBH). This effect was especially strong in secondary forest, where the gradient from the first significant tree species, *Macaranga aleuritoides* (Euphorbiaceae), to the second species, *Trema orientalis* (Cannabaceae), reflects the gradient from small to large trees occupied by ants (Figure 2.4b). Heights aboveground of the nest were strongly correlated with tree DBH.

The distribution of ant species records in trees mostly coincide with the distribution of their preferred nest sites. For example, most *Polyrhachis* species build nests on/or from the leaves, and they are thus most common on the foliage. Similarly, most *Camponotus* nest in twigs, and most *Pheidole hospes* nest in epiphytes, although not exclusively. For example, some *Camponotus* and *Colobopsis* species were strongly associated with carton nests on bark, but they actually nest in twigs (Figure 2.4a, Table 2.2). This suggests a possible parasymbiotic relationship with another ant, *Crematogaster polita*, which is the dominant species in lowland forests of New Guinea, and builds large carton nests on the bark. The most pronounced difference between the two sites was the rarity of nests under epiphytic roots in secondary forest. Interestingly, two of the more common species found in both forest types, *Crematogaster sp.* 7 and *Colobopsis vitrea*, are opportunistic in their nesting habits, occupying up to nine different nest site types in each forest plot. Such a low level of specialisation to particular nest sites, as well as to particular host trees, may represent a significant advantage, allowing those ant species to occupy a wider range of forest successional stages.

Specificity of Interactions between Ants, Nest Sites and Host Tree Species

The specificity of ant species to nest sites is higher in primary than in secondary forests, as indicated by both specificity indices: Lloyd's (primary forest: mean = 4.11; secondary forest: mean = 3.11) and HS_k (primary forest: mean = 0.59; secondary forest: mean = 0.5). However, the difference is small and not significant (Figure 2.5). Moreover, the values of both indices vary greatly between species in both forests (Table 2.2). For instance, Lloyd's index ranged from 0.06 to 10.77, indicating that some species are evenly dispersed, while others are highly aggregated with respect to nest site categories. Similarly, there was no difference in mean host specificity between primary and secondary forest. Interestingly, none of the ant species in either of the forest plots expressed specialisation to a particular tree species, except perhaps one species, *Colobopsis* aff. *sanguinifrons* in the primary forest, which showed some preference for the tree host *Teijsmanniodendron bogoriense* (Lamiaceae). This ant nested in hollow twigs in the upper canopy of this tree species. However, even

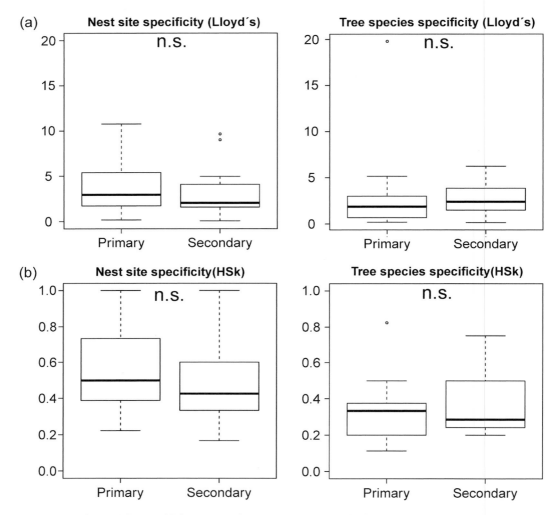

Figure 2.5. Ant species specificity to nest site types and tree species in New Guinea expressed by Lloyd's (a) and HS_k (b) indices. Box-plots show medians per ant species with 25–75 percent quartiles and whiskers as 1.5 interquartile ranges of the values. None of the differences between primary and secondary forest is significant (n.s., Kruskal-Wallis rank test, d.f. = 1, p > 0.05).

Colobopsis aff. *sanguinifrons* was not considered a specialist (HS_k < 0.9) and was found in two other tree host species from different plant families at the site (Klimes & McArthur, 2014). In contrast, several ant species (from the genera *Polyrhachis*, *Diacamma* and *Rogeria*) are specialised to particular nest sites, especially those nesting on the leaves and in epiphytes. These differences were also reflected by significantly higher specificity across all ant species to nest sites than to tree species in both specificity measures (Table 2.2; Wilcoxon signed rank test: Lloyd's, V = 980, p = 0.01; HS_k, V = 907, p < 0.001). Although this statistical comparison of the

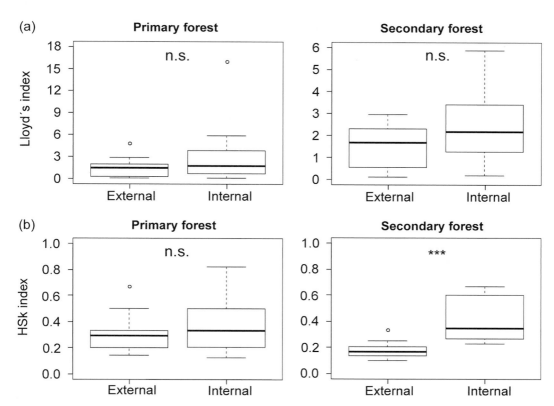

Figure 2.6.   Specificity of ant species to tree species in primary and secondary lowland forest in New Guinea for the ant communities nesting in external or internal nest sites provided by trees (see Table 2.1 for their definition) and expressed by Lloyd's (a) and HS_k (b) indices. Box-plots show medians per ant species with 25–75 percent quartiles and whiskers as 1.5 interquartile ranges of the values. Means not significantly different between the two communities are marked as n.s. (p > 0.05) and significantly as *** (p < 0.001, Kruskal-Wallis rank test, d.f. = 1).

indices should be interpreted with caution because of the very different numbers of units being compared (fewer nest site types than tree species), the lower specificity to trees than to nest sites is in agreement with the results of the variation partition-ing (Figure 2.4). As expected, the ant communities nesting internally on host trees were relatively more specialised to their hosts than communities nesting in exter-nal tree structures (Figure 2.6). However, the differences are relatively minor, and the only significant difference was in HS_k in secondary forest (mean$_{external}$ = 0.19, mean$_{internal}$ = 0.41).

The ant species with the lowest specificity values both to host tree and nest sites (HS_k < 0.2) were the invasive, tramp species *Technomyrmex brunneus* and *Anoplolepis gracilipes*. Likewise, most of the common species for both forest types were particularly generalised with respect to host tree identity and nest site location (Table 2.2).

The Importance of Host Plant and Microhabitat Diversity for Canopy Ant Communities

Although the taxonomic diversity of trees is usually strongly correlated with the species diversity of ants (Ribas et al., 2003; Basset et al., 2012; Staab et al., 2014), it was not the primary driver of variation in the diversity of ant communities in trees or their interactions with host plants in this study. Ant species richness and composition was mediated rather by overall forest structure (tree size) and by the availability of appropriate microhabitats for nesting in individual trees. Similar results have been found previously (e.g. Floren & Linsenmair, 1997; Tanaka et al., 2010; Yusah & Foster, 2016), although using much lower replication of canopy host trees than here. Canopy ants seem to be very opportunistic in regard to their choice of host plant species, and surprisingly, the specificity of those interactions is comparable for primary and secondary forest communities. This was observed despite the higher prevalence of invasive and early successional ant species in secondary forest.

Hence my results from New Guinea do not support the hypothesis that ant-plant interactions in undisturbed primary forests are more specialised than in disturbed secondary rainforests. There is no evidence of disruption of specialisation of ant-plant interactions in structurally simpler rainforests from my plot-scale data. However, the plot-scale census demonstrates that secondary forests host different and less species-diverse assemblages of trees and ants, and harbour less diverse microhabitats in which to establish nests, and hence support a more restricted range of ant-plant interactions. Moreover, the lack of some microhabitats negatively affects the taxonomic diversity and community composition of ants in secondary forests. For instance, some common genera found in primary forests (e.g. *Rogeria*, *Diacamma*, *Pheidole*) nest exclusively or preferably in epiphytes and are thus missing or rare in secondary forest. Alternatively, the lack of some microhabitats may drive some ant species to change their preferred nest site between forests. For instance, *Pheidole* sp. 7 nested mainly in epiphytes in primary forest, but was found mostly in dead branches in secondary forest. This highlights the crucial importance of nest site diversity for ant communities, which may in turn mediate interactions among ants and host tree. The positive effect of microhabitat variability on the diversity and composition of tree-dwelling ants has also been shown in other studies. For example, DaRocha et al. (2015) showed that a single old-growth tree within a cocoa monoculture was occupied by an exceptionally high number of ant species due to the presence of a rich epiphyte flora in that tree. No doubt nest sites strongly affected ant species abundance and composition in both plots; but they did not mediate the host specificity to tree taxa at the level of whole communities. This is because nesting site types were distributed rather randomly among different tree species (Figure 2.4).

Although ants did not specialise on only one nest site type in most cases, and some species were opportunistic in nesting habits, especially in secondary forest (Table 2.2), subtle differences in microhabitat characteristics (such as the size or type of twigs and branches) affected the distribution of particular ant species. For example, some ant species prefer living tree structures due to the presence of symbiotic scale insects, or they preferentially use chambers provided by myrmecophytic trees (Gullan et al., 1993; Klimes & McArthur, 2014). In contrast, an opportunist ant species might not distinguish, for example, between dry twigs, epiphytes or lianas (e.g. *Colobopsis vitrea* in this study). The importance of microhabitat diversity and availability has also been supported by experiments in simpler environments, such as plantations and within dead wood habitats (Jimenez-Soto & Philpott, 2015; Satoh et al., 2016). It is very likely that these factors are also important in more complex environments like tropical rainforests.

Some specialised interactions between ants and nest sites observed in the studied plots are highly conserved phylogenetically (e.g. *Polyrhachis* genus nesting on leaves). However, even for those expected interactions, it is interesting to know whether ant species inhabiting these common nest sites are found only on certain tree species. This does not seem to be the case, as plot-based interactions mapped here showed that ants displayed surprisingly little specialisation to host tree species, nor was host tree specificity inter-correlated with nest site species preferences in either forest type (Figure 2.4, Table 2.2). Nevertheless, there was some effect of tree size and canopy stratification on the ant species identity, including those nesting on leaves (Figure 2.4b, Klimes et al. 2015). One explanation of this pattern may be that as a tree grows the ant fauna changes dynamically through time, especially in tree taxa which are not specialised to particular ant species (Dejean et al., 2008). Furthermore, a similar shift has been demonstrated also for some myrmecophytic *Macaranga* (Euphorbiaceae) tree species in Borneo, which host only specialised ant species when small, but support a succession of other ant species including generalists, and more colonies, as they grow (Feldhaar et al., 2003). Indeed, the trees examined in this study were larger than 5 cm DBH, and they were not understorey plants, which are usually given as a model example of ant-specific myrmecophytes (Rico-Gray & Oliveira, 2007). Consequently, our plot-based census missed most of the well-known specialised interactions in New Guinea between *Colobopsis quadriceps* and *Endospermum* (Euphorbiaceae) trees (Letourneau & Barbosa, 1999). These were present in the plots but typically on trees smaller than 5 cm in DBH (author's personal observation, 2007). It is possible that the New Guinean ant fauna is also relatively less host tree-specific compared to, for example, that in the Neotropics, as New Guinea is a geologically younger region with most of the extant insect taxa having evolved relatively recently (Toussaint et al., 2014). As the plot census data presented here are unique, it is, however, difficult to compare the specificity in tree canopies between multiple regions.

As expected, ant species that build nests inside trees tended to be more host-specific than those building nests externally. Nevertheless, this effect was not strong and mostly statistically non-significant. It may be the case that internal nesting species are more specialised simply because fewer tree species have hollow structures within which ants may build a nest, rather than reflecting greater specialisation driven by intimate (myrmecophytic) ant-plant interactions. For instance, every tree species may be suitable for externally nesting ant species, but only some species (and individuals of certain size) commonly have hollow structures available for ants to nest. The size of host plants and the variety of nest sites they provide is thus much more important for canopy ant communities than the taxonomic composition of the forest or the identity of particular hosts. The importance of the environmental characteristics of host trees was therefore comparable in both primary and secondary forests.

Other Factors Influencing Ant-Plant Interactions in Rainforest Canopies

A number of factors not quantified here, such as the presence of extrafloral nectaries and hemipteran symbionts, can be very important to the maintenance of high abundances of ants in forest canopies (Davidson et al., 2003). However, they are unlikely to be major drivers of specialisation in ant-plant interactions at the community level in species-rich, tropical rainforests. This is because these plant-mediated resources are usually dominated by a few common, omnivorous ant species, although their occupancy can drive spatio-temporal relationships among ant species at the scale of individual trees and/or branches and influence species coexistence (Blüthgen et al., 2000; Dejean et al., 2007). The small or insignificant effect of nectaries as major drivers of the ant community structure was recently documented in Neotropical savannas (Camarota et al., 2015), which are well known for their diversity and high occurrence of plants with nectaries (Oliveira et al., 2002). It is likely that nectar resources have little influence on the diversity and specificity of ant-plant interactions in New Guinea as well, since nectaries are relatively rare in these rainforests and are usually occupied by common, dominant species that nest and forage in many adjacent trees, including those without nectaries (author' personal observation, 2007). Similarly, the host specificity of ants to hemipteran symbionts is relatively low in the New Guinea forests (Klimes et al., 2016) and also elsewhere in the tropics at the community level (Blüthgen et al., 2006; Staab et al., 2015). Nevertheless, since some sap-sucking herbivores, and especially extrafloral nectaries, are often restricted to particular tree species, the availability of these resources is expected to influence the specificity of ant-plant interactions, at least in some species. Yet this was not the case here because, with the possible exception of one *Colobopsis* species, all of the ants examined in both primary and secondary forest displayed very low host specificity.

Conclusions and Implications for the Conservation of Ant-Plant Interactions in the Canopy

The conversion of primary forests to secondary habitats is usually followed by a serious decline in local species richness of arboreal ants. Moreover, species composition of both ants and trees change after disturbance and facilitate invasions by exotic ant species in secondary habitats. In the lowlands of New Guinea, one of the last vast regions of pristine rainforest left on Earth, the diversity of ants and trees in secondary forests is approximately half of that harboured in primary forests at both small (20 × 20 m) and whole plot (100 × 40 m) scale. My results show that environmental predictors explain a similar proportion (~12 percent) of the variance in ant community composition in both primary and secondary forests. Most of this variance is linked to microhabitat availability (tree size and nest site variability), whereas the effect of tree host species themselves on ants is very low (~1 percent of explained variance). The canopy ant communities of primary forest are not more specialised to particular nest sites or tree species than those from the secondary forest. However, primary forests support a greater variability of microhabitats, and the rarity of some nest sites in secondary forests can result in the loss of particular ant taxa. Ant species which build their nests inside tree structures are only slightly more host-specific than those that build external nests.

These results show that almost all arboreal ant species in both primary and secondary forest are generalised with respect to host tree species, and that this level of generalisation remains stable even after forest disturbance. Similar conclusion has been recently made for ants inhabiting bird's nest ferns in Borneo (Fayle et al., 2015; Chapter 3) that suggests that this result might be general not only for host trees but also probably for smaller plants in rainforests (epiphytes, lianas). This is encouraging as it indicates that ecological interactions among ants and rainforest plants might not be seriously altered during forest disturbance, and that ants may play a comparable ecological function in both primary and secondary forests. However, the data from census plots also show that very different and less species-diverse communities of both ants and trees are involved in the ant-plant interactions in secondary forests compared to primary forests. One concern is that the high abundances of many invasive ant and plant species in disturbed rainforests may outcompete native species (Chapters 12–15). It is likely that multiple factors such as vegetation structure and ant microhabitat preferences, combined with vulnerability of disturbed and fragmented forests to ant invasions that alter interspecies interactions, are the cause of the differences in arboreal ant communities between primary and secondary forests. Therefore, in order to preserve the indigenous, species-rich ant fauna in the tropical rainforests, I suggest a strategy to protect the whole primary forest vegetation with an emphasis on maximising its microhabitat structural diversity, for example via minimising forest fragmentation and protecting old-growth large canopy trees with high load of lianas and epiphytes.

The results in terms of differences between tree communities are in agreement with other recorded vegetation changes along successional gradients in New Guinea (Whitfeld et al., 2012; Whitfeld et al., 2014) and elsewhere in the tropics (Chazdon, 2014). Whether the differences described are also typical for arboreal ant nests in primary and secondary lowland forests have yet to be elucidated, as the present findings are based on censuses from only two forest plots. More data from different tropical forests and more geographic regions are urgently needed to test whether, and to what extent, the specificity of ant-plant interactions vary in forests with different succession and disturbance histories.

References

Adis, J., Lubin, Y. D. & Montgomery, G. G. (1984). Arthropods from the canopy of inundated and terra firme forests near Manaus, Brazil, with critical considerations on the pyrethrum-fogging technique. *Studies on Neotropical Fauna and Environment,* 19, 223–36.

Agosti, D., Majer, J. D., Alonso, L. E. & Schultz, T. R. (2000). *Ants: Standard Methods for Measuring and Monitoring Biodiversity.* Washington, DC: Smithsonian Institution Press.

Basset, Y., Cizek, L., Cuenoud, P. et al. (2012). Arthropod diversity in a tropical forest. *Science,* 338, 1481–4.

Bihn, J. H., Verhaagh, M., Brändle, M. & Brandl, R. (2008). Do secondary forests act as refuges for old growth forest animals? Recovery of ant diversity in the Atlantic Forest of Brazil. *Biological Conservation,* 141, 733–43.

Blüthgen, N. & Fiedler, K. (2002). Interactions between weaver ants *Oecophylla smaragdina,* homopterans, trees and lianas in an Australian rain forest canopy. *Journal of Animal Ecology,* 71, 793–801.

Blüthgen, N., Gebauer, G. & Fiedler, K. (2003). Disentangling a rainforest food web using stable isotopes: dietary diversity in a species-rich ant community. *Oecologia,* 137, 426–35.

Blüthgen, N., Mezger, D. & Linsenmair, K. E. (2006). Ant-hemipteran trophobioses in a Bornean rainforest – diversity, specificity and monopolisation. *Insectes Sociaux,* 53, 194–203.

Blüthgen, N., Verhaagh, M., Goitia, W. et al. (2000). How plants shape the ant community in the Amazonian rainforest canopy: the key role of extrafloral nectaries and homopteran honeydew. *Oecologia,* 125, 229–40.

Butler, R. A. & Laurance, W. F. (2008). New strategies for conserving tropical forests. *Trends in Ecology & Evolution,* 23, 469–72.

Camarota, F., Powell, S., Vasconcelos, H. L., Priest, G. & Marquis, R. J. (2015). Extrafloral nectaries have a limited effect on the structure of arboreal ant communities in a Neotropical savanna. *Ecology,* 96, 231–40.

Chazdon, R. L. (2014). *Second Growth.* Chicago: University of Chicago Press.

DaRocha, W. D., Ribeiro, S. P., Neves, F. S. et al. (2015). How does bromeliad distribution structure the arboreal ant assemblage (Hymenoptera: Formicidae) on a single tree in a Brazilian Atlantic Forest agroecosystem? *Myrmecological News,* 21, 83–92.

Davidson, D. W., Cook, S. C., Snelling, R. R. & Chua, T. H. (2003). Explaining the abundance of ants in lowland tropical rainforest canopies. *Science,* 300, 969–72.

de Castro Solar, R. R., Barlow, J., Ferreira, J. et al. (2015). How pervasive is biotic homogenization in human-modified tropical forest landscapes? *Ecology Letters,* 18, 1108–18.

Dejean, A., Corbara, B., Orivel, J. & Leponce, M. (2007). Rainforest canopy ants: the implications of territoriality and predatory behavior. *Functional Ecosystems and Communities,* 1, 105–20.

Dejean, A., Djieto-Lordon, C., Cereghino, R. & Leponce, M. (2008). Ontogenetic succession and the ant mosaic: an empirical approach using pioneer trees. *Basic and Applied Ecology*, 9, 316–23.

Dejean, A., Fisher, B. L., Corbara, B. et al. (2010). Spatial distribution of dominant arboreal ants in a Malagasy coastal rainforest: gaps and presence of an invasive species. *PLoS ONE*, 5, e9319.

Ewers, R. M., Boyle, M. J. W., Gleave, R. A. et al. (2015). Logging cuts the functional importance of invertebrates in tropical rainforest. *Nature Communications*, 6, 6836.

Fayle, T. M., Edwards, D. P., Foster, W. A., Yusah, K. M. & Turner, E. C. (2015). An ant-plant by-product mutualism is robust to selective logging of rain forest and conversion to oil palm plantation. *Oecologia*, 178, 441–50.

Fayle, T. M., Turner, E. C. & Foster, W. A. (2013). Ant mosaics occur in SE Asian oil palm plantation but not rain forest and are influenced by the presence of nest-sites and non-native species. *Ecography*, 36, 1051–7.

Fayle, T. M., Turner, E. C., Snaddon, J. L. et al. (2010). Oil palm expansion into rain forest greatly reduces ant biodiversity in canopy, epiphytes and leaf-litter. *Basic and Applied Ecology*, 11, 337–45.

Feldhaar, H., Fiala, B., Hashim, R. B. & Maschwitz, U. (2003). Patterns of the *Crematogaster-Macaranga* association: The ant partner makes the difference. *Insectes Sociaux*, 50, 9–19.

Floren, A. (2005). How reliable are data on arboreal ant (Hymenoptera: Formicidae) communities collected by insecticidal fogging? *Myrmecologische Nachrichten*, 7, 91–4.

Floren, A., Freking, A., Biehl, M. & Linsenmair, K. E. (2001). Anthropogenic disturbance changes the structure of arboreal tropical ant communities. *Ecography*, 24, 547–54.

Floren, A. & Linsenmair, K. E. (1997). Diversity and recolonization dynamics of selected arthropod groups on different tree species in a lowland rainforest in Sabah, with special reference to Formicidae. In *Canopy Arthropods*, ed. N. E. Stork, J. Adis & R. K. Didham. London: Chapman & Hall, pp. 344–81.

Floren, A., Wetzel, W. & Staab, M. (2014). The contribution of canopy species to overall ant diversity (Hymenoptera: Formicidae) in temperate and tropical ecosystems. *Myrmecological News*, 19, 65–74.

Frederickson, M. E. & Gordon, D. M. (2007). The devil to pay: a cost of mutualism with *Myrmelachista* schumanni ants in 'devil's gardens' is increased herbivory on *Duroia hirsuta* trees. *Proceedings of the Royal Society B-Biological Sciences*, 274, 1117–23.

Gullan, P. J., Buckley, R. C. & Ward, P. S. (1993). Ant-tended scale insects (Hemiptera: Coccidae: *Myzolecanium*) within lowland rain forest trees in Papua New Guinea. *Journal of Tropical Ecology*, 9, 81–91.

Huxley, C. R. (1978). The ant-plants *Myrmecodia* and *Hydnophytum* (Rubiaceae), and the relationships between their morphology, ant occupants, physiology and ecology. *New Phytologist*, 80, 231–68.

Janzen, D. H. (1966). Coevolution of mutualism between ants and acacias in Central America. *Evolution*, 20, 249–75.

Jimenez-Soto, E. & Philpott, S. M. (2015). Size matters: nest colonization patterns for twig-nesting ants. *Ecology and Evolution*, 5, 3288–98.

Klimes, P., Fibich, P., Idigel, C. & Rimandai, M. (2015). Disentangling the diversity of arboreal ant communities in tropical forest trees. *PLoS ONE*, 10, e0117853.

Klimes, P., Husnik, F., Borovanska, M. & Gullan, P. J. (2016). Contrasting tri-trophic food webs between primary and secondary tropical forest: role of species ecology and phylogeny (Abstract). In *Annual Meeting of the Association for Tropical Biology and Conservation.* Montpellier: 53 rd ATBC, Le Corum, p. 220.

Klimes, P., Idigel, C., Rimandai, M. et al. (2012). Why are there more arboreal ant species in primary than in secondary tropical forests? *Journal of Animal Ecology*, 81, 1103–12.

Klimes, P. & McArthur, A. (2014). Diversity and ecology of arboricolous ant communities of *Camponotus* (Hymenoptera: Formicidae) in a New Guinea rainforest with description of four new species. *Myrmecological News,* 20, 141–58.

Lach, L., Parr, L. C. & Abbott, K. L. (2010). *Ant Ecology.* New York: Oxford University Press.

Letourneau, D. K. & Barbosa, P. (1999). Ants, stem borers, and pubescence in *Endospermum* in Papua New Guinea. *Biotropica*, 31, 295–302.

Lowman, M. D., Schowalter, T. D. & Franklin, J. F. (2012). *Methods in Forest Canopy Research.* London: University of California Press.

Novotny, V. (2010). Rain forest conservation in a tribal world: why forest dwellers prefer loggers to conservationists. *Biotropica,* 42, 546–9.

Novotny, V., Miller, S. E., Leps, J. et al. (2004). No tree an island: the plant-caterpillar food web of a secondary rain forest in New Guinea. *Ecology Letters*, 7, 1090–1100.

Oliveira, P. S., Freitas, A. V. L. & Del-Claro, K. (2002). Ant foraging on plant foliage: contrasting effects. In *The Cerrados of Brazil: Ecology and Natural History of a Neotropical Savanna*, ed. P. S. Oliveira & R. J. Marquis. New York: Columbia University Press, pp. 287–305.

R Core Team (2015). *R: A Language and Environment for Statistical Computing.* Vienna: R Foundation for Statistical Computing.

Ribas, C. R., Schoereder, J. H., Pic, M. & Soares, S. M. (2003). Tree heterogeneity, resource availability, and larger scale processes regulating arboreal ant species richness. *Austral Ecology*, 28, 305–14.

Rico-Gray, V. & Oliveira, P. S. (2007). *The Ecology and Evolution of Ant-Plant Interactions.* Chicago: University of Chicago Press.

Ryder Wilkie, K. T., Mertl, A. L. & Traniello, J. F. A. (2010). Species diversity and distribution patterns of the ants of Amazonian Ecuador. *PLoS ONE*, 5, e13146.

Satoh, T., Yoshida, T., Koyama, S. et al. (2016). Resource partitioning based on body size contributes to the species diversity of wood-boring beetles and arboreal nesting ants. *Insect Conservation and Diversity,* 9, 4–12.

Shearman, P. & Bryan, J. (2011). A bioregional analysis of the distribution of rainforest cover, deforestation and degradation in Papua New Guinea. *Austral Ecology*, 36, 9–24.

Šmilauer, P. & Lepš, J. (2014). *Multivariate Analysis of Ecological Data Using CANOCO 5.* Cambridge: Cambridge University Press.

Staab, M., Blüthgen, N. & Klein, A.-M. (2015). Tree diversity alters the structure of a tri-trophic network in a biodiversity experiment. *Oikos,* 124, 827–34.

Staab, M., Schuldt, A., Assmann, T. & Klein, A.-M. (2014). Tree diversity promotes predator but not omnivore ants in a subtropical Chinese forest. *Ecological Entomology,* 39, 637–47.

Tanaka, H. O., Yamane, S. & Itioka, T. (2010). Within-tree distribution of nest sites and foraging areas of ants on canopy trees in a tropical rainforest in Borneo. *Population Ecology,* 52, 147–57.

Toussaint, E. F. A., Hall, R., Monaghan, M. T. et al. (2014). The towering orogeny of New Guinea as a trigger for arthropod megadiversity. *Nature Communications*, 5, 4001.

Vasconcelos, H. L. & Bruna, E. M. (2012). Arthropod responses to the experimental isolation of Amazonian forest fragments. *Zoologia*, 29, 515–30.

Wardhaugh, C. W., Stork, N. E. & Edwards, W. (2013). Specialization of rainforest canopy beetles to host trees and microhabitats: not all specialists are leaf-feeding herbivores. *Biological Journal of the Linnean Society*, 109, 215–28.

Whitfeld, T. J. S., Lasky, J. R., Damas, K. et al. (2014). Species richness, forest structure, and functional diversity during succession in the New Guinea lowlands. *Biotropica*, 46, 538–48.

Whitfeld, T. J. S., Novotny, V., Miller, S. E. et al. (2012). Predicting tropical insect herbivore abundance from host plant traits and phylogeny. *Ecology*, 93, S211-S22.

Widodo, E. S., Naito, T., Mohamed, M. & Hashimoto, Y. (2004). Effects of selective logging on the arboreal ants of a Bornean rainforest. *Entomological Science*, 7, 341–9.

Woodcock, P., Edwards, D. P., Fayle, T. M. et al. (2011). The conservation value of South East Asia's highly degraded forests: evidence from leaf-litter ants. *Philosophical Transactions of the Royal Society B: Biological Sciences*, 366, 3256–64.

Woodcock, P., Edwards, D. P., Newton, R. J. et al. (2013). Impacts of intensive logging on the trophic organisation of ant communities in a biodiversity hotspot. *PLoS ONE*, 8, e60756.

Yanoviak, S. P., Silveri, C., Hamm, C. A. & Solis, M. (2012). Stem characteristics and ant body size in a Costa Rican rain forest. *Journal of Tropical Ecology*, 28, 199–204.

Yusah, K. M., Fayle, T. M., Harris, G. & Foster, W. A. (2012). Optimizing diversity assessment protocols for high canopy ants in tropical rain forest. *Biotropica*, 44, 73–81.

Yusah, K. M. & Foster, W. A. (2016). Tree size and habitat complexity affect ant communities (Hymenoptera: Formicidae) in the high canopy of Bornean rain forest. *Myrmecological News*, 23, 15–23.

3 Living Together in Novel Habitats: A Review of Land-Use Change Impacts on Mutualistic Ant-Plant Symbioses in Tropical Forests

Tom M. Fayle, Chua Wanji, Edgar C. Turner, and Kalsum M. Yusah[*]

Introduction

Mutualisms form between species when individuals provide reciprocal benefits, increasing the fitness of both partners. Ants and plants often form such mutualistic relationships, with ants providing protection from herbivory, protection from competition from other plants, seed dispersal, CO_2 and/or food, and receiving in return housing space and/or food from plants (Rico-Gray & Oliveira, 2007). Some associations are symbiotic (i.e. partners live together) while in others ants receive food benefits, but nest elsewhere. In this review we focus on ant-plant symbioses (i.e. in which entire colonies of ants inhabit plants), since these tend to be more intimate associations, sometimes have high interaction specificity, and have clearly defined partners. Although symbioses usually involve ants inhabiting plant-evolved living spaces, this is not always the case, with ants sometimes inhabiting other structures (such as the leaf litter layer in litter-collecting species) but nonetheless providing benefits to the plant in terms of protection from herbivory (Gibernau et al., 2007; Fayle et al., 2012) or nutrients (Watkins et al., 2008). For the purposes of this chapter, we exclude ants using plants as attachment points for external nests such as those inhabiting carton structures. Symbiotic ant-plant mutualisms are particularly abundant in tropical forests (Bruna et al., 2005; Feldhaar et al., 2010), where they can play important roles in structuring ecosystems (Frederickson et al., 2005; Tanaka et al., 2009).

[*] We are grateful to William A. Foster, who supervised ECT, TMF and KMY for their PhDs, during which much of the work on bird's nest ferns reviewed here was conducted, and to Nico Blüthgen, Paulo S. Oliveira and an anonymous reviewer for suggestions that improved the manuscript. TMF was funded by a Czech Science Foundation Standard Grant (16-09427S), CW by a Malaysian Ministry of Higher Education Fundamental Research Grant (FRG0373-STWN-1/2014), KMY by the Universiti Malaysia Sabah new lecturer grant scheme (SLB0071-STWN-2013), and ECT by the Isaac Newton Trust, Cambridge, PT Sinar Mas Agro Resources and Technology Tbk and the Natural Environment Research Council (NE/K016377/1). All authors are also grateful for support by the South East Asian Rainforest Research Partnership.

Figure 3.1.   Typical habitat conversion gradient for tropical forests. Note that there are two categories of continuous non-primary forest, combined here for brevity: logged forest, which is primary forest with timber selectively extracted, and secondary regrowth forest, which has regenerated following complete clearance (our definitions). The dominant agricultural habitat type varies globally, but is here depicted as oil palm plantation. Figure modified from Foster et al. (2011). Original drawings by Jake Snaddon.

Human-driven land-use causes changes to ecosystems worldwide, driven in the tropics mainly by logging of forests, clearance for expansion of agricultural land, and consequent fragmentation of remaining forest (Tilman et al., 2001; Edwards et al., 2014; Figure 3.1). Although the negative impacts of these processes on the number and identity of species are moderately well-known, changes in species interaction networks are much less studied despite being of key importance (Kaiser-Bunbury & Blüthgen, 2015). This is because network structure can determine community stability in the face of further disturbance (Dunne & Williams, 2009) and therefore affect associated ecosystem processes (Tylianakis et al., 2010).

Symbiotic ant-plant networks are abundant in tropical forests and hence are likely to be affected by habitat disturbance (Mayer et al., 2014). However, remarkably little work has been dedicated to understanding how these networks respond to human-driven land use change (Table 3.1). More specific symbioses can serve as model systems for understanding the altered selective environments in converted habitats (Laughlin & Messier, 2015), while less specific symbioses can be used as microcosms for understanding larger-scale community responses (Fayle et al., 2015b). In the following section we review studies investigating shifts in communities of ant-inhabited plants as a result of selective logging, clearance followed by secondary regrowth, forest fragmentation, and conversion of forest to agriculture. We also speculate on how other anthropogenic impacts, such as altered climate, nutrient enrichment, and invasion by non-native species, might interact with these land-use changes.

The Impacts of Logging, Forest Fragmentation, and Conversion to Agriculture on Ant-Plant Symbioses

Logging of Tropical Forest and Secondary Regrowth Following Clearance

Although it is unlikely that ant-plants are ever directly targeted for removal during commercial selective logging activities, since they tend to be epiphytes or small plants with hollow stems, felling and extraction of trees often damages the surrounding vegetation and may, therefore, indirectly affect them (Picard et al., 2012). Secondary regrowth forests, as distinct from those that have been selectively logged, also have substantially altered vegetation structure (Chazdon, 2014). Furthermore, disturbed forests differ from primary forests in having hotter, drier microclimates, and a more open vegetation structure (Hardwick et al., 2015), potentially affecting both ants and their plant hosts.

As a result of these changes, the density of ant-plants changes over time following disturbance. For example, ant-inhabited *Macaranga*, a common group of ant-plants on the island of Borneo, show an increase in density shortly after complete clearance, peaking after five years, followed by a decrease (Tanaka et al., 2007), presumably due to competition between the *Macaranga* saplings and shading by later succession species. This pattern is also seen for ant-plants in the new world tropics, where *Cecropia* in secondary regrowth increases in abundance following burning of pasture (anecdotal report; Fonseca, 1999). It is worth noting that both of these ant-plant genera, which are among the most widespread and species-rich in their respective areas, are mainly early succession pioneers that specialize on disturbed areas (Fonseca, 1999; Slik et al., 2003). In some cases, logging and regrowth has also been recorded to alter ant inhabitation. For example, *Macaranga bancana* showed lower ant inhabitation rates in secondary forest, possibly due to increased queen mortality or differences in the species of ant inhabitants (Murase et al., 2003). In Papua New Guinea, interaction networks have also been found to differ between primary forest and secondary regrowth, following clearance for food gardens, with substantial reductions in ant-inhabitation of plants (Klimeš et al., 2012; Chapter 2; note that partner benefits have not been demonstrated in this system). However, to our knowledge only one study has directly assessed the impacts of selective logging on symbiotic ant-plant mutualisms, finding no change in the relationship between epiphytic bird's nest ferns and their ant inhabitants (Fayle et al., 2015a, see also the following section). If there are differences in the occupancy and identity of ant inhabitants as a result of logging, then this could have negative impacts on plant survival (Murase et al., 2010), leading to further changes in the community.

Forest Fragmentation

Human-driven expansion of non-forest habitats often results in increasingly fragmented forest patches. This process increases the proportion of forest experiencing changes in community composition and alteration of the abiotic environment near boundaries between habitats. These 'edge effects' can penetrate far from habitat boundaries (Ewers & Didham, 2008), and hence affect a large proportion of the world's forests (Haddad et al., 2015). Fragmentation also isolates populations in the remaining habitat islands, disrupting migration and potentially leading to long-term 'extinction debt' (Laurance et al., 2011). For example, fluctuations in the size of smaller isolated populations can eventually lead to local extinction of these species from individual fragments. Fragmentation is of particular concern for species involved in obligate mutualisms, because persistence in fragments requires the presence of both partner species. Hence these populations are expected to be vulnerable to localised stochastic extinction of one partner, with recolonisation of fragments requiring simultaneous colonisation by both partners (Fortuna & Bascompte, 2006). Furthermore, co-existence between symbiont ant species in undisturbed habitats may rely on dispersal-fecundity trade-offs in combination with variation in host plant density, with species that are highly fecund but poor dispersers dominating in high plant density areas, and vice versa (Yu et al., 2001). Isolation of forest patches substantially changes the distribution of ant-plants, and hence is likely to result in extinction of ant species with poorer dispersal abilities.

Ant-plants have been documented extensively in the Biological Dynamics of Forest Fragments Project (BDFFP) in the Brazilian Amazon, in which forest fragments have been experimentally isolated since 1979 (Laurance et al., 2011). After 25 years of fragmentation, species richness of both ants and plants, overall densities of plants (Bruna et al., 2005) and network structure (Passmore et al., 2012) remain similar to those in continuous forest, suggesting that these systems are remarkably robust to the effects of change. This stability might relate to the proximity of nearby forest, which at 100 m is within the dispersal range of at least some ant species (Bruna et al., 2011), and hence would allow maintenance of sink populations in fragments. There has also been forest regrowth in the cleared areas surrounding the fragments (Laurance et al., 2011), potentially facilitating migration of ants and plants. The nature of the matrix habitat between the fragments (pasture in the case of the BDFFP), is likely to affect the persistence of ant-plant populations in these areas. This is demonstrated by the stronger impacts on ant-plant populations of fragmentation from inundation due to damming of a river, where fragments are isolated by water, rather than pasture (Emer et al., 2013). In this study from the Amazon basin the authors found a reduction in species richness of both ants and plants, and a reduction in compartmentalisation of networks in islands. Smaller and more isolated fragments were less compartmentalised (i.e. networks were not divided into groups of species, with many links within groups, but few links between groups). This is despite fragmentation having occurred only ~10 years prior to the study, and the majority of islands being ~ 100 m from the nearest mainland or large island. Interestingly, sites on the edges of continuous areas of forest were intermediate between isolated islands and non-edge forest in terms of ant-plant communities, suggesting that symbiotic ant-plant networks are susceptible to edge effects. In the longer term it is possible that the effects of fragmentation on ant-plant interactions and stochastic extinction of populations may

Table 3.1 Summary of Known Impacts of Human-Driven Habitat Change on Ant-Plant Symbiotic Networks

Habitat change type	Plant taxa	Ant taxa	Location	Habitat(s)	Habitat change	Main conclusions	Reference(s)
Logging or forest clearance with regrowth	*Teijsmanniodendron, Horsefieldia, Ficus, Macaranga*	*Anonychomyrna, Camponotus*	Papua New Guinea	Lowland rain forest	Clearance for food gardens and secondary regrowth	For trees larger than 5 cm DBH, ant inhabitation of live trees is much less common in secondary forest than primary forest.	Klimes (chapter 2); Klimes et al. (2012)
	Asplenium nidus, A. phyllitidis	Many	Malaysian Borneo	Lowland rain forest	Selective logging	Ferns and ants persist, with ants commonly inhabiting ferns, and ferns being protected by ant residents. No differences between primary and logged forest.	Fayle et al. (2015)
	Macaranga bancana	*Crematogaster*	Malaysian Borneo	Lowland rain forest	Conversion to secondary forest,* cultivated land or grassland	More saplings inhabited by non-partner *Crematogaster* species in secondary forest than primary forest.	Murase et al. (2003)
	Cecropia	*Azteca*	Brazilian Amazon	Pasture	Regrowth following burning and abandonment of pasture	Anecdotal account of forest regeneration, with *Cecropia* ant-plants dominant. Initially many, small ant-plants, with later thinning out as plants grow. *Cecropia* dominate the overstory for > 10 years, and are then replaced by later succession trees.	Fonseca (1999)
Forest fragmentation	*Hirtella*, many others	*Allomerus, Azteca*, others	Brazilian Amazon	Lowland rain forest	Experimental forest fragmentation, by pasture	No overall changes in density of plants, and little change in network structure, but some plant species become less abundant.	Bruna et al. (2005); Passmore et al. (2012)

Hirtella, Maietia, many others	*Allomerus, Pheidole,* others	Brazilian Amazon	Lowland rain forest	Forest fragmentation from dam creation	Reduction in the number of plant and ant species and colonisation rates. Increase in opportunistic species colonising.	Emer et al. (2013)
Myrmecodia, Hydnophytum	*Iridomyrmex,* others	Papua New Guinea**	Lowland rain forest, lower montane forest	Conversion to plantations and other artificial habitats	More ant-plant species in disturbed than undisturbed habitats in lowlands, opposite in highlands. More species of ant in *Myrmecodia* in undisturbed lowlands, opposite for *Hydnophytum*. More species of ant in *Myrmecodia* in disturbed highlands, very few species of ant in *Hydnophytum* in highlands (note: as very small number of species, no formal analyses conducted).	Huxley (1978)
Clearance for agriculture						
Asplenium nidus, A. phyllitidis	Many	Malaysian Borneo	Lowland rain forest	Conversion to oil palm	Ferns and ants persist across all habitats, but ant species different in oil palm. Ants still protect ferns. Lower ant abundances in ferns of a given size in oil palm.	Fayle et al. (2010); Fayle et al. (2015)
Hohenbergia, Aechmea	Many	Bahia, Brazil	Atlantic Forest	Conversion to cocoa agroforest	Introduction of agroforestry decreases interaction specificity, but epiphytes still allow maintenance of similar levels of ant diversity compared to pristine habitat.	DaRocha et al. (2016)

* Authors do not state if these areas were cleared completely and then allowed to regrow, or if they result from selective logging and subsequent regeneration.

** Papua New Guinea and nearby areas.

have wider ranging effects on the whole ecosystem, a speculation supported by the low densities of some ant-plant species in fragments (Bruna et al., 2005). However, the high degree of specificity in many ant-plant systems might protect the system from catastrophic collapse, since the impacts of extinctions of individual species are unlikely to spread through the entire ant-plant network (Passmore et al., 2012).

Conversion to Agricultural Land

Conversion of forest to agricultural land has a greater negative impact on animal and plant communities than degradation of forest (Gibson et al., 2011). In these habitats, ant-plants can usually survive only in unmanaged areas such as habitat margins, or as epiphytes on plantation trees. An example of the latter is the persistence of epiphytic bird's nest ferns (*Asplenium nidus*) in oil palm plantations in Malaysian Borneo, where ferns continue to host ants, which continue to protect the fern from herbivores (Fayle et al., 2015a, see also further pages). Persistence of ant-epiphyte symbioses in food gardens and open areas has also been reported in Papua New Guinea, with response to disturbance depending on elevation (Huxley, 1978). Partial conversion to agriculture has a less extreme impact on ant-plant symbioses. For example, cocoa agroforest, in which native shade trees are maintained, has similar overall levels of bromeliad-dwelling ant diversity to unconverted habitat, although with lower interaction specificity (DaRocha et al., 2016).

Synergy of Land-Use Impacts with Other Human-Driven Global Changes

Other anthropogenic global changes are likely to interact with the effects of differing land-use (Sala et al., 2000), with potential consequences for ant-plant symbioses. Habitat conversion has the potential to exacerbate the impacts of climate change, since increases in temperature due to logging and conversion to agriculture are often much greater than those predicted under even the most pessimistic climate change scenarios (Foster et al., 2011). Impacts of climate change on ant-plant communities can currently be extrapolated only from space-for-time surveys of ant-plants along existing climatic gradients (Mayer et al., 2014). For example, at lower altitudes in Papua New Guinea there is higher species richness of both plants and ants, and evidence for a higher level of plant protection by ants (Plowman et al., 2017). The relative importance of direct climate effects and plant protection by ants has also been investigated through transplant experiments across altitudinal gradients. In a study ranging from lowland Amazonian rain forest to montane Andean vegetation, ant-plants (*Piper immutatum*) were transported outside of their existing range both with and without their symbiotic ants (*Pheidole* sp.). Plant survival was most affected by direct climatic effects, rather than inhabitation or protection by the ant partner (Rodríguez-Castañeda et al., 2011). Extrapolating from these few studies to predict climate change impacts is challenging, because ant-plant responses will depend on multiple interacting factors, such as migration rates of the mutualistic partners, and whether ranges are defined by biotic or abiotic factors.

Nutrient enrichment may also affect ant-plant mutualisms, especially those that involve provision of nutrition from ants to plants. If plants have greater available nutrients, then ant-provided nutrients will be less valuable (Mayer et al., 2014). Such effects are likely to be greater in agricultural habitats where fertilisers are used, and in adjacent forest areas affected by fertiliser drift (Weathers et al., 2001). However, in some cases nutrient concentrations can also decrease with increasing habitat disturbance, due to depletion of the organic layer or leaching (Fernandes & Sanford, 1995; Owusu-Sekyere et al., 2006), potentially increasing the value of ant nutrient provisioning. It is therefore likely that responses are system-specific and more studies are needed for generalisations to be drawn.

Ants number among some of the most successful of invasive species, causing severe impacts on the functioning of many natural ecosystems (Lowe et al., 2000). Human-altered habitats are often highly susceptible to invasion by non-natives (King & Tschinkel, 2008) and hence ant-plant mutualisms in these habitats are likely to be affected by these newcomers (Chapters 12–15). The outcome of such interactions depends on whether (1) invasive ant species out-compete native plant ants, or (2) native ants are somehow buffered against the invaders, for example by having access to resources provided by plants that invasive ants are unable to utilise (Ness & Bronstein, 2004). As an example of the former scenario, the little fire ant, *Wasmannia auropunctata*, has been documented invading the domatia of the tree *Barteria fistulosa* in secondary forests in Gabon, and consequently reducing occupation by the native ant *Tetraponera aethiops*. This has resulted in an increase in liana coverage on the trees, as lianas are usually removed by the native ant partner (Mikissa et al., 2013). Ant-plants themselves can sometimes also become invasive species, opening up the possibility of new relationships being formed with native ants from the invaded habitat. For example, Neotropical *Cecropia* plants, which are ant-inhabited in their native ranges, thrive elsewhere, with populations in Hawaii (*C. obtusifolia*) and Peninsula Malaysia (*C. peltata*). In this case, however, plants generally do not contain ants, despite abundant non-specialist ant partners inhabiting *Cecropia* in its native range. This may be because access holes into domatia have not been made by the plant's regular ant partner and also because an absence of specialist herbivores has ensured that lack of protection is not a significant cost to the plants (Putz & Holbrook, 1988; Wetterer, 1997). In general, it seems likely that the degree of interaction specificity will influence the manner in which non-native species of ants and plants interact. With accelerating habitat change, movement of products around the world, and the impacts of climate change taking effect, we are likely to see the formation of further new combinations of ant and plant partners in the future. Understanding the costs and benefits for partners in these novel symbioses is likely to be a fruitful future research direction, informing both core ecological knowledge as well as habitat management strategies for biodiversity and ecosystem services.

Figure 3.2.   Bird's nest fern (*Asplenium nidus*) in the high canopy of lowland Dipterocarp rain forest in Malaysian Borneo. The largest ferns reach 200 kg wet weight (Ellwood & Foster, 2004) and can support diverse arthropod communities, including multiple colonies of co-existing ants. Inset photograph shows a colony of ant belonging to the genus *Diacamma*, one of many species that excavate nesting cavities in the root mass of these ferns. Main photograph credit Chien C. Lee; inset Tom Fayle. (A black-and-white version of this figure will appear in some formats. For the color version, please refer to the plate section.)

The Interaction between Epiphytic Bird's Nest Ferns and Ants as a Model System

The interaction between epiphytic bird's nest ferns (*Asplenium* spp.) and their ant symbionts serves as a useful model system for exploring impacts of habitat change on mutualistic interactions. Here we review the current state of research regarding these ferns and their ant symbionts.

Bird's nest ferns are common throughout the old world tropics (Holttum, 1976). They are litter intercepting epiphytes (Figure 3.2; Fayle et al., 2008), probably deriving the majority of their nutrient requirements from the decomposition of falling leaves that are collected in a broad rosette of fronds (Turner et al., 2007). In lowland Dipterocarp rain forest in Malaysian Borneo, there are two common species of bird's nest fern: *A. phyllitidis* and *A. nidus* (Fayle et al., 2009). *A. phyllitidis* is restricted to more shaded areas, where the continuous canopy layer provides more living space for this species. *A. nidus* is more abundant in areas that are open at ground level and where there are higher densities of emergent trees, since both of these areas provide the open habitat that this species requires. This leads to a vertical stratification, with *A. phyllitidis* being found only below 30 m, but *A. nidus* being found at all heights in the canopy, up to 60 m in the tallest emergent trees. Both species collect leaf litter, and the resulting mass of decomposing organic material, held together by the fern's root mass, is damp and cool, with temperature being buffered compared to that in the surrounding canopy (Turner & Foster, 2006; Freiberg & Turton, 2007). This refuge from the hot, dry rain forest canopy is an attractive habitat for a range of animals (mainly arthropods), the most abundant of which are the Coleoptera, Isoptera, Collembola, Acari, Diptera and Formicidae (Floater, 1995; Rodgers & Kitching, 1998; Walter et al., 1998; Ellwood et al., 2002, 2009; Karasawa & Hijii, 2006a, 2006b; 2006c; Turner & Foster, 2009; Rodgers & Kitching, 2011). As a result of this, the ferns can substantially increase the overall arthropod biomass that an area of canopy supports (Ellwood & Foster, 2004). Furthermore, bird's nest ferns occasionally provide nesting sites for birds (Thorstrom & Roland, 2000; Roland et al., 2005) and stingless bees (N. Blüthgen, personal communication, 2016), roosts for bats (Hodgkison et al., 2003) and habitats for frogs (Scheffers et al., 2013; Scheffers et al., 2014) and earthworms (Richardson et al., 2006). The ferns also co-occur with other epiphytic plant species, which can use the fern's mossy core as a substrate (T. M. Fayle personal observation, 2006), although it is not clear if these aggregations are 'ant gardens', in which ants have planted seeds to strengthen nest structure. Marasmioid fungi, which play a role in the decomposition of leaf litter (Snaddon et al., 2012), are also found in 36 per cent of the ferns in the litter held in the fern rosette (30 of the 83 ferns from Fayle et al. (2012), and the ferns also support communities of fungi and bacteria (Donald et al., 2017).

The most abundant animal group found in bird's nest ferns are the ants, comprising on average 86 per cent of individuals, and 91 per cent of biomass of all arthropods in primary forest ferns in Borneo (Turner & Foster, 2009), although in larger ferns termites are sometimes even more abundant than ants (Ellwood et al., 2002). Multiple ant colonies can co-exist within the litter-root mass (note that ferns do not grow domatia for ants), with larger ferns supporting more ant colonies; up to 12 resident ant species in larger ferns (Fayle et al., 2012). There is considerable ant species turnover between ferns, with at least 71 species across 27 genera using the ferns as nesting sites in primary forest. The identity of these ant species depends weakly on the height of the fern within the rain forest canopy, and on the size of fern, but once these factors are taken into account, there is no difference in ant composition or species richness between the two fern species, *A. nidus* and *A. phyllitidis*. Furthermore, some ant species found in leaf litter on the forest floor also inhabit the ferns (Fayle et al., 2015a). This indicates that the symbiotic

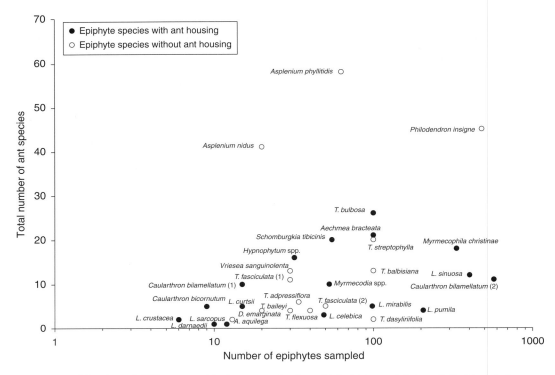

Figure 3.3.   Number of different symbiotic ant species found inhabiting epiphytes plotted in relation to sampling intensity and presence of ant housing. Some genus names are abbreviated for clarity: *Tillandisa, Leucanopteris, Dimerandra, Aechmea*. Two species are represented twice, denoted numerically in brackets. Data from publications for which both sampling intensity and number of ant species were reported for ant-epiphyte systems in habitats unmodified by humans (Huxley, 1978; Fisher & Zimmerman, 1988; Gay & Hensen, 1992; Dejean et al., 1995; Blüthgen et al., 2000; Stuntz et al., 2002; Dejean et al., 2003; Gibernau et al., 2007; Dutra & Wetterer, 2008; Fayle et al., 2012; Talaga et al., 2015). Figure reproduced and updated from supplementary online material of Fayle et al. (2012), with permission from John Wiley and Sons.

relationship is non-specific. This is a similar pattern to that observed for some ant-inhabited bromeliads (Blüthgen et al., 2000), where interactions have low specificity compared to a range of other systems (Bluthgen et al., 2007). This low specificity results in the ferns supporting more ant species than epiphytes that grow structures adapted for housing ants, although many other species lacking housing also have low ant diversity (Figure 3.3). The diverse ants inhabiting bird's nest ferns compete with one another for nesting space within the ferns (Ellwood et al., 2016), with species that have more similar body sizes competing most strongly (Fayle et al., 2015b). This competition controls fern-dwelling ant species abundance distributions.

Both ferns and ants receive by-product benefits from their symbiosis. The ants protect the fern from herbivory (Fayle et al., 2012), although this seems to be a result of normal foraging behaviour, with resident ants failing to aggressively defend ferns from disturbance (T. M. Fayle, personal observation), as would be expected in a protection mutualism. However, the presence of one ant species in the genus *Monomorium* has a negative impact on herbivory rates (Fayle et al., 2015a). An unidentified species in the same genus has also been observed to actively protect *Asplenium nidus* in India, while tending to coccids that mimic the fern's sori (clusters of spore-containing bodies) (Patra et al., 2008). Despite this protective behaviour, this species of *Monomorium* is not particularly common (15/83 ferns; 18 per cent) and the protective effect from herbivores remains even when this species is removed from analyses, indicating that multiple ant species provide this by-product service to the ferns. The lack of a tight mutualistic relationship is probably because there is little incentive for resident ants to promote fern growth, since larger ferns support more species of ants, rather than larger colonies of particular species (Fayle et al., 2012). This failure on the part of the fern to direct benefits towards more beneficial ant species probably arises because ferns are constrained to maintain a leaf litter layer and a soil root mass, which can be inhabited by a wide range of ant species as well as other taxa. Such a situation can be contrasted to those in which plants create pre-formed domatia, in which the increased intimacy of the interaction creates greater opportunities for partner selection and punishment (Edwards et al., 2006). Furthermore, the ferns have not been observed to provide food to their ant inhabitants, and *Asplenium* are not recorded as ever having foliar nectaries (www.extrafloralnectaries.org/). Hence, although ferns and ants receive by-product benefits from the symbiosis, neither partner has adaptively increased investment in the relationship, resulting in a two-way by-product mutualism. This interaction can be seen as an old world parallel to ant-bromeliad interactions in the Neotropics, with both groups being highly abundant, comprising some leaf-litter-collecting species, and showing low specificity of ant inhabitants (Blüthgen et al., 2000).

Throughout the tropics, but particularly in SE Asia, expansion of oil palm plantation following logging is a major driver of forest clearance (Wilcove et al., 2013). Surprisingly, bird's nest fern populations are resilient to habitat change, with abundances decreasing in logged forest, but increasing in oil palm plantation (90, 53 and 117 ferns per hectare in primary forest, logged forest and oil palm plantation respectively (Turner, 2005; see also Padmawathe et al., 2004). However, only the high canopy species *A. nidus* survives in oil palm plantations, perhaps due to its pre-adaptation to hot and dry environments (Fayle et al., 2011). Despite substantial reductions in total arthropod abundance (67.2 per cent decrease) and biomass (87.5 per cent decrease) between ferns in primary forest and those in oil palm plantation (Turner & Foster, 2009), the numbers of species of ants per fern do not change (Fayle et al., 2010). This is in contrast to leaf litter and canopy communities more broadly, which both show substantial reductions in ant species richness. However, in oil palm plantations, a completely different set of ant species inhabit the ferns. The oil palm fern ants show stronger species segregation (consistent with the existence of interspecific competition) than those in primary forest. This pattern is not driven by the presence of non-native ant species (Fayle et al., 2013), with analyses in which non-native species are removed showing even stronger patterns of

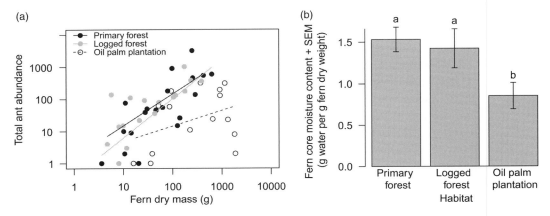

Figure 3.4.   (a) The volume of suitable nesting space for ants in ferns differs between habitats, with ferns in oil palm plantation supporting lower total abundances of ants per unit dry weight than ferns from forest habitats. (b) One explanation for this is that ferns in oil palm plantations have significantly lower moisture content than those in either primary or logged forest. Standard error bars are shown. Reproduced with permission from Fayle et al. (2015a).

species segregation. This effect is even more pronounced for ants in the ferns than in the rest of the canopy. The degree of specificity of the interaction remains low in logged forest and in oil palm plantation, with oil palm showing even greater overlap between fern-dwelling and litter ants than the other two habitats (Fayle et al., 2015a). Furthermore, the positive relationship between the fern size and the number of ant species observed in primary forest ferns (Fayle et al., 2012) persists in both logged forest and in oil palm plantation (Fayle et al., 2015a), and there is also no relationship in these habitats between the size of colonies of individual ant species and fern size. This indicates that there is little opportunity for partner fidelity feedbacks in human-modified habitats. Hence, neither ferns nor their resident ants invest in partner fitness, since for the ants, this would not result in benefits being fed back to that colony, and for plants there remains no opportunity to direct benefits to better partners. Interestingly, the relationship between total ant abundance (not that for any particular colony) and fern size differs between oil palm plantation and logged or primary forest, with a given increase in fern size resulting in a much smaller increase in total ant abundance in oil palm plantation (Figure 3.4a). This is probably because the hotter, drier microclimate in oil palm plantations (Turner & Foster, 2006) results in a lower moisture content in oil palm ferns (Figure 3.4b), leading to a reduction in the habitable volume of the fern. Non-native species in oil palm plantation, which are common in the ferns, play a significant role in driving the relationships between fern size and ant species richness/abundance. This indicates that the persistence of this two-way by-product mutualism in oil palm plantations depends to some extent on non-native species. The result also raises the question as to whether more generalist interactions are more robust to habitat change.

Future Research Directions

As the review has demonstrated, this is an area with a paucity of studies. However, ant-plant symbioses offer useful model systems for understanding network responses to disturbance and shifts in costs and benefits for symbiotic partners. Fruitful work could be conducted in a range of different directions.

Differential Responses of Specialised and Generalised Species to Habitat Change

Generalist species are predicted to be better able to persist in human-modified habitats than specialist species, because they are less likely to suffer total loss of all partner species and because they are likely to form new connections more easily. Ant-plant symbiotic systems present an opportunity to test this prediction. For example, a similar pattern has already been found in terms of spatial turnover of ant-EFN bearing plant interactions within one habitat type, with a central core of generalists (those species interacting with many other species) remaining unchanged over larger spatial scales (Dáttilo et al., 2013). With regard to forest fragmentation impacts on networks involving more specialised species, impacts are observed to be greater where ants and plants cannot cross matrix habitats (Bruna et al., 2005; Passmore et al., 2012; Emer et al., 2013). Furthermore, for a less specialised interaction, the symbiosis persists, even in plantation habitats (Fayle et al., 2015a), partly because non-native species are able to take the place of native ant partners. It is also possible that in disturbed habitats there might be some 'rewiring' of the network, with persisting species forming novel connections with each other (in addition to interacting with newly arrived species). Hence we predict that the responses of specialist and generalist mutualists will depend on (1) landscape connectivity with source populations of ants and plants, (2) whether non-native species can take the place of native partners for less specialised interactions and (3) the degree to which the network 'rewires' itself following disturbance.

Impacts of Abiotic Changes on Costs and Benefits of Interactions

Shifts in the abiotic environment that occur during habitat conversion, such as changes in temperature and nutrient availability, are expected to alter the outcomes of mutualistic interactions, specifically in relation to the value of investing in partners. For example, if converted habitats are more nutrient-poor, then the value of hosting plant-feeding ants will increase; if a hotter habitat means that a smaller volume of the plant is habitable, this may break the relationship between ant colony size and plant size, reducing the value for ants of investing in plant growth (for an example specifically relating to bird's nest ferns see Figure 3.4 and the section 'Synergy of Land-Use Impacts with Other Human-Driven Global Changes). In converted habitats, if species persist, they do so in an adaptive landscape very different from the one in which they evolved (Laughlin & Messier, 2015). Hence robustness to habitat change will depend on species' abilities to respond plastically over short time periods. It would be worthwhile measuring costs and benefits for partners directly in relation to changes in various abiotic variables along habitat disturbance gradients. Such measurements could allow better prediction of persistence of species involved in mutualisms. Over longer time periods, tracking evolutionary changes in mutualistic behaviours in converted habitats would also be of interest.

Impacts of Changes in the Biotic Environment

Symbiotic ant and plant species experience novel biotic environments as a result of human-induced changes, both in terms of their partner species and other species that impact on the interaction. For example, non-native *Cecropia peltata* (that are ant-inhabited in their native range) in Peninsula Malaysia experience less herbivory than plants in their native range, despite lacking ant inhabitants (Putz & Holbrook, 1988), perhaps due to a release from specialist herbivores. This represents a radical change in the benefits of ant-inhabitation. A similar pattern is observed when large mammalian herbivores are excluded from *Acacia* ant-plants in Kenya, with the benefits of ant-inhabitation being reduced (Palmer et al., 2008). Hence, even in supposedly pristine habitats, previous mammalian herbivore extinctions might leave mutualistic partners behaving sub-optimally. It would be worthwhile exploring how costs and benefits vary across habitat disturbance gradients both with partner identity and in relation to the presence of other interacting taxa, such as herbivores.

Conclusion

The world's tropical forests are changing rapidly as a result of human disturbance. This not only causes species extinctions at local and global scales, and shifts in species composition, but also drives a re-organisation of interactions between those species that persist. Understanding the nature of these novel interaction networks is vital if we are to maintain ecosystem functioning in human-modified landscapes. Here we have described how mutualistic symbioses between ants and plants are altered when humans exploit tropical forests, although a lack of studies makes generalisation of results challenging. Ant symbioses with bird's nest ferns serve as a useful model system for exploring the impacts of habitat change on non-specific mutualistic interactions. Future research might profitably compare responses to habitat change for mutualistic species with a range of degrees of interaction specificity, and assess the way that costs and benefits of the interaction change in relation to shifts in both abiotic and biotic environments.

References

Blüthgen, N., Menzel, F., Hovestadt, T., Fiala, B. and Bluthgen, N. (2007). Specialization, constraints, and conflicting interests in mutualistic networks. *Current Biology*, 17, 341–346.
Blüthgen, N., Verhaagh, M., Goitía, W. and Blüthgen, N. (2000). Ant nests in tank bromeliads – an example of non-specific interaction. *Insect. Soc.*, 47, 313–316.

Bruna, E. M., Izzo, T. J., Inouye, B. D., Uriarte, M. and Vasconcelos, H. L. (2011). Asymmetric dispersal and colonization success of Amazonian plant-ants queens. *PLoS ONE*, 6, e22937.

Bruna, E. M., Vasconcelos, H. L. and Heredia, S. (2005). The effect of habitat fragmentation on communities of mutualists: Amazonian ants and their host plants. *Biological Conservation*, 124, 209–216.

Chazdon, R. L. (2014). *Second Growth: The Promise of Tropical Forest Regeneration in an Age of Deforestation*. Chicago: University of Chicago Press.

DaRocha, W. D., Neves, F. S., Dáttilo, W. and Delabie, J. H. C. (2016). Epiphytic bromeliads as key components for maintenance of ant diversity and ant–bromeliad interactions in agroforestry system canopies. *Forest Ecology and Management*, 372, 128–136.

Dáttilo, W., Guimarães, P. R. and Izzo, T. J. (2013). Spatial structure of ant-plant mutualistic networks. *Oikos*, 122, 1643–1648.

Dejean, A., Durou, S., Olmsted, I., Snelling, R. R. and Orivel, J. (2003). Nest site selection by ants in a flooded Mexican mangrove, with special reference to the epiphytic orchid *Myrmecophila christinae*. *Journal of Tropical Ecology*, 19, 325–331.

Dejean, A., Olmsted, I. and Snelling, R. R. (1995). Tree-epiphyte-ant relationships in the low inundated forest of Sian Ka'an biosphere reserve, Quintana Roo, Mexico. *Biotropica*, 27, 57–70.

Donald, J., Maxfield, P., Murray, D. and Ellwood, M. D. F. (2017) How tropical epiphytes at the Eden Project contribute to rainforest canopy science. *Sibbaldia: The Journal of Botanic Garden Horticulture*, 14, 55–68.

Dunne, J. A. and Williams, R. J. (2009). Cascading extinctions and community collapse in model food webs. *Philosophical Transactions of the Royal Society B-Biological Sciences*, 364, 1711–1723.

Dutra, D. and Wetterer, J. K. (2008). Ants in myrmecophytic orchids of Trinidad (Hymenoptera: Formicidae). *Sociobiology*, 51, 249–254.

Edwards, D. P., Hassall, M., Sutherland, W. J. and Yu, D. W. (2006). Selection for protection in an ant-plant mutualism: host sanctions, host modularity, and the principal-agent game. *Proceedings of the Royal Society B: Biological Sciences*, 273, 595–602.

Edwards, D. P., Tobias, J. A., Sheil, D., Meijaard, E. and Laurance, W. F. (2014). Maintaining ecosystem function and services in logged tropical forests. *Trends in Ecology & Evolution*, 29, 511–520.

Ellwood, M. D. F., Blüthgen, N., Fayle, T. M., Foster, W. A. and Menzel, F. (2016). Analysis of pairwise interactions reveals unexpected patterns in tropical ant communities *Acta Oecologica*, 75, 24–34.

Ellwood, M. D. F. and Foster, W. A. (2004). Doubling the estimate of invertebrate biomass in a rainforest canopy? *Nature*, 429, 549–551.

Ellwood, M. D. F., Jones, D. T. and Foster, W. A. (2002). Canopy ferns in lowland dipterocarp forest support a prolific abundance of ants, termites and other invertebrates. *Biotropica*, 34, 575–583.

Ellwood, M. D. F., Manica, A. and Foster, W. A. (2009). Stochastic and deterministic processes jointly structure tropical arthropod communities. *Ecology Letters*, 12, 277–284.

Emer, C., Venticinque, E. and Fonseca, C. R. (2013). Effects of dam-induced landscape fragmentation on Amazonian ant-plant mutualistic networks. *Conservation Biology*, 27, 763–773.

Ewers, R. M. and Didham, R. K. (2008). Pervasive impact of large-scale edge effects on a beetle community. *Proceedings of the National Academy of Sciences*, 105, 5426–5429.

Fayle, T. M., Chung, A. Y., Dumbrell, A. J., Eggleton, P. and Foster, W. A. (2009). The effect of rain forest canopy architecture on the distribution of epiphytic ferns (*Asplenium* spp.) in Sabah, Malaysia. *Biotropica*, 41, 676–681.

Fayle, T. M., Dumbrell, A. J., Turner, E. C. and Foster, W. A. (2011). Distributional patterns of epiphytic ferns are explained by the presence of cryptic species. *Biotropica*, 43, 6–7.

Fayle, T. M., Edwards, D. P., Foster, W. A., Yusah, K. M. and Turner, E. C. (2015a). An ant-plant by-product mutualism is robust to selective logging of rain forest and conversion to oil palm plantation. *Oecologia*, 178, 441–450.

Fayle, T. M., Edwards, D. P., Turner, E. C. et al. (2012). Public goods, public services, and by-product mutualism in an ant-fern symbiosis. *Oikos*, 121, 1279–1286.

Fayle, T. M., Eggleton, P., Manica, A., Yusah, K. M. and Foster, W. A. (2015b). Experimentally testing and assessing the predictive power of species assembly rules for tropical canopy ants. *Ecology Letters*, 18, 254–262.

Fayle, T. M., Ellwood, M. D. F., Turner, E. C. et al. (2008). Bird's nest ferns: islands of bio-diversity in the rainforest canopy. *Antenna*, 32(1), 34–37.

Fayle, T. M., Turner, E. C. and Foster, W. A. (2013). Ant mosaics occur in SE Asian oil palm plantation but not rain forest and are influenced by the presence of nest-sites and non-native species. *Ecography*, 36, 1051–1057.

Fayle, T. M., Turner, E. C., Snaddon, J. L. et al. (2010). Oil palm expansion into rain forest greatly reduces ant biodiversity in canopy, epiphytes and leaf-litter. *Basic and Applied Ecology*, 11, 337–345.

Feldhaar, H., Gadau, J. and Fiala, B. (2010). Speciation in obligately plant-associated crema-togaster ants: host distribution rather than adaption towards specific hosts drives the process. In: Glaubrecht, M (ed.) *Evolution in Action*. Berlin Heidelberg: Springer, pp. 193–213.

Fernandes, D. N. and Sanford, R. L. (1995). Effects of recent land-use practices on soil nutrients and succession under tropical wet forest in Costa Rica. *Conservation Biology*, 9, 915–922.

Fisher, B. L. and Zimmerman, J. K. (1988). Ant/orchid associations in the Barro Colorado National Monument, Panama. *Lindleyana*, 3, 12–16.

Floater, G. J. (1995). Effect of epiphytes on the abundance and species richness of litter-dwelling insects in a Seychelles cloud forest. *Tropical Ecology*, 36, 203–212.

Fonseca, C. R. (1999). Amazonian ant-plant interactions and the nesting space limitation hypothesis. *Journal of Tropical Ecology*, 15, 807–825.

Fortuna, M. A. and Bascompte, J. (2006). Habitat loss and the structure of plant–animal mutualistic networks. *Ecology Letters*, 9, 281–286.

Foster, W. A., Snaddon, J. L., Turner, E. C. et al. (2011). Establishing the evidence base for maintaining biodiversity and ecosystem function in the oil palm landscapes of South East Asia. *Philosophical Transactions of the Royal Society B: Biological Sciences*, 366, 3277–3291.

Frederickson, M. E., Greene, M. J. and Gordon, D. M. (2005). 'Devil's gardens' bedevilled by ants. *Nature*, 437, 495–496.

Freiberg, M. and Turton, S. M. (2007). Importance of drought on the distribution of the birds nest fern, *Asplenium nidus*, in the canopy of a lowland tropical rainforest in north-eastern Australia. *Austral Ecology*, 32, 70–76.

Gay, H. and Hensen, R. (1992). Ant specificity and behaviour in mutualisms with epiphytes: the case of Lecanopteris (Polypodiaceae). *Biological Journal of the Linnean Society*, 47, 261–284.

Gibernau, M., Orivel, J., Delabie, J. H. C., Barabe, D. and Dejean, A. (2007). An asymmetrical relationship between an arboreal ponerine ant and a trash-basket epiphyte (Araceae). *Biological Journal of the Linnean Society*, 91, 341–346.

Gibson, L., Lee, T. M., Lian Pin Koh et al. (2011). Primary forests are irreplaceable for sustaining tropical biodiversity. *Nature*, 478, 378–381.

Haddad, N. M., Brudvig, L. A., Clobert, J. et al. (2015). Habitat fragmentation and its lasting impact on Earth's ecosystems. *Science Advances*, 1, e1500052.

Hardwick, S. R., Toumi, R., Pfeifer, M. et al. (2015). The relationship between leaf area index and microclimate in tropical forest and oil palm plantation: Forest disturbance drives changes in microclimate. *Agricultural and Forest Meteorology*, 201, 187–195.

Hodgkison, R., Balding, S. T., Akbar, Z. and Kunz, T. H. (2003). Roosting ecology and social organization of the spotted-winged fruit bat, *Balionycteris maculata* (Chiroptera: Pteropodidae), in a Malaysian lowland dipterocarp forest. *Journal of Tropical Ecology*, 19, 667–676.

Holttum, R. E. (1976). *Asplenium* Linn., sect. *Thamnopteris* Presl. *Gardens' Bulletin, Singapore*, 27, 143–154.

Huxley, C. R. (1978). The ant-plants *Myrmecodia* and *Hydnophytum* (Rubiaceae), and the relationships between their morphology, ant occupants, physiology and ecology. *New Phytologist*, 80, 213–268.

Kaiser-Bunbury, C. N. and Blüthgen, N. (2015). Integrating network ecology with applied conservation: a synthesis and guide to implementation. *AoB Plants*, 7: plv076. doi:10.1093/aobpla/plv076

Karasawa, S. and Hijii, N. (2006a). Determinants of litter accumulation and the abundance of litter-associated microarthropods in bird's nest ferns (*Asplenium nidus* complex) in the forest of Yambaru on Okinawa Island, southern Japan. *Journal of Forest Research*, 11, 313–318.

(2006b). Does the existence of bird's nest ferns enhance the diversity of oribatid (Acari: Oribatida) communities in a subtropical forest? *Biodiversity and Conservation*, 15, 4533–4553.

(2006c). Effects of distribution and structural traits of bird's nest ferns (*Asplenium nidus*) on oribatid (Acari: Oribatida) communities in a subtropical Japanese forest. *Journal of Tropical Ecology*, 22, 213–222.

King, J. R. and Tschinkel, W. R. (2008). Experimental evidence that human impacts drive fire ant invasions and ecological change. *Proceeding of the National Academy of Sciences*, 105, 20339–20343.

Klimes, P., Idigel, C., Rimandai, M. et al. (2012). Why are there more arboreal ant species in primary than secondary tropical forests? *Journal of Animal Ecology*, 81, 1103–1112.

Laughlin, D. C. and Messier, J. (2015). Fitness of multidimensional phenotypes in dynamic adaptive landscapes. *Trends in Ecology & Evolution*, 30, 487–496.

Laurance, W. F., Camargo, J. L. C., Luizão, R. C. C. et al. (2011). The fate of Amazonian forest fragments: a 32-year investigation. *Biological Conservation*, 144, 56–67.

Lowe, S., Browne, M., Boudjelas, S. and De Poorter, M. (2000). *100 of the World's Worst Invasive Alien Species: A Selection from the Global Invasive Species Database*. The Invasive Species Specialist Group (ISSG), Auckland, New Zealand, 12 pp.

Mayer, V. E., Frederickson, M. E., McKey, D. and Blatrix, R. (2014). Current issues in the evolutionary ecology of ant-plant symbioses. *New Phytologist*, 202, 749–764.

Mikissa, J. B., Jeffery, K., Fresneau, D. and Mercier, J. L. (2013). Impact of an invasive alien ant, Wasmannia auropunctata Roger, on a specialised plant–ant mutualism, Barteria fistulosa Mast. and Tetraponera aethiops F. Smith., in a Gabon forest. *Ecological Entomology*, 38, 580–584.

Murase, K., Itioka, T., Nomura, M. and Yamane, S. (2003). Intraspecific variation in the status of ant symbiosis on a myrmecophyte, Macaranga bancana, between primary and secondary forests in Borneo. *Population Ecology*, 45, 221–226.

Murase, K., Yamane, S., Itino, T. and Itioka, T. (2010). Multiple factors maintaining high species-specificity in Macaranga-Crematogaster (Hymenoptera: Formicidae) myrmecophytism: higher mortality in mismatched ant-seedling pairs. *Sociobiology*, 55, 883–898.

Ness, J. and Bronstein, J. (2004). The effects of invasive ants on prospective ant mutualists. *Biological Invasions*, 6, 445–461.

Owusu-Sekyere, E., Cobbina, J. and Wakatsuki, T. (2006). Nutrient cycling in primary, secondary forests and cocoa plantation in the Ashanti Region, Ghana. *West African Journal of Applied Ecology*, 9, 1–9. http://dx.doi.org/10.4314/wajae.v9i1.45680

Padmawathe, R., Qureshi, Q. and Rawat, G. S. (2004). Effects of selective logging on vascular epiphyte diversity in a moist lowland forest of Eastern Himalaya, India. *Biological Conservation*, 119, 81–92.

Palmer, T. M., Stanton, M. L., Young, T. P. et al. (2008). Breakdown of an ant-plant mutualism follows the loss of large herbivores from an African Savanna. *Science*, 319, 192–195.

Passmore, H. A., Bruna, E. M., Heredia, S. M. and Vasconcelos, H. L. (2012). Resilient networks of ant-plant mutualists in Amazonian forest fragments. *PLoS ONE*, 7, e40803.

Patra, B., Bera, S. and Hickey, R. J. (2008). Soral crypsis: protective mimicry of a coccid on an Indian fern. *Journal of Integrative Plant Biology*, 50, 653–658.

Picard, N., Gourlet-Fleury, S. and Forni, É. (2012). Estimating damage from selective logging and implications for tropical forest management. *Canadian Journal of Forest Research*, 42, 605–613.

Plowman, N. S., Hood, A. S. C., Moses, J., Redmond, C., Novotny, V., Klimes, P. and Fayle, T. M. (2017). Network reorganization and breakdown of an ant-plant protection mutualism with elevation. *Proceedings of the Royal Society B: Biological Sciences* 284: 20162564. http://dx.doi.org/10.1098/rspb.2016.2564.

Putz, F. E. and Holbrook, N. M. (1988). Further observations on the dissolution of mutualism between Cecropia and its ants: the Malaysian case. *Oikos*, 53, 121–125.

Richardson, B. A., Borges, S. and Richardson, M. J. (2006). Differences between epigeic earthworm populations in tank bromeliads from Puerto Rico and Dominica. *Caribbean Journal of Science*, 42, 380–385.

Rico-Gray, V. and Oliveira, P. S. (2007). *The Ecology and Evolution of Ant-Plant Interactions*. Chicago: University of Chicago Press.

Rodgers, D. J. and Kitching, R. L. (1998). Vertical stratification of rainforest collembolan (Collembola: Insecta) assemblages: description of ecological patterns and hypotheses concerning their generation. *Ecography*, 21, 392–400.

(2011). Rainforest Collembola and the insularity of epiphyte microhabitats. *Insect Conservation and Diversity*, 4, 99–106.

Rodríguez-Castañeda, G., Forkner, R. E., Tepe, E. J., Gentry, G. L. and Dyer, L. A. (2011). Weighing defensive and nutritive roles of ant mutualists across a tropical altitudinal gradient. *Biotropica*, 43, 343–350.

Roland, L.-A. R. d., Rabearivony, J., Razafimanjato, G., Robenarimangason, H. and Thorstrom, R. (2005). Breeding biology and diet of Banded Kestrels Falco zoniventris on Masoala Peninsula, Madagascar. *Ostrich*, 76, 32–36.

Sala, O. E., Stuart Chapin, F., III et al. (2000). Global biodiversity scenarios for the year 2100. *Science*, 287, 1770–1774.

Scheffers, B. R., Edwards, D. P., Diesmos, A., Williams, S. E. and Evans, T. A. (2013). Microhabitats reduce animal's exposure to climate extremes. *Global Change Biology*, n/a-n/a.

Scheffers, B. R., Phillips, B. L. and Shoo, L. P. (2014). Asplenium bird's nest ferns in rainforest canopies are climate-contingent refuges for frogs. *Global Ecology and Conservation*, 2, 37–46.

Slik, F. J. W., Keßler, P. J. A. and Welzen, P. C. v. (2003). Macaranga and Mallotus species (Euphorbiaceae) as indicators for disturbance in the mixed lowland dipterocarp forest of East Kalimantan (Indonesia). *Ecological Indicators*, 2, 311–324.

Snaddon, J. L., Turner, E. C., Fayle, T. M. et al. (2012). Biodiversity hanging by a thread: the importance of fungal-litter trapping systems in tropical rainforests. *Biology Letters*, 8, 397–400.

Stuntz, S., Ziegler, C., Simon, U. and Zotz, G. (2002). Diversity and structure of the arthropod fauna within three canopy epiphyte species in central Panama. *Journal of Tropical Ecology*, 18, 161–176.

Talaga, S., Dézerald, O., Carteron, A. et al. (2015). Tank bromeliads as natural microcosms: a facultative association with ants influences the aquatic invertebrate community structure. *C. R. Biol.*, 338, 696–700.

Tanaka, H., Inui, Y. and Itioka, T. (2009). Anti-herbivore effects of an ant species, Crematogaster difformis, inhabiting myrmecophytic epiphytes in the canopy of a tropical lowland rainforest in Borneo. *Ecol. Res.*, 24, 1393–1397.

Tanaka, H. O., Yamane, S., Nakashizuka, T., Momose, K. and Itioka, T. (2007). Effects of deforestation on mutualistic interactions of ants with plants and hemipterans in tropical rainforest of Borneo. *Asian Myrmecology*, 1, 31–50.

Thorstrom, R. and Roland, L.-A. R. d. (2000). First nest description, breeding behaviour and distribution of the Madagascar Serpent-Eagle Eutriorchis astur. *Ibis*, 142, 217–224.

Tilman, D., Fargione, J., Wolff, B. et al. (2001). Forecasting agriculturally driven global environmental change. *Science*, 292, 281–284.

Turner, E. C. (2005). The ecology of the Bird's Nest Fern (*Asplenium* spp.) in unlogged and managed habitats in Sabah, Malaysia. PhD, University of Cambridge, Cambridge.

Turner, E. C. and Foster, W. A. (2006). Assessing the influence of Bird's nest ferns (*Asplenium* spp.) on the local microclimate across a range of habitat disturbances in Sabah, Malaysia. *Selbyana*, 27, 195–200.

(2009). The impact of forest conversion to oil palm on arthropod abundance and biomass in Sabah, Malaysia. *Journal of Tropical Ecology*, 25, 23–30.

Turner, E. C., Snaddon, J. L., Johnson, H. R. and Foster, W. A. (2007). The impact of bird's nest ferns on stemflow nutrient concentration in a primary rain forest, Sabah, Malaysia. *Journal of Tropical Ecology*, 23, 721–724.

Tylianakis, J. M., Laliberté, E., Nielsen, A. and Bascompte, J. (2010). Conservation of species interaction networks. *Biological Conservation*, 143, 2270–2279.

Walter, D. E., Seeman, O., Rodgers, D. and Kitching, R. L. (1998). Mites in the mist: how unique is a rainforest canopy knockdown fauna? *Australian Journal of Ecology*, 23, 501–508.

Watkins, J. E., Cardelús, C. L. and Mack, M. C. (2008). Ants mediate nitrogen relations of an epiphytic fern. *New Phytologist*, 180, 5–8.

Weathers, K. C., Cadenasso, M. L. and Pickett, S. T. (2001). Forest edges as nutrient and pollutant concentrators: potential synergisms between fragmentation, forest canopies, and the atmosphere. *Conservation Biology*, 15, 1506–1514.

Wetterer, J. K. (1997). Ants on *Cecropia* in Hawaii. *Biotropica*, 29, 128–132.

Wilcove, D. S., Giam, X., Edwards, D. P., Fisher, B. and Koh, L. P. (2013). Navjot's nightmare revisited: logging, agriculture, and biodiversity in Southeast Asia. *Trends in Ecology & Evolution*, 28, 531–540.

Yu, D. W., Wilson, H. B. and Pierce, N. E. (2001). An empirical model of species coexistence in a spatially structured environment. *Ecology*, 82, 1761–1771.

4 Ecology of Leaf-Cutting Ants in Human-Modified Landscapes

Marcelo Tabarelli, Felipe F. S. Siqueira, Julia Backé,
Rainer Wirth, and Inara R. Leal[*]

Introduction

Natural landscapes have been rapidly converted into human-modified landscapes (HML) globally, and future population growth, particularly across the tropics, will probably speed up this trend (Laurance et al., 2014). In the tropical region, HML refer to vegetation mosaics consisting of one or several components, such as old-growth and edge-affected forest remnants, secondary forest patches of varying ages and managed plantations of exotic species (Tabarelli et al., 2010). These components are usually immersed in open-habitat matrix consisting of pasture lands or agricultural fields. Such changes in the nature of forest habitats and spatial reorganisation of landscape components occur in parallel to other threats for biodiversity persistence in HML, such as hunting, logging, plant collection, biocide/fertiliser spillover and fire (Tabarelli et al., 2004; Laurance et al., 2014). In this emerging ecological scenario, some native species benefit and tend to proliferate, i.e. the 'winner species' (sensu Tabarelli et al. 2012), with cascading impacts on biological dynamics still to be investigated.

Leaf-cutting ants (hereafter LCA) offer prime examples for such winner species and represent one of the most characteristic and conspicuous groups of social insects inhabiting the majority of neotropical and subtropical habitats, from semi-arid grasslands to humid forests, and pasture lands to agricultural fields (Fowler & Claver, 1991; Farji-Brener & Ruggiero, 1994; Weber, 1966). Strictly, they currently

[*] We thank Paulo Oliveira and Suzanne Koptur for the invitation to write this chapter. We also thank the financial support of the collaborating projects CAPES/DFG (process 007/01) and CAPES/DAAD (grants 257/07, BEX 8836/11–6), CNPq-DFG (grant 490450/2013-0) and the following agencies: CNPq (grants 540322/01, 471904/2004-0, 305970/2004–6, 304346/2007-1, 473529/2007-6, 403770/2012-2, 470480/2013-0), FACEPE (grants 0738-2.05/12, APQ-0138-2.05/14), ICMbio (grants) and Schimper Foundation (grant 1959/1–2). Conservação Internacional do Brasil (CI-Brasil). Centro de Estudos Ambientais do Nordeste (CEPAN) and Usina Serra Grande provided infrastructure and logistic support during field work. Finally, this review would not have been possible without the invaluable practical and intellectual contributions of Ana G. D. Bieber, Christoph Dohm, Clarissa Knoechelmann, Manoel V. de Araújo Jr, Michele M. Corrêa, Olivier P. G. Darrault, Paulo S. D. da Silva, Pille Urbas, Poliana F. Falcão, Sebastian T. Meyer, Veralucia S. Barbosa, Walkiria R. de Almeida and many other graduate students not mentioned here.

consist of 40 species within the genera *Atta* and *Acromyrmex*, in the Myrmicinae subfamily (Schultz & Brady, 2008). Some species have narrow geographic distributions or are habitat specialists (e.g. *Atta robusta*, Teixeira et al., 2003), while others are generalists with broad distributional range (e.g. *Atta sexdens*, Fowler et al., 1989). Despite such diversity, LCA share a distinct set of morphological, metabolic and natural history characters, together with the unique habit of cutting fresh leaf material to cultivate a symbiotic fungus that serves as their main food source (Weber, 1966; Mueller et al., 1998; De Fine Licht & Boomsma, 2010; Hölldobler & Wilson, 2011).

Precisely, LCA are able to harvest tremendous quantities and types of plant materials (Wirth et al., 2003; Herz et al., 2007; Costa et al., 2008; Falcão et al., 2011). In the case of *Atta* species, plant consumption per colony can range from ca. 70 to 500 kg (dry weight) per year, making *Atta* ants the preeminent herbivore of Neotropical forests and savannas (Wirth et al., 2003; Herz et al., 2007; Costa et al., 2008 and references therein). Accordingly, LCA herbivory impacts plant fitness, demography and community structure (Wirth et al., 2003; Leal et al., 2014a; Corrêa et al., 2016). In addition to herbivory, LCA affect their environment via construction and maintenance of colossal nests. *Atta* nests, for instance, may achieve up to 8,000 subterranean interconnected chambers (Moreira et al., 2004a, 2004b), with more than 20 m³ or 40 tons of soil removed to the surface (reviewed by Farji-Brener & Illes, 2000; Hölldobler & Wilson, 2011). Nest construction/maintenance may result in large mounds of soil above their surface reaching up to 250 m² in area (Cherrett, 1989). These nest-related activities can alter canopy cover, light regime, litter cover and soil attributes (chemical and physical attributes), while LCA colonies are active for 8–20 years (Fowler et al., 1986; Meyer et al., 2009), and may even persist long after colony death (Farji-Brener & Illes, 2000; Bieber et al., 2011). Such tangible impacts cascade across multiple levels of biological organisation (from plant population to ecosystem level) and spatial scales, such as the alteration of plant assemblages in nest-impacted areas (hundreds of m²) and foraging zones that often reach as much as 2 ha (Corrêa et al., 2010; 2016).

Therefore, LCA represent key herbivores acting as ecosystem engineers as recently summarised (Leal et al., 2014a; Farji-Brener & Werenkraut, 2015; Chapter 18). The prominent ecological role played by this group is highlighted by the following features. First, LCA show that some herbivorous insects are able to generate ecologically important disturbance regimes via non-trophic activities. Second, impacts of LCA can be observed at multiple spatio-temporal scales and levels of biological organisation. Third, ecosystem-level effects by LCA include ecosystem engineering capable not only of altering the abundance of other organisms, but also the successional trajectory of vegetation. Finally, impacts of leaf-cutting ants are context-dependent, species-specific, and synergistically reinforced by anthropogenic interferences (Leal et al., 2014a).

In fact, it is not a novelty that some LCA species operate as agricultural pests (for review see Montoya-Lerma et al., 2012), but only recently consistent scientific attention has been devoted to the response of LCA to human disturbances and the potential consequences for vegetation dynamics and ecosystem processes in HML. Briefly, there has been accumulating evidence that some LCA species proliferate across HML, such as highly fragmented agro-mosaics, with tangible impacts on the successional trajectory experienced by remaining native vegetation. With the continuous expansion of agricultural frontiers, HML are expected to dominate in tropical regions (Laurance et al., 2014). This global trend has fuelled the scientific agenda and research initiatives focused on the biological dynamic, biodiversity persistence and provision of ecosystem services in HML (see Melo et al., 2013; Arroyo-Rodriguez et al., 2017). In this context, species loss, biotic homogenisation, community-level impoverishment, secondarisation, biological invasion and the disruption of plant-animal interactions have been reported (Girão et al., 2007; Santos et al., 2008; Wirth et al., 2008; Lôbo et al., 2011; Leal et al., 2014b, 2015). Yet, the way ant-plant interactions are altered and contribute to new biological arrangements persists as a wide avenue of future research. Indeed, ant species represent a substantial fraction of tropical biodiversity (Basset et al., 2012) and engage in multiple relationships (i.e. ecosystem functions), such as animal predation, herbivory, pollination, seed predation and dispersal (Beattie, 1985; Hölldobler & Wilson, 1990; Leal et al., 2014b; 2015).

In this chapter we present recent advances in the field of LCA ecology, focusing on the forces driving both their proliferation and their impacts on the successional trajectory experienced by the native vegetation in HML, particularly in the context of forested tropical ecosystems (Figure 4.1). Major portions of our chapter rest upon a recent and comprehensive review about the multiple impacts of LCA on neotropical vegetation (Leal et al., 2014a). Surprisingly, however, the bulk of available knowledge is currently derived from studies in tropical rain forests, grassland and savanna ecosystems, while the situation in tropical dry forests (TDFs) has been largely neglected (but see Barrera et al., 2015), despite the even higher threat status and conservation priority of TDFs compared to other tropical biomes (Ceballos & García, 1995; Sampaio, 1995; Leal et al., 2005; Grau et al., 2008; Santos et al., 2011). Therefore, we extend this review by placing additional focus on TDFs, especially in the context of chronic anthropogenic disturbances (sensu Singh, 1998). We provide preliminary information about the multiple LCA-derived effects on the successional trajectory experienced by the Caatinga vegetation (Figure 4.1), the largest TDF in South America (Sampaio, 1995; Leal et al., 2005; Pennington et al., 2009). We hope that this chapter will inspire a more comprehensive assessment of the forces driving the nature of human-modified tropical landscapes, such as the LCA-driven impacts, as a crucial step to guarantee the full potential of anthropogenic landscapes in terms of their biodiversity conservation and environmental services.

Figure 4.1.   Leaf-cutting ants (LCA) in human-modified tropical forest landscapes in Atlantic Forest landscape in Serra Grande (Alagoas, northeast Brazil) (a–f) and in Caatinga dry tropical forest in the Catimbau National Park (Pernambuco, northeast Brazil) (g–k). Undisturbed

LCA Basic Ecology and Responses to Human Disturbances

LCA bear a wide set of key life history attributes, which enable them to operate as a dominant herbivore group across both pristine/natural and HML. Most prominently, we refer to a generalist diet and an extensive, highly flexible foraging system. The fungus-farming lifestyle allows a high degree of polyphagy with up to 50 per cent of the local forest flora (Vasconcelos & Fowler, 1990; Wirth et al., 2003) and a rich variety of plant matter harvested (e.g. leaves, twigs, bark, flowers, fruits, seed endosperm; Shepherd, 1985; Wirth et al., 2003; Falcão et al., 2011; Figure 4.1). The accessibility of LCA to such a wide resource base is believed to enable/facilitate their function as large-scale herbivores (De Fine Licht et al., 2013). Their enormous rates of biomass consumption are achieved through complex and highly organised foraging/transportation devices, usually in the form of thoroughly cleared, up to 30 cm wide trunk trails, sometimes extending more than 250 m from the nest into the foraging territories (Hölldobler & Wilson, 2011), which cover areas often as large as 1–2 ha (Wirth et al., 2003; Urbas et al., 2007; Silva et al., 2009, 2013). These foraging systems appear to be extremely flexible, allowing LCA to profit from the continuous emergence/recruitment of palatable resources (Kost et al., 2005; Silva et al., 2013), such as plant species benefited and proliferating in response to human disturbances, as, for example, the creation of forest edges and roads (Vasconcelos et al., 2006; Leal et al., 2014a).

In fact, the human-induced spread of early successional plant species (Laurance et al., 2006; Lôbo et al., 2011; Tabarelli et al., 2012) and the concomitant release of LCA from bottom-up regulation have been identified as key mechanisms supporting their proliferation in HML (Urbas et al., 2007; Wirth et al., 2008; Falcão et al., 2011; Figure 4.2a). Exhibiting colony density increments up to 20 times (Meyer et al., 2009; Dohm et al., 2011), these insects are probably among the most 'successful' species in anthropogenically modified tropical landscapes (Wirth et al., 2007; Tabarelli et al., 2012). In the case of tropical rain forests (particularly in Atlantic

---

continuous forest (a); hyper-fragmented landscape with forest fragments in a matrix of sugar cane plantations (b); typical forest edge zone along harvested sugar cane field (c); nest clearing by *Atta cephalotes* with central nest mound (note that the ants remove both understory vegetation and leaf litter from the immediate nest surface) (d); *A. cephalotes* foraging worker ants on a horizontal trunk and (e) typical leaf damage by LCA on pioneer vegetation (*Cecropia* sp.). (f) Catimbau National Park aerial view showing mosaic of degenerated (foreground) and conserved Caatinga vegetation (right background) (g); giant *Atta opaciceps* nest (ca. 250 m^2 surface area) located on a roadside in degenerated Caatinga scrubland (h); surface of *Acromyrmex balzani* nest (i), the most abundant LCA in the Catimbau National Park; the photo depicts external refuse dump (1), loose excavated soil (2) and nest entrance with ventilation turrets (3); *Atta sexdens* worker carrying a flower of *Poincianella pyramidalis* (Fabaceae) (j), one of the most abundant food plants and *Atta opaciceps* harvesting cladode and fruit of *Tacinga inamoena* (Cactaceae) (k). Photo credits: A. Gambarini (a and b); I. R. Leal (c and g); R. Wirth (d, e, f and i); F. M. P. Oliveira (h and k); B. Büdel (j).

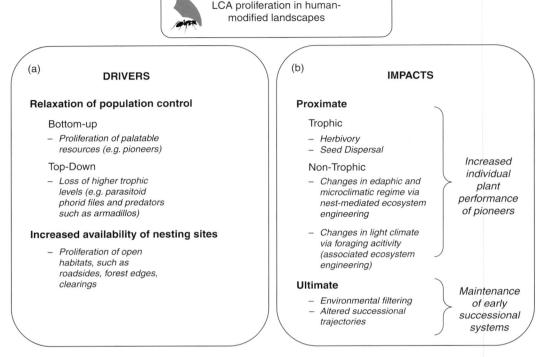

Figure 4.2.   Drivers (a) and impacts (b) of leaf-cutting ant proliferation in human-modified ecosystems. See the text for further details and contextual explanations.

and Amazon forests), the conversion of natural landscapes implies the replacement of old-growth forest stands by edge-affected habitats, such as forest edges and small forest fragments (Broadbent et al., 2008; Tabarelli et al., 2008). Thereby, in highly fragmented landscapes, only edge-affected habitats tend to persist (Joly et al., 2015). This is of crucial significance to LCA, as these habitats tend to be dominated by colonising, successional, or pioneer plants species (Laurance et al., 2006; Tabarelli et al., 2012). In edge-affected habitats, this group of plants can account for over 80 per cent of all stems and species, while old-growth stands support less than 20 per cent (Laurance et al., 2006; Santos et al., 2008). With reduced levels of both chemical and structural defences against herbivores, pioneer plants are more palatable (Coley, 1980; Coley et al., 1985) and thus preferred by LCA (Farji-Brener, 2001; Wirth et al., 2003; Urbas et al., 2007; Falcão et al., 2011). Such decoupling from population control is not restricted to bottom-up regulation alone, but has also been demonstrated for top-down control by natural enemies (Figure 4.2a). Edge-affected habitats and HML tend to support lower abundance of LCA predators due to habitat loss and hunting (Terborgh et al., 2001; Wirth et al., 2008), and the colonies suffer reduced attack rates by parasitoid phorid flies, because of the drier microclimatic conditions compared to non-disturbed areas (Almeida et al., 2008).

Apart from cascading benefits via the disruption of trophic interactions, LCA may also directly profit from human alterations, for example, the increased availability of nesting sites (Figure 4.2a). The mere opening of forested land facilitates LCA colonisation as evidenced by nesting site preferences of founder queens for open habitats, such as pastures (Vasconcelos, 1990) and roads/roadsides (Farji-Brener, 1996; Vasconcelos et al., 2006; Vieira-Neto & Vasconcelos, 2010).

Impacts on Vegetation Dynamics

Traditionally, the role played by LCA on vegetation dynamics has been concentrated in natural tropical rain forest landscapes such as those on Barro Colorado Island in Panama (Wirth et al., 2003) and at the La Selva Biological Station in Costa Rica (Garrettson et al., 1998). In this ecological context LCA can promote plant species persistence and diversity by creating the environmental heterogeneity required for plant niche partitioning. We refer, for instance, to ant nests as recruitment sites and sunflecks on the ground resulting from ant-related foliage removal in the canopy. However, investigation of LCA ecology and effects in HML, such as the highly fragmented Atlantic Forest landscapes in northeast Brazil (Urbas et al., 2007; Wirth et al., 2007; Almeida et al., 2008; Meyer et al., 2009, 2011a, 2011b, 2013; Silva et al., 2009, 2012, 2013; Corrêa et al., 2010, 2016; Facão et al., 2011), has amplified our understanding, providing evidence for distinct ecological roles in this context.

LCA herbivory, seed dispersal and nest-related activities affecting soil conditions and light regime can impose multiple effects on plant individuals, populations and assemblages that collectively can alter the successional trajectory experienced by forest edges and small forest fragments (see Leal et al., 2014a, for a review; Figure 4.2b). Briefly, forest area-based herbivory rates of LCA vary from 2.1 per cent of the available foliage area in an undisturbed late-successional forest in Panama (0.52 *Atta colombica* colonies/ha, Herz et al., 2007) to 36 per cent at the edge of human-modified Atlantic Forest (2.79 *A. cephalotes* colonies/ha, Urbas et al., 2007; Wirth et al., 2007; Meyer et al., 2009). The latter value greatly exceeds the overall rate of herbivory estimated for tropical forests (5–15 per cent, Schowalter et al., 1986; Landsberg & Ohmart, 1989; Coley & Barone, 1996). As LCA concentrate their foraging on pioneer plant species (Wirth et al., 2003; Falcão et al., 2011), this ecological group may suffer greatly from negative impacts of LCA herbivory on plant reproductive success and population dynamics (Leal et al., 2014a). On the other hand, there is compelling evidence suggesting positive net effects from LCA presence on pioneers. First, pioneers may suffer disproportionately lower fitness losses than shade-tolerant food plants due to their increased tolerance to defoliation, especially under high-light environments (Rosenthal & Kotanen, 1994). In addition, seed dispersal activities favour pioneer species directly since pioneer flora contain a large set of small-seeded species bearing fleshy fruits such as those from the Melastomataceae family (Silva et al., 2012; Costa et al., 2013; Leal et al., 2014a; Chapters 6, 7). LCA also favour pioneer species indirectly, for example, via increased light availability resulting from herbivory (called 'associated ecosystem engineering' in Figure 4.2b; Corrêa et al., 2016), and other non-trophic effects like the alterations in microclimatic conditions due to nest construction and maintenance (Corrêa et al., 2010; Meyer et al., 2011a, 2011b). Correspondingly, *A. cephalotes* nests increase light levels in approximately 0.6 per cent of the total area in the interior of Atlantic Forest, but up to 6 per cent in forest edge zones, where colonies are aggregated (Meyer et al., 2011a).

Even more strikingly, the nest-driven decline in litter cover and nutrient status of topsoil layers, usually a local phenomenon spanning ca. 0.6 ha per colony, were estimated to affect the entire forest edge (Meyer et al., 2013). Such shifts may impose environmental filtering for: (1) light-sensitive, shade-tolerant species; (2) plant species bearing ant-targeted seedlings; and (3) plant species whose seeds require undisturbed habitats for better germination (Corrêa et al., 2010; Meyer et al., 2011b). On the contrary, such disturbance favours some pioneer species (Figure 4.2b), which proliferate and support increased LCA abundance by offering palatable foliage. Accordingly, LCA foraging areas tend to be smaller in edge-affected habitats (Urbas et al., 2007). In synthesis, there is a feedback between forest fragmentation and disturbance that promotes pioneer proliferation, which, in turn, supports higher ant colony density (Figure 4.2a, b). LCA therefore amplify environmental homogeneity and favour the persistence of early successional systems in HML (Figure 4.2b).

The Neglected Response/Role of Leaf-Cutting Ants in Chronically Disturbed Dry Tropical Forests

As already mentioned, LCA are present in nearly all tropical/subtropical ecosystems in the Americas, but research on their impact has been concentrated mainly on grasslands, savannas and tropical rain forests. However, in addition to humid forests as the Atlantic and Amazon, there has been evidence for the proliferation of LCA in HML of TDF, such as the Caatinga vegetation in northeast Brazil. Apparently, LCA species such as *A. laevigata*, *A. opaciceps* and *A. sexdens* benefit from the proliferation of pioneer species (e.g. *Croton argyrophyllus*; Euphorbiaceae), which dominate regenerating forest stands following the abandonment of agricultural fields (i.e. slash-and-burn agriculture), pasture lands and roadside vegetation. In fact, colony density along Caatinga vegetation paralleling roads is fivefold higher than farther spots (Siqueira, unpublished data; Figure 4.3a), a trend in line with previous findings from tropical rain forests (Dohm et al., 2011), Cerrado savanna (Vasconcelos et al., 2006) and arid steppe of Patagonia (Farji-Brener, 1996). Furthermore, quantitative area-based colony surveys in a forest mosaic of various successional stages (fallow vegetation) with interspersed pasture lands revealed a previously undocumented prevalence of LCA in the Caatinga, as reflected by a species-rich community (*Atta sexdens, A. opaciceps, A. laevigata, Acromyrmex rugosus, Ac. balzani*) and

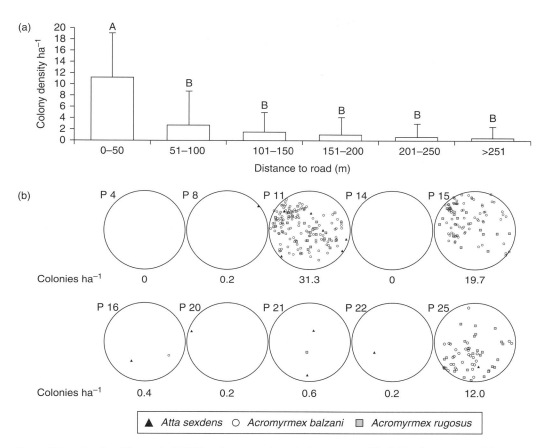

Figure 4.3. Leaf-cutting ants (LCA) as key organisms in human-modified Caatinga tropical dry forest (TDF). Mean colony density (+ SD) of *Atta* leaf-cutting ants (*A. laevigata*, *A. opaciceps* and *A. sexdens*) in six distance zones parallel to roads passing through disturbed and conserved areas of Caatinga in Catimbau National Park in Pernambuco (NE-Brazil) (a). Different letters indicate significant differences ($P < 0.05$) among distance zones; the number of sample areas was 50 for each distance zone up to 300 m into the Caatinga (Siqueira, unpublished data). Quantitative plot-based surveys of leaf-cutting ant communities in 10 plots of 4.32 ha spread across the eastern section of the Catimbau National Park showing a pronounced spatial aggregation with local hyper-abundance of nests (b) (Modified from Backé, 2015).

hyper-abundant nests (7 col ha^{-1} ± 11.9, mean ± SD; Backé 2015; Figure 4.3b). The corresponding colony maps indicated pronounced spatial underdispersion as some sites were completely devoid of colonies, while others exhibited spatial aggregation with up to 31.3 col ha^{-1} (Figure 4.3b). Among the factors driving nest density we identified (1) road proximity, (2) vegetation cover, (3) chronic disturbance and (4) rainfall (Backé, 2015 and Siqueira, unpublished data). Proliferation of LCA in the context of shifting cultivation opens a new research agenda for the investigation of the mechanism supporting persistence and proliferation. In terms of LCA ecology, it is noteworthy that there is no foliage available during 6–8 months every year (the dry season) and that prolonged droughts (2–3 years long) are frequent (Sampaio, 1995).

In addition to nest hyper-abundance in particular spots (where total nest area of LCA occupies up to 48.5 per cent of the forest ground), edaphic engineering and intense plant harvesting in early regenerating stands reinforce the chance that LCA can affect Caatinga vegetation dynamics. We refer to soil movement and deposition on nest mounds and, more importantly, the presence of external nest refuse dumps composed of organic waste in nests of all locally occurring LCA species (including *Atta laevigata* and *A. sexdens*, which are otherwise known as internal refuse dumpers; Siqueira, unpublished data), both altering soil conditions around the nests and probably patterns of seedling recruitment and growth. In contrast to HML in tropical rain forests, in which matrix lands are usually devoted to commodity production, the Caatinga landscapes are spatially configured as vegetation mosaics due to shifting cultivation for subsistence production and livestock farming (Sampaio, 1995). We refer to mosaics consisting of small patches of both agriculture/pasture lands and regenerating forest stands of varying age, i.e. fallow vegetation ranging from few years old to decades old (Sobrinho et al., 2016; Figure 4.1). In some landscapes, agriculture and pasture stands compose the matrix, while in others old-growth forest and secondary forest stands represent the dominant matrix habitat.

In this ecological context of the Caatinga HML, it is reasonable to expect that the most important LCA impacts refer to those affecting secondary succession/forest recovery or regeneration dynamics. Preliminary results and field observation suggest that (1) LCA and other key organisms such as biological soil crusts (BSC) are favoured by human disturbances, particularly the presence of fallow spots, and (2) these organisms may interact and impose distinct/differential impacts on forest regeneration, here briefly presented as working hypotheses (Figure 4.4). One of the possibilities is that LCA facilitate forest regeneration by a variety of mechanisms: (1) seed and nutrient input from adjacent forest stands, (2) enhanced seedling recruitment in nest-related spots enriched by external refuses, and (3) offering favourable spots for BSC development, such as inactive nest mounds. More precisely, a five-fold increase in BSC cover on inactive LCA nests suggests that nests may serve as nuclei of regeneration (F.F.S. Siqueira, unpublished data). Note that BSC have been reported to benefit plant recruitment and growth, particularly by incrementing soil fertility via nitrogen and carbon inputs (Belnap & Lange, 2001). Such progressive succession can result in the recovery of old-growth forest stands or scrub vegetation depending on the duration of the fallow period (Figure 4.4), although assemblages are expected to support only a subset of LCA-promoted plant species.

In contrast, as human disturbance intensifies (i.e. reduced fallow period, increased plant collection by human populations and livestock pressure), LCA may retard forest regeneration, contribute to maintain regeneration in its initial stages (shrub-dominated or scrub vegetation) or speed up the desertification process (Figure 4.4). Briefly, these possibilities may result from the negative impacts on plant individuals and populations caused by intense herbivory, seed harvesting and deposition

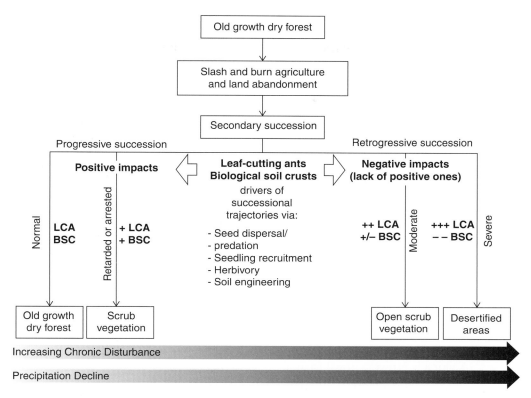

Figure 4.4. Leaf-cutting ants (LCA) and biological soil crusts (BSC) drive secondary succession along chronic disturbance and precipitation gradients through herbivory, seed dispersal/predation, and soil engineering, which in turn impacts germination and seedling recruitment/performance. Declines in the abundance and positive effects of LCA and BSC with land use intensity and precipitation lead to differential outcomes of successional trajectories. Increasing and decreasing abundance of organisms is indicated by the number of + or – symbols, respectively (+/- refers to less complex BSC communities).

on nests and nutrient storage on deep soil layers by LCA and the collapse of BSC cover. In addition, active and inactive nest mounds may consist of unsuitable sites for seedling recruitment as observed in the Atlantic Forest (Corrêa et al., 2010; Bieber et al., 2011; Meyer et al., 2011a, 2011b, 2013). This retrogressive succession can be intensified as precipitation declines (following projected climate trends) and (1) negatively affects the establishment/performance of BSC and vascular plants, particularly stress-sensitive species, and (2) turns herbivory-related foliage loss more deleterious to plant fitness (Figure 4.4).

Similar to the case of tropical rain forests, we postulate a synergism between human disturbances, particularly shifting cultivation, increased abundance of LCA and altered patterns of Caatinga vegetation dynamics. Caatinga vegetation supports a socioecological system in which human populations are highly dependent of forest products (e.g. firewood, timber, fruits and fodder) and services such as capture, storage and the recovery of nutrients stocks via forest regeneration during fallow periods (Kauffman et al., 1993; Pereira et al., 2003; Leal et al., 2005; Ramos & Albuquerque, 2012). As the predominant forest habitat, regenerating stands represent a key component of HML in the Caatinga region. This implies that LCA can affect the socioecological sustainability of the Caatinga by influencing the rates of forest recovery and the structure of plant assemblages. The working hypotheses postulating an anthropogenic-biogenic synergism between human populations and LCA with multiple possibilities in terms of impacts on vegetation dynamics support the notion that LCA operate as key engineers across a variety of ecosystems (see also Farji-Brener & Silva, 1995, 1996; Vasconcelos et al., 2006; Farji-Brener & Ghermandi, 2008; Farji-Brener et al., 2010; Vieira-Neto & Vasconcelos, 2010) for data on LCA in savannas and steppes). However, like the documented impacts of LCA on plant individuals, populations and assemblages, their role in species assembly and vegetation dynamics are presumably context-dependent (Leal et al., 2014a) and largely mediated by the nature of human disturbances imposed on this landscape, including climate change.

Future Research

A globally growing threat to food production and natural resources, such as timber and forage, tends to locally disturb practically all ecosystems inhabited by LCA, from steppes to tropical forests (Harvey et al., 2008; Melo et al., 2013). In addition to local disturbances, such as habitat loss and fragmentation, CO_2 increase and nitrogen fertilisation, altered rainfall patterns may favour several ecological plant groups exploited by LCA such as native/exotic pioneers, fast-growing and successional species (Santos et al., 2014). In other words, the natural relationship between LCA and their environment is expected to change everywhere with a myriad of potential impacts on the biological dynamics of HML.

In this context, the LCA-driven successional trajectories proposed here as among the ultimate impacts on neotropical ecosystems represent working hypotheses towards a better understanding of the role played by LCA in HML. To test these hypotheses, future research needs to examine (1) additional life-history traits making LCA successful in disturbed areas, (2) the nature of novel, emergent or transition ecosystems resulting from the proliferation of LCA, particularly trophic cascades mediated by LCA-driven vegetation shifts, (3) the effects of inactive nests on plants, (4) the mechanisms behind soil modifications around active ant nests, (5) the fate of nutrients deposited into subterranean fungus and refuse chambers (i.e. the potential role of nests as carbon sinks), and (6) the generality of both patterns and explanatory mechanisms via cross-taxa and cross-ecosystem comparisons. Exclusion-based experiments are probably required, particularly in the case of simultaneous occurrence of acute (e.g. forest replacement by large areas of commodity production) and chronic disturbances (e.g. shifting agriculture, plant collection, livestock production), as it is almost impossible to discriminate the multiple drivers and ecological roles of LCA without controlled experiments. Such ecological roles are far from negligible: LCA are not only agricultural pests, but also key ecological players in HML.

References

Almeida, W. R., Wirth, R. and Leal, I. R. (2008). Edge-mediated reduction of phorid parasitism on leaf-cutting ants in a Brazilian Atlantic Forest. *Entomologia Experimentalis et Applicata*, 129, 251–257.

Arroyo-Rodríguez, V., Melo, F. P. L., Martínez-Ramos, M., Bongers, F., Chazdon, R. L., Meave, J. A., Norden, N., Santos, B. A., Leal, I. R. and Tabarelli, M. (2017). Multiple successional pathways in human-modified tropical landscapes: new insights from forest succession, forest fragmentation and landscape ecology research. *Biological Reviews*, 92, 326–340.

Backé, J. (2015). Lebensgemeinschaften von Blattschneider-ameisen und ihre Rolle als Ökosystemingenieure unter dem Einfluss von Klimawandel und menschlicher Störung in der brasilianischen Caatinga. MSc thesis, University of Kaiserslautern, Germany.

Barrera, C. A., Buffa, L. M., Valladares, G. (2015). Do leaf-cutting ants benefit from forest fragmentation? Insights from community and species-specific responses in a fragmented dry forest. *Insect Conservation and Diversity*, 8, 456–463.

Basset, Y., Cizek, L., Cuenoud, P. et al. (2012). Arthropod diversity in a tropical forest. *Science*, 338, 1481–1484.

Beattie, A. J. (1985). *The Evolutionary Ecology of Ant-Plant Mutualisms*. Cambridge: Cambridge University Press.

Belnap, J. and Lange, O. L. (2001). Biological soil crusts: structure, function, and management. *Ecological Studies*, 150, 2nd edition. Berlin, Heidelberg, New York: Springer.

Bieber, A. G. D., Oliveira, M. A., Wirth, R., Tabarelli, M. and Leal, I. R. (2011). Do abandoned nests of leaf-cutting ants enhance plant recruitment in the Atlantic Forest? *Austral Ecology*, 36, 220–232.

Broadbent, E.N., Asner, G.P., Keller, M., Knapp, D.E., Oliveira, P.J.C. and Silva, J.N. (2008). Forest fragmentation and edge effects from deforestation and selective logging in the Brazilian Amazon. *Biological Conservation*, 141, 1745–1757.

Ceballos, G. and García, A. (1995). Conserving neotropical biodiversity: the role of dry forests in western Mexico. *Conservation Biology*, 9, 1349–1356.

Cherrett, J. M. (1989). Leaf-cutting ants. *In Ecosystems of the World* (eds. H. Lieth and M. J. A. Werger). Amsterdam: Elsevier, pp. 473–486.

Coley, P. D. (1980). Effects of leaf age and plant life history patterns on herbivory. *Nature*, 284, 545–546.

Coley, P. D., and Barone, J. A. (1996). Herbivory and plant defenses in tropical forests. *Annual Review of Ecology and Systematics*, 27, 305–335.

Coley, P. D., Bryant, J. P. and Chapin III, F. S. (1985). Resource availability and plant antiherbivore defense. *Science*, 230, 895–899.

Corrêa, M. M., Silva, P. S. D., Wirth, R., Tabarelli, M. and Leal, I. R. (2010). How leaf-cutting ants impact forests: drastic nest effects on light environment and plant assemblages. *Oecologia*, 162, 103–115.

Corrêa, M. M., Silva, P. S. D., Wirth, R., Tabarelli, M. and Leal, I. R. (2016). Foraging activity of leaf-cutting ants changes light availability and plant assemblage in Atlantic Forest. *Ecological Entomology*, 41, 442–450.

Costa, A. N., Vasconcelos, H. L., Vieira-Neto, E. H. M. and Bruna, E. M. (2008). Do herbivores exert top-down effects in Neotropical savannas? *Journal of Vegetation Science*, 19, 849–854.

Costa, U. A. S., Pinto, S. R. R., Silva, F. A., Oliveira, M., Agra, D. B., Marques, E. and Leal, I. R. (2013). O papel das formigas como dispersores secundários de sementes na Floresta Atlântica Nordestina. In *Serra Grande: uma floresta de idéias* (eds. M. Tabarelli, A. V. Aguiar Neto, I. R. Leal and A. V. Lopes). Recife: Editora Universitária da UFPE, pp. 415–438.

De Fine Licht, H. H. and Boomsma, J. J. (2010). Forage collection, substrate preparation, and diet composition in fungus-growing ants. *Ecological Entomology*, 35, 259–269.

De Fine Licht, H. H., Schiott, M., Rogowska-Wrzesinska, A., Nygaard, S., Roepstorff, P. and Boomsma, J. J. (2013). Laccase detoxification mediates the nutritional alliance between leaf-cutting ants and fungus-garden symbionts. *Proceedings of the National Academy of Sciences*, 110, 583–587.

Dohm, C., Leal, I. R., Tabarelli, M., Meyer, S. T. and Wirth, R. (2011). Leaf-cutting ants proliferate in the Amazon: an expected response to forest edge? *Journal of Tropical Ecology*, 27, 645–649.

Falcão, P. F., Pinto, S. R. R., Wirth, R. and Leal, I. R. (2011). Edge-induced narrowing of dietary diversity in leaf-cutting ants. *Bulletin of Entomological Research*, 101, 305–311.

Farji-Brener, A. G. (1996). Posibles vías de expansión de la hormiga cortadora de hojas *Acromyrmex lobicornis* hacia la Patagonia. *Ecología Austral*, 6, 144–150.

(2001). Why are leaf-cutting ants more common in early secondary forests than in old-growth tropical forests? An evaluation of the palatable forage hypothesis. *Oikos*, 92, 169–177.

Farji-Brener, A. G. and Ghermandi, L. (2008). Leaf-cutting ant nests near roads increase fitness of exotic plant species in natural protected areas. *Proceedings of the Royal Society B –Biological Sciences*, 275, 1431–1440.

Farji-Brener, A. G. and Illes, A. E. (2000). Do leaf-cutting ant nests make 'bottom-up' gaps in Neotropical rain forests? A critical review of the evidence. *Ecology Letters*, 3, 219–227.

Farji-Brener, A. G., Lescano, N. and Ghermandi, L. (2010). Ecological engineering by a native leaf-cutting ant increases the performance of exotic plant species. *Oecologia*, 163, 163–169.

Farji-Brener, A. G. and Ruggiero, A. (1994). Leaf-cutting ants (*Atta* and *Acromyrmex*) inhabitating Argentina: patterns in species richness and geographical range sizes. *Journal of Biogeography*, 21, 391–399.

Farji-Brener, A. G. and Silva, J. (1995). Leaf-cutting ants and forest groves in a tropical parkland savanna of Venezuela: facilitated succession? *Journal of Tropical Ecology*, 11, 651–669.

(1996). Leaf-cutter ants' (*Atta laevigata*) aid to the establishment success of *Tapirira velutinifolia* (Anacardiaceae) seedlings in a parkland savanna? *Journal of Tropical Ecology*, 12, 163–168.

Farji-Brener, A. G. and Werenkraut, V. (2015). A meta-analysis of leaf-cutting ant nest effects on soil fertility and plant performance. *Ecological Entomology*, 40,150–158.

Fowler, H. G. and Claver, S. (1991). Leaf-cutter ant assemblies: effects of latitude, vegetation, and behavior. In *Ant-Plant Interactions* (eds. C. R. Huxley and D. F. Cutler). Oxford: Oxford University Press, pp. 51–59.

Fowler, H. G., Pagani, M. I., Silva, O. A., Forti, L. C. and Sales, N. B. (1989). A pest is a pest is a pest? The dilemma of neotropical leaf-cutting ants: keystone taxa of natural ecosystems. *Environmental Management*, 13, 671–675.

Fowler, H. G., Pereira da Silva, V., Forti, L. C., Saes, N. B. (1986). Population dynamics of leaf-cutting ants: a brief review. In *Fire Ants and Leaf-Cutting Ants: Ecology and Management* (eds. C. S. Lofgren and R. K. Vander Meer). Boulder: Westview Press, pp. 123–145.

Garrettson, M., Stetzel, J. F., Halpern, B. S., Hearn, D. J., Lucey, B. T. and McKone, M. J. (1998). Diversity and abundance of understory plants on active and abandoned nests of leaf-cutting ants (*Atta cephalotes*) in a Costa Rican rain forest. *Journal of Tropical Ecology*, 14, 17–26.

Girão, L. C., Lopes, A. V., Tabarelli, M. and Bruna, E. M. (2007). Changes in tree reproductive traits reduce functional diversity in a fragmented Atlantic Forest landscape. *PLoS ONE*, 2, e908.

Grau, H. R., Gasparri, N. I. and Aide, T. M. (2008). Balancing food production and nature conservation in the neotropical dry forests of northern Argentina. *Global Change Biology*, 14, 985–997.

Harvey, C. A., Komar, O., Chazdon, R. et al. (2008). Integrating agricultural landscapes with biodiversity conservation in the Mesoamerican hotspot. *Conservation Biology*, 22, 8–15.

Herz, H., Beyschlag, W. and Hölldobler, B. (2007). Herbivory rate of leaf-cutting ants in a tropical moist forest in Panama at the population and ecosystem scales. *Biotropica*, 39, 482–488.

Hölldobler, B. and Wilson, E. O. (1990). *The Ants*. Cambridge, MA: Harvard University Press. (2011). *The Leafcutter Ants: Civilization by Instinct*. London: W.W. Norton.

Joly, C. A., Metzger, J. P. and Tabarelli, M. (2014). Experiences from the Brazilian Atlantic Forest: ecological findings and conservation initiatives. *New Phytologist*, 204, 459–473.

Kauffman, B., Sanford Jr., R. L., Cummings, D. L., Salcedo, I. H. and Sampaio, E. V. S. B. (1993). Biomass and nutrient dynamics associated with slash fires in neotropical dry forests. *Ecology*, 74, 140–151.

Kost, C., Gama de Oliveira, E., Knoch, T. and Wirth, R. (2005). Temporal and spatial patterns, plasticity, and ontogeny of foraging trails in leaf-cutting ants. *Journal of Tropical Ecology*, 21, 677–688.

Landsberg, J. and Ohmart, C. (1989). Levels of insect defoliation in forests: patterns and concepts. *Trends in Ecology & Evolution*, 4, 96–100.

Laurance, W. F., Nascimento, H. E. M., Laurance, S. G. et al. (2006). Rain forest fragmentation and the proliferation of successional trees. *Ecology*, 87, 469–482.

Laurance, W. F., Sayer, J. and Cassman, K. G. (2014). Agricultural expansion and its impacts on tropical nature. *Trends in Ecology & Evolution*, 29, 107–116.

Leal, I. R., Silva, J. M. C., Tabarelli, M. and Lacher, T. E. (2005). Changing the course of biodiversity conservation in the Caatinga of Northeastern Brazil. *Conservation Biology*, 19, 701–706.

Leal, I. R., Wirth, R., Tabarelli, M. (2014a). The multiple impacts of leaf-cutting ants and their novel ecological role in human-modified neotropical forests. *Biotropica*, 46, 516–528.

Leal, L. C., Andersen, A. N. and Leal, I. R. (2014b). Anthropogenic disturbance reduces seed dispersal services for myrmecochorous plants in the Brazilian Caatinga. *Oecologia*, 174, 173–181.

Leal, L. C., Andersen, A. N., Leal, I. R. (2015). Disturbance winners or losers? Plants bearing extrafloral nectaries in Brazilian Caatinga. *Biotropica*, 47, 468–474.

Lôbo, D., Leão, T., Melo, F. P. L., Santos, A. M. M. and Tabarelli, M. (2011). Forest fragmentation drives Atlantic Forest of northeastern Brazil to biotic homogenization. *Diversity and Distribution*, 17, 287–296.

Melo, F. P. L., Arroyo-Rodríguez, V., Fahrig, L., Martínez-Ramos, M. and Tabarelli, M. (2013). On the hope for biodiversity-friendly tropical landscapes. *Trends in Ecology and Evolution*, 28, 461–468.

Meyer, S. T., Leal, I. R., Tabarelli, M. and Wirth, R. (2011a). Ecosystem engineering by leaf-cutting ants: nests of *Atta cephalotes* drastically alter forest structure and microclimate. *Ecological Entomology*, 36, 14–24.

(2011b). Performance and fate of tree seedlings on and around nests of the leaf-cutting ant *Atta cephalotes*: ecological filters in a fragmented forest. *Austral Ecology*, 36, 779–790.

Meyer, S. T., Leal, I. R. and Wirth, R. (2009). Persisting hyper-abundance of keystone herbivores (*Atta* spp.) at the edge of an old Brazilian Atlantic Forest fragment. *Biotropica*, 41, 711–716.

Meyer, S. T., Neubauer, M., Sayer, E. J., Leal, I. R., Tabarelli, M. and Wirth, R. (2013). Leaf-cutting ants as ecosystem engineers: topsoil and litter perturbations around *Atta cephalotes* nests reduce nutrient availability. *Ecological Entomology*, 38, 497–504.

Montoya-Lerma, J., Giraldo-Echeverri, C., Armbrecht, I., Farji-Brener, A. G. and Calle, Z. (2012). Leaf-cutting ants revisited: towards rational management and control. *International Journal of Pest Management*, 58, 225–247.

Moreira, A. A, Forti, L. C., Andrade, A. P. P., Boaretto, M. A. C. and Lopes J. F. S. (2004a). Nest architecture of *Atta laevigata* (F. Smith 1858) (Hymenoptera: Formicidae). *Studies on Neotropical Fauna and Environment*, 39, 109–116.

Moreira, A. A., Forti, L. C., Boaretto, M. A. C., Andrade, A. P. P., Lopes, J. F. S. and Ramos V. M. (2004b). External and internal structure of *Atta bisphaerica* Forel (Hymenoptera: Formicidae) nests. *Journal of Applied Entomology*, 128, 204–211.

Mueller, U. G., Rehner, S. A. and Schultz, T. R. (1998). The evolution of agriculture in ants. *Science*, 281, 2034–2038.

Pennington, R.T., Lavin, M. and Oliveira-Filho, A. (2009). Woody plant diversity, evolution, and ecology in the tropics: perspectives from seasonally dry tropical forests. *Annual Review of Ecology, Evolution and Systematics*, 40, 437–57.

Pereira, I. M., Andrade, L. A., Sampaio, E. V. S. B. and Barbosa, M. R. V. (2003). Use-history effects on structure and flora of Caatinga. *Biotropica*, 35, 154–165.

Ramos, M. A. and Albuquerque, U. P. (2012). The domestic use of firewood in rural communities of the Caatinga: how seasonality interferes with patterns of firewood collection. *Biomass and Bioenergy*, 39, 147–158.

Rosenthal, J. P. and Kotanen, P. M. (1994). Terrestrial plant tolerance to herbivory. *Trends in Ecology and Evolution*, 9,145–148.

Sampaio, E. V. S. B. (1995). Overview of the Brazilian Caatinga. In *Seasonally Dry Forests* (eds. S. H. Bullock, H. A. Mooney and E. Medina). Cambridge: Cambridge University Press, pp. 35–58.

Santos, B. A., Peres, C. A., Oliveira, M. A., Grillo, A. S., Alves-Costa, C. P. and Tabarelli, M. (2008). Drastic erosion in functional attributes of tree assemblages in Atlantic Forest fragments of northeastern Brazil. *Biological Conservation*, 141, 249–260.

Santos, B. A., Tabarelli, M., Melo, F.P. L. et al. (2014). Phylogenetic impoverishment of Amazonian tree communities in an experimentally fragmented forest landscape. *PLoS One*, 9, e113109.

Santos, J. C., Leal, I. R., Almeida-Cortez, J. S., Fernandes, G. W. and Tabarelli, M. (2011). Caatinga: the scientific anonymity experienced by a dry tropical forest. *Tropical Conservation Science*, 3, 276–286.

Schowalter, T. D., Hargrove, W. W. and Crossley Jr, D. A. (1986). Herbivory in forested ecosystems. *Annual Review of Entomology*, 31, 177–196.

Schultz, T. R. and Brady, S. G. (2008). Major evolutionary transitions in ant agriculture. *Proceedings of the National Academy of Sciences*, 105, 5435–5440.

Shepherd, J. D. (1985). Adjusting foraging effort to resources in adjacent colonies of the leaf-cutting ant, *Atta colombica*. *Biotropica*, 17, 245–252.

Silva, P. S. D., Bieber, A. G. D., Knoch, T. A., Tabarelli, M., Leal I. R. and Wirth, R. (2013). Foraging in highly dynamic environments: leaf-cutting ants adjust foraging trail networks to pioneer plant availability. *Entomologia Experimentalis et Applicata*, 147, 110–119.

Silva, P. S. D., Bieber, A. G. D., Leal, I. R., Wirth, R. and Tabarelli, M. (2009). Decreasing abundance of leaf-cutting ants across a chronosequence of advancing Atlantic Forest regeneration. *Journal of Tropical Ecology*, 25, 223–227.

Silva, P. S. D., Leal, I. R., Wirth, R., Melo, P. F. L. and Tabarelli, M. (2012). Leaf-cutting ants alter seedling assemblages across second-growth stands of Brazilian Atlantic Forest. *Journal of Tropical Ecology*, 28, 361–368.

Singh, S. P. (1998). Chronic disturbance, a principal cause of environmental degradation in developing countries. *Environmental Conservation*, 25, 1–2.

Sobrinho, M. S., Tabarelli, M., Machado, I. C., Sfair, J., Bruna, E. M. and Lopes, A. V. (2016). Land use, fallow period and the recovery of a Caatinga forest. *Biotropica*, doi: 10.1111/btp.12334.

Tabarelli, M., Aguiar, A. V., Ribeiro, M. C., Metzger, J. P. and Peres C. A. (2010). Prospects for biodiversity conservation in the Atlantic Forest: lessons from aging human-modified landscapes. *Biological Conservation*, 143, 2328–2340.

Tabarelli, M., Lopes, A. V. and Peres, C. A. (2008). Edge-effects drive tropical forest fragments towards an early-successional system. *Biotropica*, 40, 657–661.

Tabarelli, M., Peres, C. A. and Melo, F. P. L. (2012). The 'few winners and many losers' paradigm revisited: emerging prospects for tropical forest biodiversity. *Biological Conservation*, 155, 136–140.

Tabarelli, M., Silva, J. M. C. and Gascon, C. (2004). Forest fragmentation, synergisms and the impoverishment of Neotropical forests. *Biodiversity and Conservation*, 13, 1419–1425.

Teixeira, M. C., Schroeder J. H. and Mayhé-Nunes, A. J. (2003). Geographic distribution of *Atta robusta* Borgmeier (Hymenoptera: Formicidae). *Neotropical Entomology*, 32, 719–721.

Terborgh, J., Lopez, L., Nuñez, V. P. et al. (2001). Ecological meltdown in predator-free forest fragments. *Science*, 294, 1923–1926.

Urbas, P., Araújo Jr., M. V., Leal, I. R. and Wirth, R. (2007). Cutting more from cut forests: edge effects on foraging and herbivory of leaf-cutting ants in Brazil. *Biotropica*, 39, 489–495.

Vasconcelos, H. L. (1990). Habitat selection by the queens of the leaf-cutting ant *Atta sexdens* L. in Brazil. *Journal of Tropical Ecology*, 6, 249–252.

Vasconcelos, H. L. and Fowler, H. G. (1990). Foraging and fungal substrate selection by leaf-cutting ants. In *Applied Myrmecology – A World Perspective,* (eds. R. K. Vander Meer, K. Jaffe, and A. Cedeno). Boulder: Westview Press, pp. 411–419.

Vasconcelos, H. L., Vieira-Neto, E. H. M., Mundim, F. M. and Bruna, E. M. (2006). Roads alter the colonization dynamics of a keystone herbivore in Neotropical savannas. *Biotropica*, 38, 661–666.

Vieira-Neto, E. H. M. and Vasconcelos, H. L. (2010). Developmental changes in factors limiting colony survival and growth of the leaf-cutter ant *Atta laevigata*. *Ecography*, 33, 538–544.

Weber, N. A. (1966). Fungus growing ants. *Science*, 153, 587–604.

Wirth, R., Beyschlag, W., Herz, H., Ryel, R. J. and Hölldobler, B. (2003). Herbivory of leaf-cutter ants: a case study of Atta colombica in the tropical rainforest of Panama. *Ecological Studies*, 164. New York: Springer.

Wirth, R., Leal, I. R. and Tabarelli, M. (2008). Plant-herbivore interactions at the forest edge. *Progress in Botany*, 69, 423–448.

Wirth, R., Meyer, S. T., Almeida, W. R., Araujo Jr, M. V., Barbosa, V. S. and Leal, I. R. (2007). Increasing densities of leaf-cutting ants (*Atta* spp.) with proximity to the edge in a Brazilian Atlantic Forest. *Journal of Tropical Ecology*, 23, 501–505.

Part II

Ant-Seed Interactions and Man-Induced Disturbance

5 Global Change Impacts on Ant-Mediated Seed Dispersal in Eastern North American Forests

Robert J. Warren II, Joshua R. King, Lacy D. Chick, and Mark A. Bradford

Introduction

Global change alters the distributions and interactions of species through large-scale fragmentation and alteration of forests, as well as rapid increases in both the mean and variability of temperatures, both augmenting the introduction of exotic species. The impacts of these changes are not ubiquitous, but rather context-dependent and may be considered drivers or secondary forces for which shifts in communities, and therefore species interactions, may occur.

Ant-mediated seed dispersal (myrmecochory) is a climate-dependent interaction. Both ants and plants rely on temperature to cue seasonal activity, such as the timing of life history events, and changes in temperature prompt changes in ant community composition, behavior, and interactions with other ants. Climatic changes disrupt the synchrony between ant-dispersed (myrmecochorous) plant seed release and the initiation of ant foraging in the spring. Moreover, changes in habitat structure change the template by which plants and ants interact, in that habitat fragmentation changes ant dispersal behavior away from forest edges so that these become formidable barriers for myrmecochorous plants. In addition, habitat fragmentation increases exotic ant invasion, as does warming, and both can disrupt ant-plant seed dispersal interactions.

Here, we examine the impacts of three major agents of global change – forest fragmentation, invasive species, and climate change – on the interactions between woodland myrmecochores and ants in eastern North America (NA). The ant-plant interaction may be unbalanced and nonspecific, in that many plants rely on a few ant species, therefore, seed dispersal depends strongly on the presence of particular ant species (and primarily of one genus in eastern NA). Thus, we focus on the main seed-dispersing genus of ants in eastern NA deciduous forests, *Aphaenogaster*, to investigate potential global change impacts on the interaction. Alone, and in concert, global change drivers pose great challenges for ant-mediated seed dispersal in eastern NA.

Ant-Mediated Seed Dispersal

Myrmecochory is a climate-mediated interaction that occurs worldwide and is employed by at least 11,000 plant species (Gorb & Gorb, 2003; Lengyel et al., 2009; Rico-Gray & Oliveira, 2007). Plants in the myrmecochore guild (sensu *stricto*) produce lipid-rich external appendages on the seed surface (elaiosomes) that induce solitary foraging, nongranivorous ants to retrieve seeds back to their colony (Marshall et al., 1979). Ant workers then remove elaiosomes and feed them to larvae, and the seeds are discarded undamaged – either within nest midden or near the ant nest (Canner et al., 2012; Handel, 1976; Servigne & Detrain, 2008).

Whereas many ants show interest in myrmecochore seeds, only ~100 ant species worldwide provide successful dispersal services (Gove et al., 2007; Ness et al., 2009; Warren II & Giladi, 2014). In eastern NA, ant species in the *Aphaenogaster rudis-texana-miamiana complex* (De Marco & Cognato, 2016) (hereafter *A. rudis*) are the dominant seed dispersers (Ness et al., 2009; Lubertazzi, 2012). They are the most common and abundant ants in eastern deciduous forests north of Georgia (King et al., 2013), and they retrieve ~75 percent of available seeds (Ness et al., 2009; Warren II et al., 2010) from >80 myrmecochore plant species (Table 5.1). Primary seed dispersal by these species occurs at the scale of 1–2 m, and can reach further with secondary dispersal (Canner et al., 2012), but in the absence of ants, myrmecochore seed movement decreases to centimeters (Gomez & Espadaler, 2013; Warren II et al., 2010; Zelikova et al., 2008).

The benefit from myrmecochore interactions appear quite asymmetrical, with inconsistent findings regarding the contribution of elaiosomes toward whole-colony nutrition and demographic structure (Bono & Heithaus, 2002; Brew et al., 1989; Caut et al., 2013; Clark & King, 2012; Fischer et al., 2008; Gammans et al., 2005; Keller & Passera, 1989; Marussich, 2006; Morales & Heithaus, 1998; Warren II et al., 2015b). There is little to no data on the contribution of elaiosomes toward colony fitness or persistence. Moreover, ants disperse myrmecochore "cheaters" that provide faux elaiosomes, with seed retrieval not dependent on elaiosome nutrition (Pfeiffer et al., 2010; Turner & Frederickson, 2013). Finally, ant presence, abundance, and distribution do not appear dependent on myrmecochore abundance or even presence (Bronstein et al., 2006; Mitchell et al., 2002; Ness et al., 2009). As such, disruption in ant-plant seed interactions should primarily impact plant participants because the ants seem to persist as effectively with or without myrmecochores.

Given that the plants appear strongly dependent on the ants for population persistence, it is perhaps not surprising that they have adaptations that appear to "choose" ant partners that are effective seed dispersers. Specifically plants use traits such as elaiosome size, and chemistry, and seed phenology and presentation, to attract ants that quickly move seeds to protected microhabitat with minimal harm (see Giladi, 2006; Warren II & Giladi, 2014 and references therein). For example, seed-dispersing ants share several similar traits, such as highly transient colonies, relatively larger body and mandible size and locally abundant foraging workers adept at rapidly discovering and exploiting food resources (Warren II & Giladi, 2014). These traits

Table 5.1 Myrmecochorous Plants in Eastern North America along with Seed Set Phenology (Early, Middle or Late Summer) and Distribution (Southern, Northern, Throughout)

Species	Family	Seed set	Distribution	
Allium tricoccum Aiton	Alliaceae	Late	Throughout	
Anemone acutiloba (DC.) G. Lawson	Ranunculaceae	Middle	Throughout	
Anemone americana (DC.) H. Hara	Ranunculaceae	Early	Throughout	
Anemone quinquefolia L.	Ranunculaceae	Middle	Throughout	
Asarum arifolium Michx.	Aristolochiaceae	Late	Southern	*
Asarum canadense L.	Aristolochiaceae	Late	Throughout	
Asarum contractum (H.L.Blomq.) Barringer.	Aristolochiaceae	Late	Southern	*
Asarum minus Ashe	Aristolochiaceae	Late	Southern	*
Asarum rhombiformis Gaddy	Aristolochiaceae	Late	Southern	*
Asarum shuttleworthii Britten & Baker f.	Aristolochiaceae	Late	Southern	*
Asarum virginicum L.	Aristolochiaceae	Late	Southern	*
Carex jamesii Schwein.	Cyperaceae	Late	Throughout	
Carex laxiculmis Schwein.	Cyperaceae	Late	Throughout	
Chrysogonum virginianum L.	Asteraceae	Middle	Throughout	
Claytonia virginica L.	Portulacaceae	Middle	Throughout	
Corydalis flavula (Raf.) DC.	Fumariaceae	Early	Throughout	*
Dicentra cucullaria (L.) Bernh.	Fumariaceae	Middle	Throughout	
Erythronium albidum Nutt.	Liliaceae	Early	Northern	*
Erythronium americanum Ker Gawl.	Liliaceae	Early	Throughout	*
Erythronium umbilicatum Parks & Hardin	Liliaceae	Early	Southern	**
Galium circaezans Michx.	Rubiaceae	Late	Throughout	
Prosartes lanuginosa (Michx.) D. Don	Liliaceae	Late	Throughout	
Prosartes maculata (Buckley) A. Gray	Liliaceae	Late	Southern	*
Sanguinaria canadensis L.	Papaveraceae	Middle	Throughout	
Scleria triglomerata Michx.	Cyperaceae	Late	Throughout	
Tiarella cordifolia L.	Saxifragaceae	Middle	Throughout	
Trillium catesbaei Elliott	Melanthiaceae	Middle	Southern	*
Trillium cernuum L.	Melanthiaceae	Middle	Northern	
Trillium cuneatum Raf.	Melanthiaceae	Early	Southern	†*
Trillium decipiens J. D. Freeman	Melanthiaceae	Early	Southern	†*
Trillium decumbens Harbison	Melanthiaceae	Early	Southern	†*
Trillium erectum L.	Melanthiaceae	Late	Throughout	
Trillium grandiflorum (Michx.) Salisb.	Melanthiaceae	Late	Throughout	
Trillium luteum (Muhl.) Harbison	Melanthiaceae	Middle	Southern	*
Trillium maculatum Raf.	Melanthiaceae	Early	Southern	†*
Trillium nivale Riddell	Melanthiaceae	Early	Northern	*

(*continued*)

Table 5.1 *(cont.)*

Species	Family	Seed set	Distribution	
Trillium persistens Duncan	Melanthiaceae	Early	Southern	†*
Trillium pusillum Michx.	Melanthiaceae	Middle	Southern	*
Trillium reliquum J. D. Freeman	Melanthiaceae	Early	Southern	†*
Trillium rugelii Rendle	Melanthiaceae	Middle	Southern	*
Trillium sessile L.	Melanthiaceae	Middle	Throughout	
Trillium simile Gleason	Melanthiaceae	Middle	Southern	*
Trillium stamineum Harbison	Melanthiaceae	Middle	Southern	*
Trillium sulcatum Patrick	Melanthiaceae	Middle	Southern	*
Trillium underwoodii Small	Melanthiaceae	Early	Southern	†*
Trillium undulatum Willd.	Melanthiaceae	Late	Throughout	
Trillium vaseyi Harbison	Melanthiaceae	Late	Southern	*
Trillium viride Beck	Melanthiaceae	Middle	Northern	
Uvularia floridana Chapm.	Colchicaceae	Middle	Southern	*
Uvularia grandiflora Sm.	Colchicaceae	Late	Throughout	
Uvularia perfoliata L.	Colchicaceae	Late	Throughout	
Uvularia puberula Michx.	Colchicaceae	Late	Throughout	
Uvularia sessilifolia L.	Colchicaceae	Late	Throughout	
Viola blanda Willd.	Violaceae	Middle	Throughout	
Viola brittoniana Pollard	Violaceae	Early	Throughout	*
Viola canadensis L.	Violaceae	Middle	Throughout	
Viola cucullata Aiton	Violaceae	Middle	Throughout	
Viola hastata Michx.	Violaceae	Early	Throughout	*
Viola lanceolata L.	Violaceae	Early	Throughout	*
Viola macloskeyi Lloyd	Violaceae	Early	Throughout	*
Viola pedata L.	Violaceae	Middle	Throughout	
Viola pubescens Aiton	Violaceae	Early	Throughout	*
Viola renifolia A. Gray	Violaceae	Middle	Northern	
Viola rostrata Pursh	Violaceae	Early	Throughout	*
Viola rotundifolia Michx.	Violaceae	Early	Throughout	*
Viola sagittata Aiton	Violaceae	Early	Throughout	*
Viola septemloba Leconte	Violaceae	Early	Southern	†*
Viola sororia Willd.	Violaceae	Early	Throughout	*
Viola striata Aiton	Violaceae	Early	Throughout	*
Viola triloba Schwein.	Violaceae	Early	Throughout	*
Viola villosa Walter	Violaceae	Early	Southern	†*
Viola walteri House	Violaceae	Early	Southern	†*

Myrmecochorous plants that set seed early (†) or have primarily southern distributions (*) may be at greatest risk for disrupted ant dispersal due to a greater influx of exotic ants in the southern region and a greater chance of regional extirpation with climate change and habitat fragmentation.

Source: Data compiled from eFloras (2016); Radford et al. (1968); Weakley (2015); USDA (2016).

contribute toward the rapid discovery, and retrieval, of myrmecochore seeds. These behaviors likely benefit plants in many ways, including the reduction of negative density-dependence effects caused by the accumulation of seeds at maternal plants and at ant nests, placement of seeds in establishment-friendly microhabitats, and seed protection from predators and fire (see Giladi, 2006; Warren II & Giladi, 2014 and references therein). As a result, temporal or spatial mismatches between plants and targeted ants cause plant dispersal failure. For example, climate warming can cause a temporal mismatch between seed set and ant foraging activity, and high soil moisture – though often needed by the plants – can cause spatial mismatches because effective seed dispersing ants are often restricted to drier soils by more aggressive and non-seed-dispersing ants (Warren II et al., 2012). Such mismatches can be magnified by the loss or reduction of key seed-dispersing ants or by changes in their behavior due to habitat fragmentation, invasive ant species that are either "poor dispersers" or seed predators, and/or climate change (Ness, 2004; Ness & Morin, 2008; Rodriguez-Cabal et al., 2012; Warren II et al., 2011a, 2015a, 2015b; Warren II & Bradford, 2013). When ant-mediated seed dispersal fails, seedlings aggregate near parental plants (Warren II & Bradford, 2013; Warren II et al., 2015a) and gene flow is limited in the plant populations (Beattie, 1978; Kalisz et al., 1999; Zhou et al., 2007), causing inbreeding, pathogen accumulation, and, ultimately, may lead to population failure.

Here, we explore the effects of global change on ant-mediated seed dispersal in eastern NA. We consider changes in forest fragmentation, the introduction of exotic invasive ants, warming temperatures, and the interactions among these global change drivers. Given that the foraging behavior of worker ants drive the interaction, we focus on how these changes impact the key seed disperser in eastern NA, *A. rudis,* as a specific case helping to understand and predict effects on ants, and how those effects may cascade through to the plants they disperse.

Forest Fragmentation

A primary driver of forest loss in eastern NA is development for urban and, especially, residential land use (Drummond & Loveland, 2010; Chapter 1). Low-density rural home development (exurbanization) is the fastest-growing form of land use in the United States, and it is the greatest source of forest fragmentation (Brown et al., 2005; Drummond & Loveland, 2010; Theobald, 2005). Habitat conversion, fragmentation and the introduction and spread of invasive species represent root causes of extinction and loss of ecosystem function.

Most plant diversity in eastern NA deciduous forests occurs in the herbaceous layer (Gilliam, 2007; Whigham, 2004). As a result of habitat alteration and destruction (e.g., logging, urbanization), woodland herbs communities have declined throughout their range, with recovery time estimated in centuries (Duffy & Meier, 1992; Whigham, 2004). Secondary forests (once logged) rarely contain the same herbaceous diversity and composition as primary forests (never logged), even with long periods of post-logging recovery (Duffy & Meier, 1992; Harrelson & Matlack, 2006; Jackson et al., 2009; Matlack, 1994a; Meier et al., 1995; but see also Ford et al., 2000; Gilliam, 2002). Whereas these long-term changes may be attributed to enduring changes in soil and forest microhabitats (e.g., Flinn & Vellend, 2005; Harrelson & Matlack, 2006; Matlack, 1993; Peterson & Campbell, 1993), a fairly consistent pattern emerges in the herbaceous communities: ant-dispersed plants are absent from recovering forests (Cain et al., 1998; Harrelson & Matlack, 2006; Sorrells & Warren II, 2011). Given that myrmecochores contribute an estimated 30–70 percent of the herbaceous richness in these woodlands (Beattie & Culver, 1981; Beattie & Hughes, 2002; Handel et al., 1981; Ness et al., 2009), their absence or reduction poses considerable threat to forest community diversity.

One of the major reasons that ant-dispersed herbs are absent from many disturbed and re-establishing forests is that ants carry seeds much shorter distances (1–2 m) than those traveled by vertebrate- or wind-dispersed seeds (Brunet & von Oheimb, 1998; Cain et al., 1998; Matlack, 1994a). Woodland myrmecochore absence often is attributed to poor dispersal (Jacquemyn & Brys, 2008; Matlack, 1994a; McLachlan & Bazely, 2001; Meier et al., 1995; Mitchell et al., 2002; Sorrells & Warren II, 2011). Short dispersal distances mean that myrmecochore populations can advance only 1–2 m year^{-1} (Gomez & Espadaler, 2013), making population recovery times consistent with the observed lag of 100–200 years between forest disturbance and the reestablishment of original woodland plant communities (Duffy & Meier, 1992; Harrelson & Matlack, 2006; Jackson et al., 2009; Matlack, 1994a; Meier et al., 1995).

The direct effect of forest disturbance on woodland plants is extirpation; however, the indirect effect may be a change in ant presence, abundance, and/or behavior (Ness, 2004; Ness & Morin, 2008; Rodriguez-Cabal et al., 2012; Warren II et al., 2015a, 2015b). *Aphaenogaster* ants are highly abundant in eastern NA forests, particularly mature forests (Wike et al., 2010), and uncommon in open habitats (Heithaus & Humes, 2003; King et al., 2013; Lubertazzi, 2012; Mitchell et al., 2002; Warren II et al., 2015b; Wike et al., 2010). *Aphaenogaster* ants can disperse 50–1500 m year^{-1} through aerial mating flights (Hölldobler & Wilson, 1990; Talbot, 1966), so that they can find and colonize even small forest patches relatively quickly (Mitchell et al., 2002). *Aphaenogaster* abundance at forest edges varies greatly, however (Banschbach et al., 2012; Mitchell et al., 2002; Ness & Morin, 2008; Pudlo et al., 1980; Warren II et al., 2015b), likely because they only forage at edges (Ness, 2004; Warren II et al., 2015b). Hence, forest edges may inhibit myrmecochore seed dispersal between forest patches because seeds move only toward the forest interior where the ants locate their nests (Matlack, 1994b; Mitchell et al., 2002; Ness, 2004; Warren II et al., 2015b). Indeed, Warren II et al., (2015b) found that myrmecochorous plant diversity declined with proximity to forest edge whereas nonmyrmecochore woodland plants were unaffected.

Woodland herb diversity depends not only on suitable habitat for plants, but also suitable habitat for the ants that disperse their seeds (Warren II & Bradford, 2013; Warren II et al., 2010, 2011a, 2014). As such, myrmecochores can be absent from habitats where the abiotic conditions are suitable for their persistence, but unfavorable for their dispersers or dispersal behavior.

Invasive Ants

Invasive and native ants often do not mix well, with native ant diversity and abundance generally diminished in the presence of many invasive ants (Holway et al., 2002; Chapters 13–15). Invasive ants exhibit some common traits, such as mammoth population sizes and fierce interspecific aggression, which seem to give them advantage in monopolizing food resources to the detriment of native species (McGlynn, 1999; Sanders et al., 2003). Invasive ants might substitute as seed dispersers in invaded habitats, but with few exceptions (Prior et al., 2015), ant invasion reduces or eliminates ant-mediated seed dispersal worldwide (Carney et al., 2003; Christian, 2001; Zettler et al., 2001).

Potentially the greatest direct threat to *A. rudis* in eastern NA is *Brachyponera chinensis* (Emery, formerly *Pachycondyla chinensis*; "Chinese needle ant"), a forest invader that has become prevalent in southeastern US forests in recent years (Bednar, 2010; Bednar & Silverman, 2011; Guenard & Dunn, 2010; Nelder et al., 2006; Rodriguez-Cabal et al., 2012; Warren II et al., 2015a). *Aphaenogaster rudis* and *B. chinensis* appear to share ecological niche requirements in temperature, moisture, coarse woody material (CWM) as nesting habitat and termites as preferred food (Bednar, 2010; Bednar & Silverman, 2011; Warren II et al., 2015a). Rodriguez-Cabal et al. (2012) found 96 percent fewer *A. rudis* workers in *B. chinensis*-invaded plots, with no similar impact on other large forest ants, and Warren II et al. (2015a) observed *A. rudis* displacement from artificial nests by *B. chinensis* in forest plots similar to the observed displacement from naturally occurring CWM in the invaded habitat.

In eastern NA, the dominant seed-dispersing ant, *A. rudis*, thrives in mesic forest habitat (King et al., 2013; Lubertazzi, 2012; Ness et al., 2009; Warren II et al., 2012). Hence, interactions with *Solenopsis invicta* (Buren; "red imported fire ant"), mostly a hot, open habitat invader (King & Tschinkel, 2016), are limited to forest edges at the southeastern end of the *A. rudis* NA range (Ness, 2004). Given that *Linepithema humile* (Mayr, formerly *Iridomyrmex humilis*; "Argentine ant") appears to have more flexible microhabitat requirements, greater pressure might occur between *L. humile* and A. *rudis*, but *L. humile* is not (yet) widespread in eastern deciduous forest habitats, and in one instance of overlap in a mixed pine-hardwood forest, *A. rudis* appeared unaffected by *L. humile* invasion (Rowles & Silverman, 2010). *Myrmica rubra* (L.; "northern fire ant") appears limited to highly mesic forests (Garnas, 2004; Groden et al., 2005), a microhabitat where *A. rudis* already gets displaced by native ant species (Warren II et al., 2010, 2012), but limited data suggests a negative relationship between *M. rubra* and *A. rudis* where they do co-occur (Garnas, 2004, Warren, personal observation).

Three things can happen when an invasive ant displaces a native seed-dispersing ant: (1) dispersal fails, (2) the exotic replaces the native as an effective seed disperser or (3) the exotic replaces the native as a seed disperser, but provides inferior dispersal. Most current data on these effects is based on seed bait stations, but caution must be exercised when interpreting these studies as seed retrieval, in itself, is not a reliable indicator of effective dispersal (Prior et al., 2015; Warren II & Giladi, 2014; Warren II et al., 2015a). Given that most invasive ants are omnivorous, they usually show interest and even will pick up elaiosome-bearing seeds, but deliver poor or no dispersal services (Ness & Bronstein, 2004; Rodriguez-Cabal et al., 2009, 2012). For example, in NA, *S. invicta* removes 50–100 percent of elaiosome-bearing seeds (*Trillium* and *Viola* spp.), but the post-removal fate of those seeds generally differs greatly from that when removed by *A. rudis* (Ness, 2004; Stuble et al., 2010; Zettler et al., 2001). *Aphaenogaster* ants are quick to discover elaiosome-bearing seeds, pick them up, whole, and retrieve them to their nest, where the elaiosome is removed without harming the discarded seed. In contrast, though they remove as many seeds as *A. rudis*, *S. invicta* generally transports them a lesser distance, damages or destroys them, and discards them in locations unfavorable for germination or vulnerable to seed predators (Ness, 2004; Stuble et al., 2010; Zettler et al., 2001).

Whereas the impact of *S. invicta* on ant-mediated seed dispersal in eastern NA may be limited by forest edge, *B. chinensis* and *M. rubra* venture much further into the intact forest habitat where *A. rudis* and myrmecochorous plants thrive. *Brachyponera chinensis* is associated with precipitous declines in *A. rudis* abundance as well as declines in the abundance and dispersion of *Asarum arifolium* [(Michx.) Small 1903, formerly *Hexastylis arifolia*], a common myrmecochorous plant in the southeastern United States (Rodriguez-Cabal et al., 2012; Warren II et al., 2015a). Conversely, myrmecochorous plant dispersal and germination may be greater with invasive *M. rubra* than native *A. rudis*, particularly for an invasive myrmecochorous plant (Prior et al., 2015). The difference between *B. chinensis* and *M. rubra* as seed-dispersers in their invaded range may trace back to traits evolved in their home ranges. *Brachyponera chinensis* shows some interest in seeds in its home range, but that interest is low, and may be geared more toward granivory than dispersal (Eguchi, 2004; Ohnishi et al., 2008). In contrast, *M. rubra* shows great interest in elaiosome-bearing seeds and is a seed disperser in its native range (Bulow-Olsen, 1984; Gorb & Gorb, 2000; Prior et al., 2015; Servigne & Detrain, 2008). As a result, *B. chinensis* invasion resulted in a decrease in seed removal whereas *M. rubra* was associated with an increase (Prior et al., 2015; Rodriguez-Cabal et al., 2012). However, we note that in seed bait station experiments conducted in Buffalo, NY (US), in May 2015, *M. rubra* only removed 13 percent of *S. canadensis* seeds in a heavily invaded area.

The data on *S. invicta*, *B. chinensis*, and *M. rubra* indicate that the effects of invasive ants on seed dispersal may be contingent on system and plant species, but overall appear detrimental to myrmecochorous plants in eastern NA (Ness, 2004; Prior et al., 2015; Rodriguez-Cabal et al., 2012; Servigne & Detrain, 2008; Stuble et al., 2010; Warren II et al., 2015a; Zettler et al., 2001). Similarly, though there is little information on the impact of *L. humile* on eastern NA myrmecochores, its effects on seed dispersal worldwide suggest similar detrimental effects, particularly in handling after collection (Bond & Slingsby, 1984; Carney et al., 2003; Christian, 2001; Gomez & Oliveras, 2003; Rodriguez-Cabal et al., 2009; Rowles & O'Dowd, 2009). The direct effect of exotic ant invasion into woodland habitat appears to be the displacement of native ants or the disruption of their foraging behaviors, including the key seed dispersing *Aphaenogaster* ants. The indirect effect generally appears to be the disruption of seed-dispersing interactions. Plant benefits from ant-mediated seed-dispersal include the movement of seeds away from negative density-dependent effects caused by seed accumulation near parents, and the placement of seeds in ant nests protected from seed predators, and rich in nutrients and moisture (Chapter 7). The invasion of *L. humile, S. invicta, M. rubra* and, particularly, *B. chinensis* in eastern NA may reduce or negate most, if not all, of these benefits.

Climate Change

Temperature strongly influences ant communities. Ant survival, reproduction, foraging, dispersal, physiology, and behavior vary with ambient temperatures (Andrew et al., 2013; Chen et al., 2002; Diamond et al., 2012; Kuriachan & Vinson, 2000; Pelini et al., 2011; Sanders et al., 2007). For example, a temperature shift of ~2°C corresponds with considerable changes in ant behavior and community composition (Arnan et al., 2007; Lessard et al., 2010; Warren II et al., 2011a; Warren II & Bradford, 2013; Warren II & Chick, 2013; Wittman et al., 2010). Given that much of NA may warm several degrees by mid-century, major shifts in ant behaviors and communities can be expected. As with other species, ants likely will respond to changing temperatures either spatially (relocation) or temporally (phenology) – and these changes will impact their interactions with other species, including the plants they disperse (Singer & Parmesan, 2010; Urban et al., 2012; Walther et al., 2002; Warren II & Chick, 2013).

Many ants thrive in climates with temperatures well below their maximum tolerance so that rising minimum temperatures may be the most effective (Diamond et al., 2012; Stuble et al., 2014; Warren II et al., 2011a; Warren II & Chick, 2013). For example, in the southern Appalachian Mountains (US), the warm-adapted, low-elevation *Aphaenogaster* species migrated upward with warming minimum temperatures and replaced the cold-adapted, high-elevation *Aphaenogaster* species (Warren II & Chick, 2013). A changing ant community means that effective seed-dispersing ants may join or leave the local assemblage. If the interaction with plants was diffuse – that is, many ant species effectively disperse seeds – then the functional relationships would remain intact. However, even though numerous species of ants interact with numerous species of plants, many destroy the seed or provide poor dispersal services so that only a few species effectively transport seeds (Warren II & Giladi, 2014). In eastern NA, ants in the genus *Aphaenogaster* are the only effective seed dispersers, and each of the 6–12 species in the cryptic *Aphaenogaster* genera appear to have unique temperature optima that determines their range boundaries (Warren II et al., 2011b). With changing temperatures, those range boundaries appear to be shifting with warm-adapted *Aphaenogaster* species moving upward or northward (Warren II & Bradford, 2013; Warren II & Chick, 2013).

The seasonal patterning (phenology) of ant behavior is closely linked with temperature – foraging increases with warming (Fellers, 1989; Pelini et al., 2011; Pelini et al., 2012; Stuble et al., 2013; Warren II et al., 2011a; Warren II & Bradford, 2013; Warren II et al., 2014; Zelikova et al., 2008). Concomitantly, temperature controls plant phenology (Dahlgren et al., 2007; Rathcke & Lacey, 1985; Warren II et al., 2011a), and ant-dispersed plants release their seeds relatively early in spring in concordance with ant foraging (Warren II & Giladi, 2014). The timing of myrmecochore seed release seemingly corresponds with peak early spring foraging by *A. rudis* (see Warren II & Giladi (2014) and references therein). If both plant and ant follow the same temperature cues, the interaction is portable and resistant to climate change (Warren II et al., 2011a). However, Warren II and Bradford (2013) found that a shift from the cold- to warm-adapted *Aphaenogaster* species resulted in phenological asynchrony with early blooming myrmecochores. The warm-adapted *Aphaenogaster* required temperatures 6 °C higher to break winter dormancy and thus began foraging approximately four weeks later than the cold-adapted species (Warren II & Bradford, 2013). As a result, *Aphaenogaster* foraging failed to synchronize with early blooming *Anemone americana* (DC.) Ker Gawl., and the plant seedlings aggregated around the parents – a similar pattern to what is found when ants are experimentally or naturally excluded from mymecochores (Warren II et al., 2010; Zelikova et al., 2011). A key area of climate change research necessary to predict ant impacts is to experimentally quantify how changing temperatures impact seed dispersing ant colony growth and maintenance (e.g.,Porter & Tschinkel, 1993) in addition to foraging behavior.

Global Change Synergy

Individually, habitat fragmentation, species invasion, and climate change are recognized drivers of ecosystem change; interacting together, they pose a great threat to biodiversity (Didham et al., 2007; Mantyka-Pringle et al., 2012; Staudt et al., 2013; see Figure 5.1). For example, anthropogenic forest disturbance creates edge habitat that appears more suitable for exotic invasive ants than the native seed-dispersing ants that prefer intact forest (Holway et al., 2002; King & Tschinkel, 2008, 2013; King et al., 2009, 2013; Lubertazzi, 2012; Suarez et al., 1998). A warming climate lifts the minimum temperature barriers that may prevent some southeastern US exotic ants from invading northern forests (Bertelsmeier et al., 2013, 2008; Hellmann et al., 2008). And a warming climate may leave ant-dispersed plants stranded in fragmented forests unable to track their suitable habitat because the woodland ants do not disperse seeds across forest edges (Jules & Rathcke, 1999; Ness, 2004; Ness & Morin, 2008; Warren II et al., 2015b).

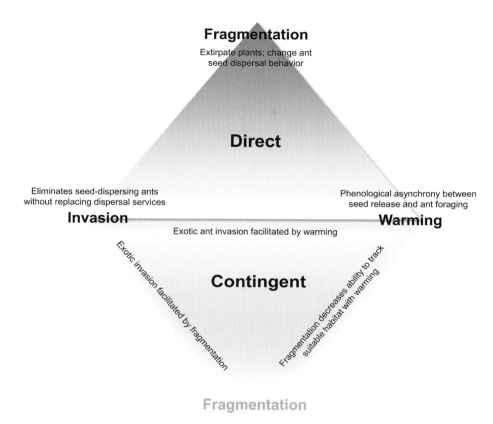

Figure 5.1. Conceptual graphic of direct and contingent effects of global change on ant-mediated seed dispersal in eastern North America.

It is hard to devise an alternative hypothesis by which habitat fragmentation, species invasion, and climate change increase or preserve interactions between woodland herbs and seed-dispersing ants in eastern NA. Although *M. rubra* might substitute for *A. rudis* as an effective disperser of myrmecochorous plants (Prior et al., 2015), most exotic ants clearly do not (Bond & Slingsby, 1984; Carney et al., 2003; Christian, 2001; Rodriguez-Cabal et al., 2009, 2012; Warren II et al., 2015a). Woodland plants may also tolerate projected warming changes and not be pushed against forest edges in fragmented patches, but they may nevertheless be out of synch with the ant species that move into the patches, native or exotic (Rodriguez-Cabal et al., 2012; Warren II & Bradford, 2013; Warren II et al., 2015a). It is also possible that edge invaders such as *S. invicta* and *L. humile* may rarely pose a great threat to interior forest ant species, yet *B. chinensis* appears to be able to push far into forest habitat, and continued warming may allow it and other invasive ants greater access to northeastern US forests (Bertelsmeier et al., 2013; Rice & Silverman, 2013).

Ant-mediated seed dispersal alleviates maladaptive density-dependent conditions, and delivers seeds to benign sites with lessened seed predation and greater fertility (Giladi, 2006; Handel & Beattie, 1990; Warren II & Giladi, 2014). As a result, failed myrmecochore dispersal results in greater seedling mortality and reduced population genetic flow (Beattie, 1978; Boyd, 2001; Kalisz et al., 1999; Kjellsson, 1991; Zhou et al., 2007). Alternate modes of dispersal have been suggested for myrmecochorous plants (Bale et al., 2003; Myers et al., 2004); however, these have not been linked with plant population viability as has ant-mediated dispersal. Hence, a failure in ant-mediated dispersal may lead to the slow extirpation of myrmecochore populations in many regions. Rather than acting as physiological barriers for these woodland herbs, the global changes we consider here may act as biotic barriers. That is, as increased fragmentation and climate warming prompts changes in native and exotic ant species' ranges and behaviors, the data reviewed here suggest that ant-mediated seed dispersal will fail more in the southern than in the northern portion of the range. That is, rather than a shift northward *en masse*, a more likely scenario is that the northern populations persist and the southern populations are steadily extirpated due to the additive multiple threats, a process that appears to already be occurring (Warren II & Bradford, 2013; Warren II & Chick, 2013). Myrmecochore species that only occur at the southern end of the range and set seed early in spring (Table 5.1, genera such as *Asarum*, *Trillium* and *Viola*) may very well be those species whose extinction is imminent as forests continue to be fragmented, climate warms, and novel exotic ant species establish in their ranges.

References

Andrew, N. R., Hart, R. A., Jung, M.-P., Hemmings, Z. and Terblanche, J. S. (2013). Can temperate insects take the heat? A case study of the physiological and behavioural responses in a common ant, *Iridomyrmex purpureus* (Formicidae), with potential climate change. *Journal of Insect Physiology*, 59, 870–880.

Arnan, X., Rodrigo, A. and Retana, J. (2007). Uncoupling the effects of shade and food resources of vegetation on Mediterranean ants: an experimental approach at the community level. *Ecography*, 30, 161–172.

Bale, M. T., Zettler, J. A., Robinson, B. A., Spira, T. P. and Allen, C. R. (2003). Yellow jackets may be an underestimated component of an ant-seed mutualism. *Southeastern Naturalist*, 2, 609–614.

Banschbach, V. S., Yeamans, R., Brunelle, A., Gulka, A. and Holmes, M. (2012). Edge effects on community and social structure of northern temperate deciduous forest ants. *Psyche*, 2012, 548260.

Beattie, A. J. (1978). Plant-animal interactions affecting gene flow in *Viola*. In *The Pollination of Flowers by Insects*, ed. Richards, A. J. London: Academic Press, pp. 151–164.

Beattie, A. J. and Culver, D. C. (1981). The guild of myrmecochores in the herbaceous flora of West Virginia forests. *Ecology*, 62, 107–115.

Beattie, A. J. and Hughes, L. (2002). Ant-plant interactions. In *Plant-Animal Interactions: An Evolutionary Approach,* ed. Herrera, C. M. and Pellmyr, O. Oxford: Blackwell Science, pp. 211–235.

Bednar, D. M. (2010). *Pachycondyla (=Brachyponera) chinensis Predation on Reticuletermes virginicus and Competition with Aphaenogaster rudis*. Master of Science Thesis, North Carolina State University, USA.

Bednar, D. M. and Silverman, J. (2011). Use of termites, *Reticulitermes virginicus*, as a springboard in the invasive success of a predatory ant, *Pachycondyla (=Brachyponera) chinensis*. *Insectes Sociaux*, 58, 459–467.

Bertelsmeier, C., Guenard, B. and Courchamp, F. (2013). Climate change may boost the invasion of the Asian needle ant. *PLoS ONE*, 8, e75438.

Bertelsmeier, C., Luque, G. M., Hoffmann, B. D. and Courchamp, F. (2015). Worldwide ant invasions under climate change. *Biodiversity Conservation*, 24, 117–128.

Bond, W. and Slingsby, P. (1984). Ant-plant mutalism: the Argentine ant (*Iridomyrmex Humilis*) and myrmecochorous Proteaceae. *Ecology*, 65, 1031–1037.

Bono, J. M. and Heithaus, E. R. (2002). Population consequences of changes in ant-seed mutualism in *Sanguinaria canadensis*. *Insectes Sociaux*, 49, 320–325.

Boyd, R. (2001). Ecological benefits of myrmecochory for the endangered chaparral shrub *Fremontodendron decumbens* (Sterculiaceae). *American Journal of Botany*, 88, 234–241.

Brew, C. R., O'Dowd, D. J. and Rae, I. D. (1989). Seed dispersal by ants: behaviour-releasing compounds in elaiosomes. *Oecologia*, 80, 490–497.

Bronstein, J. L., Alarcon, R. and Geber, M. (2006). The evolution of plant-insect mutualisms. *New Phytologist*, 172, 412–428.

Brook, B. W., Sodhi, N. S. and Bradshaw, C. J. A. (2008). Synergies among extinction drivers under global change. *Trends in Ecology & Evolution*, 23, 453–460.

Brown, D. G., Johnson, K. M., Loveland, T. R. and Theobald, D. M. (2005). Rural land-use trends in the conterminous United States. *Ecological Applications*, 15, 1851–1863.

Brunet, J. and von Oheimb, G. (1998). Migration of vascular plants to secondary woodlands in southern Sweden. *Journal of Ecology*, 86, 429–438.

Bulow-Olsen, A. (1984). Diplochory in *Viola*: a possible relation between seed dispersal and soil seed bank. *American Midland Naturalist*, 112, 251–260.

Cain, M. L., Damman, H. and Muir, A. (1998). Seed dispersal and the Holocene migration of woodland herbs. *Ecological Monographs*, 68, 325–347.

Canner, J. E., Dunn, R. R., Giladi, I. and Gross, K. (2012). Redispersal of seeds by a keystone ant augments the spread of common wildflowers. *Acta Oecologica*, 40, 31–39.

Carney, S. E., Byerley, M. B. and Holway, D. A. (2003). Invasive Argentine ants (*Linepithema humile*) do not replace native ants as seed dispersers of *Dendromecon rigida* (Papaveraceae) in California, USA. *Oecologia*, 135, 576–582.

Caut, S., Jowers, M. J., Cerda, X. and Boulay, R. (2013). Questioning the mutual benefits of myrmecochory: a stable isotope-based experimental approach. *Ecological Entomology*, 38, 390–399.

Chen, Y., Hansen, L. D. and Brown, J. J. (2002). Nesting sites of the carpenter ant, *Camponotus vicinus* (Mayr) (Hymenoptera: Formicidae) in northern Idaho. *Environmental Entomology*, 31, 1037–1042.

Christian, C. E. (2001). Consequences of biological invasions reveal importance of mutualism for plant communities. *Nature*, 413, 576–582.

Clark, R. E. and King, J. R. (2012). The ant, *Aphaenogaster picea*, benefits from plant elaiosomes when insect prey is scarce. *Environmental Entomology*, 41, 1405–1408.

Dahlgren, J. P., von Zeipel, H. and Ehrlen, J. (2007). Variation in vegetative and flowering phenology in a forest herb caused by environmental heterogeneity. *American Journal of Botany*, 94, 1570–1576.

De Marco, B. and Cognato, A. I. (2016). A multiple gene phylogeny reveals polyphyly among eastern North American *Aphaenogaster* species (Hymenoptera: Formicidae). *Zoologica Scripta*, DOI: 10.1111/zsc.12168.

Diamond, S. E., Nichols, L. M., McCoy, N., Hirsch, C., Pelini, S. L., Sanders, N. J., Ellison, A. M., Gotelli, N. J. and Dunn, R. R. (2012). A physiological trait-based approach to predicting the responses of species to experimental climate warming. *Ecology*, 93, 2305–2312.

Didham, R. K., Tylianakis, J. M., Gemmell, N. J., Rand, T. A. and Ewers, R. M. (2007). Interactive effects of habitat modification and species invasion on native species decline. *Trends in Ecology & Evolution*, 22, 489–496.

Drummond, M. A. and Loveland, T. R. (2010). Land-use pressure and a transition to forest-cover loss in the eastern United States. *Bioscience*, 60, 286–298.

Duffy, D. C. and Meier, A. J. (1992). Do Appalachian herbaceous understories ever recover from clearcutting? *Conservation Biology*, 6, 196–201.

Eguchi, K. (2004). A survey on seed predation by omnivorous ants in the warm-temperate zone of Japan (Insecta, Hymnoptera, Formicidae). *New Entomologist*, 53, 7–18.

Fellers, J. H. (1989). Daily and seasonal activity in woodland ants. *Oecologia*, 78, 69–76.

Fischer, R. C., Richter, A., Hadacek, F. and Mayer, V. (2008). Chemical differences between seeds and elaiosomes indicate an adaptation to nutritional needs of ants. *Oecologia*, 155, 539–547.

Flinn, K. M. and Vellend, M. (2005). Recovery of forest plant communities in post-agricultural landscapes. *Frontiers in Ecology and the Environment*, 3, 243–250.

Ford, W. M., Odom, R. H., Hale, P. E. and Chapman, B. R. (2000). Stand-age, stand characteristics, and landform effects on understory herbaceous communities in southern Appalachian cove-hardwoods. *Biological Conservation*, 93, 237–246.

Gammans, N., Bullock, J. J. and Schonrogge, K. (2005). Ant benefits in a seed dispersal mutualism. *Oecologia*, 146, 43–49.

Garnas, J. (2004). European fire ants on Mount Desert Island, Maine: population structure, mechanisms of competition and community impacts of *Myrmica rubra* L. (Hymenoptera: Formicidae). *Ecology and Environmental Sciences*. Orono, Maine: University of Maine.

Giladi, I. (2006). Choosing benefits or partners: a review of the evidence for the evolution of myrmecochory. *Oikos*, 112, 481–492.

Gilliam, F. S. (2002). Effects of harvesting on herbaceous layer diversity of a central Appalachian hardwood forest in West Virginia, USA. *Forest Ecology and Management*, 155, 33–43.

(2007). The ecological significance of the herbaceous layer in temperate forest ecosystems. *Bioscience*, 57, 845–858.

Gomez, C. and Espadaler, X. (2013). An update of the world survey of myrmecochorous dispersal distances. *Ecography*, 36, 1193–1201.

Gomez, C. and Oliveras, J. (2003). Can the Argentine ant (*Linepithema humile* Mayr) replace native ants in myrmecochory? *Acta Oecologia*, 24, 47–53.

Gorb, E. and Gorb, S. (2000). Effects of seed aggregation on the removal rates of elaiosome-bearing *Chelidonium majus* and *Viola odorata* seeds carried by *Formica polyctena* ants. *Ecological Research*, 15, 187–192.

(2003). *Seed Dispersal by Ants in a Deciduous Forest Ecosystem*. Dordrecht, The Netherlands: Kluwer Academic Publishers.

Gove, A. D., Majer, J. D. and Dunn, R. R. (2007). A keystone ant species promotes seed dispersal in "diffuse" mutualism. *Oecologia*, 153, 687–697.

Groden, E., Drummond, F. A., Garnas, J. and Franceour, A. (2005). Distribution of an invasive ant, *Myrmica rubra* (Hymenoptera: Formicidae), in Maine. *Journal of Economic Entomology*, 98, 1774–1784.

Guenard, B. and Dunn, R. R. (2010). A new (old), invasive ant in the hardwood forests of eastern North America and its potentially widespread impacts. *PLoS ONE*, 5, e11614.

Handel, S. N. (1976). Ecology of *Carex pedunculata* (Cyperaceae), a new North American myrmecochore. *American Journal of Botany*, 63, 1071–1079.

Handel, S. N. and Beattie, A. J. (1990). Seed dispersal by ants. *Scientific American*, 263, 76–83.

Handel, S. N., Fisch, S. B. and Schatz, G. E. (1981). Ants disperse a majority of herbs in a mesic forest community in New-York state. *Bulletin of the Torrey Botanical Club*, 108, 430–437.

Harrelson, S. M. and Matlack, G. R. (2006). Influence of stand age and physical environment on the herb composition of second-growth forest, Strouds Run, Ohio, USA. *Journal of Biogeography*, 33, 1139–1149.

Heithaus, E. R. and Humes, M. (2003). Variation in communities of seed-dispersing ants in habitats with different disturbance in Knox County, Ohio. *Ohio Journal of Science*, 103, 89–97.

Hellmann, J. J., Byers, J. E., Bierwagen, B. G. and Dukes, J. S. (2008). Five potential consequences of climate change for invasive species. *Conservation Biology*, 22, 534–543.

Hölldobler, B. and Wilson, E. O. (1990). *The Ants*. Cambridge, MA: Belknap.

Holway, D. A., Lach, L., Suarez, A. V., Tsutsui, N. D. and Case, T. J. (2002). The causes and consequences of ant invasions. *Annual Review of Ecology and Systematics*, 33, 181–233.

Jackson, B. C., Pitillo, J. D., Allen, H. L., Wentworth, T. R., Bullock, B. P. and Loftis, D. L. (2009). Species diversity and composition in old growth and second growth rich coves of the Southern Appalachian Mountains. *Castanea*, 74, 27–38.

Jacquemyn, H. and Brys, R. (2008). Effects of stand age on the demography of a temperate forest herb in post-agricultural forests. *Ecology*, 89, 3480–3489.

Jules, E. S. and Rathcke, B. J. (1999). Mechanisms of reduced *Trillium* recruitment along edges of old-growth forest. *Conservation Biology*, 13, 784–793.

Kalisz, S., Hanzawa, F. M., Tonsor, S. J., Thiede, D. A. and Voigt, S. (1999). Ant-mediated seed dispersal alters pattern of relatedness in a population of *Trillium grandiflorum*. *Ecology*, 80, 2620–2634.

Keller, L. and Passera, L. (1989). Size and fat-content of gynes in relation to the mode of colony founding in ants (Hymenoptera; Formicidae). *Oecologia*, 80, 236–240.

King, J. R. and Tschinkel, W. R. (2008). Experimental evidence that human impacts drive fire ant invasions and ecological change. *Proceedings of the National Academy of Sciences*, 105, 20339–20343.

    (2013). Experimental evidence for weak effects of fire ants in a naturally invaded pine-savanna ecosystem. *Ecological Entomology*, 38, 68–75.

    (2016). Experimental evidence that dispersal drives ant community assembly in human-altered ecosystems. *Ecology*, 97, 236–249.

King, J. R., Tschinkel, W. R. and Ross, K. G. (2009). A case study of human exacerbation of the invasive species problem: transport and establishment of polygyne fire ants in Tallahassee, Florida, USA. *Biological Invasions*, 11, 373–377.

King, J. R., Warren II, R. J. and Bradford, M. A. (2013). Social insects dominate eastern US temperate hardwood forest macroinvertebrate communities in warmer regions. *PLoS ONE*, 8, e75843.

Kjellsson, G. (1991). Seed fate in an ant-dispersed sedge, *Carex pilulifera* L.: recruitment and seedling survival in tests of models for spatial dispersion. *Oecologia*, 88, 435–443.

Kuriachan, I. and Vinson, S. B. (2000). A queen's worker attractiveness influences her movement in polygynous colonies of the red imported fire ant (Hymenoptera: Formicidae) in response to adverse temperature *Environmental Entomology*, 29, 943–949.

Lengyel, S., Gove, A. D., Latimer, A. M., Majer, J. D. and Dunn, R. B. (2009). Ants sow the seeds of global diversification in flowering plants. *PLoS ONE*, 4, e5480.

Lessard, J. P., Sackett, T. E., Reynolds, W. N., Fowler, D. A. and Sanders, N. J. (2010). Determinants of the detrital arthropod community structure: effects of temperature and resources along an environmental gradient. *Oikos*, 120, 333–343.

Lubertazzi, D. (2012). The biology and natural history of *Aphaenogaster rudis*. *Psyche*, 2012, 752815.

Mantyka-Pringle, C. S., Martin, T. G. and Rhodes, J. R. (2012). Interactions between climate and habitat loss effects on biodiversity: a systematic review and meta-analysis. *Global Change Biology*, 18, 1239–1252.

Marshall, D. L., Beattie, A. J. and Bollenbacher, W. E. (1979). Evidence for diglycerides as attractants in an ant-seed interaction. *Journal of Chemical Ecology*, 5, 335–344.

Marussich, W. A. (2006). Testing myrmecochory from the ant's perspective: the effects of *Datura wrightii* and *D. discolor* on queen survival and brood production in *Pogonomyrmex californicus*. *Insectes Sociaux*, 53, 403–411.

Matlack, G. R. (1993). Microenvironment variation within and among forest edge sites in the Eastern United-States. *Biological Conservation*, 66, 185–194.

(1994a). Plant-species migration in a mixed-history forest landscape in Eastern North-America. *Ecology*, 75, 1491–1502.

(1994b). Vegetation dynamics of the forest edges-trends in space and successional time. *Journal of Ecology*, 82, 113–123.

McGlynn, T. P. (1999). Non-native ants are smaller than related native ants. *American Naturalist*, 6, 690–699.

McLachlan, S. M. and Bazely, D. R. (2001). Recovery patterns of understory herbs and their use as indicators of deciduous forest regeneration. *Conservation Biology*, 15, 98–110.

Meier, A. J., Bratton, S. P. and Duffy, D. C. (1995). Possible ecological mechanisms for loss of vernal-herb diversity in logged eastern deciduous forests. *Ecological Applications*, 5, 935–946.

Mitchell, C. E., Turner, M. G. and Pearson, S. M. (2002). Effects of historical land use and forest patch size on myrmecochores and ant communities. *Ecological Applications*, 12, 1364–1377.

Morales, M. A. and Heithaus, E. R. (1998). Food from seed-dispersal mutualism shifts sex ratios in colonies of the ant *Aphaenogaster rudis*. *Ecology*, 79, 734–739.

Myers, J. A., Vellend, M., Gardescu, S. and Marks, P. L. (2004). Seed dispersal by white-tailed deer: implications for long-distance dispersal, invasion and migration of plants in eastern North America. *Oecologia*, 139, 35–44.

Nelder, M. P., Paysen, E. S., Zungoli, P. A. and Benson, E. P. (2006). Emergence of the introduced ant *Pachycondyla chinensis* (Formicidae: Ponerinae) as a public health threat in the southeastern United States. *Journal of Medical Entomology*, 43, 1094–1098.

Ness, J. H. (2004). Forest edges and fire ants alter the seed shadow of an ant-dispersed plant. *Oecologia*, 138, 228–454.

Ness, J. H. and Bronstein, J. L. (2004). The effects of invasive ants on the prospective ant mutualists. *Biological Invasions*, 6, 445–461.

Ness, J. H. and Morin, D. F. (2008). Forest edges and landscape history shape interactions between plants, seed-dispersing ants and seed predators. *Biological Conservation*, 141, 838–847.

Ness, J. H., Morin, D. F. and Giladi, I. (2009). Uncommon specialization in a mutualism between a temperate herbaceous plant guild and an ant: Are *Aphaenogaster* ants keystone mutualists? *Oikos*, 12, 1793–1804.

Ohnishi, Y., Suzuki, N., Katayama, N. and Teranishi, S. (2008). Seasonally different modes of seed dispersal in the prostrate annual, *Chamaesyce maculata* (L.) Small (Euphorbiaceae), with multiple overlapping generations. *Ecological Research*, 23, 299–305.

Pelini, S. L., Boudreau, M., McCoy, N., Ellison, A. M., Gotelli, N. J., Sanders, N. J. and Dunn, R. R. (2011). Effects of short-term warming on low and high latitude forest ant communities. *Ecosphere*, 2, 1–12.

Pelini, S. L., Diamond, S. E., MacLean, H. J., Ellison, A. M., Gotelli, N. J., Sanders, N. J. and Dunn, R. R. (2012). Common garden experiments reveal uncommon responses across temperatures, locations, and species of ants. *Ecology and Evolution*, 2, 3009–3015.

Peterson, C. J. and Campbell, J. E. (1993). Microsite differences and temporal change in plant communities of treefall pits and mounds in an old-growth forests. *Bulletin of the Torrey Botanical Club*, 120, 451–460.

Pfeiffer, M., Huttenlocher, H. and Ayasse, M. (2010). Myrmecochorous plants use chemical mimicry to cheat seed-dispersing ants. *Functional Ecology*, 24, 545–555.

Porter, S. D. and Tschinkel, W. R. (1993). Fire ant thermal preferences: behavioral control of growth and metabolism. *Behavioral Ecology and Sociobiology*, 32, 321–329.

Prior, K. M., Robinson, J. M., Meadly Dunphy, S. A. and Frederickson, M. E. (2015). Mutualism between co-introduced species facilitates invasion and alters plant community structure. *Proceedings of the Royal Society B-Biological Sciences*, 282.

Pudlo, R. J., Beattie, A. J. and Culver, D. C. (1980). Population consequences of changes in ant-seed mutualism in *Sanguinaria canadensis*. *Oecologia*, 146, 32–37.

Rathcke, B. and Lacey, E.P. (1985). Phenological patterns of terrestrial plants. *Annual Review of Ecology and Systematics*, 16, 179–214.

Rice, E. S. and Silverman, J. (2013). Propagule pressure and climate contribute to the displacement of *Linepithema humile* by *Pachycondyla chinensis*. *PLoS ONE*, 8, 856281.

Rico-Gray, V. and Oliveira, P. S. (2007). *The Ecology and Evolution of Ant-Plant Interactions*. Chicago: University of Chicago Press.

Rodriguez-Cabal, M. A., Stuble, K. L., Guenard, B., Dunn, R. R. and Sanders, N. J. (2012). Disruption of ant-seed dispersal mutualisms by the invasive Asian needle ant (*Pachycondyla chinensis*). *Biological Invasions*, 14, 557–565.

Rodriguez-Cabal, M. A., Stuble, K. L., Nunez, M. A. and Sanders, N. J. (2009). Quantitative analysis of the effects of the exotic Argentine ant on seed-dispersal mutualisms. *Biology Letters*, 5, 499–502.

Rowles, A. D. and O'Dowd, D. J. (2009). New mutualism for old: indirect disruption and direct facilitation of seed dispersal following Argentine ant invasion. *Oecologia*, 158, 709–716.

Rowles, A. D. and Silverman, J. (2010). Argentine ant invasion associated with loblolly pines in the Southeastern United States: minimal impacts but seasonally sustained. *Environmental Entomology*, 39, 1141–1150.

Sanders, N. J., Gotelli, N. J., Heller, N. E. and Gordon, D. M. (2003). Community disassembly by an invasive species. *Proceedings of the National Academy of Sciences*, 100, 2474–2477.

Sanders, N. J., Lessard, J. P., Fitzpatrick, M. C. and Dunn, R. R. (2007). Temperature, but not productivity or geometry, predicts elevational diversity gradients in ants across spatial grains. *Global Ecology and Biogeography*, 16, 640–649.

Servigne, P. and Detrain, C. (2008). Ant-seed interactions: combined effects of ant and plant species on seed removal patterns. *Insectes Sociaux*, 55, 220–230.

Singer, M. C. and Parmesan, C. (2010). Phenological asynchrony between herbivorous insects and their hosts: signal of climate change or pre-existing adaptive strategy? *Proceedings of the Royal Society B-Biological Sciences*, 365, 3161–3176.

Sorrells, J. S. and Warren II, R. J. (2011). Ant-dispersed herb colonization lags behind forest re-establishment. *Journal of the Torrey Botanical Society*, 138, 77–84.

Staudt, A., Leidner, A. K., Howard, J., Brauman, K. A., Dukes, J. S., Hansen, L. J., Paukert, C., Sabo, J. and Solorzano, L. A. (2013). The added complications of climate change: understanding and managing biodiversity and ecosystems. *Frontiers in Ecology and the Environment*, 11, 494–501.

Stuble, K. L., Kirkman, L. K. and Carroll, C. R. (2010). Are red imported fire ants facilitators of native seed dispersal? *Biological Invasions*, 12, 1661–1669.

Stuble, K. L., Patterson, C. M., Rodriguez-Cabal, M. A., Ribbons, R. R., Dunn, R. R. and Sanders, N. J. (2014). Ant-mediated seed dispersal in a warmed world. *PeerJ*, 2, e286.

Stuble, K. L., Pelini, S. L., Diamon, S. E., Fowler, D. A., Dunn, R. R. and Sanders, N. J. (2013). Foraging by forest ants under experimental warming: a test at two sites. *Ecology and Evolution*, 3, 482–491.

Suarez, A. V., Bolger, D. T. and Case, T. J. (1998). Effects of fragmentation and invasion on native ant communities in a coastal Southern California. *Ecology*, 79, 2041–2056.

Talbot, M. (1966). Flights of the ant *Aphaenogaster treatae*. *Kansas Entomological Society*, 39, 67–77.

Theobald, D. M. (2005). Landscape patterns of exurban growth in the USA from 1980 to 2020. *Ecology and Society*, 10, 32.

Turner, K. M. and Frederickson, M. E. (2013). Signals can trump rewards in attracting seed-dispersing ants. *PLoS ONE*, 8, e71871.

Urban, M. C., Tewksbury, J. J. and Sheldon, K. S. (2012). On a collision course: competition and dispersal differences create no-analogue communities and cause extinctions during climate change. *Proceedings of the Royal Society B-Biological Sciences*, 279, 2072–2080.

Walther, G. R., Post, E., Convey, P., Menzel, A., Parmesan, C., Beebee, T. J. C., Fromentin, J. M., Hoegh-Guldberg, O. and Bairlein, F. (2002). Ecological responses to recent climate change. *Nature*, 416, 389–395.

Warren II, R. J., Bahn, V. and Bradford, M. A. (2011a). Temperature cues phenological synchrony in ant-mediated seed dispersal. *Global Change Biology*, 17, 2444–2454.

Warren II, R. J. and Bradford, M. A. (2013). Mutualism fails when climate response differs between interacting species. *Global Change Biology*, 20, 466–474.

Warren II, R. J. and Chick, L. (2013). Upward ant distribution shift corresponds with minimum, not maximum, temperature tolerance. *Global Change Biology*, 19, 2082–2088.

Warren II, R. J. and Giladi, I. (2014). Ant-mediated seed dispersal: a few ant species (Hymenoptera: Formicidae) benefit many plants. *Myrmecological News*, 20, 129–140.

Warren II, R. J., Giladi, I. and Bradford, M.A. (2010). Ant-mediated seed dispersal does not facilitate niche expansion. *Journal of Ecology*, 98, 1178–1185.

(2012). Environmental heterogeneity and interspecific interactions influence occupancy be key seed-dispersing ants. *Environmental Entomology*, 41, 463–468.

(2014). Competition as a mechanism structuring mutualisms. *Journal of Ecology*, 102, 486–495.

Warren II, R. J., McAfee, P. and Bahn, V. (2011b). Ecological differentiation among key plant mutualists from a cryptic ant guild. *Insectes Sociaux*, 58, 505–512.

Warren II, R. J., McMillan, A., King, J. R., Chick, L. and Bradford, M. A. (2015a). Forest invader replaces predation but not dispersal services by a keystone species. *Biological Invasions*, 23, 3153–3162.

Warren II, R. J., Pearson, S., Henry, S., Rossouw, K., Love, J.P., Olejniczak, M., Elliott, K. and Bradford, M.A. (2015b). Cryptic indirect effects of exurban edges on a woodland community. *Ecosphere*, 6, 218.

Whigham, D. E. (2004). Ecology of woodland herbs in temperate deciduous forests. *Annual Review of Ecology Evolution and Systematics*, 35, 583–621.

Wike, L., Martin, F. D., Paller, M. H. and Nelson, E. A. (2010). Impact of forest seral stage on use of ant communities for rapid assessment of terrestrial ecosystem health. *Journal of Insect Science*, 10, 1–16.

Wittman, S. E., Sanders, N. J., Ellison, A. M., Jules, E. S., Ratchford, J. S. and Gotelli, N. J. (2010). Species interactions and thermal constraints on ant community structure. *Oikos*, 119, 551–559.

Zelikova, T. J., Dunn, R. R. and Sanders, N. J. (2008). Variation in seed dispersal along an elevational gradient in Great Smoky Mountains National Park. *Acta Oecologica*, 34, 155–162.

Zelikova, T. J., Sanders, D. and Dunn, R. R. (2011). The mixed effects of experimental ant removal on seedling distribution, belowground invertebrates, and soil nutrients. *Ecosphere*, 2, 1–14.

Zettler, J. A., Spira, T. P. and Allen, C. R. (2001). Ant-seed mutualisms: can red imported fire ants source the relationship? *Biological Conservation*, 101, 249–253.

Zhou, H., Chen, J. and Chen, F. (2007). Ant-mediated seed dispersal contributes to the local spatial pattern and genetic structure of *Globba lancangensis* (Zingiberaceae). *Journal of Heredity*, 98, 317–324.

6 Effects of Human Disturbance and Climate Change on Myrmecochory in Brazilian Caatinga

Inara R. Leal, Laura C. Leal, Fernanda M. P. Oliveira,
Gabriela B. Arcoverde, and Alan N. Andersen[*]

Introduction

Myrmecochory refers to specialized seed dispersal by ants and is one of the world's major seed-dispersal syndromes, involving more than 11,000 angiosperm species (4.5 percent of the global total) from 334 genera and 77 families (Lengyel et al., 2009). Myrmecochorous plants have specially adapted seeds that possess a food appendage designed for ant attraction and transport (van der Pijl, 1982; Beattie, 1985). The appendages are typically arils, ariloides or caruncles (Gorb & Gorb, 2003), and are collectively referred to as elaiosomes because they are all rich in fatty acids. Generalized omnivorous ants are attracted to the elaiosomes and use them as handles for seed transportation (Beattie, 1985). Once seeds reach the nest, ants eat the elaiosome and typically discard the seed intact, in nest galleries or externally on nest middens or nearby refuse dumps, where they can potentially germinate and establish (Beattie, 1985; Hughes & Westoby, 1992; Manzaneda & Rey, 2012).

Myrmecochory is particularly common among temperate forest herbs in the northern Hemisphere (Beattie & Culver, 1981; Gorb & Gorb, 2003) and sclerophyll shrubs of Mediterranean-climate landscapes in South Africa, southern Australia and southern Europe (Bond & Slingsby, 1983; Westoby et al., 1991; Garrido et al.,

[*] We thank Paulo Oliveira and Suzanne Koptur for the invitation to write this chapter. Our studies on myrmecochory in Caatinga have been supported by the "Conselho Nacional de Desenvolvimento Científico e Tecnológico" (CNPq, processes: DCR 300582/1998-6, Universal 477290/2009-4 and 470480/2013-0, PELD 403770/2012-2, CNPQ-DFG 490450/2013-0), "Coordenação de Aperfeiçoamento de Pessoal de Nível Superior" (CAPES, processes: Estágio Sênior 2009/09-9 and 2411-14-8, "Sandwich" Doctorate 1650-12-2, PVE 88881.030482/2013-01) and "Fundação de Amparo à Ciência e Tecnologia do Estado de Pernambuco" (FACEPE, processes: APQ 0140-2.05/08 and 0738-2.05/12, PRONEX 0138-2.05/14). We would like to thank the landowners for giving us permission to work on their properties in the Xingó region, Parnamirim municipality and Catimbau region. Finally, we also thank all our students and colleagues who have assisted with field work and participated in fruitful discussions: Antônio F. M. Oliveira, Carlos H. F. Silva, Elâine M. S. Ribeiro, Felipe F. S. Siqueira, José D. Ribeiro-Neto, Kátia F. Rito, Kelaine Demetrio, Marcelo Tabarelli, Marcos V. Meiado, Rainer Wirth, Talita Câmara and Xavier Arnan.

2002). In the past decade, the Brazilian Caatinga has been recognized as another hotspot of myrmecochory (Leal et al., 2007, 2015). Caatinga is a mosaic of xerophytic, deciduous, semiarid thorn scrubs and seasonally dry forests that covers 730,000 km^2 in northeastern Brazil (Instituto Brasileiro de Geografia e Estatística, 1985; Sampaio, 1995). The soils form a complex mosaic, ranging from nutrient-rich clays to nutrient-poor sands (Sampaio, 1995). Although the seeds of most Caatinga plants are dispersed abiotically (Griz & Machado, 2001; Tabarelli et al., 2003), a large number of species from many families rely on the seed dispersal services by ants (e.g. 25 percent of local woody flora in Leal et al., 2007). Myrmecochory is especially prevalent in the Euphorbiaceae, the second largest plant family in the Caatinga flora (Moro et al., 2014), where about 70 percent of its species – woody plants from the genera *Cnidoscolus*, *Croton*, *Jatropha* and *Manihot* – have their caruncle-bearing seeds dispersed exclusively by ants (Leal et al., 2015). Although we have improved our understanding of myrmecochory in Caatinga over recent years (Leal et al., 2007, 2014a, 2015; Lôbo et al., 2011), there has been no systematic analysis of the impact of human disturbance on this dispersal mode (but see Leal et al., 2014b).

Like other ecosystems around the world, Caatinga has experienced a gradual but persistent degradation process varying from relatively minor biomass reduction to complete desertification (Leal et al., 2005; Ministério do Meio Ambiente, 2011; Ribeiro et al., 2015, 2016; Schulz et al., 2016). Much of this degradation is a consequence of the very dense (i.e. 26 inhabitants per km^2, Medeiros et al., 2012) and low-income rural populations of the region (Ab'Sáber, 1999), which are highly dependent on forest resources for their livelihoods (Leal et al., 2005; Gariglio et al., 2010; Sunderland et al., 2015). Activities such as firewood collection, exploitation of non-timber products, hunting, grazing by livestock and intentional introduction of invasive plant species (Singh, 1998; Martorell & Peters, 2005; Davidar et al., 2010) have had a negative influence on biodiversity, including a replacement of disturbance-sensitive species by disturbance-adapted species (Leal et al., 2005; Ribeiro et al., 2015, 2016; Oliveira et al., 2017; Ribeiro-Neto et al., 2016). Such chronic anthropogenic disturbance and its consequences are typical of developing countries and fast-growing global economies (Singh, 1998; Martorell & Peters, 2005; Laurance & Peres, 2006; Davidar et al., 2010; Ribeiro et al., 2015). Effective seed dispersal is critical for vegetation recovery after such disturbance (Farwig & Berens, 2012; Marini et al., 2012), and so any diminishing of seed-dispersal services by ants could have particularly significant consequences for a large component of the Caatinga flora.

In addition to chronic anthropogenic disturbance, the Caatinga biota is threatened by climate change. Predictions of the International Panel of Climate Change indicate that the Caatinga region will face an increase in temperature of 1.8–4°C, and a reduction in rainfall of 22 percent by 2100 from a 2000 baseline (Magrin et al., 2014). The range of climatic variation is also predicted to increase (Schär et al., 2004), including a higher frequency of extreme weather events; this might have greater ecological consequences than just the predicted shift in average conditions (Jentsch et al., 2009). Climate change could further intensify the negative effects of disturbance. There is increasing evidence that climate change and disturbance can have complex and sometimes synergistic effects on biodiversity (Travis, 2003; Ponce-Reyes et al., 2013;

García-Valdés et al., 2015), with warm and arid environments (such as Caatinga) likely to be at greatest risk (Anderson-Teixeira et al., 2013; Gibb et al., 2015).

In this chapter, we present a synthesis of our research on myrmecochory in Caatinga vegetation, with a particular focus on the impacts of chronic anthropogenic disturbance and climate change on this dispersal mode. First, we describe ant-diaspore interactions occurring in Caatinga, including the plant and ant species involved, diaspore types, ant behavior toward diaspores and the quality of dispersal services provided by ants. We then focus on true myrmecochores (mainly species of the Euphorbiaceae), describing species composition, diaspore traits and their influence on seed removal, seed germination and seedling survival. Finally, we examine the effect of chronic anthropogenic disturbance on the quality of the dispersal service provided by ants, and present an assessment of how myrmecochory in Caatinga might be affected by climate change. We conclude with some future directions of research on myrmecochory in Caatinga.

Interactions between Ants and Diaspores in Brazilian Caatinga

Diaspore Species

Ants from Caatinga vegetation interact with diaspores of a wide variety of plant species, including non-myrmecochores (i.e. those without elaiosome) as well as myrmecochores (Table 6.1). The non-myrmecochorous diaspores are fleshy fruits that belong to many plant families, including Anacardiaceae, Annonaceae, Boraginaceae, Cactaceae, Malpighiaceae, Rhamnaceae, Rubiaceae, Sapotaceae and Simaroubaceae. Most of these diaspore species are transported by ants, but some of the largest ones are not, and are just "cleaned" (i.e. fruit pulp is removed) in situ (Table 6.1). The interactions between ants and non-myrmecochorous diaspores in Caatinga represent secondary dispersal, similar to that described in other neotropical ecosystems such as rainforests (Levey & Byrne, 1993; Pizo & Oliveira, 2000; Passos & Oliveira, 2003) and savannas (Leal & Oliveira, 1998; Christianini et al., 2007), where a large variety of primarily vertebrate-dispersed plant species are secondarily dispersed by ants once the seeds reach the ground (Chapter 7).

The majority of true myrmecochores belongs to the Euphorbiaceae and possesses carunculate seeds. Ant dispersal has been directly observed in 16 carunculate euphorb species (Table 6.1, Figure 6.1); another 100 Caatinga euphorb species have carunculate seeds and are therefore presumably also myrmecochorous (Leal et al., 2015). Other species bearing elaiosomes categorized as true arils and sarcotestas (sensu Gorb & Gorb, 2003) also occur in Caatinga (Table 6.1). Myrmecochorous diaspores reach the ground by being passively dropped by parent plants or via ballistic dispersal; ants have never been observed climbing the vegetation to access fruits or seeds.

Ant Species

Diaspores are manipulated by different ant species in three main ways (Table 6.2, Figure 6.1): (1) large ant species (e.g. of *Camponotus*, *Dinoponera*, *Ectatomma* and

Table 6.1. Plant Species with Diaspores Interacting with Ants on the Ground in Caatinga Vegetation, Northeastern Brazil

Plant family and species	Fruit type	Fruit length (mm)	Seed length (mm)	Elaiosome type	Interaction	
					cleaning	removal
Anacardiaceae						
Myracrodruon urundeuva Allemão	Drupe	14	4	–		X
Schinus terebinthifolius Raddi	Drupe	6	5	–		X
Spondias tuberosa Arruda	Drupe	48	28	–	X	
Anonaceae						
Annona coriacea Mart.	Symcarp	130	28	–	X	
Boraginaceae						
Varronia globosa Jacq.	Drupe	5	3	–	X	X
Varronia leucocephala (Moricand) J. S. Mill.	Drupe	5	3	–	X	X
Burseraceae						
Commiphora leptophloeos (Mart.) J.B. Gillett	Capsule	15	10	Aril	X	X
Cactaceae						
Cereus jamacaru DC.	Berry	50	1.5	–		X
Melocactus bahiensis (Britton & Rose) Luetzelb.	Berry	15	1.2	–		X
Melocactus zehntneri (Britton & Rose) Luetzelb.	Berry	15	1.2	–		X
Pilosocereus gounellei (F. A. C. Weber) Byles & Rowley	Berry	50	1.5	–		X
Pilosocereus pachycladus F. Ritter	Berry	45	1.3	–		X
Tacinga inamoena (K.Schum.) N.P.Taylor & Stuppy	Berry	30	2	–		X
Tacinga palmadora (Britton & Rose) N.P.Taylor & Stuppy	Berry	30	2	–		X
Capparaceae						
Colicodendron yco Mart.	Capsule	60	15	Sarcotesta	X	X
Cynophalla flexuosa (L.) J.Presl	Capsule	50	10	Sarcotesta	X	X

(continued)

Table 6.1 (*cont.*)

Plant family and species	Fruit type	Fruit length (mm)	Seed length (mm)	Elaiosome type	Interaction	
					cleaning	removal
Celastraceae						
Maytenus rigida Mart.	Capsule	8	5	Aril	X	X
Euphorbiaceae						
Cnidoscolus pubescens Pohl	Capsule	24	12.5	Caruncle	X	X
Cnidoscolus quercifolius Pohl.	Capsule	25	13.5	Caruncle	X	X
Cnidoscolus urens (L.) Arthur	Capsule	16	7.2	Caruncle	X	X
Croton adamantinus Müll.Arg.	Capsule	8	4	Caruncle	X	X
Croton argyrophyllus Kunth	Capsule	6.5	4.8	Caruncle	X	X
Croton blanchetianus Baill.	Capsule	5.5	3	Caruncle	X	X
Croton campestris A.St.-Hil.	Capsule	10	4.3	Caruncle	X	X
Croton grewioides Baill.	Capsule	5	2.8	Caruncle	X	X
Croton heliotropiifolius Kunth	Capsule	7	4.5	Caruncle	X	X
Croton nepetaefolius Baill.	Capsule	10	5	Caruncle	X	X
Croton sonderianus Müll. Arg.	Capsule	10	5	Caruncle	X	X
Jatropha gossypiifolia L.	Capsule	14	7.3	Caruncle	X	X
Jatropha mollissima Pohl. (Baill.)	Capsule	25	12	Caruncle	X	X
Jatropha mutabilis Pohl. (Baill.)	Capsule	24	12	Caruncle	X	X
Jatropha ribifolia (Pohl) Baill.	Capsule	13	7	Caruncle	X	X
Manihot carthaginensis subsp. *glaziovii* (Müll. Arg.) Allem	Capsule	22	11	Caruncle	X	X
Malpighiaceae						
Byrsonima gardneriana A.Juss.	Drupe	10	7	–	X	X
Rhamnaceae						
Ziziphus joazeiro Mart.	Drupe	18	14	–	X	X

Table 6.1 (*cont.*)

Plant family and species	Fruit type	Fruit length (mm)	Seed length (mm)	Elaiosome type	Interaction	
					cleaning	removal
Rubiaceae						
Guettarda angelica Mart. ex Müll.Arg.	Drupe	6	4	–	X	X
Sapotaceae						
Manilkara rufula (Miq.) H.J.Lam	Berry	25	12	–	X	
Sideroxylon obtusifolium (Roem. & Schult.) T.D.Penn.	Berry	13.5	9	–	X	X
Simaroubaceae						
Simaba ferruginea A.St.-Hil.	Drupe	35	30	–	X	

Fruit type, fruit, and seed size according to van Roosmalen (1985), Andrade-Lima (1989), Lorenzi (1998) and Barroso et al. (1999). Elaiosome type *sensu* Gorb & Gorb (2003).

Source: From Leal et al. (2007); Lôbo et al. (2011); Leal et al. (2014a, 2014b, 2015, and unpublished data).

Odontomachus) often individually remove the whole diaspore to their nests; (2) species of *Acromyrmex*, *Atta*, *Crematogaster*, *Cyphomyrmex*, *Dorymyrmex*, *Pheidole* and *Trachymyrmex* recruit nest mates and cooperatively transport diaspores or parts of the elaiosome to nests and (3) species of *Solenopsis* and some of *Pheidole* commonly recruit nestmates and remove elaiosome in situ and do not transport the diaspores.

Although many ant species are attracted to diaspores, the dispersal service provided by them is highly variable and related to behavioral and morphological differences between the ant species (Hughes & Westoby, 1992; Gove et al., 2007; Manzaneda & Rey, 2008; 2012; Ness et al., 2009). In particular, large-bodied, solitary foraging ants typically offer superior dispersal services because individual workers can quickly collect seeds, transport them over relatively long distances and deposit them isolated or in small groups in external nest refuse piles (Ness et al., 2004; Gove et al., 2007; Aranda-Rickert & Fracchia, 2012). Conversely, small-bodied, recruit-foraging ants are typically low-quality dispersers because they often feed on the elaiosome in situ without transporting the diaspore, or use mass-recruiting system to remove diaspores, usually only over very short distances, and deposit them in large groups in nest refuse (Andersen & Morrison, 1998; Lôbo et al., 2011). The most important high-quality disperser ant species in Caatinga are those of *Dinoponera* (which are able to remove diaspores >20 m), *Ectatomma* and *Camponotus*, whereas species of *Solenopsis* and *Pheidole* are the most common low-quality dispersers (Leal et al., 2014a). Leaf-cutting ants (species of *Atta* and *Acromyrmex*) are also classified as low-quality dispersers despite their relatively

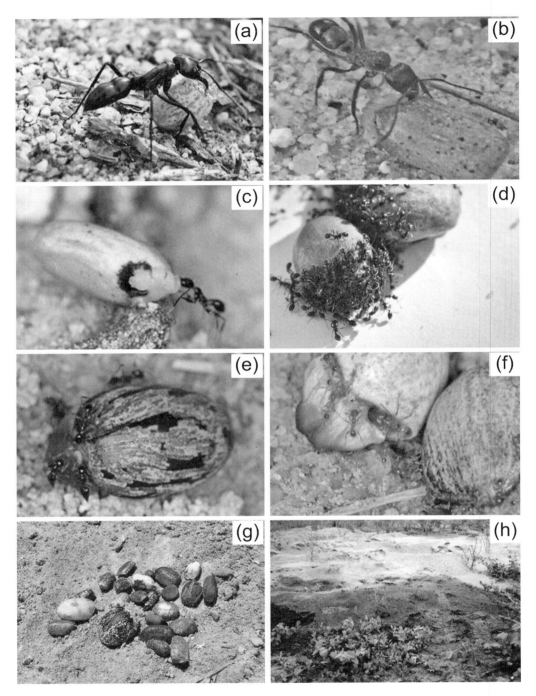

Figure 6.1. Interactions between ants and seeds in Caatinga vegetation: *Dinoponera quadriceps* carrying a *Jatropha mutabilis* seed (a), *Ectatomma muticum* carrying a *Jatropha ribifolia* seed (b), *Pheidole fallax* carrying a *Jatropha gossypiifolia* seed (c), *Solenopsis tridens* removing elaiosomes of *Jatropha mutabilis* seeds in situ (d), *Crematogaster* sp. removing the elaiosome of a *Cnidoscolus quercifolius* seed in situ (e), *Solenopsis* sp. removing elaiosomes of *Jatropha mollissima* seeds in situ (f), Euphorbiaceae seeds from different species on an *Atta laevigata* nest (g), and seedlings growing on an *Atta opaciceps* nest (h). Photo credits: L.C. Leal (a, b); I. R. Leal (c, e, f); F. M. P. Oliveira (d, g, h).

Table 6.2 Ant Species Recorded Interacting with Diaspores on the Ground in Caatinga Vegetation, Northeast Brazil

Ant subfamily and species	Ant behavior	Quality
Dolichoderinae		
Dorymyrmex bruneus (Forel)	RT	L
Dorymyrmex thoracicus (Gallardo)	RT	L
Dorymyrmex goeldii (Forel)	RT	L
Ectatomminae		
Ectatomma muticum (Mayr)	I	H
Formicinae		
Camponotus blandus (Smith)	I	H
Camponotus crassus (Mayr)	I	H
Myrmicinae		
Atta laevigata (Smith)	RT	L
Atta opaciceps (Borgmeier)	RT	L
Atta sexdens (L.)	RT	L
Acromyrmex balzani (Emery)	RT	L
Acromyrmex rugosus (Smith)	RT	L
Crematogaster sp. 1	RT	L
Crematogaster sp. 2	RT	L
Cyphomyrmex rimosus (Spinola)	RT	L
Pheidole fallax (Mayr)	RT	L
Pheidole radoszkowskii (Mayr)	RT	L
Pheidole sp. 1	RT	L
Pheidole sp. 2	RT	L
Pheidole sp. 3	RT	L
Pheidole sp. 4	RT	L
Solenopsis tridens (Forel)	RC	L
Solenopsis virulens (Smith)	RC	L
Solenopsis sp. 1	RC	L
Solenopsis sp. 2	RC	L
Solenopsis sp. 3	RC	L
Trachymyrmex sp. 1	RT	L
Trachymyrmex sp. 2	RT	L
Ponerinae		
Dinoponera quadriceps (Kempf)	IT	H
Odontomachus haematodus (Linnaeus)	IT	H

Ant behavior: I = individual transport of diaspores to nests; RT = recruitment of nest mates and transport of diaspores or parts of them to the nest; and RC = recruitment of workers and removal of the fruit pulp or elaiosome without diaspore transport. Quality of seed dispersal service: H = high-quality seed dispersers: large-bodied, subordinate predatory or omnivore ants that transport diaspores to significant distances and deposit them isolated or in small groups in nest refuses; L = low-quality seed dispersers: small-bodied, dominant or granivore ants that feed on the elaiosome in situ without transporting the diaspore, or transport diaspores to short distances and deposit them in large groups in nest refuses, or cut or bury all seedlings growing on nests.

Source: From Leal et al. (2007); Lôbo et al. (2011); Leal et al. (2014a, 2014b, 2015, and unpublished data).

large body size and ability to transport seeds long distances, because they usually cut or bury all seedlings growing on or near their nests (Table 6.2).

Despite myrmecochory being such a common dispersal syndrome around the world, there have been relatively few studies of the fate of dispersed seeds in terms of germination and seedling establishment (e.g. Hanzawa et al., 1988; Hughes & Westoby, 1992; Manzaneda & Rey, 2012). In Caatinga, we have observed that seeds deposited by *Dinoponera quadriceps* near its nest entrances experience a higher likelihood of escaping from post-dispersal predation by insects compared with control seeds, and seedling abundance is more than twice as high around nest entrances compared with control areas (Arcoverde, 2012). Although *D. quadriceps* is not involved in active seed and seedling protection, seed predators and folivores are likely to be deterred by resident ants, thus enhancing the prospects of seeds and seedlings (Arcoverde, 2012). This same pattern was observed in neotropical Restinga vegetation, where ponerine ants promote higher seed germination and seedling survival of non-myrmecochorous plants in ant nests as compared to control areas (Passos & Oliveira, 2002, 2004; Chapter 7). Therefore it suggests that some disperser ant species not only disproportionally benefit myrmecochorous and non-myrmecochorous plants during the seed transport phase, but may also extend their positive influence on myrmecochorous reproductive success into the post-dispersal phase.

Euphorbs: The Dominant Myrmecochores in Caatinga

Euphorbiaceae is by far the most important family of myrmecochores in Caatinga. Leal et al. (2015) identified 116 (68 percent) out of the 186 euphorb species known from this ecosystem as myrmecochorous, including species of *Croton* (58 species), *Manihot* (20), *Euphorbia* (10), *Cnidoscolus* (9), *Sebastiania* (9), *Jatropha* (5), *Mabea* (2), *Stilingia* (2) and *Ricinus* (1). All of them have carunculate seeds that are diplochorous, with ballistic discharge of seeds from explosive dehiscent capsules followed by dispersal by ants.

The interactions of ants with euphorb seeds have been described in detail, including removal rate and distance, seed fate, seed germination and seedling growth rate (Leal et al., 2007, 2014a). The rate of seed removal varies among sites, but was frequently very high (e.g. >70 percent after 12 hours) for all seven species studied by Leal et al. (2007). Approximately 80 percent of removed seeds are transported to ant nests. Dispersal distances can exceed 25 m, but mean distance is much shorter, varying from 1.2 m to 5.3 m; even the lower figure substantially exceeds that of ballistic dispersal (see Leal et al., 2007; Leal et al., 2014a). Within the nest, ants remove the elaiosome and retain the cleaned seeds inside the nest or more frequently deposit them in the vicinity of the nest entrance, including refuse piles and nest mounds (Leal et al., 2007). Although seeds become clustered close to nest entrances, they experience a higher chance of escaping from predation compared with seeds deposited in areas away from nests (Lôbo et al., 2011; Arcoverde, 2012).

Ant attraction to euphorb seeds is mediated by the elaiosome, as the removal rate of elaiosome-bearing seeds was at least two-fold higher than seeds without elaiosome for all the seven myrmecochore species evaluated by Leal et al. (2007). Elaiosomes served as handles, without which the hard and smooth seed coats would not allow seed transport by ants. Interspecific variation in elaiosome size and chemical composition also influence seed removal by ants. Low-quality disperser ants were equally attracted to the elaiosomes of all study species, while high-quality dispersers such as *Dinoponera quadriceps* and *Ectatomma muticum* showed a strong preference for diaspores of *Jatropha mollissima*, which had the highest elaiosome mass and especially proportional mass (Leal et al., 2014a). This finding suggests that myrmecochorous plants can preferentially target high-quality seed-disperser ants through the evolution of particular elaiosome traits (Leal et al., 2014a).

The deposition of transported seeds on ant nest improves seed germination for most species evaluated in Leal et al. (2007). This may be a peculiarity of Euphorbiaceae and other caruncle-bearing seed species, as their elaiosome (also referred to as micropylar aril) covers the micropyle, the structure responsible for seed imbibition (Gorb & Gorb, 2003). Caruncle removal by ants may facilitate seed imbibition, and consequently enhance seed germination (Leal et al., 2007). In the case of *Cnidoscolus quercifolius*, a higher germination rate in nest soil is followed by an increased seedling growth rate, due to higher nutrient content and cation exchange capacity of nest soil (Leal et al., 2007). In addition, soil penetrability is three times higher in ant nests than in random sites (Leal et al., 2007). Thus, ant nests may provide a deep, soft, nutrient-enriched and moist substrate that promotes seed germination and enhances seedling performance, particularly at sites covered by shallow and rocky soils, which are very common in Caatinga (Leal et al., 2007).

Effects of Chronic Disturbance

Overgrazing by livestock and the continual extraction of forest products are important drivers of chronic anthropogenic disturbances in Caatinga, which can lead to the gradual extirpation of woody plant species (Ribeiro et al., 2015, 2016) and alteration of vegetation structure (Ribeiro et al., 2015, 2016), increased species similarity in sites under higher level of disturbance (i.e. biotic homogenization) for both plant and ant assemblages (Ribeiro-Neto et al., 2016), and impoverished stocks of soil nutrient (Schulz et al., 2016). Leal et al. (2014b) investigated the effects of chronic anthropogenic disturbance on myrmecochory in Caatinga vegetation near Parnamirim municipality (8°5'S; 39°34'W; 393 m asl) in the state of Pernambuco. They used five surrogates of chronic disturbances (density of people and livestock, and proximity to urban center, houses and roads), which were combined in a disturbance index through a principal component analysis. All these surrogates have been shown to individually have negative effect on species richness and composition for

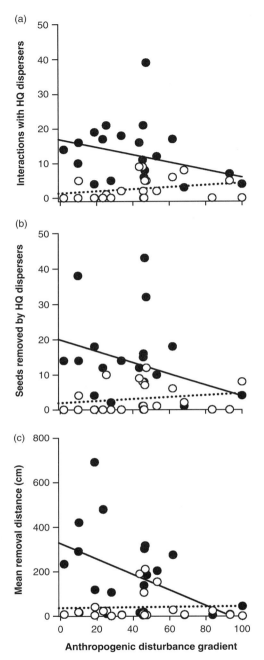

Figure 6.2. Number of interactions (a), number of removals (b) and mean removal distance (c) of seeds from *Jatropha mollissima* (black circles) and *Croton sonderianus* (open circles) by high-quality (HQ) seed-disperser ant species over an anthropogenic disturbance gradient in Brazilian Caatinga (modified from Leal et al., 2014b). Zero and 100 represent the extreme values of disturbance that represents sites under the lowest and highest human disturbance pressure, respectively. General linear models indicate that number of interactions (F = 9.12, df = 1, p < 0.001), number of removals (F = 6.83, df = 1, p<0.001) and removal distances (F = 7.94, df = 1, p < 0.0001) were negatively affected by disturbance for *Jatropha mollissima*, but not affected for *Croton sonderianus* (number of interactions: F = 5.82, df = 1, p = 0.09; number of removals: F = 3.81, df = 1; p = 0,45; removal distance: F = 0.64, df = 1, p = 0.42).

both plant and ant assemblages (Ribeiro et al., 2015; Oliveira et al., 2017; Ribeiro-Neto et al., 2016).

Their first important result is that overall ant disperser composition varied markedly with disturbance. The abundance of the two key high-quality disperser ants (*Dinoponera quadriceps* and *Ectatomma muticum*) decreased with disturbance, whereas a range of low-quality dispersers (species of *Pheidole*, *Solenopsis*, *Camponotus* and *Crematogaster*) were positively associated with disturbance (Leal et al., 2014b). *Dinoponera quadriceps*, like many other large predatory ants (Hoffmann & Andersen, 2003), is very sensitive to habitat disturbance (Leal et al., 2012). This species provides a key ecological service to myrmecochorous plants in Caatinga, accounting for 97 percent of all removals farther than 2 m, and 100 percent of all removals beyond 5 m (Leal et al., 2014b).

These disturbance-mediated changes in seed disperser composition had important effects on ant dispersal services in terms of the number of ant-seed interactions, number of removals and removal distances (Figure 6.2). Of the myrmecochore species assessed, *Jatropha mollissima* bears the most attractive elaiosome to high-quality disperser ants (Leal et al., 2014a), and thus the dispersal services it receives were particularly affected by disturbance (Figure 6.2). Conversely, the interaction with high-quality seed dispersers was not affected by disturbance for *Croton sonderianus*, because this species has the least attractive elaiosome (Figure 6.2; Leal et al., 2014a).

We also investigated the influence of chronic anthropogenic disturbance on two post-dispersal services provided by *Dinoponera quadriceps*: protecting seeds from predators and promoting seedling establishment (Arcoverde, 2012). We measured insect seed predation and seedling richness and densities on *D. quadriceps* nest refuse piles and in adjacent control areas along the same chronic disturbance gradient investigated by Leal et al. (2014b). Although ant nests presented lower seed predation, and higher seedling richness and densities compared to control areas, no relationship with anthropogenic disturbance was detected (Figure 6.3). This result indicates that although disturbance markedly reduces rates of seed removal by *D. quadriceps* and the distances seeds are removed, post-dispersal services are maintained even in areas under higher level of chronic disturbance pressure.

Effects of Climate Change

Few studies have addressed the influence of climate change on myrmecochory. Warren II et al. (2011) demonstrated that warming temperatures act as the primary phenological cue for plant fruiting, but that the key seed disperser, *Aphaenogaster rudis*, fails to emerge during early fruit set (Chapter 5). Under a future climate there might therefore be a temporal mismatch between seed fall and the foraging activity of the key seed disperser. However, Stuble et al. (2014) argue that ant-seed-dispersal mutualisms may be more robust to increased temperature than currently assumed. There are no published studies of the interaction between disturbance and climate change. To fill this gap, we

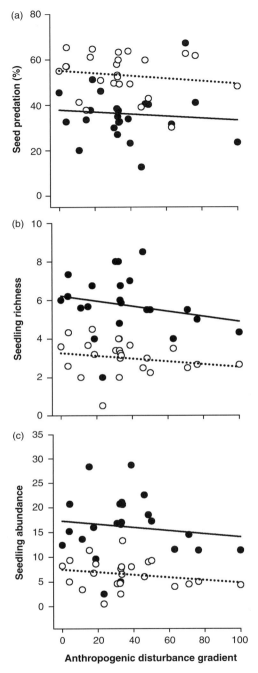

Figure 6.3. Seed predation (a), seedling richness (b) and seedling abundance (c) on *Dinoponera quadriceps* nests (black circles) and control areas (open circles) over an anthropogenic disturbance gradient in Brazilian Caatinga (modified from Arcoverde, 2012). Zero and 100 represent sites under the lowest and highest human disturbance pressure, respectively. General Linear Models indicate that *Dinoponera* nests presented lower values of seed predation (F = 57.08, df = 1, p < 0.001) and higher values of seedling richness (F = 74.30, df = 1, p < 0.001) and abundance (F = 84.52, df = 1, p < 0.001) as compared to control areas, but anthropogenic disturbance did not affect these variables (seed predation: F = 0.37, df = 1, p = 0.543; seedling richness: F = 1.93, df = 1, p = 0.167; seedling abundance: F = 0.661, df = 1, p = 0.416).

are currently conducting a Long-Term Ecological Project (see www.peldcatimbau.org) investigating the interactive effects of chronic disturbance and decreasing rainfall on myrmecochory in Caatinga vegetation of Catimbau National Park (8°30'S; 37°20'W; 350–1100 m asl). The Park is an ideal system for investigating such interactive effects, as it has a substantial resident population that continues to use natural resources and has a steep gradient in mean annual rainfall from 1100 to 480 mm. We have estimated chronic anthropogenic disturbance using surrogates of human use (proximity to urban center, houses and roads) and direct metrics measured on our plots (number of livestock feces, length of livestock trails and number of stumps) combined in a disturbance index through a principal component analysis. For the rainfall gradient, we used mean annual precipitation within each plot obtained from the updated WorldClim global climate data repository (www.worldclim.org) with a 1-km resolution rainfall.

Preliminary results suggest that disturbance and rainfall have different effects on different myrmecochore species (Oliveira, unpublished data). For example, decreasing rainfall has a stronger negative effect on seed removal rate and removal distances of *Jatropha mutabilis* than does increasing chronic anthropogenic disturbance, whereas neither had an effect on either seed removal or removal distance for *Croton nepetaefolius* (Figure 6.4). As was the case in the Parnamirim study, *Dinoponera quadriceps* was the most important seed disperser in Catimbau, being responsible for 62 percent of removal events with a mean removal distance of 7.3 m (Oliveira, unpublished data). *Dinoponera quadriceps* reduced its dispersal services in dryer and more disturbed areas for *Jatropha mutabilis* diaspores (the species with the highest absolute and proportional elaiosome biomass), but it remained unaltered for *Croton nepetaefolius* (the species with the smallest absolute and proportional elaiosome biomass, Oliveira, unpublished data). Its abundance declined markedly with decreasing rainfall, but was not affected by disturbance (Figure 6.5, Arcoverde, unpublished data). However, its removal rates were lower at sites with lower rainfall and higher disturbance (Figure 6.4a, b), which suggests that disturbance changes ant behavior toward diaspores. We found no interaction between disturbance and rainfall (Figures 6.4 and 6.5).

Our findings suggest that myrmecochory in Caatinga is sensitive to variations in both rainfall and chronic disturbance, although rainfall has a stronger impact. This is consistent with effects on other taxa (plants, ants and other insects) and other ant-plant interactions (extrafloral nectary-bearing plants and ant attendants) we are investigating in our LTER project. Thus, the rapid climatic change in the region will probably have stronger effects on Caatinga biodiversity, functions and services than continued chronic disturbance.

Conclusions and Future Directions

The studies described here have provided important information on the distribution of myrmecochorous species in Brazilian Caatinga, their interactions with ants and the influence of chronic anthropogenic disturbance and decreasing rainfall on this

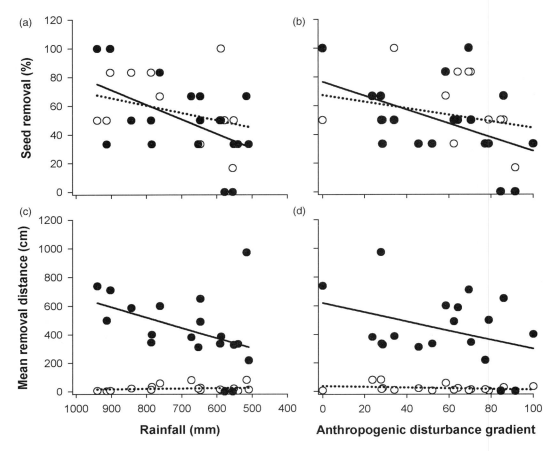

Figure 6.4. Removal rate (a, b) and mean removal distance (c, d) of seeds from *Jatropha mutabilis* (black circles) and *Croton nepetaefolius* (open circles) by seed disperser ant species over rainfall and anthropogenic disturbance gradients in Brazilian Caatinga (Oliveira, unpublished data). Zero and 100 represent sites under the lowest and highest human disturbance pressure, respectively. General Linear Models indicate that removal rate of *Jatropha mutabilis* seeds was affected by both rainfall and disturbance (whole model: $F_{3,15}$ = 5.67, p < 0.01, rainfall: p < 0.01, disturbance: p < 0.01, interaction rainfall*disturbance: p = 0.85), while removal distance was related only to rainfall (whole model: $F_{3,15}$ = 3.26, p = 0.05, rainfall: p= 0.04, disturbance: p= 0.07, interaction rainfall*disturbance: p = 0.23). For *Croton nepetaefolius* results were not significant for either seed removal (whole model: $F_{3,15}$ = 2.70, p = 0.08, rainfall: p= 0.14, disturbance: p= 0.20, interaction rainfall*disturbance: p = 0.08) or removal distance (whole model: $F_{3,15}$ = 1.36, p = 0.29, rainfall: p= 0.58, disturbance: p= 0.23, interaction rainfall*disturbance: p = 0.17). In all models the interaction between rainfall and disturbance was not significant.

dispersal syndrome. All these results together indicate that, despite a high diversity of seed dispersing ant species, Caatinga has only a limited number of high-quality seed dispersers, and these are functionally very different from other disperser species in terms of distance dispersal and response to chronic anthropogenic disturbance.

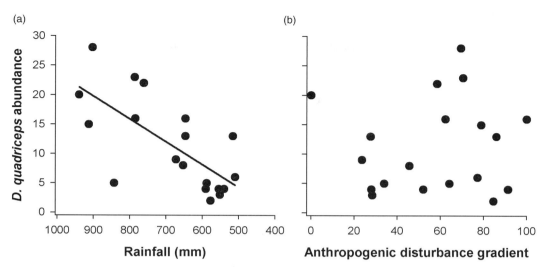

Figure 6.5.   *Dinoponera quadriceps* abundance over rainfall (a) and anthropogenic disturbance (b) gradients in Brazilian Caatinga (Arcoverde, unpublished data). Zero and 100 represent sites under the lowest and highest human disturbance pressure, respectively. General Linear Model indicate that *Dinoponera quadriceps* abundance decreased with decreasing rainfall, but it was not affected by disturbance; interaction between rainfall and disturbance was not significant (whole model: $F_{3,15} = 4.99$, $p = 0.01$, rainfall: $p < 0.01$, disturbance: $p = 0.82$, interaction rainfall*disturbance: $p = 0.76$).

In particular, the species responsible for most long-distance dispersal is highly sensitive to disturbance, and so dispersal distance decreases markedly with increasing disturbance. However, the post-dispersal services of this species are maintained in disturbed areas, decreasing seed predation and increasing seedling recruitment in the nest vicinity. Rainfall appears to have a stronger effect on removal rates and distances than does disturbance, both declining with increasing aridity. Increasing disturbance and declining rainfall have important longer-term implications for recruitment of myrmecochorous plants, and therefore future vegetation composition and structure.

Our studies have also provided important insights into myrmecochory more generally. Despite the prominence of myrmecochory as a dispersal syndrome globally, the benefits to plants in having their seed dispersed by ants remain unclear. Potential benefits include directed dispersal to nutrient-enriched microsites (Beattie, 1985; Hanzawa et al., 1988), protection from predators (Heithaus, 1981; Smith et al., 1989) or fire (Berg, 1975; Hughes & Westoby, 1992) and distance dispersal (Andersen, 1988; Higashi et al., 1989; Boyd, 2001). A notable feature of myrmecochory is that it occurs almost exclusively in plants of small stature (herbs and shrubs), and we have shown that this is also the case for myrmecochory in Caatinga euphorbs (Leal et al., 2015). We point out that distance dispersal is the only proposed benefit where plant stature is relevant, which strongly suggests that the primary benefit of myrmecochory to plants is related to distance dispersal, and in particular the avoidance of parental and sibling competition (Leal et al., 2015).

The apparent primacy of distance dispersal as a benefit of myrmecochory to plants places a premium on large-bodied disperser ant species, given that dispersal distances are so strongly related to body size (Ness et al., 2004). Notably, these high-quality seed dispersers are especially sensitive to habitat disturbance and apparently to increased aridity. This represents a double jeopardy for Caatinga myrmecochores, given high human pressure and projected lower rainfall under global climate change.

A priority for future research is to examine in more detail the interactive effects of increasing disturbance and aridity on the seed-dispersal services provided by ants in Caatinga not observed so far. We also need a better understanding of the consequences of such effects for plant populations. For example, a feasible scenario is that the creation of open areas by disturbance has a negative impact on high-quality seed-disperser ants, and that this has a negative impact on larger-seeded plant species that tend to be larger in stature, thus maintaining habitat openness. Finally, we also need a better understanding of the growth and survival of seedlings located near ant nest entrances, disentangling the effects of changed soil conditions and protection from herbivores. Such future research would benefit our under-standing of myrmecochory not only in Caatinga, but also in other arid environ-ments where there are just a few high-quality ant dispersers that are highly sensitive to habitat disturbance and climate change.

References

Ab'Sáber, A. N. (1999). Dossiê Nordeste seco. *Estudos Avançados*, 13, 1–59.

Ahrends, A., Burgess, N. D., Milledge, S. A. H., Bulling, M. T., Fisher, B., Smart, J. C. R., Clarke, G. P., Mhoro, B. E. and Lewis, S. L. (2010). Predictable waves of sequential forest degradation and biodiversity loss spreading from an African city. *Proceedings of the National Academy of Sciences of the United States of America*, 107, 14556–14561.

Andersen, A. N. (1988). Dispersal distance as a benefit of myrmecochory. *Oecologia*, 75, 507–511.

Andersen, A. N. and Morrison, S. (1998). Myrmecochory in Australia's seasonal tropics: effects of disturbance on distance dispersal. *Australian Journal of Ecology*, 23, 483–491.

Anderson-Teixeira, K. J., Miller, A. D., Mohan, J. E., Hudiburg, T. W., Duval, B. D. and DeLucia, E. H. (2013). Altered dynamics of forest recovery under a changing climate. *Global Change Biology*, 19, 2001–2021.

Andrade-Lima, D. (1989). *Plantas da Caatinga*. Rio de Janeiro: Academia Brasileira de Ciências.

Aranda-Rickert, A. and Fracchia, S. (2012). Are subordinate ants the best seed dispersers? Linking dominance hierarchies and seed dispersal ability in myrmecochory interaction. *Arthropod-Plant Interactions*, 6, 297–306.

Arcoverde, G. B. (2012). Efeitos de perturbações antrópicas na proteção de sementes e estabelecimento de plântulas em ninhos de Dinoponera quadriceps Santschi (Hymenoptera: Formicidae) no semi-árido nordestino. Master thesis, Universidade Federal de Pernambuco, Recife.

Barroso, G. M., Morim, M. P., Peixoto, A. L. and Ichaso, C. L. F. (1999). *Frutos e sementes: morfologia aplicada à sistemática de dicotiledôneas.* Viçosa: Universidade Federal de Viçosa.

Beattie, A. J. (1985). *The evolutionary ecology of ant-plant mutualisms.* Cambridge: Cambridge University Press.

Beattie, A. J. and Culver, D. C. (1981). The guild of myrmecochores in the herbaceous flora of West Virginia forests. *Ecology*, 62, 107–115.

Berg, R. Y. (1975). Myrmecochorous plants in Australia and their dispersal by ants. *Australian Journal of Botany*, 62, 714–722.

Bond, W. and Slingsby, P. (1983). Seed dispersal by ants in Cape shrublands and its evolutionary implications. *South Africa Journal of Science*, 79, 231–233.

Boyd, R. S. (2001). Ecological benefits of myrmecochory for the endangered chaparral shrub *Fremontodendron decumbens* (Sterculiaceae). *American Journal of Botany*, 88, 234–241.

Christianini, A. V, Mayhé-Nunes, A. J. and Oliveira P. S. (2007). The role of ants in the removal of non-myrmecochorous diaspores and seed germination in a Neotropical savanna. *Journal of Tropical Ecology*, 23, 343–351.

Davidar, P., Sahoo, S., Mammen, P. C. et al. (2010). Assessing the extent and causes of forest degradation in India: where do we stand? *Biological Conservation*, 143, 2937–2944.

Farwig, N. and Berens, D. G. (2012). Imagine a world without seed dispersers: a review of threats, consequences and future directions. *Basic and Applied Ecology*, 13, 109–115.

García-Valdés, R., Svenning, J. C., Zavala, M. A., Purves, D. W. and Araújo, M. B. (2015). Evaluating the combined effects of climate and land-use change on tree species distributions. *Journal of Applied Ecology*, 52, 902–912.

Gariglio, M. A., Sampaio, E. V. S. B., Cestaro, L. A. and Kageyama, P. Y. (2010). *Uso sustentável e conservação dos recursos florestais da caatinga.* Brasília: Serviço Florestal Brasileiro.

Garrido, J. L., Rey, P. J., Cerdá, X. and Herrera, C. M. (2002). Geographical variation in diaspore traits of an ant-dispersed plant (*Helleborus foetidus*): are ant community composition and diaspore traits correlated? *Journal of Ecology*, 90, 446–455.

Gibb, H., Sanders, N. J., Dunn, R. R. et al. (2015). Climate mediates the effects of disturbance on ant assemblage structure. *Proceedings of the Royal Society of London B: Biological Sciences*, 282, 20150418.

Gorb, E. and Gorb, S. (2003). *Seed dispersal by ants in a deciduous forest ecosystem. Mechanisms, strategies, adaptation.* Dordrecht: Kluwer Academic Publishers.

Gove, A. D., Majer, J. D. and Dunn, R. R. (2007). A keystone ant species promotes seed dispersal in a "diffuse" mutualism. *Oecologia*, 153, 687–697.

Griz, L. M. S. and Machado, I. C. (2001). Fruiting phenology and seed dispersal syndromes in Caatinga, a tropical dry Forest in the Northeast of Brazil. *Journal of Tropical Ecology*, 17, 303–321.

Hanzawa, F. M., Beattie, A. J. and Culver, D. C. (1988). Directed dispersal: demographic analysis of an ant-seed mutualism. *American Naturalist*, 131, 1–13.

Heithaus, E. R. (1981). Seed predation by rodents on three ant-dispersed plants. *Ecology*, 62, 136–145.

Higashi, S., Tsuyuzaki, S. and Ohara, I. F. (1989). Adaptive advantages of ant-dispersed seeds in the myrmecochorous plant *Trillium tschonoskii* (Liliaceae). *Oikos*, 54, 389–394.

Hoffmann, B. D. and Andersen, A. N. (2003) Responses of ants to disturbance in Australia with particular reference to functional groups. *Austral Ecology*, 28, 444–464.

Hughes, L. and Westoby, M. (1992). Fate of seeds adapted for dispersal by ants in Australian sclerophyll vegetation. *Ecology*, 73, 1285–1299.

Instituto Brasileiro de Geografia e Estatística, 1985 (1985). *Atlas Nacional do Brasil: Região Nordeste*. Rio de Janeiro: IBGE.

Jentsch, A., Kreyling, J., Boettcher-Treschkow, J. and Beierkuhnlein, C. (2009). Beyond gradual warming: extreme weather events alter flower phenology of European grassland and heath species. *Global Change Biology*, 15, 837–849.

Laurance, W. F. and Peres, C. A. (2006). *Emerging threats to tropical forests*. Chicago: University of Chicago Press.

Leal, I. R., Filgueiras, B. K. C., Gomes, J. P. and Andersen, A. N. (2012). Effects of habitat fragmentation on ant richness and functional composition in Atlantic Forest of northeastern Brazil. *Biodiversity and Conservation*, 21, 1687–1701.

Leal, I. R., Leal, L. C. and Andersen, A. N. (2015). The benefits of myrmecochory: a matter of stature. *Biotropica*, 47, 281–285.

Leal, I. R. and Oliveira, P. S. (1998). Interactions between fungus-growing ants (Attini), fruits and seeds in cerrado vegetation in Southeast Brazil. *Biotropica*, 30, 170–178.

Leal, I. R., Silva, J. M. C., Tabarelli, M. and Lacher, T. E. (2005). Changing the course of biodiversity conservation in the Caatinga of Northeastern Brazil. *Conservation Biology*, 19, 701–706.

Leal, I. R., Wirth, R. and Tabarelli, M. (2007). Seed dispersal by ants in the semi-arid Caatinga of North-east Brazil. *Annals of Botany*, 99, 885–894.

Leal, L. C., Andersen, A. N. and Leal, I. R. (2014b). Anthropogenic disturbance reduces seed dispersal services for myrmecochorous plants in the Brazilian Caatinga. *Oecologia*, 174, 173–181.

Leal, L. C., Lima-Neto, M. C., Oliveira, A. F. M., Andersen, A. N. and Leal, I. R. (2014a). Myrmecochores can target high-quality disperser ants: variation in elaiosome traits and ant preferences for myrmecochorous Euphorbiaceae in Brazilian Caatinga. *Oecologia*, 174, 493–500.

Lengyel, S., Gove, A. D., Latimer, A. M., Majer, J. D. and Dunn, R. R. (2009). Ants sow the seeds of global diversification in flowering plants. *PLoS ONE*, 4(5), e5480.

Levey, D. J. and Byrne, M. M. (1993). Complex ant-plant interactions: rain forest ants as secondary dispersers and post-dispersal seed predators. *Ecology*, 74, 1802–1812.

Lôbo, D., Tabarelli, M. and Leal, I. R. (2011). Realocation of *Croton sonderianus* (Euphorbiaceae) seeds by *Pheidole fallax* Mayr (Formicidae): a case of post-dispersal seed protection by ants. *Neotropical Entomology*, 40, 440–444.

Lorenzi, H. (1998). *Árvores brasileiras: manual de identificação e cultivo de plantas arbóreas nativas do Brasil*. Nova Odessa: Editora Plantarum.

Magrin, G. O., Marengo, J. A., Boulanger, J. P. et al. (2014). Central and South America. In *Climate change 2014: impacts, adaptation, and vulnerability. Part B: regional aspects. contribution of working group II to the fifth assessment report of the intergovernmental panel on climate change*, eds. V. R. Barros, C. B. Field, D. J. Dokken, M. D. Mastrandrea, K. J. Mach, T. E. Bilir, M. Chatterjee, K. L. Ebi, Y. O. Estrada, R. C. Genova, B. Girma, E. S. Kissel, A. N. Levy, S. MacCracken, P. R. Mastrandrea and L. L. White. Cambridge: Cambridge University Press, pp. 1499–1566.

Manzaneda, A. J. and Rey, P. J. (2008). Geographic variation in seed removal of a myrmecochorous herb: influence of variation in functional guild and species composition of disperser assemblage through spatial and temporal scale. *Ecography*, 31, 583–591.

Manzaneda A. J. and Rey P. J. (2012). Geographical and interspecific variation and the nutrient-enrichment hypothesis as an adaptive advantage of myrmecochory. *Ecography*, 35, 322–332.

Marini, L., Bruun, H. H., Heikkinen, R. K., Helm, A., Honnay, O., Krauss, J., Kuhn, I., Lindborg, R., Partel, M. and Bommarco, R. (2012). Traits related to species persistence and dispersal explain changes in plant communities subjected to habitat loss. *Diversity and Distributions*, 18, 890–908.

Martorell, C. and Peters, E. M. (2005). The measurement of chronic disturbance and its effects on the threatened cactus *Mammilaria pectinifera*. *Biological Conservation*, 124, 199–207.

Medeiros, S. S., Cavalcante, A. M. B., Perez Marin, A. M., Tinôco, L. B. M., Salcedo, I. H. and Pinto, T. F. (2012). *Sinopse do censo demográfico para o semiárido brasileiro*. Campina Grande: Instituto Nacional de Seminário.

Ministério do Meio Ambiente (2011). *Monitoramento do desmatamento nos biomas brasileiros por satélite: monitoramento do bioma Caatinga de 2008 a 2009*. Brasília: Ministério do Meio Ambiente.

Moro, M. F., Lughadha, E. N., Filer, D. L., Araújo, F. S. and Martins, F. R. (2014). A catalogue of the vascular plants of the Caatinga Phytogeographical Domain: A synthesis of floristic and phytosociological surveys. *Phytotaxa*, 160, 1–30.

Ness, J. H., Bronstein, J., Andersen, A. N. and Holland, J. N. (2004). Ant body size predicts dispersal distance of ant-adapted seeds: implications of small-ant invasions. *Ecology*, 85, 1244–1250.

Ness, J. H., Morin, F. and Giladi, I. (2009). Uncommon specialization in mutualism between a temperate herbaceous plant guild and the ant: are *Aphaenogaster* ants keystone mutualistics? *Oikos*, 118, 1793–1804.

Oliveira, F. M. P., Ribeiro-Neto, J. D., Andersen, A. N. and Leal, I. R. (2017). Chronic anthropogenic disturbance as a secondary driver of ant community structure: interactions with soil type in Brazilian Caatinga. *Environmental Conservation*, 44, 115–123.

Passos, L. and Oliveira, P. S. (2002). Ants affect the distribution and performance of *Clusia criuva* seedlings, a primarily bird-dispersed rainforest tree. *Journal of Ecology*, 90, 517–528.

(2003). Interactions between ants, fruits and seeds in a restinga forest in south-eastern Brazil. *Journal of Tropical Ecology*, 19, 261–270.

(2004). Interaction between ants and fruits of *Guapira opposite* (Nyctaginaceae) in a Brazilian sandy plain rainforest: ant effects on seeds and seedlings. *Oecologia*, 139, 376–382.

Pizo, M. A. and Oliveira, P. S. (2000). The use of fruits and seeds by ants in the Atlantic forest of southeast Brazil. *Biotropica*, 32, 851–861.

Ponce-Reyes, R., Nicholson, E., Baxter, P.W.J., Fuller, R.A. and Possingham, H. (2013). Extinction risk in cloud forest fragments under climate change and habitat loss. *Diversity and Distributions*, 19, 518–529.

Ribeiro, E. M. S., Arroyo-Rodríguez, V., Santos, B. A., Tabarelli, M. and Leal, I. R. (2015). Chronic anthropogenic disturbance drives the biological impoverishment of the Brazilian Caatinga vegetation. *Journal of Applied Ecology*, 52, 611–620.

Ribeiro, E. M. S., Arroyo-Rodríguez, V., Santos, B. A., Tabarelli, M. & Leal, I. R. (2016). Phylogenetic impoverishment of plant communities following chronic human disturbances in the Brazilian Caatinga. *Ecology*, 97, 1583–1592.

Ribeiro-Neto, J. D., Arnan, X., Tabarelli, M. and Leal, I. R. (2016). Chronic anthropogenic disturbance causes homogenization of plant and ant communities in the Brazilian Caatinga. *Biodiversity and Conservation*, 25, 943–956.

Sampaio, E. V. S. B. (1995). Overview of the Brazilian Caatinga. In *Seasonal dry tropical forests*, eds. S. H. Bullock, H. A. Mooney and E. Medina. Cambridge: Cambridge University Press, pp. 35–63.

Schär, C., Vidale, P. L., Lüthi, D., Frei, C., Häberli, C., Liniger, M. A. and Appenzeller, C. (2004). The role of increasing temperature variability in European summer heatwaves. *Nature*, 427, 332–336.

Schulz, K., Voigt, K., Beusch, C., Almeida-Cortez, J. S., Kowarik, I., Walz, A. and Cierjacks, A. (2016). Grazing deteriorates the soil carbon stocks of Caatinga forest ecosystems in Brazil. *Forest Ecology and Management*, 367, 62–70.

Singh, S. P. (1998). Chronic disturbance, a principal cause of environmental degradation in developing countries. *Environmental Conservation*, 25, 1–2.

Smith, B. H., Forman, P. D. and Boyd, A. E. (1989). Spatial patterns of seed dispersal and predation of 2 myrmecochorous forest herbs. *Ecology*, 70, 1649–1656.

Stuble, K. L., Patterson, C. M., Rodriguez-Cabal, M. A., Ribbons, R. R., Dunn, R. R. and Sanders, N. J. (2014). Ant-mediated seed dispersal in a warmed world. *Peer J.*, 2, e286.

Sunderland, T., Apgaua, D., Baldauf, C. et al. (2015). Global dry forests: a prologue. *International Forestry Review*, 17, 1–9.

Tabarelli, M., Vicente, A. and Barbosa, D. C. A. (2003) Variation of seed dispersal spectrum of woody plants across a rainfall gradient in northeastern Brazil. *Journal of Arid Environments*, 53, 197–210.

Travis, J. M. J. (2003). Climate change and habitat destruction: a deadly anthropogenic cocktail. *Proceedings of the Royal Society of London B: Biological Sciences*, 270, 467–473.

van der Pijl, L. (1982). *Principles of dispersal in higher plants*. Berlin: Springer Verlag.

van Roosmalen, M. G. M. (1985). *Fruits of the Guianan flora*. Utrecht: Institute of Systematic Botany.

Warren II, R. J., Bahn, V. and Bradford, M. A. (2011). Temperature cues phenological synchrony in ant-mediated seed dispersal. *Global Change Biology*, 17, 2444–2454.

Westoby, M., French, K., Hugdes, L., Rice, B. and Rodgerson, L. (1991). Why do more plant species use ants for dispersal on infertile compared with fertile soils? *Australian Journal of Ecology*, 16, 445–455.

7   Anthropogenic Disturbances Affect the Interactions between Ants and Fleshy Fruits in Two Neotropical Biodiversity Hotspots

Paulo S. Oliveira, Alexander V. Christianini, Ana G. D. Bieber, and Marco A. Pizo*

Introduction

The main ecological function of fleshy fruits is probably advertisement, such that they look or smell attractive enough to entice a particular type of animal (frugivore) to eat them (Schaefer & Ruxton, 2011). By feeding on the nutritious fleshy portion of fruits and seeds (pericarp, pulp, aril), mobile frugivores frequently detach them from the parent plant and deposit the seeds someplace else. As such, seed dispersal is essentially a mutualism in which nutrition of a frugivore is exchanged for mobility of a plant's offspring (Jordano, 2000). The tropical region not only harbours a great diversity of plant and animal species, but also contains a huge variety of fruit traits (colour, morphology and chemistry) associated with many types of seed dispersal strategies (e.g. Galetti et al., 2011). Seed dispersal by animals clearly predominates, and it is estimated that nearly 90 per cent of tropical forest eudicots have fleshy fruits/seeds (i.e. diaspores, sensu van der Pijl, 1969) that rely on vertebrate frugivores for dissemination (Fleming et al., 1987; Levey et al., 1994). Individual plant species, however, are rarely adapted to seed dispersal by only one particular species of frugivore (Howe & Smallwood, 1982). Indeed, because most plant species are involved with numerous dispersers, a clear picture of the seed rain of a given plant species frequently entails understanding seed movements mediated by dispersal agents not only from multiple animal species, but also from several animal groups as well (e.g. Vander Wall & Longland, 2005).

* We thank André Freitas, Javier Ibarra, Sebastián Sendoya and Paulo Silva for help with the illustrations, and most especially Luísa Mota for the drawings and Luciana Passos for the colour photos of the ant-*Clusia* system. Comments from Nádia Barbosa and two reviewers substantially improved the manuscript. Our studies were supported by the National Council for Scientific and Technological Development (CNPq), the São Paulo Research Foundation (FAPESP), the São Paulo Forestry Institute (IF), and the Research Funds of Campinas State University (FAEPEX).

Detailed field observations of animal behaviour and seed fate have made increasingly apparent that seed movements can involve sequential steps, each performed by a distinct dispersal agent, in a process known as diplochory or two-phase dispersal (Vander Wall & Longland, 2005). Phase one dispersal refers to the initial movement of the seed away from the parent plant, which can result in avoidance of density-dependent seed and seedling mortality near the parent (Janzen, 1970). Phase two often involves subsequent seed movements that may result in seed deposition in predictable, favourable microsites where seedling establishment is increased (i.e. directed dispersal). Two-phase dispersal systems can be regarded as special cases of secondary dispersal, which deserve special attention since they combine two dispersal modes that can offer more benefits to plants (Vander Wall & Longland, 2005; Camargo et al., 2016).

Ants as Participants in Two-Phase Dispersal Systems

Despite the high abundance of frugivores in tropical habitats, large amounts of fruits fall to the ground spontaneously (Jordano, 2000; Laman, 1996). In addition, many fleshy fruits can reach the ground dropped by vertebrate frugivores while foraging, or in their faeces (Roberts & Heithaus, 1986; Byrne & Levey, 1993; Kaspari, 1993; Pizo & Oliveira, 1999, 2000). This huge amount of fallen, nutritious fleshy fruits coupled with the remarkable diversity and density of tropical ground-dwelling ants (e.g. Longino et al., 2002) set up the ecological scenario for the observed prominence of ant-fruit interactions on the floor of tropical ecosystems (Figure 7.1), including forests and savannas (Levey & Byrne, 1993; Leal & Oliveira, 1998; Böhning-Gaese et al., 1999; Pizo & Oliveira, 2000; Pizo, 2008; Passos & Oliveira, 2003; Christianini et al., 2012). Field observations and experiments have demonstrated that ants actively participate in phase two of a number of primarily vertebrate-dispersed plants, providing a range of benefits to seeds and seedlings that may include protection of seeds against pathogens and predators, increased germination success and directed dispersal to nutrient-rich ant nests where seedling establishment is improved (see Beattie 1985; Rico-Gray & Oliveira, 2007; and included references). Reciprocal benefits to ant colonies from consumption of fallen lipid-rich fleshy seeds have recently been demonstrated through improved larval development, thus confirming the mutualistic nature of this interaction (Bottcher & Oliveira, 2014).

As opposed to true myrmecochory (i.e. specialized dispersal by ants) in which a distinct ant-attractive seed appendage (elaiosome) induces transportation (Chapters 5 and 6), interactions between ants and fallen fleshy diaspores (non-myrmecochorous, primarily vertebrate-dispersed) are more generalized and may involve a diversity of ants attending a broad range of fruit/seed types differing in morphology and chemical composition of the edible portion (but see Pizo & Oliveira, 2001; Christianini et al., 2012).

ANT SPECIES

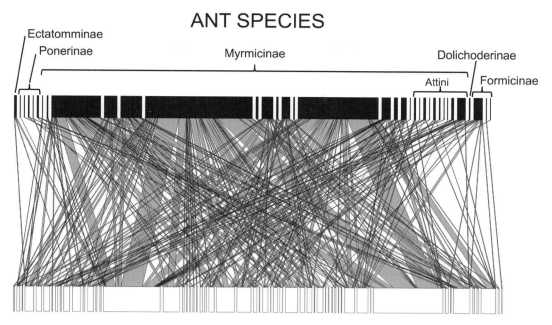

PLANT SPECIES

Figure 7.1.   Network representation of the interactions between ants and primarily vertebrate-dispersed diaspores on the floor of a lowland Brazilian rainforest. Each node is species-specific and identifies an ant (black) or a plant (white); node size (squares) is proportional to species' contribution to network structure and line thickness indicates the frequency of the interaction. Ant species in the subfamily Myrmicinae are by far the most frequently recorded interacting with fallen fleshy diaspores. Based on data from Pizo and Oliveira (2000).

Habitat Fragmentation and Seed Dispersal Processes

Owing mostly to human-induced habitat alterations, plant-disperser interactions are currently susceptible to disruption in many tropical ecosystems, and the magnitude of this interference for seed dispersal and plant communities remains uncertain (McConkey et al., 2012). Habitat loss and fragmentation caused by agriculture, cattle ranching, urban expansion, roads and different types of commercial interests have changed plant and animal abundances in many tropical environments (Laurance et al., 2010). Animal population declines are aggravated by hunting activities aimed at certain groups that are important seed dispersers – birds, ungulates and primates are at particularly high risk in tropical forests (Endo et al., 2010) and savannas (El Bizri et al., 2015).

Because the majority of tropical plant species have animal-dispersed fleshy diaspores (Levey et al., 1994), the loss of frugivorous vertebrates could drastically reduce seed dispersal and recruitment and thus negatively affect plant populations and communities in tropical fragmented biomes (McConkey et al., 2012). In addition to affecting movement behaviour of vertebrate frugivores (e.g. Pinto & Keitt, 2008), habitat fragmentation can also affect the occurrence patterns of insect species participating in two-phase dispersal systems such as dung-beetles and large ponerine ants, which tend to avoid habitat edges due to biotic or abiotic conditions (Santos-Heredia, et al. 2011; Christianini & Oliveira, 2013).

The future of tropical biodiversity depends largely on the extent to which natural ecological processes can be maintained in isolated habitat fragments. In the past few decades it has become increasingly evident that invertebrates are crucial for the maintenance of viable biological communities through their numerous interspecific interactions (Janzen, 1977), including seed dispersal (Christianini et al., 2014). Increased emphasis has thus been placed on the knowledge of invertebrate biology and natural history, threats to their ecological role in human-disturbed habitats, as well as on the incorporation of the so-called little things into conservation programmes (Wilson, 1987).

This chapter examines how human-induced habitat change can affect the interactions between ants and fallen fleshy diaspores in two Brazilian biodiversity hotspots, the Atlantic rainforest and the cerrado savanna (Myers et al., 2000). Since habitat loss and fragmentation can have pervasive impacts on the components of multi-step seed dispersal systems, the outcomes of these interactions are also expected to be affected (McConkey et al., 2012). We focus on plant-disperser interactions from an integrated perspective, taking into account both plant (diaspore traits) and animal attributes (behaviour of distinct dispersal agents) that contribute to patterns of seed fate and seedling establishment. We argue that despite the negative effects of anthropogenic disturbance on ant-mediated dispersal systems, endurance of many ant species in fragmented habitats will likely make them increasingly more relevant for plant regeneration processes in vertebrate-impoverished environments.

Two Threatened Neotropical Hotspots: The Cerrado Savanna and the Atlantic Rainforest

Positioned between Amazonia and the Atlantic forests, the so-called *cerrados* comprise the largest savanna region in South America ($\approx$ 2 million km^2), covering nearly 22 per cent of the land surface of Brazil and small parts of Bolivia and Paraguay (Oliveira & Marquis, 2002). The cerrado domain is characterized by a mosaic of contrasting vegetation types, including open grasslands, woody savannas and dense semideciduous forests. The cerrado itself refers only to the woody savannas, which encompass a continuum of vegetation physiognomies determined by the density of woody plants, ranging from extensive grasslands with scattered shrubs to forest woodlands with a scanty understory of herbs and subshrubs (Oliveira-Filho & Ratter, 2002). The cerrados host an enormous biodiversity with about 10,000 species of plants and over 1,250 species of vertebrates (Myers et al., 2000). The diversity of invertebrates is uncertain but insects have been estimated to be around

90,000 species (Dias, 1992). Continuous deforestation for cash crops such as soybean and corn, cattle ranching, and charcoal production makes the cerrado one of the most threatened and overexploited natural landscapes in Brazil; the remaining stands are in a highly fragmented state and less than 20 per cent can be considered intact (Klink & Machado, 2005).

The Atlantic rainforest originally extended over nearly 1.5 million km², mainly along the Brazilian coast (92 per cent), but also reaching some areas of Paraguay and Argentina (Ribeiro et al., 2009). The Atlantic forest domain encompasses a range of vegetation types, namely, the rain, cloud, coastal sandy and semideciduous forests, as well as their associated non-forest ecosystems such as wetlands and scrublands that may grow on particular substrates (Eisenlohr & Oliveira-Filho, 2015). The biome is recognized as a main global biodiversity hotspot, hosting nearly 20,000 plant species and over 1,360 vertebrate species (Myers et al., 2000). Because of its location along the highly urbanized and populated Brazilian Atlantic coast (≈ 120 million people), this biome has suffered unparalleled levels of habitat loss and other anthropogenic disturbances since the sixteenth century with the European colonization and initial coastal agricultural settlements (Ribeiro et al., 2009). High levels of deforestation continue in most regions, particularly around metropolitan areas where habitat loss and fragmentation is often associated with hunting, logging, agriculture, pastures, and fire, which have led to the near extinction of a large proportion of Atlantic forest biodiversity. Only about 11–16 per cent of the original forest is estimated to remain in Brazil, in the form of an extremely fragmented landscape (Ribeiro et al., 2009).

Natural History of Ant-Diaspore Interactions

Ant Behaviour

Ground-dwelling ants (and some arboreal species as well) may exploit fallen fleshy diaspores from a variety of plant life forms, including trees, shrubs, palms, herbs, lianas, epiphytes, hemiepiphytes and parasites (Pizo et al., 2005; Christianini et al. 2012). Irrespective of the size (≈ 0.02–30 g) or type of reward (fruit pulp or seed aril), diaspores are usually discovered by ants in the leaf litter in less than 10 minutes. Our surveys using tuna and honey baits indicate that the use of fallen diaspores by the litter-foraging ant community (excluding fungus-growers) is opportunistic and matches the relative abundance of ant species on the floor (Pizo & Oliveira, 2000). Once a diaspore is discovered by ants, the outcome of the interaction will largely depend on the size of the diaspore relative to the size of the ant, as well as on the chemical composition of the diaspore's fleshy portion. These two factors will influence the type of ant attending the diaspore, its foraging behaviour towards the food reward and, ultimately, the ant-induced effects on seeds and seedlings.

Ants treat fallen fleshy diaspores in different ways. They can remove the entire diaspore, tear off pieces of the fleshy part (pulp, aril) or collect liquids from it. The following behavioural categories are seen, depending on the size of the ant: (1) large ants in the subfamily Ponerinae (*Pachycondyla*, *Odontomachus*, *Dinoponera*) and in the Myrmicinae tribe Attini (*Atta*, *Acromyrmex*) often individually remove diaspores of up to 1 g to their nests; (2) small- to medium-sized myrmicine ants (*Pheidole*, *Crematogaster*) usually recruit up to 100 workers that feed on the diaspore on the spot, although they sometimes remove diaspores ≤0.05 g; (3) *Solenopsis* species often cover the diaspore with soil before collecting liquid and solid food from the fleshy part. Large fruits (>>1 g) are normally consumed on the spot, irrespective of the size of the ant. If broken open, small-seeded heavy fruits (>10 g) may have their seeds and attached pulp removed by large ponerine ants, as well as by the large fungus-growers *Atta* and *Acromyrmex* (see Leal & Oliveira 1998; Pizo & Oliveira, 2000; Passos & Oliveira, 2003; Christianini et al., 2012).

Diaspore Traits Mediating Ant Attendance

In addition to the diaspore/ant size ratio, a key factor mediating the outcome of the interaction with ants is the chemical composition of the fleshy reward attached to the seed, in particular the lipid content. Analogous to what has been shown for true myrmecochoric seeds bearing lipid-rich elaiosomes (Hughes & Westoby, 1992), the interaction between ants and non-myrmecochorous diaspores are largely mediated by the fatty acids of the fleshy portion (fruit pulp or seed aril), whose composition is indeed very similar to that found in the seed elaiosomes of true myrmecochores (Beattie 1985; Hughes et al., 1994; Pizo & Oliveira, 2001).

Our field work in Atlantic rainforest and cerrado savanna has shown that ant response to these diaspore features (size and lipid content) will largely determine ant-derived effects on seeds and seedlings. Predominantly carnivorous ponerine ants hunt for arthropod prey and use nutritious diaspores as a secondary food source (e.g. Horvitz & Beattie, 1980; Fourcassié & Oliveira, 2002). These large ants prefer protein- and lipid-rich fleshy diaspores that they transport to their nests (≈ 10 m away), feed the fleshy portion to larvae and then discard the viable seeds in the nest vicinity (Passos & Oliveira, 2004; Christianini & Oliveira, 2010, 2013; Bottcher & Oliveira, 2014). Around the nest entrance, more suitable physical (increased penetrability) and chemical (nutrient-enriched) soil properties may facilitate seedling establishment and growth compared to surrounding environment (Horvitz, 1981; Passos & Oliveira, 2002, 2004). Directed dispersal to ant nests can also render an additional benefit to seeds if removal of fallen diaspores decreases the risk of seed predation by rodents and insects beneath parent plants (Pizo 1997; Pizo & Oliveira, 1998). Lipid-rich diaspores too large (>3 g) to be transported by individual ants often have the piecemeal removal of the fleshy material by recruited ants on the spot (Pizo & Oliveira, 2001). In such cases, the seeds can be entirely cleaned from pulp or aril in less than 24 h, reducing fungi infestation of seeds and increasing germination success (Pizo et al., 2005, and included references).

Lipid-poor diaspores are rarely attended by ponerines and attract more frequently myrmicine ants, most especially fungus-growing attines (Pizo & Oliveira, 2001). A single leaf-cutter ant colony can remove thousands of plant diaspores in a few days to its nest at distances of up to 100 m (Roberts & Heithaus, 1986; Dalling & Wirth, 1998; Wirth et al., 2003; Christianini & Oliveira, 2009). Some seeds may be lost on the way back to the nest, and many are often discarded around nest entrances or in deep middens. Although traditionally considered pests, the activity of attine ants (including leaf-cutters) at fallen fleshy diaspores has already been shown to positively affect seed germination through seed cleaning (Oliveira et al., 1995; Leal & Oliveira, 1998), by reshaping the seed rain generated by primary dispersers, and through seedling/juvenile distribution (Farji-Brener & Ghermandi, 2004; Christianini & Oliveira, 2009). Because leaf-cutter ants often change light and soil conditions near nests (Meyer et al., 2011), they may influence the germination and growth of small, often dormant seeds of pioneer plants (Dalling & Wirth, 1998; Christianini & Oliveira, 2009), and simultaneously decrease the abundance of preferred harvested species due to severe leaf damage (Corrêa et al., 2010). Since leaf-cutter ants often benefit from human disturbance, their role as ecosystem engineers mediating plant regeneration processes in human-modified landscapes can be relevant (Chapters 4 and 18).

In the following sections we report distinctive case studies in cerrado and Atlantic rainforest that document ant-induced effects on seed fate and seedling establishment of non-myrmecochorous plants. Potential impacts of habitat disturbance and fragmentation on ant-diaspore interactions are inferred from field experiments investigating edge effects on ant-derived services to seeds and seedlings in cerrado, and from experiments using lipid-rich synthetic fruits to evaluate patterns of ant attendance in undisturbed and disturbed Atlantic forest sites.

Ant-Diaspore Interactions in Cerrado Savanna

Similar to what has been reported for tropical forests (Fleming et al., 1987), most shrubs and trees from the cerrado produce fruits and seeds adapted for dispersal by vertebrate frugivores (Gottsberger & Silberbauer-Gottsberger, 1983). Despite the rarity of true myrmecochorous plants in cerrado, interactions between ants and fleshy diaspores are prominent in this savanna. According to a 15-month survey in a cerrado reserve in southeastern Brazil, at least 71 species of ants interact with fallen diaspores from 38 species of plants (Christianini et al., 2012). Of course not all of these interactions would lead to seed dispersal and some ants have a granivorous diet, which may constrain recruitment of certain plants (Ferreira et al., 2011). More recently, however, detailed studies reveal an overlooked positive role of ants in plant regeneration in the cerrado.

Case Studies

Xylopia aromatica (Annonaceae)

Fallen fleshy diaspores in cerrado are frequently attended by ponerine ants in the genera *Dinoponera*, *Pachycondyla* and *Odontomachus* (Figure 7.2a). These large ants present a series of traits that make them high-quality dispersers, such as being solitary predators/scavengers that promptly retrieve fleshy fruits and seeds to their nests, discarding the intact viable seeds at refuse dumps (Fourcassié & Oliveira, 2002; Passos & Oliveira, 2002). Both in Atlantic forests and cerrado, large 'poneromorph' ants (sensu Bolton, 2003) are particularly attracted to fallen seeds coated by a lipid-rich aril (Pizo & Oliveira, 2001; Christianini & Oliveira, 2010). Although ponerine ants are not the most common ants in interaction with fallen diaspores in cerrado, they directly influence the regeneration of non-myrmecochorous plants producing lipid-rich arillate seeds, such as the bird-dispersed tree *Xylopia aromatica* (33 per cent of lipids in the fleshy portion). The seeds are readily removed by at least eight species of birds that take away around 32 per cent of the seed crop from the canopy of *Xylopia* trees (Christianini & Oliveira, 2010; Figure 7.2a). Another 25 per cent of the seed crop falls beneath the parent canopy as ripe seeds or after being dropped by birds (ca. 43 per cent is unripe or is preyed on before dispersal). Ants from 30 species interact with the fallen seeds (Figure 7.2b), and in less than 24 h they remove over 80 per cent of the arillate seeds to their nests. *Xylopia* seedlings were more common near the nests of *Pachycondyla*, *Odontomachus* and *Dinoponera* than at control plots away from ant nests. These large ants retrieve the diaspores of *Xylopia* to their nests at distances ca. twofold farther than smaller ants such as *Pheidole* (mean of 1.6 m vs. 0.6 m). In this dispersal system, while birds have the premier role of disseminating seeds long distances (tens to hundreds of meters), ants may reshape the seed rain at small spatial scales (centimeters to few meters), changing local seed densities and providing a fine-tuning to dispersal after the coarse dispersal by birds (see Horvitz & Le Corff, 1993). Thus, birds and ants act as complementary dispersers of *Xylopia* seeds at different spatial scales (Figure 7.2c). Similar dispersal systems should be common for other lipid-rich non-myrmecochorous diaspores in the cerrado (see Christianini et al., 2007, 2012).

Miconia rubiginosa (Melastomataceae)

Studies by Leal & Oliveira (1998) in cerrado have indicated a widespread use of non-myrmecochorous plant diaspores as substrates for fungus culturing by attine ants. Controlled experiments indicated that these ants could increase seed germination by removing the fleshy pulp covering the seeds, decreasing pathogen infestation (Leal and Oliveira, 1998). But what is the role of these ants in plant regeneration? That question was addressed by a comparative study of the role of birds and ants in seed dispersal of the melastome *Miconia rubiginosa* (Christianini & Oliveira, 2009). Birds from 13 species feed on the sugar-rich fruits of *Miconia*

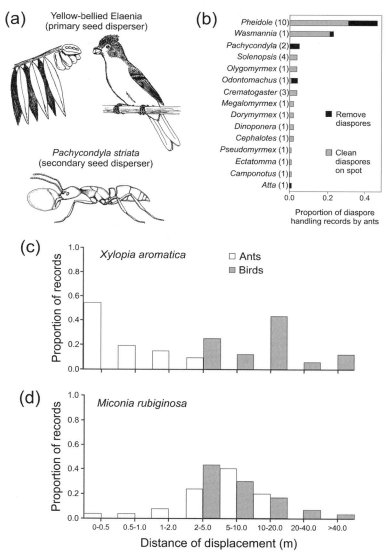

Figure 7.2. Birds and ants as complementary seed dispersers of fleshy diaspores in cerrado savanna. (a) The Yellow-bellied Elaenia (*Elaenia flavogaster*) consume the lipid-rich arillate seeds of *Xylopia aromatica* in the plant canopy, whereas fallen seeds are carried to the nest by large ponerines such as *Pachycondyla striata*. (b) Most ground-dwelling ants only inspect and clean *X. aromatica* seeds on spot, but large poneromorphs readily remove them (number of species in brackets). Birds and ants remove seeds at different scales. (c) While birds may remove *X. aromatica* seeds to at least tens of meters away from parent trees, ants usually act as short-distance dispersers (mostly < 2 m) and reshape the seed shadow locally. (d) The sugar-rich fruit pulp of *Miconia rubiginosa* are very attractive to leaf-cutting ants (*Atta* spp.), which can remove thousands of fallen fruits and may even rival some short-distance bird dispersers. Adapted from Christianini and Oliveira (2009) with permission from Springer (d), and from Christianini and Oliveira (2010) with permission from John Wiley and Sons (b, c). Drawings: Luisa Mota.

and remove nearly 25 per cent of the seed crop away from parent plants. Bird activity in the canopy also drops up to 19 per cent of the ripe fruit crop beneath the parent plants, where the chances of plant recruitment are small. However, ants (12 species) are often attracted to fallen fruits and seeds embedded in bird droppings. Some ants (especially the leaf-cutters *Atta sexdens* and *A. laevigata*) remove these fallen diaspores to their nests up to ca. 20 m away (Figure 7.2d). Seed samples taken from refuse piles around *Atta* nest entrances and from bird droppings revealed that neither birds nor ants affected seed germination compared to controls (seeds directly removed from mature fruits). In the field, saplings of *M. rubiginosa* were more common around leaf-cutter ant nests than around adult *Miconia* or non-*Miconia* trees, suggesting that the area surrounding the ant nest is adequate for recruitment. In this system, leaf-cutter ants can thus act as seed rescuers by removing seeds from beneath parent plants, and also reshape the seed rain produced by birds (Christianini & Oliveira, 2009).

The two-phase dispersal systems here described for *X. aromatica* and *M. rubiginosa* strongly suggest that different ant groups may complement the role of vertebrate frugivores for the regeneration of non-myrmecochorous plants in cerrado.

Edge Effects Affecting Ant-Diaspore Interactions in Cerrado: The Case of Erythroxylum pelleterianum (Erythroxylaceae)

The rapid conversion of vast cerrado areas in recent decades has produced highly fragmented landscapes of the savanna, most of them with habitat remnants immersed in a matrix of pastures and croplands (Klink & Machado, 2005). While edge effects have long been considered among the main drivers of change in tropical forest remnants (Laurance et al., 2010), only more recently their consequences have been investigated in cerrado savanna (Mendonça et al., 2015). Since changes in abiotic conditions may influence ant distribution and activity in the cerrado (see Vasconcelos et al., 2006; Brandão et al., 2011), it is likely that edge effects influence the interactions between ants and fallen non-myrmecochorous diaspores. We tested this idea in a study with the common shrub *Erythroxylum pelleterianum*.

In a fragment of cerrado in southeastern Brazil, at least five species of birds feed on fruits of *Erythroxylum* and act as primary dispersers, but around 45 per cent of the fruits produced reach the ground beneath the parent plant (Christianini & Oliveira, unpublished data). Remains of the lipid-rich (68 per cent in dry mass) fruit pulp of *Erythroxylum* attached to the seeds attract several ant species. Small ants, such as *Pheidole*, *Solenopsis* and *Wasmannia*, usually clean the seed on the spot or remove diaspores short distances (often <1 m), while large ants such as the poneromorphs *Pachycondyla*, *Odontomachus* and *Dinoponera* readily remove the fallen diaspores to their nests up to 7 m away. These large poneromorphs remove the fleshy portion of the diaspores to feed larval nestmates and discard the intact seed thereafter in a midden where it germinates. Diaspore removal rates by ants are higher near fragment edges, but small-sized granivorous Myrmicinae ants (e.g. *Pheidole*, *Wasmannia*) account for the vast majority (80 per cent) of the interactions observed in the hotter and drier conditions at edges. On the other hand, in the comparatively mild interior of cerrado, interactions between ants and fallen *Erythroxylum* fruits are more evenly distributed among several ant groups, with a substantial contribution (30 per cent) from the large, seed-dispersing ponerines *Odontomachus*, *Pachycondyla* and *Dinoponera*. Therefore, due to more variable removal distances, seed treatments and places of seed deposition, *Erythroxylum* seeds can reach more diversified fates in the interior compared to cerrado edges.

Ant-derived benefits to seedlings of *E. pelleterianum* differed between cerrado locations. While seedlings growing near nests of *O. chelifer* survived better than control seedlings in the cerrado interior, no such effect was seen at edges (Figure 7.3a). The enhanced performance of seedlings growing in ant nests in the cerrado interior is likely due to combined effects of increased soil fertility (Christianini & Oliveira, unpublished data) and possibly protection against insect herbivores by patrolling ants (see further pages). The lower residence time of colonies of *Odontomachus chelifer* at edges compared to the interior of cerrado (Figure 7.3b) apparently does not allow a significant accumulation of nest debris, increasing the chance of seedling mortality near edges due to abiotic and biotic stressors (Christianini & Oliveira, 2013).

The increased fragmentation of cerrado landscapes and the consequent pervasive creation of edges are likely to negatively affect the regeneration of plants producing lipid-rich diaspores that attain enhanced dispersal in the interior of cerrado as a result of the interaction with large poneromorph ants. As our case study with *E. pelleterianum* suggests, poneromorph ants are more susceptible to the changes in abiotic conditions at cerrado edges. Nest persistence is lower at edges, which cascades to a decrease of ant-derived benefits to seedlings and affects the regeneration of cerrado. However, vegetation changes at the edges of cerrado may not be so predictable. The proliferation of leaf-cutter ants near edges of fragments (Vasconcelos et al., 2006) may also increase the removal of sugar-rich diaspores, benefiting the regeneration of melastome species (see also Lima et al., 2013). The net effect of *Atta* species on plant regeneration processes is not straightforward given that leaf-cutter ants also frequently defoliate young plants for fungus culturing (Corrêa et al., 2010). Changes in the interactions between ants and diaspores coupled with increased disturbance by fire and competition with invasive African grasses in the edges of cerrado (Mendonça et al., 2015) are likely to drive the small fragments of cerrado to an impoverished state in the long term.

Ant-Diaspore Interactions in the Atlantic Rainforest

In Atlantic rainforests, trees may present a pattern of continuous fruit production and fallen fleshy fruits can be available for ants on a year-round basis, comprising a total of up to 400 kg ha^{-1}yr^{-1} (Morellato, 1992). Since fleshy fruits and seeds may

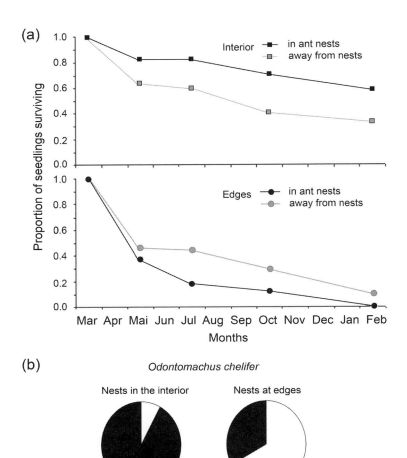

Figure 7.3.   Edge effects alter the outcome of ant-diaspore interactions in cerrado. (a) In the interior of cerrado, seedlings of *Erythroxylum pelleterianum* benefit from growing near ant nests compared to control plots without nests (* $P < 0.05$), but no such effect is detected at edges. (b) Nests of the large ponerine *Odontomachus chelifer*, an important local seed disperser, have higher residence time in the cerrado interior compared to edge habitats. Adapted from Christianini & Oliveira (2013) with permission from Springer (a).

account for an important part of the diet of ant colonies, and a wide assemblage of ant species in different subfamilies is known to feed on them on a regular basis in Neotropical forests (Figure 7.1), it is highly unlikely that any fleshy diaspore that falls to the forest floor will pass undetected by litter-foraging ants (see Horvitz & Beattie, 1980; Pizo & Oliveira, 2000; Passos & Oliveira, 2003).

Case Studies

Guapira opposita (Nyctaginaceae)

Fleshy fruits are usually poor in protein, but the fruit pulp of the common Atlantic forest tree *Guapira opposita* has one of the highest protein contents (28.4 per cent) so far reported in the literature (Jordano, 2000), and a minor amount (0.5 per cent) of lipids (Passos & Oliveira, 2004). A total of 24 bird species have been recorded visiting fruiting trees of *G. opposita* in lowland and sandy Atlantic forests (Figure 7.4a; M. A. Pizo personal observations, see also Galetti et al., 2000). If not carried away by frugivorous birds, mature fruits can be dropped with bits of pulp attached after manipulation by birds in the canopy, and many can still fall spontaneously beneath parent trees with the entire pulp intact. Although fruits of *G. opposita* lack any morphological specialization for dispersal by ants, once the fruits reach the forest floor they are rapidly discovered by ground-dwelling ants (11 species). The large ponerines *Odontomachus chelifer* and *Pachycondyla striata* are the main seed vectors (Figure 7.4b) and together account for >50 per cent of the interactions between ants and fallen *Guapira* fruits, frequently displacing them up to 4 m to their nests (Passos & Oliveira, 2004). Field observations suggest that colonies of these large, primarily carnivorous ponerines complement their protein intake by collecting *G. opposita* fruits on the floor of coastal sandy forests, where arthropod prey can be scarce (Pizo et al., 2005). Indeed, behavioural observations with captive colonies of ponerines revealed that entire fruits are transported inside the nest, where bits of protein-rich pulp are fed to larvae and worker nestmates. Deposition of cleaned, viable seeds in the surrounding area outside the nest, and increased number of seedlings and juveniles at this location, further confirmed that the interaction between ponerine ants and *G. opposita* fruits illustrate a case of directed dispersal rather than granivory (Figure 7.4c). Soil analyses indicate that the association of *G. opposita* seedlings with nests of *O. chelifer* is beneficial compared to background soils because nest microsites are richer in phosphorous, potassium and calcium, and the ants also increase soil penetrability, all of which likely improve seed germination and seedling growth (see also Horvitz, 1981; Levey & Byrne, 1993; Passos & Oliveira, 2002, 2003). An additional potential benefit of growing around nests of ponerines is that hunting activity by these primarily carnivorous ants might confer some anti-herbivore protection to neighbouring seedlings (Figure 7.4d), analogous to ant-garden epiphytes (see Davidson & Epstein, 1989).

Clusia criuva (Clusiaceae)

Clusia criuva is a common tree in sandy Atlantic forests producing fruit capsules that expose five diaspores each containing up to 17 seeds enveloped by a red aril with an extremely high (83.4 per cent) lipid content (Figure7.5a; Passos & Oliveira, 2002). A total of 14 bird species have been recorded ingesting and dispersing > 80 per cent of diaspores in the canopy and subsequently defecating intact seeds, thus acting as

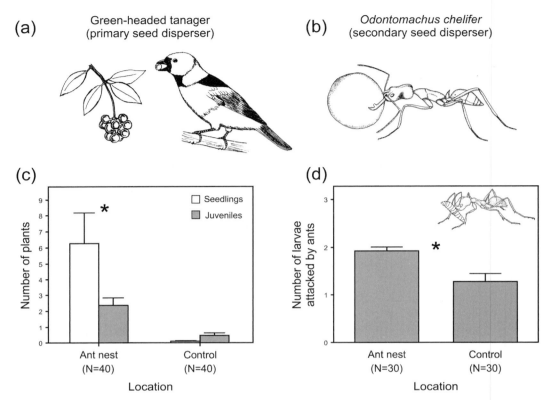

(a) Green-headed tanager (primary seed disperser)

(b) *Odontomachus chelifer* (secondary seed disperser)

Figure 7.4.   The dual seed dispersal system of *Guapira opposita* and ant-derived benefits to seedlings in sandy plain Atlantic forest. (a) Fruits are dispersed from the canopy by at least 24 bird species, including the Green-headed tanager *Tangara seledon*. (b) Large ponerine ants carry fallen fruits to the nest where the protein-rich pulp is fed to larvae; viable seeds are discarded around the nest. (c) Quantity of seedlings and juveniles around nests of *Odontomachus chelifer* and in control plots without nests. (d) Aggression by ants to dipteran larvae placed on seedlings growing in *O. chelifer* nests, and in adjacent plots (two larvae per location). Data are means + 1 SE. Adapted from Passos and Oliveira (2004) with permission from Springer (c, d). Drawings: Luisa Mota.

legitimate seed dispersers (Pizo, personal observations; Galetti et al., 2000). Arillate seeds can fall to the ground within entire fresh diaspores, dropped by birds, or embedded in bird feces containing plenty of undigested aril (Figure 7.5b, c). Litter-foraging ants promptly respond to this lipid-rich resource on the ground, removing 89 per cent of the fallen diaspores and 98 per cent of the seeds embedded in bird droppings in 24 h. The large seed vectors *Pachycondyla striata* and *Odontomachus chelifer* together account for 34 per cent of the ant-diaspore records on the forest floor, and are able to individually transport newly fallen or feces-embedded seeds into their nests up to 10 m away (Figure 7.5d). Observations with captive ponerine colonies revealed that the lipid-rich aril is fed to larvae whereas intact and viable seeds are discarded in the nest surroundings, where seedlings are more frequently

Figure 7.5. Seed dispersal and ant-mediated seedling establishment in the Atlantic forest tree, *Clusia criuva*. (a) Open mature fruit exposing diaspores (each with up to 17 seeds) coated by a red, lipid-rich aril. (b) *Turdus albicollis* (White-necked thrush), the main frugivore and seed disperser in the tree canopy (Photo: Octavio C. Salles). (c) Bird dropping beneath a fruiting tree containing embedded seeds with bits of red aril attached. (d) Worker of *Odontomachus chelifer* transporting a fallen diaspore to its nest, where the aril will be fed to larvae and the viable seeds discarded nearby. (e) Seedlings of *C. criuva* clumped in the vicinity of a nest of *O. chelifer* (tagged 'N'). After Passos & Oliveira (2002), photo (d) reproduced with permission from John Wiley and Sons. (A black-and-white version of this figure will appear in some formats. For the color version, please refer to the plate section.)

found compared to control areas without nests (Figure 7.5e). Increased early seedling survival through time nearby *Pachycondyla striata* nests correlated with higher concentrations of total nitrogen and phosphorus compared to control plots. These results confirm the beneficial effects from secondary, directed dispersal by ants on the distribution and survival of seedlings of primarily bird-dispersed *C. criuva* trees.

Some plant species may have diaspores with special structures for two-step dispersal systems sequentially involving vertebrate frugivores and ants (Vander Wall & Longland, 2005, and included references). The cases described in this chapter, however, show that lack of morphological specialization for ant-dispersal does not prevent secondary, ant-induced seed movements when protein- and lipid-rich fleshy diaspores attract both birds and ants (Pizo & Oliveira, 1998; Böhning-Gaese et al., 1999; Passos & Oliveira, 2002, 2004; Christianini & Oliveira 2010). Indeed, while distances of seed displacement by large *Odontomachus* and *Pachycondyla* ants may be small compared to primary bird dispersers (Figure 7.2c; see Horvitz & Le Corff, 1993), ant-induced seed movements can produce non-random spatial recruitment patterns, which are clearly beneficial at the early stages of plant development.

Effect of Habitat Fragmentation on Ant-Fruit Interactions: An Experiment with Synthetic Fruits

Since particular functional groups of ants, including large predatory *Odontomachus* and *Pachycondyla* species, have been shown to be sensitive to habitat fragmentation in Atlantic rainforests (Leal et al., 2012), we examined whether anthropogenic disturbance would affect patterns of ant attendance and ant-induced displacement of fruits. Field experiments were performed in human-created fragments (90–145 ha) *versus* undisturbed forest sites. We used synthetic fruits (8-mm diameter; 0.2-g weight) consisting of one plastic 'seed' (3-mm diameter) coated by a whitish 'pulp' rich in lipids (75 per cent; Bieber et al. 2014), which simulated the size and composition of fleshy fruits of many species commonly found on the floor of Atlantic rainforests (Pizo & Oliveira, 2000; Passos & Oliveira, 2003). This method allowed us to mimic fallen fleshy fruits in identical conditions (i.e. ripeness, size, quantity) and compare ant attendance among eight study sites simultaneously (Bieber et al. 2014).

Overall, 51 ant species were attracted to the lipid-rich synthetic fruits – while large ponerines individually transported entire 'fruits' up to ≈1.5 m, smaller myrmicines either sequentially removed the pulp on spot or carried entire 'fruits' in groups of recruited workers (Figure 7.6a, b). Synthetic fruits in undisturbed forests attracted a higher number of species per sampling station than those in disturbed forest fragments (Figure 7.6c). Most notably, the large ponerines *Odontomachus chelifer* and *Pachycondyla striata* (≈1.5 cm) were more common in undisturbed than in disturbed forest sites, whereas large *Pheidole* species (≥3 mm) did not differ between areas (Figure 7.6d). As a consequence of the contrasting response of these two ant groups to forest disturbance, the frequency of stations with synthetic fruits removed by ants was greater at undisturbed compared to disturbed forest sites (Figure 7.6e).

The experimental simulation with lipid-rich synthetic fruits confirmed that the ongoing process of fragmentation of the Atlantic rainforest may negatively affect potentially mutualistic interactions between ground-dwelling ants and fallen fleshy fruits. Although large *Pheidole* species may provide services to fallen diaspores in fragments through seed cleaning and displacement, our results strongly suggest that fallen fleshy fruits in disturbed forest fragments have decreased chances of interacting with large ant species that move seeds further and to favourable recruitment sites (Figure 7.6d; see also Guimarães & Cogni, 2002; Zwiener et al., 2012; Almeida et al., 2013).

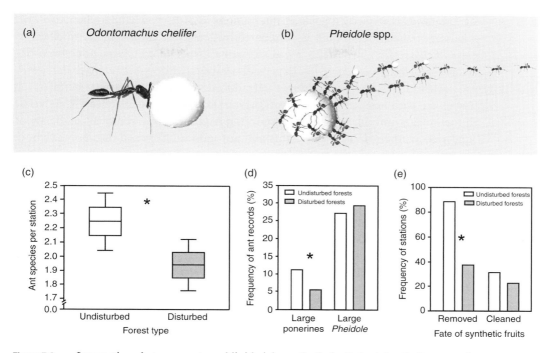

Figure 7.6. Interactions between ants and lipid-rich synthetic fruits in Atlantic forest, and patterns of ant attendance in undisturbed and disturbed forest sites. (a) Worker of *Odontomachus chelifer* (≈1.5 cm) inspects a synthetic fruit before carrying it to the nest. (b) Recruited workers of *Pheidole* species sequentially remove the 'pulp' from a synthetic fruit on spot; the 'seed' will be entirely cleaned in <5 h; large *Pheidole* (≥3 mm) can cooperatively transport entire 'fruits'. (c) Mean number of ant species (± 1 SE; whiskers delimit the 95 per cent confidence interval) attending 'fruits' is significantly higher in undisturbed than in disturbed forests. (d) Frequency of particular ant groups at sampling stations with synthetic fruits; large ponerines are more frequent in undisturbed than in disturbed forests whereas large *Pheidole* do not differ between forest types. (e) Occurrence of beneficial behaviours at sampling stations with synthetic fruits in the two forest types; frequency of 'fruit' removal decreases markedly in disturbed forests whereas occurrence of 'seed' cleaning by ants is not altered by disturbance. Asterisks denote $P < 0.05$. Adapted from Bieber et al. (2014). Artwork: Paulo S. D. Silva.

Ant-Diaspore Interactions in the Anthropocene

Our data on ant-diaspore interactions in cerrado savanna and Atlantic forest show that ground-dwelling ants play a key role in the regeneration of many plants adapted for vertebrate seed dispersal in these two highly disturbed Brazilian biomes. This type of mutualistic, non-specialized interaction has also been recognized as relevant for the regeneration of plants in other tropical areas around the world (e.g. Byrne & Levey, 1993; Böhning-Gaese et al., 1999; Leal et al., 2007; Dausmann et al., 2008). Our second conclusion is that ant-diaspore interactions can be negatively affected by habitat disturbance. In general, interactions in disturbed habitats included only a subset of the ant community, dominated by generalist and small myrmicine species (*Pheidole* spp.), which provide less effective dispersal than large poneromorphs (*Odontomachus*, *Pachycondyla*). In cerrado savanna, although *Erythroxyulum* seeds are removed in greater numbers at cerrado edges, seedlings survive better when associated with nests of ponerine ants, which show higher persistence through time in the moister interior of the cerrado. In the Atlantic forest, removal frequency of lipid-rich synthetic diaspores by ants and ant-induced dispersal distances are both considerably reduced at disturbed fragments.

The fast pace of habitat conversion, fragmentation and defaunation of vertebrate frugivores around the world has raised questions about the chances of regeneration for many plant species in the remnant patches of native habitats (McConkey et al., 2012). Seed dispersal is an issue of increasing concern not only due to its importance for the local persistence of populations, but also due to its role in community dynamics, evolution and the maintenance of healthy plant populations in an era of climate change (Mokany et al., 2015). Global climate change is already affecting plant phenology and reshaping the distributional range of suitable habitat for many species (Travis et al., 2013; Morellato et al., 2016). Without dispersal, many plant populations are likely to be confined to less favourable habitats through time, with increasing chances of extinction.

Decades ago Redford (1992) indicated that large tracts of tropical native habitats in relatively good conservation conditions were becoming empty and quiet forests due to the absence of many vertebrates that were continuously harvested due to commercial or subsistence hunting. Pervasive defaunation is currently a recognized phenomenon (Dirzo et al., 2014), and in vertebrate-depleted habitats seeds of many plants are attaining only truncated distances of dispersal close to parental plants due to the virtual absence of large vertebrate frugivores (Terborgh et al., 2008). Changes in seed size have already been documented in Atlantic rainforest and are attributed to the disruption of seed dispersal interactions with large frugivores (Galetti et al., 2013). In spite of the undisputable importance of vertebrates for ecosystem functioning (Estes et al., 2011), many of their relevant services are shared with invertebrates, as demonstrated in this chapter for ant-mediated, two-phase seed dispersal systems. Depletion of large vertebrates in disturbed habitats may be partially compensated by small animals, which may disperse and prey on a larger fraction of seeds compared to normal undisturbed conditions (Wright, 2003; Dausmann et al. 2008). It is thus likely that plant regeneration will increasingly rely on the small rodents, birds and invertebrates for seed dispersal in human-impacted habitat patches.

As many other animals, ants are also subject to the negative effects of habitat loss, disturbance and fragmentation (Laurance et al., 2002; this chapter). Many ant species, however, are relatively resistant or have higher resilience to such disturbances than large vertebrates. For instance, small ants such as *Pheidole* spp. are more common in interactions with fallen diaspores near edges than in the interior of cerrado (Christianini & Oliveira, 2013), and seem less affected by forest disturbance than large ponerines (Figure 7.6d; Bieber et al., 2014). Moreover, ants show fast population and species richness recoveries after fire in cerrado (Vasconcelos et al., 2009), and leaf-cutters can even increase in abundance after disturbance in forests and savannas (Chapter 4). Ants may positively influence the recovery of vegetation near disturbed tracts of habitats (Gallegos et al., 2014), and can be used to assist restoration processes in degraded lands (Henao-Gallego et al., 2012). The remarkable abundance of ants in terrestrial ecosystems, coupled with their pervasive interactions with plants, suggest that these 'little things' (Wilson, 1987) will have an increasingly enhanced role in two-step dispersal systems and in plant regeneration processes (Christianini et al., 2014), particularly in frugivore-impoverished tropical habitats.

References

Almeida, F. S, Mayhé-Nunes, A. J. & Queiroz, J. M. (2013). The importance of Poneromorph ants for seed dispersal in altered environments. *Sociobiology*, 60, 229–35.

Beattie, A. J. (1985). *The Evolutionary Ecology of Ant-Plant Mutualisms*. Cambridge: Cambridge University Press.

Bieber, A. G. D., Silva, P. S. D., Sendoya, S. F. & Oliveira, P. S. (2014). Assessing the impact of deforestation of the Atlantic rainforest on ant-fruit interactions: a field experiment using synthetic fruits. *PLoS ONE*, 9, e90369.

Böhning-Gaese, K., Gaese, B. H. & Rabemanantsoa, S. B. (1999). Importance of primary and secondary seed dispersal in the Malagasy tree *Commiphora guillaumini*. *Ecology*, 80, 821–32.

Bolton, B. (2003). Synopsis and classification of Formicidae. *Memoirs of the American Entomological Institute*, 71, 1–370.

Bottcher, C. & Oliveira, P.S. (2014). Consumption of lipid-rich seed arils improves larval development in a Neotropical primarily carnivorous ant, *Odontomachus chelifer* (Ponerinae). *Journal of Tropical Ecology*, 30, 621–4.

Brandão, C. R. F., Silva, R. R. & Feitosa, R. M. (2011). Cerrado ground-dwelling ants (Hymenoptera: Formicidae) as indicators of edge effects. *Zoologia (Curitiba)*, 28, 379–87.

Byrne, M. M. & Levey, D. J. (1993). Removal of seeds from frugivore defecations by ants in a Costa Rican rain forest. *Vegetatio*, 107/108, 363–74.

Camargo, P. H. S. A., Martins, M. M., Feitosa, R. M. & Christianini, A. V. (2016). Bird and ant synergy increases the seed dispersal effectiveness of an ornithochoric shrub. *Oecologia*, 181, 507–18.

Christianini, A. V., Mayhé-Nunes, A. J. & Oliveira, P. S. (2007). The role of ants in the removal of non-myrmecochorous diaspores and seed germination in a Neotropical savanna. *Journal of Tropical Ecology*, 23, 343–51.

 (2012). Exploitation of fallen diaspores by ants: are there ant-plant partner choices? *Biotropica*, 44, 360–7.

Christianini, A. V. & Oliveira, P. S. (2009). The relevance of ants as seed rescuers of a primarily bird-dispersed tree in the Neotropical cerrado savanna. *Oecologia*, 160, 735–45.

(2010). Birds and ants provide complementary seed dispersal in a Neotropical savanna. *Journal of Ecology*, 98, 573–82.

(2013). Edge effects decrease ant-derived benefits to seedlings in a Neotropical savanna. *Arthropod-Plant Interactions*, 7, 191–9.

Christianini, A. V., Oliveira, P. S, Bruna, E. M. & Vasconcelos, H. L. (2014). Fauna in decline: Meek shall inherit. *Science*, 345, 1129.

Corrêa, M. M., Silva, P. S. D., Wirth, R., Tabarelli, M. & Leal, I. R. (2010). How leaf-cutting ants impact forests: drastic nest effects on light environment and plant assemblages. *Oecologia*, 162, 103–15.

Dalling, J. W. & Wirth, R. (1998). Dispersal of *Miconia argentea* seeds by the leaf-cutting ant *Atta colombica*. *Journal of Tropical Ecology*, 14, 705–10.

Dausmann, K. H., Glos, J., Linsenmair, K. E. & Ganzhorn, J. U. (2008). Improved recruitment of a lemur-dispersed tree in Malagasy dry forests after the demise of vertebrates in forest fragments. *Oecologia*, 157, 307–16.

Davidson, D. W. & Epstein, W. W. (1989). Epiphytic associations with ants. In U. Lüttge, ed., *Vascular plants as epiphytes*. Berlin: Springer, pp. 200–33.

Dias, B. F. S. (1992). *Alternativas de Desenvolvimento dos Cerrados: Manejo e Conservação dos Recursos Naturais Renováveis*. Brasília: Funatura.

Dirzo, R., Young, H. S., Galetti et al. (2014). Defaunation in the Anthropocene. *Science*, 345, 401–6.

Eisenlohr, P. V. & Oliveira-Filho, A. T. (2015). Revisiting patterns of tree species composition and their driving forces in the Atlantic forests of Southeastern Brazil. *Biotropica*, 47, 689–701.

El Bizri, H. R., Morcatty, T. Q., Lima, J. J. S. & Valsecchi, J. (2015). The thrill of the chase: uncovering illegal sport hunting in Brazil through YouTube™ posts. *Ecology and Society*, 20, 30.

Endo, W., Peres, C. A., Salas, et al. (2010). Game vertebrate densities in hunted and non-hunted forest sites in Manu National Park, Peru. *Biotropica*, 42, 251–61.

Estes, J. A., Terborgh, J., Brashares, J. S., et al. (2011). Trophic downgrading on planet Earth. *Science*, 333, 301–6.

Farji-Brener, A. G. & Ghermandi, L. (2004). Seedling recruitment in a semi-arid Patagonian steppe: facilitative effects of refuse dumps of leaf-cutting ants. *Journal of Vegetation Science*, 15, 823–30.

Ferreira, A. V., Bruna, E. M. & Vasconcelos, H. L. (2011). Seed predators limit plant recruitment in Neotropical savannas. *Oikos*, 120, 1013–22.

Fleming, T. H., Breitwisch, R. & Whitesides, G. H. (1987). Patterns of tropical vertebrate frugivore diversity. *Annual Review of Ecology and Systematics*, 18, 91–109.

Fourcassié, V. & Oliveira, P. S. (2002). Foraging ecology of the giant Amazonian ant *Dinoponera gigantea* (Hymenoptera, Formicidae, Ponerinae): activity schedule, diet, and spatial foraging patterns. *Journal of Natural History*, 36, 2211–27.

Galetti, M., Guevara, R., Côrtes, M. C., et al. (2013). Functional extinction of birds drives rapid evolutionary changes in seed size. *Science*, 340, 1086–90.

Galetti, M., Laps, R. & Pizo, M. A. (2000). Frugivory by toucans (Ramphastidae) at two altitudes in the Atlantic forest of Brazil. *Biotropica*, 32, 842–50.

Galetti, M., Pizo, M. A. & Morellato, L. P. C. (2011). Diversity of functional traits of fleshy fruits in a species-rich Atlantic rain forest. *Biota Neotropica*, 11, 181–93.

Gallegos, S.C., Hensen, I. & Schleuning, M. (2014). Secondary dispersal by ants promotes forest regeneration after deforestation. *Journal of Ecology*, 102, 659–66.

Gottsberger, G. & Silberbauer-Gottsberger, I. (1983). Dispersal and distribution in the cerrado vegetation of Brazil. *Sonderbänd des Naturwissenschaftlichen Vereins in Hamburg*, 7, 315–52.

Guimarães, P. R., & Cogni, R. (2002). Seed cleaning of *Cupania vernalis* (Sapindaceae) by ants: edge effect in a highland forest in south-east Brazil. *Journal of Tropical Ecology* 18, 303–7.

Henao-Gallego, N., Escobar-Ramírez, S., Calle, Z., Montoya-Lerma, J. & Armbrecht, I. (2012). An artificial aril designed to induce seed hauling by ants for ecological rehabilitation purposes. *Restoration Ecology*, 20, 555–60.

Horvitz, C. C. (1981). Analysis of how ant behaviors affect germination in a tropical myrmecochore *Calathea-Microcephala* (P and E) Koernicke (Marantaceae) – microsite selection and aril removal by Neotropical ants, *Odontomachus*, *Pachycondyla*, and *Solenopsis* (Formicidae). *Oecologia*, 51, 47–52.

Horvitz, C. C. & Beattie, A. J. (1980). Ant dispersal of *Calathea* (Marantaceae) seeds by carnivorous ponerines (Formicidae) in a tropical rain forest. *American Journal of Botany*, 67, 321–6.

Horvitz, C. C. & Le Corff, J. (1993). Spatial scale and dispersion pattern of ant- and bird-dispersed herbs in two tropical lowland rain forests. *Vegetatio*, 107, 351–62.

Howe, H. F. & Smallwood, J. (1982). Ecology of seed dispersal. *Annual Review of Ecology and Systematics*, 13, 201–28.

Hughes, L. & Westoby, M. (1992). Effect of diaspore characteristics on removal of seeds adapted for dispersal by ants. *Ecology*, 73, 1300–12.

Hughes, L., Westoby, M. & Jurado, E. (1994). Convergence of elaiosomes and insect prey: evidence from ant foraging behaviour and fatty acid composition. *Functional Ecology*, 8, 358–65.

Janzen, D. H. (1970). Herbivores and the number of tree species in tropical forests. *American Naturalist*, 104, 501–29.

(1977). Promising directions of study in tropical animal-plant interactions. *Annals of the Missouri Botanical Garden*, 64, 706–36.

Jordano, P. (2000). Fruits and frugivory. In M. Fenner, ed., *Seeds: The Ecology of Regeneration in Plant Communities*. Wallingford: CAB International, pp. 125–65.

Kaspari, M. (1993). Removal of seeds from Neotropical frugivore droppings: ant responses to seed number. *Oecologia*, 95, 81–8.

Klink, C. A. & Machado, R. B. (2005). Conservation of the Brazilian Cerrado. *Conservation Biology*, 19, 707–13.

Laman, T. G. (1996). *Ficus* seed shadows in a Bornean rain forest. *Oecologia*, 107, 347–55.

Laurance, W. F., Camargo, J. L. C., Luizão, R. C. C. et al. (2010). The fate of Amazonian forest fragments: a 32-year investigation. *Biological Conservation*, 144, 56–67.

Laurance, W. F., Lovejoy, T. E., Vasconcelos, H. L. et al. (2002). Ecosystem decay of Amazonian forest fragments: a 22-year investigation. *Conservation Biology*, 16, 605–18.

Leal, I. R., Filgueiras, B. K. C., Gomes, J. P., Iannuzzi, L. & Andersen A. N. (2012). Effects of habitat fragmentation on ant richness and functional composition in Brazilian Atlantic forest. *Biodiversity and Conservation*, 21, 1687–701.

Leal, I. R. & Oliveira, P. S. (1998). Interactions between fungus-growing ants (Attini), fruits and seeds in cerrado vegetation in Southeast Brazil. *Biotropica*, 30, 170–8.

Leal, I. R., Wirth, R. & Tabarelli, M. (2007). Seed dispersal by ants in the semi-arid Caatinga of north-east Brazil. *Annals of Botany,* 99, 885–94.

Levey, D. J. & Byrne, M. M. (1993). Complex ant–plant interactions: rain forest ants as secondary dispersers and post-dispersal seed predators. *Ecology*, 74, 1802–12.

Levey, D. J., Moermond, T. C. & Denslow, J. S. (1994). Frugivory: An overview. In L. A. McDade, K. S. Bawa, H. A. Hespenheide & G. S. Hartshorn, eds., *La Selva: Ecology and Natural History of a Neotropical Rain Forest.* Chicago: University of Chicago Press, pp. 282–94.

Lima, M. H. C., Oliveira, E. G. & Silveira, F. A. O. (2013). Interactions between ants and non-myrmecochorous fruits in *Miconia* (Melastomataceae) in a Neotropical savanna. *Biotropica,* 45, 217–23.

Longino, J. T, Coddington, J. & Colwell, R. K. (2002). The ant fauna of a tropical rain forest: estimating species richness three different ways. *Ecology*, 83, 689–702.

McConkey, K.R., Prasad, S., Corlett, R.T. et al. (2012). Seed dispersal in changing landscapes. *Biological Conservation*, 146, 1–13.

Mendonça, A. H., Russo, C., Melo, A. C. G. & Durigan, G. (2015). Edge effects in savanna fragments: a case study in the cerrado. *Plant Ecology & Diversity*, 8, 493–503.

Meyer, S. T., Leal, I. R., Tabarelli, M. & Wirth, R. (2011). Ecosystem engineering by leaf-cutting ants: nests of *Atta cephalotes* drastically alter forest structure and microclimate. *Ecological Entomology*, 36, 14–24.

Mokany, K., Prasad, S. & Westcott, D. A. (2015). Impacts of climate change and management responses in tropical forests depend on complex frugivore-mediated seed dispersal. *Global Ecology and Biogeography*, 24, 685–94.

Morellato, L. P. C. (1992). Nutrient cycling in two south-east Brazilian forests. I Litterfall and litter standing crop. *Journal of Tropical Ecology*, 8, 205–15.

Morellato, L. P. C., Alberton, B., Alvarado, S. T. et al. (2016). Linking plant phenology to conservation biology. *Biological Conservation*, 195, 60–72.

Myers, N., Mittermeier, R. A., Mittermeier, C. G., da Fonseca, G. A. B. & Kent, J. (2000). Biodiversity hotspots for conservation priorities. *Nature*, 403, 853–8.

Oliveira, P. S., Galetti, M., Pedroni, F. & Morellato, L. P. C. (1995). Seed cleaning by *Mycocepurus goeldii* ants (Attini) facilitates germination in *Hymenaea courbaril* (Caesalpiniaceae). *Biotropica*, 27, 518–22.

Oliveira, P. S. & Marquis, R. J. (eds.) (2002). *The Cerrados of Brazil: Ecology and Natural History of a Neotropical Savanna.* New York: Columbia University Press.

Oliveira-Filho, A. T. & Ratter, J. A. (2002). Vegetation physiognomies and woody flora of the Cerrado biome. In P. S. Oliveira & R. J. Marquis, eds., *The Cerrados of Brazil: Ecology and Natural History of a Neotropical Savanna.* New York: Columbia University Press, pp. 91–120.

Passos, L. & Oliveira, P. S. (2002). Ants affect the distribution and performance of *Clusia criuva* seedlings, a primarily bird-dispersed rainforest tree. *Journal of Ecology,* 90, 517–28.

(2003). Interactions between ants, fruits and seeds in a restinga forest in south-eastern Brazil. *Journal of Tropical Ecology,* 19, 261–70.

(2004). Interaction between ants and fruits of *Guapira opposita* (Nyctaginaceae) in a Brazilian sandy plain rainforest: ant effects on seeds and seedlings. *Oecologia*, 139, 376–82.

Pinto, N. & Keitt, T. H. (2008). Scale-dependent responses to forest cover displayed by frugivore bats. *Oikos*, 117, 1725–31.

Pizo, M. A. (1997). Seed dispersal and predation in two populations of *Cabralea canjerana* (Meliaceae) in the Atlantic forest of southeastern Brazil. *Journal of Tropical Ecology*, 13, 559–78.

Pizo, M. A. (2008). The use of seeds by a twig-dwelling ant on the floor of a tropical rain forest. *Biotropica*, 40, 119–21.

Pizo, M. A. & Oliveira P. S. (1998). Interaction between ants and seeds of a nonmyrmecochorous Neotropical tree, *Cabralea canjerana* (Meliaceae), in the Atlantic forest of southeast Brazil. *American Journal of Botany*, 85, 669–74.

Pizo, M. A. & Oliveira P. S. (1999). Removal of seeds from vertebrate faeces by ants: effects of seed species and deposition site. *Canadian Journal of Zoology*, 77, 1595–1602.

Pizo, M. A. & Oliveira, P. S. (2000). The use of fruits and seeds by ants in the Atlantic forest of southeast Brazil. *Biotropica*, 32, 851–61.

Pizo, M. A. & Oliveira, P. S. (2001). Size and lipid content of nonmyrmecochorous diaspores: effects on the interaction with litter-foraging ants in the Atlantic rain forest of Brazil. *Plant Ecology*, 157, 37–52.

Pizo, M. A., Passos, L. & Oliveira, P. S. (2005). Ants as seed dispersers of fleshy diaspores in Brazilian Atlantic forests. In P.-M. Forget, J. E. Lambert, P. E. Hulme and S. B. Vander Wall, eds., *Seed Fate: Predation and Secondary Dispersal*. Wallingford: CABI Publishing, pp. 315–29.

Redford, K. H. (1992). The empty forest. *BioScience*, 42, 412–22.

Ribeiro, M. C., Metzger, J. P., Martensen, A. C., Ponzoni, F. J. & Hirota, M. M. (2009). The Brazilian Atlantic Forest: how much is left, and how is the remaining forest distributed? Implications for conservation. *Biological Conservation,* 142, 1141–53.

Rico-Gray, V. & Oliveira, P. S. (2007). *The Ecology and Evolution of Ant-Plant Interactions*. Chicago: University of Chicago Press.

Roberts, J. T. & Heithaus, R. (1986). Ants rearrange the vertebrate-generated seed shadow of a Neotropical fig tree. *Ecology*, 67, 1046–51.

Santos-Heredia, C., Andresen, E. & Stevenson, P. (2011). Secondary seed dispersal by dung beetles in an Amazonian forest fragment of Colombia: influence of dung type and edge effect. *Integrative Zoology*, 6, 399–408.

Schaefer, H. M. & Ruxton, G. D. (2011). *Plant-Animal Communication*. Oxford: Oxford University Press.

Terborgh, J., Nuñez-Iturri, G., Pitman, N. C. A. et al. (2008). Tree recruitment in an empty forest. *Ecology,* 89, 1757–68.

Travis, J. M. J., Delgado, M., Bocedi, G. et al. (2013). Dispersal and species response to climate change. *Oikos*, 122, 1532–40.

van der Pijl, L. (1969). *Principles of Seed Dispersal in Higher Plants*. Berlin: Springer-Verlag.

Vander Wall, S. B. & Longland, W. S. (2005). Diplochory and the evolution of seed dispersal. In P.-M. Forget, J. E. Lambert, P. E. Hulme and S. B. Vander Wall, eds., *Seed Fate: Predation and Secondary Dispersal*. Wallingford: CABI Publishing, pp. 297–314.

Vasconcelos, H. L., Pacheco, R., Silva, R. C. et al. (2009). Dynamics of the leaf-litter arthropod fauna following fire in a Neotropical woodland savanna. *PLoS ONE*, 4, e7762.

Vasconcelos, H. L., Vieira Neto, E. M. H., Mundim, F. M. R. & Bruna, E. M. (2006). Roads alter the colonization dynamics of a keystone herbivore in Neotropical savannas. *Biotropica*, 38, 661–5.

Wilson, E. O. (1987). The little things that run the world. *Conservation Biology*, 1, 344–6.

Wirth, R., Herz, H., Ryel, R., Beyschlag, W. & Hölldobler, B. (2003). *Herbivory of Leaf-Cutting Ants – A Case Study on Atta Colombica in the Tropical Rainforest of Panama.* Berlin: Springer.

Wright, S. J. (2003). The myriad consequences of hunting for vertebrates and plants in tropical forests. *Perspectives in Plant Ecology, Evolution and Systematics*, 6, 73–86.

Zwiener, V. P., Bihn, J. H. & Marques, M. C. M. (2012). Ant-diaspore interactions during secondary succession in the Atlantic forest of Brazil. *Revista de Biologia Tropical*, 60, 933–42.

Part III

Ant-Plant Protection Systems under Variable Habitat Conditions

8 Plasticity and Efficacy of Defense Strategies against Herbivory in Ant-Visited Plants Growing in Variable Abiotic Conditions

Akira Yamawo

Ant-Plant Mutualism as a Biotic Defense in Plants

Many plants have evolved indirect biotic defenses that attract the natural enemies of herbivores, including ants (Koptur, 1992; Rico-Gray & Oliveira, 2007; Heil, 2015). Ant-plant protection mutualism exists in a diverse array of plant species. These plants often provide a food type, such as extrafloral nectar or lipid-rich food bodies (FBs), to visiting ants. In return, the visiting ants protect the plants against herbivores. Extrafloral nectaries (EFNs) are plant glands that secrete sugar, water and amino acids (Heil et al., 2000; Ness, 2003). EFNs have been described in virtually all above-ground plant tissues, including leaves, petioles, stipules, cotyledons, fruits and stems, of approximately 4,000 species worldwide, with examples in 745 genera from 108 families (Weber & Keeler, 2013). FBs contain carbohydrates, lipids and proteins and emerge at the leaf and stem surface (Heil et al., 1998; Fischer et al., 2002). Although fewer plants produce FBs compared to EFNs, FBs are also known from a diverse range of unrelated plant taxa and can be a significant contribution to the dietary needs of ant colonies (Schupp & Feener, 1991; Heil, Hilpert, Krüger et al., 2004; Webber et al., 2007).

In addition to indirect defenses, plants have evolved direct defense traits for avoiding herbivory, including both morphological and chemical defenses (reviewed in Howe & Westley, 1988; Walters, 2011; Mithöfer & Boland, 2012). Ant-visited plants express both direct and indirect defense mechanisms (Agrawal & Spiller, 2004; Rudgers et al., 2004; Kobayashi et al., 2008; Yamawo et al., 2014). For example, *Paulownia tomentosa* (Paulowniaceae) and *Mallotus japonicus* (Euphorbiaceae) employ trichomes as a physical defense as well as EFNs and FBs (Figure 8.1; see also Kobayashi et al., 2008; Yamawo et al., 2014). A series of studies on the tallow tree (*Triadica sebifera*) also reported that this plant exhibits not only a chemical defense but also an indirect defense via ants (Huang et al., 2010; Wang et al., 2013). Some studies suggest that plants involved in obligate ant-plant mutualisms may also have direct defenses (e.g. Folgarait & Davidson, 1994; Murase et al., 2003).

Figure 8.1. (a) Leaf of *Mallotus japonicus* with ants, *Pristomyrmex punctatus*. (b) *P. punctatus* collecting extrafloral nectar of *M. japonicus*. (c) *Pheidole noda* attacking caterpillar of *Parasa lepida* (Lepidoptera: Limacodidae) on leaf of *M. japonicus*. Photos by Akira Yamawo (a, c) and Hiromi Mukai (b). (A black-and-white version of this figure will appear in some formats. For the color version, please refer to the plate section.)

Ant-visited plant species often occur in disturbed sites and successional habitats, accompanied by spatial and temporal variation in biotic and abiotic habitat conditions (Bentley, 1977; O'Dowd, 1982; Schupp & Feener, 1991) because most of these plants are pioneer species (Koptur, 1992). Vegetation succession and disturbance are likely to cause variations in abiotic factors such as light, nutritional conditions and soil moisture (Chen et al., 2014; Yamawo et al., 2014). The relative costs and benefits of defenses are likely affected by such abiotic conditions, and ant-visited plants must cope with these conditions via phenotypic plasticity.

Here we review recent advances in the study of plasticity and efficacy of defense strategies against herbivory in ant-visited plants growing under variable abiotic conditions. First, we explore the relative costs of ant defense in comparison with other defensive traits. Second, we summarize the plasticity and efficacy of defense strategies of ant-visited plants in relation to abiotic habitat conditions. Finally, we highlight the major gaps in our current knowledge and discuss future research objectives.

Allocation and Ecological Cost of Indirect Defense by Ants

Many elements of biotic condition vary over space and time, for example, in the abundance and diversity of herbivores (Rudgers & Strauss, 2004; Nogueira et al., 2012). Therefore, plants have evolved plastic mechanisms to minimize the costs of defense. Many studies using induction experiments on indirect defense traits have suggested the presence of costs in indirect defense by ants (review in Heil, 2010).

Plants produce FBs and EF nectar and adjust the sugar and amino acid content in nectar in response to leaf damage (Smith et al., 1990; Heil et al., 2000; Ness, 2003; Pulice & Packer, 2008; review in Heil, 2015), root damage (Wäckers & Bezemer, 2003), volatile organic compounds caused by leaf damage (Kost & Heil, 2005, 2006; Choh et al., 2006; Li et al., 2012) or applied jasmonic acid (JA) (Heil et al., 2001; Kost & Heil, 2008; Heil, 2015). These inductions increase patrolling by defending ants (Bentley, 1977; Stephenson, 1982; Heil et al., 2001; Ness, 2003) and thereby reduce leaf damage (Heil et al., 2001; Kost & Heil, 2008). For example, *Catalpa bignonioides* (Bignoniaceae) expressed high sugar content in EF nectar after leaf damage by *Ceratomia catalpae* (Lepidoptera: Sphingidae), following which ant attendance at caterpillar-attacked leaves increased two- to threefold within 24 hours of herbivory relative to attendance at neighboring undamaged leaves (Ness, 2003). The induction of EFN secretions is complete within six days after leaf damage (Wäckers et al., 2001; Ness, 2003).

In addition to herbivores, the abundance and species composition of ants are important factors in the efficacy of indirect defense, and also vary depending on space and time (Cogni et al., 2003; Holland et al., 2010; Lach et al., 2010; Falcão et al., 2014; Yamawo et al., 2014). Therefore, plants change their investment in defense traits in response to ant community dynamics (Table 8.1). Some obligate ant-plant mutualism systems reduce their production of FBs and EF nectar when they are not well defended by their attendant ants (Risch & Rickson, 1981; Heil et al., 1997, 2009; Heil, 2013). In *Acacia* species (Leguminosae), the plasticity of EF nectar production shifted the competitive balance between mutualistic and parasitic ants. Such selective production of a greater amount of nectar thereby favors the maintenance of protection mutualists on plants (Heil, 2013). In facultative ant-plant mutualisms, plants produce EF nectar more often in the presence of ant workers. Heil et al. (2000) found that the attendance of ants increased the production of EF nectar in *Macaranga tanarius* (Euphorbiaceae), and similar results were reported for *Inga* species (Leguminosae) (Bixenmann et al., 2011) and *M. japonicus* (Yamawo et al., 2015). Plants may screen ant defenders through nectar removal (Heil et al., 2000, 2009, 2013). Thus, because plants adjust their production of rewards, many studies have suggested that indirect defense by ants is costly for plants.

Indeed, several studies have demonstrated the cost (perhaps the allocation cost) of mutualism with ants in obligate ant-plant systems. According to the results of an experiment in which exterminating ant colonies actually increased growth, reproduction or both aspects in host trees in the absence of herbivores, hosting ants is costly to *Acacia drepanolobium* (Leguminosae) and *Cordia nodosa* (Boraginaceae) (Stanton & Palmer, 2011; Frederickson et al., 2012; Chapter 10). The production cost of FBs was found to be approximately 5 percent of the above-ground biomass production in saplings of *Macaranga triloba* (Euphorbiaceae) (Heil et al., 1997). Cases of ants injuring plants were also identified in some cases of obligate ant-plant mutualism as direct costs (Izzo & Vasconcelos, 2002; Palmer et al., 2008). Nonetheless, three recent meta-analyses have reported that protective ants very

Table 8.1 Effects of Ant Attendance in the Defense Expression of Ant-Visited Plants

Plant species	Plant family	Type of mutualism	Type of stimulus	Ant species	Response	Region	Source
Piper cenocladum	Piperaceae	Obligate	Presence of ants	*Pheidole bicornis*	Production of FBs and reduction of chemical substances	Central America	Risch & Rickson (1981); Dyer et al. (2001)
Macaranga triloba	Euphorbiaceae	Obligate	Presence of ants	Ants	Increased production of FBs	South Asia	Heil et al. (1997)
Macaranga tanarius	Euphorbiaceae	Facultative	Presence of insects including ant workers	Insects	Increased production of EF nectar	South Asia	Heil et al. (2000)
Acacia species	Leguminosae	Obligate	Difference of ant species	*Pseudomyrmex* spp.	Increased production of EF nectar for mutualistic ant species	Central America	Heil et al. (2009); Heil (2013)
Inga species	Leguminosae	Obligate	Presence of ants	Ants	Increased production of EF nectar	Central America	Bixenmann et al. (2011)
Mallotus japonicus	Euphorbiaceae	Facultative	Presence of ants	*Pheidole noda*	Increased production of EF nectar growth and reduction of chemical substances	East Asia	Yamawo et al. (2015)
Cecropia glaziovii	Urticaceae	Obligate	Colonies of ants	*Azteca muelleri*	None	Central America	Oliveira et al. (2015)

FB – food body; EF – extrafloral.

rarely have negative effects on their plant partners (Chamberlain & Holland, 2009; Rosumek et al., 2009; Trager et al., 2010). In facultative mutualism, the expression of indirect defense traits, such as the formation of EFNs and secretion of EF nectar, has little effect on plant fitness and growth. EF nectar production in *Chamaecrista fasciculata* (Leguminosae) was found to not reduce plant growth or seed production (Rutter & Rausher, 2004). In addition, ant attendance of *C. fasciculata* had no influence on plant growth (Barton, 1986). The implications of these findings are that the allocation cost of indirect defense traits are minor. In fact, the cost of EFN production was found to be as little as approximately 1 percent of the total energy invested in leaves in *Ochroma pyramidale* (Malvaceae), *Vicia sativa* var. *angustifolia* (Leguminosae) and *M. japonicus* (O'Dowd, 1979; Katayama & Suzuki, 2011; Yamawo et al., 2015).

A few studies have compared the allocation costs between indirect and direct defense traits in facultative ant-plant mutualism systems. For instance, Katayama and Suzuki (2011) reported a smaller allocation cost of EF nectar corresponding to < 0.1 percent of total plant biomass than that of the chemical defense substance corresponding to > 20 percent of total plant biomass in *Vicia sativa* var. *angustifolia* (Leguminosae). The growth of *M. japonicus* was negatively related with the expression of direct defense traits, such as the densities of trichomes, pellucid dots and leaf phenolic concentration, but it was not related to indirect defense traits such as sugar concentration in EF nectar and the number of FBs (Yamawo et al., 2015). Such a trade-off suggests that the expression of direct defense traits is more costly than that of indirect traits. The estimated total amount of sugars secreted by EFNs and FBs in *M. japonicus* corresponded to < 0.1 percent of total plant biomass, whereas that of phenolic compounds represented approximately 10 percent of total plant biomass (Yamawo et al., 2015). In addition, ant-attended plants had greater height, leaf biomass and total biomass (Figure 8.2). On the basis of these findings, the authors proposed that ants provide low-cost indirect defenses that allow plants to re-allocate their energy from direct defenses to growth (Figure 8.3). Rapid growth in pioneer trees should increase their survival under light competitive conditions. The preferential use of indirect defense is found on other plant species. Decreased production of chemical defense substances in response to attendant ants was reported in *Piper cenocladum* (Piperaceae) (Dyer et al., 2001; but see Oliveira et al., 2015). *Cordia nodosa* defended by ants was found to induce chemical (phenolics) and physical (leaf toughness, trichomes) defenses when ant defense failed (Frederickson et al., 2013). Thus, the plasticity of plant defense strategies in response to the ant-environment supports the optimal defense theory: plants have evolved their defense traits to maximize their fitness, both ecologically and physiologically (McKey, 1974). These results indicate that, in some facultative ant-plant mutualisms, the allocation costs of displaying indirect defense traits are lower than those of displaying direct defense traits. However, the relative costs of direct and indirect defenses in other species, including species employing the obligate ant-plant mutualism system, remain unresolved.

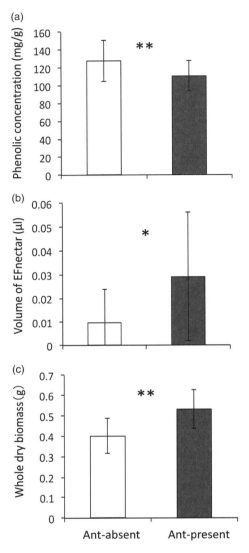

Figure 8.2.   Defense traits and dry biomass of *Mallotus japonicus* growing under the ant-absent and ant-present conditions (*n* = 15 each). (a) Phenolic concentration, (b) Volume of EF nectar and (c) Whole dry biomass. Error bars represent standard deviation. EF – extrafloral. Asterisk indicates significant differences (GLM, *$P < 0.05$, ** $P < 0.01$). Figure is based on data from Yamawo et al. (2015).

Ecological costs are also found in indirect defenses by ants, resulting from the negative effects of defenses on interactions between a plant and other organisms that could affect fitness under natural conditions. The occupation of EFNs by nectar-robbing insects could prevent mutualistic ants from exhibiting their defensive effects (Beattie, 1985; Koptur, 1992; Heil et al., 2004). In addition, visitations

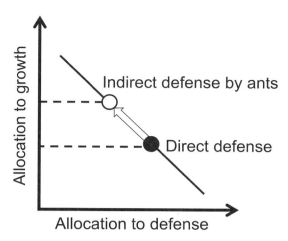

Figure 8.3.   Schematic illustration of inferred relationship in the trade-off between growth and defenses in *Mallotus japonicus* (Yamawo et al., 2015). White arrow indicates the shift of defense strategy in response to ant attendance.

by aggressive ants reduce the frequency of plant-pollinator interactions and hence plant reproduction (Ness, 2006; Malé et al., 2012; LeVan et al., 2014; Miller, 2014). Reduction of direct defenses has been observed in obligate ant-plant mutualisms (Heil et al., 2004; Eck et al., 2001, but see Steward & Keeler, 1988; Letourneau & Barbosa, 1999; Heil, Delsinne, Hilpert et al., 2002) and in ant-visited, EFN-bearing *Inga* spp. over an elevational gradient (Koptur, 1985). A phylogenetically controlled analysis of 31 wild cotton species (*Gossypium*), however, observed trade-offs between direct defensive traits (leaf trichomes and toxic leaf glands), but did not find a trade-off between EFNs and toxic glands/trichomes (Rudgers et al., 2004). On the basis of a meta-analysis of 25 studies, Koricheva and Romero (2012) found a significant negative correlation between plant allocation to direct and ant-mediated defenses. However, these trade-offs were significant only in plants that offered more costly rewards to ants, such as food bodies and/or domatia, and not in those that offered EFNs. These results suggest that plants have evolved EFNs in conjunction with direct defense traits (Steward & Keeler, 1988; Rudgers et al., 2004; Kobayashi et al., 2008; Yamawo et al., 2012b) due to the low production costs of EFNs. Thus, while some studies have demonstrated the ecological costs of ant defenses, the differences in relative cost among defenses, including direct and indirect costs, remain unclear.

Plasticity of Defense Strategies in Response to Abiotic Conditions

Plants alter the intensity of their defenses in response to abiotic habitat conditions such as soil nutrients, soil water and light conditions (Bryant et al., 1987; Folgarait & Davidson, 1994, 1995; Hemming & Lindroth, 1999; Stamp, 2003). The growth-differentiation balance (GDB) hypothesis (Herms & Mattson, 1992) and resource availability (RA) hypothesis (Coley et al., 1985) both integrate the trade-offs between growth and defense traits with the responses of net assimilation and growth rates according to the availability of resources. These hypotheses predict that rapidly growing plants under high nutrient conditions will have low levels of defense because the production of new leaves is supported by an export of photosynthate from source leaves, leaving few resources for the synthesis of defenses. Results of several studies support the GDB and RA hypotheses (e.g. Bryant et al., 1987; Folgarait & Davidson, 1995; Hemming & Lindroth, 1999; Stamp, 2003). The carbon-nutrient balance (CNB) hypothesis assumes that a plant will invest preferentially in carbon-based defenses when nutrients limit growth more than photosynthesis, such as high light conditions. However, in situations where photosynthesis is limited by factors other than nutrients, these will be allocated to defense traits (Bryant et al., 1983). Results overall are equivocal (Stamp, 2003), although many studies have examined the CNB hypothesis. Nevertheless, several studies do support the CNB hypothesis (Folgarait & Davidson, 1994; Stamp, 2003).

Numerous experimental studies under common garden or artificial experimental conditions have demonstrated the plasticity of direct and indirect defense traits in ant-visited plants in response to abiotic conditions (Table 8.2). Soil fertilization, or increased soil nutrient contents, was found to increase the production of EFNs and FBs (Folgarait & Davidson, 1995; Heil, Hilpert, Fiala et al., 2002; Wagner & Fleur Nicklen, 2010). For example, *Mallotus japonicus* increased the production of EFNs and FBs and decreased the densities of pellucid dots and trichomes (chemical and physical defense traits, respectively) in response to soil nutrient fertilization (Yamawo et al., 2012b; see color plates section). Folgarait and Davidson (1995) and Heil et al. (2002) reported positive effects of nutrient supply on the production of FBs in *Cecropia* species and *M. triloba*, while *Acacia constricta* grown near an ant nest demonstrated more EFNs per leaf than plants growing further away, as the nests provide soil nutrients for plants (Wagner & Fleur Nicklen, 2010). The induction of EFNs also depends on soil nutrient levels (Mondor & Addicott, 2003), and similar studies have reported the importance of water availability to soil nutrient absorption. Water availability was found to limit the production of EFNs and EF nectar (Yamawo et al., 2012b; Newman & Wagner, 2013; Newman et al., 2015). For example, in quaking aspen, *Populus tremuloides* (Salicaceae), water stress reduced the sugar concentration or volume of EF nectar as well as its induction (Newman et al., 2015). Yamawo et al. (2012b) demonstrated that, under low soil moisture conditions, young *M. japonicus* decreased the number of EFNs and volume of EF nectar, leading to less ant attendance, while increasing densities of trichomes and pellucid dots. The reduction of water availability directly limits the production of EF nectar because nectar contains water; thus, the GDB and RA hypotheses do not support indirect defense by ants, whereas several other studies on direct defenses do support these hypotheses.

Light availability also decisively influences indirect defenses. The production of EF nectar depends on carbon fixation by the plant (Xu & Chen, 2015) and sugar

Table 8.2 Effects of Environmental Factors on Defense Expression in Ant-Visited Plants

Environmental factors	Plant species	Type of mutualism	Response	Region	Source
Light intensity	*Cecropia* species	Obligate	Increase in carbon-rich FB production and decrease in lipid-rich FBs	Central America	Folgarait & Davidson (1994)
Light intensity	*Mallotus japonicus*	Facultative	Increase in size of EFNs and decrease in FB production	East Asia	Yamawo & Hada (2010); Yamawo et al. (2014)
Light intensity	*Phaseolus lunatus*	Facultative	Depends on light environment and herbivory	South America	Radhika et al. (2010)
Light intensity	*Inga* species	Facultative	Increase in EF nectar production	Central America	Bixenmann et al. (2011)
Light quality	*Passiflora edulis*	Facultative	Decrease in EF nectar induction by simulated herbivory	South America	Izaguirre et al. (2013)
Light intensity	*Senna mexicana* var. *chapmanii*	Facultative	Increase in EF nectar production	North America	Jones & Koptur (2015b)
Soil nutrients	*Cecropia* species	Obligate	Increase in carbon-rich and lipid-rich FB production	Central America	Folgarait & Davidson (1995)
Soil nutrients	*Macaranga triloba*	Obligate	Increase in FB production	Southeast Asia	Heil et al. (2001)
Soil nutrients	*Vicia faba*	Facultative	Depends on soil nutrient levels and leaf damage	World wide	Mondor & Addicott (2003); Mondor et al. (2006)
Soil nutrients	*Cordia alliodora*	Obligate	Increase in number of ants in domatia	Central America	Trager & Bruna (2006)
Soil nutrients	*Acacia constricta*	Facultative	Increase in EF nectar production	North America	Wagner & Fleur Nicklen (2010)
Soil nutrients	*Mallotus japonicus*	Facultative	Increase in EFN and FB production, decrease in trichome (physical defense trait) and pellucid dots (chemical defense trait)	East Asia	Yamawo et al. (2012b); Yamawo et al. (2014)
Soil nutrients	*Ricinus communis*	Facultative	No-effect	World wide	De Sibio & Rossi (2016)
Soil water	*Mallotus japonicus*	Facultative	Increase in EFN and FB production, decrease in trichome (physical defense trait) and pellucid dots (chemical defense trait)	East Asia	Yamawo et al. (2012a); Yamawo et al. (2014)
Soil water	*Populus tremuloides*	Facultative	Increase in EFNs	North America	Newman & Wagner (2013); Newman et al. (2015)

Note: FB – food body; EFN – extrafloral nectary; EF – extrafloral.

flux in the phloem (Millán-Cañongo et al., 2014). Consequently, in several EFN-bearing plants that have a mutualistic association with ants, abundant light results in increased production of carbon-based biotic defense traits, including EFNs and EF nectar, resulting in increased attraction of ants (Folgarait & Davidson, 1994; Yamawo & Hada, 2010; Bixenmann et al., 2011; Jones & Koptur, 2015b). The induction of EF nectar is also found under high light conditions (Radhika et al., 2010). Plants that produce both carbon- and nitrogen-based rewards change the balance of these in relation to light conditions. For instance, Folgarait and Davidson (1994) demonstrated that in *Cecropia* (Urticaceae), an ant-visited plant species with an obligate ant-plant mutualism, Müllerian bodies rich in glycogen are produced as a reward for ants under high light conditions, whereas lipid-rich pearl bodies are produced under low light conditions. Yamawo and Hada (2010) also found a negative relationship between two indirect defenses in *M. japonicus*; under low light conditions, the size of EFNs that had a positive relationship to productivity was reduced, while the production of pearl bodies increased. Similarly, inductions of EF nectar by JA, JA-Ile (JA conjugated with amino acid isoleucine) or leaf damage treatments are large under high light conditions (Radhika et al., 2010). These results support the CNB hypothesis (Bryant et al., 1983) and suggest that it can powerfully predict different rewards, such as EFN and FB development, among individuals. On the other hand, diurnal patterns of EF nectar production, which peak during the night, were observed in *Macaranga tanarius* and *Phaseolus lunatus* (Leguminosae) (Heil et al., 2000; Radhika et al., 2010). Intensity of extrafloral nectar induction depends on the time of leaf damage (Jones & Koptur, 2015a). These diurnal patterns of EF nectar production may adapt to consumer and pollinator activity (Heil et al., 2000; Raine et al., 2002) rather than to the effect of light intensity.

In addition to light availability, light quality is a strong regulator of indirect defenses (Radhika et al., 2010; Izaguirre et al., 2013 but see Jones & Koptur, 2015b). For example, Radhika et al. (2010) exposed no-treatment and JA or JA-Ile treatment *P. lunatus* plants to different ratios of red (R) to far-red (FR) radiation and measured EF nectar production. In response to increases in the ratio of R to FR radiation to 10:90 and 50:50 from 0:100, EF nectar production significantly increased in JA or JA-Ile treatment plants. In *Passiflora edulis* (Passifloraceae), low red:far red light ratios strongly suppressed the induction of EF nectar response triggered by simulated herbivory or methyl jasmonate application, and plants grew more (elongation of the stem and petiole length) (Izaguirre et al., 2013). Light quality is a constant strong regulator of indirect defenses; however, directions of regulation differ among plant species, and low R:FR ratio reflects the presence of light competitor plants. These differences among plant species may depend on variations in response to competitor plants; although light conditions influence the expression of direct defenses in numerous plant species (Walters, 2011), a change in direct defenses was not found (Folgarait & Davidson, 1994; Yamawo & Hada, 2010). Thus, although the effect of light quality on indirect defenses depends on species, intensity of light changes the types of indirect defenses, namely carbon-based EFN or nutrition-based FBs in ant-visited plant.

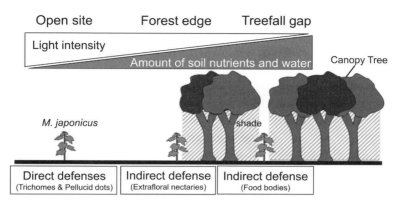

Figure 8.4.   Schematic illustration of habitat conditions and defense strategy of *Mallotus japonicus* in relation to light intensity (information from Yamawo et al., 2014).

This plasticity in defense traits should be related to variations in abiotic habitat conditions in the field, and reflect the variations of defense strategies in plants. For example, a few studies have identified an effect of light availability on ant-plant mutualism in the field. Kersch and Fonseca (2005) conducted a field experiment to determine the modulation effect of the availability of light and nutrients on the interaction strength between *Inga vera* (Mimosoideae) and its associated ants. Although they did not estimate defense traits, they found that ants had a positive effect on plant growth under high light conditions. In contrast, soil nutrient conditions did not affect ant-plant protection mutualisms. de la Fuente and Marquis (1999) compared ant and herbivore abundance and the growth of ant visited and ant excluded *Stryphnodendron macrostachyum* (Leguminosae) saplings grown in sunlight and shaded microhabitats of a plantation in Costa Rica over a seven-month period. Plants in shaded conditions were more frequently visited and protected by ants than those in sunlight conditions, probably due to lower herbivore attack in the sunlight condition. Leaf water content was also higher in the shade than in sunlight; however, the chemical defense substances of *S. macrostachyum*, including saponins and tannins, were not different under varied light conditions.

Yamawo et al. (2014) reported on the plasticity and efficacy of defense strategies against herbivory in ant-visited plants growing in variable abiotic habitat conditions in the field (Figure 8.4); *Mallotus japonicus* saplings expressed different defense strategies in three habitats in relation to abiotic conditions. Trichomes and pellucid dots were expressed to a greater degree at open sites with high light intensity and low soil nutrients or moisture than at forest edges and treefall gaps with low light intensity and high soil nutrients or moisture. In fact, leaves collected from the open sites were consumed less by the generalist herbivore *Spodoptera litura* than were those from other habitats. This result indicates that direct defenses function to a greater degree at open sites in *M. japonicus*. At open sites, *M. japonicus* favors carbon-based physical and chemical defenses, such as trichomes and pellucid dots, because of high light intensity (Bryant et al., 1983), and low expression of direct defense traits at forest edges and treefall gaps may be attributed to high production costs. In these habitats, *M. japonicus* must allocate more of its resources to vertical growth to outcompete other trees that are also attempting to fill the gap in the canopy. Under restricted light conditions, the expression of physical and chemical defense traits is costly for plants that are not shade tolerant, such as herbs and pioneer tree species (Kitajima et al., 2012). Plants exhibiting a high expression of chemical defenses often possess low competitive ability (Ballhorn et al., 2014). Therefore, at forest edges and treefall gaps, plants that are not shade tolerant are expected to decrease their expression of direct defense traits and allocate more resources to their own growth (Dalling & Hubbell, 2002; Tripathi & Raghubanshi, 2013). Although EFNs are a carbon-based defense trait, the production of EF nectar is inhibited in dry environments such as open sites. Instead of physical and chemical defenses, under the nutrient- and water-rich conditions of forest edges and treefall gaps, plants depend on indirect defenses because they have the resources necessary to produce EF nectar and FBs. The adoption of low-cost indirect defenses contributes not only to defensive functions, but also allows increased resource allocation to growth (Yamawo et al., 2014, 2015). Furthermore, food bodies are abundant in treefall gaps (Figure 8.4).

Chloroplasts/plastids play an important role in lipid metabolism. Mutations in several genes that encode chloroplast/plastid-localized proteins alter salicylic acid (SA) synthesis and plant defense signaling (Wildermuth et al., 2001; Slaymaker et al., 2002), suggesting that the SA pathway is important for food body production. SA also leads to an increase in tri-unsaturated fatty acids (Belkhadi et al., 2010), and its pathway depends on a low red to far-red (R:FR) ratio (Ballaré et al., 2012). The R:FR ratio decreases more than ultraviolet during canopy closure, which consequently regulates the JA pathway (Fournier et al., 2004). Thus, the SA pathway is active in treefall gaps. Indeed, because plants express nitrogen-based defenses such as food bodies (Walters, 2011), coherent variation in defense strategies in ant-visited plants appears to be in relation to abiotic habitat conditions. However, such empirical studies are limited to a few species, and studies should be extended to other species or systems to reveal the general pattern of phenotype plasticity in the defenses of ant-visited plants.

Interestingly, a similar pattern of variations in defense relationships to habitat conditions were found among *Cecropia* and *Macaranga* species which included both obligate and facultative systems. In *Cecropia* and *Macaranga* species, obligate ant-plant mutualism using food bodies was intensive in species growing in treefall gaps and on the forest floor under low light conditions (Davidson & Fisher, 1991; Davidson & McKey, 1993; Itioka, 2005). In contrast, in these species, facultative ant-plant mutualism using EFNs and direct defenses was intensive in species growing in open sites, forest edges and treefall gaps. Thus, specialization-generalization in an ant-plant mutualism system may depend on abiotic habitat conditions as well as the diversity of habitats. Further investigation of relationships between abiotic habitat conditions and intra- interspecific variations in plant defenses would be useful for understanding the evolution of ant-plant mutualism systems.

Conclusions and Future Research Objectives

We reviewed the study of the costs and plasticity of plant defense strategies involving direct defenses and indirect defense by ants and identified gaps in our knowledge. Even after accepting that ant-visited plants often demonstrate other direct defenses, many researchers have focused only on ant-plant relationships, as they appear to have concentrated on the evolution and maintenance mechanisms in systems of mutualism rather than plant defense. Under such a context, the relative costs of direct plant defense and indirect defenses by ants have tended to be overlooked, with the exception of a few studies. The relative costs of different ant-plant mutualism systems or in unrelated species should be investigated to understand the evolution of indirect defense by ants. Plasticity in response to JA treatment (simulated leaf damage) or ant visitation that does not influence resource availability and relative effects of direct and indirect defense may facilitate the comparisons of costs. Ultimately, a complete understanding of the relative costs of defenses will be accomplished only by revealing the molecular mechanisms involved in both direct and indirect defenses.

Moreover, our review found that many ant-visited plants alter the expression of indirect defense traits, or the balance of direct and indirect defenses, in relation to abiotic habitat conditions. Such plasticity in defenses suggests that a defense system involving ants has evolved through interaction with other defense systems within the heterogeneity of abiotic habitat conditions. Despite the variations in defense strategy in relation to abiotic habitat conditions found not only in intraspecific plasticity but also in interspecific differences – including facultative and obligate ant plant mutualism – studies of the relationship between abiotic habitat conditions and defense strategies in ant-visited plants are limited. The study of the defense strategies of ant-visited plants using multiple defense systems may accelerate understanding the ecology and evolution of ant-plant protection mutualisms. We should be focusing our attention on the differences in costs in relation to abiotic habitat conditions. The classical theories of plant defense strategies such as the optimal defense, CNB, GDB and RA hypotheses are useful for understanding the general patterns of ant-plant mutualism.

In summary, we should pay attention to the general and phylogeny dependent patterns of variations in direct and indirect defenses of ant-visited plants in relation to abiotic habitat conditions, in addition to biotic conditions, and the differences in these costs. Revealing general patterns in both variations in abiotic habitat conditions and ant-plant mutualism may enable the prediction of effects of artificial disturbance or climate change on ecology as well as the evolution of ant-plant mutualism. Forest disturbances, such as forest fragmentation due to road construction and defoliation, alter habitat conditions and affect ant diversity and abundance

(Philpott et al., 2010). Light intensity and water availability increase and decrease with disturbance, respectively, and their relationship to soil nutrients is complex (Ellsworth & Reich, 1992; García et al., 2006). Such changes in habitat conditions may alter ant-plant mutualism systems, including facultative and obligate systems (Maschwitz et al., 1996; Murase et al., 2003). Thus, the study of ant-plant mutualism in the context of plant defense strategies using multiple defense traits in relation to abiotic habitat conditions is useful for understanding the dynamics of ant-plant mutualism in a changing world.

References

Agrawal, A. A. and Spiller, D. A. (2004). Polymorphic buttonwood: effects of disturbance on resistance to herbivores in green and silver morphs of a Bahamian shrub. *American Journal of Botany*, 91, 1990–1997.

Ballaré, C. L., Mazza, C. A., Austin, A. T. and Pierik, R. (2012). Canopy light and plant health. *Plant Physiology*, 160, 145–155.

Ballhorn, D. J., Godschalx, A. L., Smart, S. M., Kautz, S. and Schädler, M. (2014). Chemical defense lowers plant competitiveness. *Oecologia*, 176, 811–824.

Barton, A. M. (1986). Spatial variation in the effect of ants on extrafloral nectary plant. *Ecology*, 67, 495–504.

Beattie, A. J. (1985). *The Evolutionary Ecology of Ant-Plant Mutualisms*. New York: Cambridge University Press.

Belkhadi, A., Hediji, H., Abbes, Z. et al. (2010). Effects of exogenous salicylic acid pre-treatment on cadmium toxicity and leaf lipid content in *Linum usitatissimum* L. *Ecotoxicology and Environmental Safety*, 73, 1004–1011.

Bentley, B. L. (1977). Extrafloral nectaries and protection by pugnacious bodyguards. *Annual Review of Ecology and Systematics*, 8, 407–427.

Bixenmann, R. J., Coley, P. D. and Kursar, T. A. (2011). Is extrafloral nectar production induced by herbivores or ants in a tropical facultative ant-plant mutualism? *Oecologia*, 165, 417–425.

Bryant, J. P., Chapin III, F. S. and Klein, D. R. (1983). Carbon/nutrient balance of boreal plants in relation to vertebrate herbivory. *Oikos*, 40, 357–368.

Bryant, J. P., Chapin III, F. S., Reichardt, P. B. and Clausen, T. P. (1987). Response of winter chemical defense in Alaska paper birch and green alder to manipulation of plant carbon/nutrient balance. *Oecologia*, 72, 510–514.

Chamberlain, S. A. and Holland, J. N. (2009). Quantitative synthesis of context dependency in ant-plant protection mutualisms. *Ecology*, 90, 2384–2392.

Chen, X., Adams, B., Bergeron, C., Sabo, A. and Hooper-Bùi, L. (2014). Ant community structure and response to disturbances on coastal dunes of Gulf of Mexico. *Journal of Insect Conservation*, 19, 1–13.

Choh, Y., Kugimiya, S. and Takabayashi, J. (2006). Induced production of extrafloral nectar in intact lima bean plants in response to volatiles from spider mite-infested conspecific plants as a possible indirect defense against spider mites. *Oecologia*, 147, 455–460.

Cogni, R., Freitas, A. V. and Oliveira, P. S. (2003). Interhabitat differences in ant activity on plant foliage: ants at extrafloral nectaries of *Hibiscus pernambucensis* in sandy and mangrove forests. *Entomologia Experimentalis et Applicata*, 107, 125–131.

Coley, P. D., Bryant, J. P. and Chapin, F. S. (1985). Resource availability and plant antiherbi-vore defense. *Science*, 230, 895–899.

Dalling, J. W. and Hubbell, S. P. (2002). Seed size, growth rate and gap microsite condi-tions as determinants of recruitment success for pioneer species. *Journal of Ecology*, 90, 557–568.

Davidson, D. W. and Fisher, B. L. (1991). Symbiosis of ants with *Cecropia* as a function of light regime. *In Ant-Plant Interactions,* Huxley, C. R. and Cutler, D. F. (eds.). Oxford: Oxford University Press, pp. 289–309.

Davidson, D. W. and McKey, D. (1993). Ant-plant symbioses: stalking the chuyachaqui. *Trends in Ecology and Evolution*, 8, 326–332.

de la Fuente, M. A. S. and Marquis, R. J. (1999). The role of ant-tended extrafloral nec-taries in the protection and benefit of a Neotropical rainforest tree. *Oecologia*, 118, 192–202.

De Sibio, P. R. and Rossi, M. N. (2016). Interaction effect between herbivory and plant fertilization on extrafloral nectar production and on seed traits: An experimental study with *Ricinus communis* (Euphorbiaceae). *Journal of Economic Entomology*, 109, 1612–1618.

Dyer, L. A., Dodson, C. D., Beihoffer, J. and Letourneau, D. K. (2001). Trade-offs in anti-herbivore defenses in *Piper cenocladum*: ant mutualists versus plant secondary metabo-lites. *Journal of Chemical Ecology*, 27, 581–592.

Eck, G., Fiala, B., Linsenmair, K. E., Hashim, R. B. and Proksch, P. (2001). Trade-off between chemical and biotic antiherbivore defense in the South East Asian plant genus *Macaranga*. *Journal of Chemical Ecology*, 27, 1979–1996.

Ellsworth, D. S. and Reich, P. B. (1992). Water relations and gas exchange of *Acer saccharum* seedlings in contrasting natural light and water regimes. *Tree Physiology*, 10, 1–20.

Falcão, J. C. F., Dáttilo, W. and Izzo, T. J. (2014). Temporal variation in extrafloral nec-tar secretion in different ontogenic stages of the fruits of *Alibertia verrucosa* S. Moore (Rubiaceae) in a Neotropical savanna. *Journal of Plant Interactions*, 9, 137–142.

Fischer, R. C., Richter, A., Wanek, W. and Mayer, V. (2002). Plants feed ants: food bodies of myrmecophytic *Piper* and their significance for the interaction with *Pheidole bicornis* ants. *Oecologia*, 133, 186–192.

Folgarait, P. J. and Davidson, D. W. (1994). Antiherbivore defenses of myrmecophytic *Cecropia* under different light regimes. *Oikos*, 71, 305–320.

  (1995). Myrmecophytic *Cecropia*: antiherbivore defenses under different nutrient treat-ments. *Oecologia*, 104, 189–206.

Fournier, A. R., Gosselin, A., Proctor, J. T. et al. (2004). Relationship between understory light and growth of forest-grown American ginseng (*Panax quinquefolius* L.). *Journal of the American Society for Horticultural Science*, 129, 425–432.

Frederickson, M. E., Ravenscraft, A., Arcila Hernández, L. M. et al. (2013). What happens when ants fail at plant defence? *Cordia nodosa* dynamically adjusts its investment in both direct and indirect resistance traits in response to herbivore damage. *Journal of Ecology*, 101, 400–409.

Frederickson, M. E., Ravenscraft, A., Miller, G. A. et al. (2012). The direct and ecological costs of an ant-plant symbiosis. *The American Naturalist*, 179, 768–778.

García, L. V., Maltez-Mouro, S., Pérez-Ramos, I. M., Freitas, H. and Marañón, T. (2006). Counteracting gradients of light and soil nutrients in the understorey of Mediterranean oak forests. *Web Ecology*, 6, 67–74.

Heil, M. (2010). Plastic defense expression in plants. *Evolutionary Ecology* 24, 555–569.

(2013). Let the best one stay: screening of ant defenders by *Acacia* host plants functions independently of partner choice or host sanctions. *Journal of Ecology*, 101, 684–688.

(2015). Extrafloral nectar at the plant-insect interface: a spotlight on chemical ecology, phenotypic plasticity, and food webs. *Annual Review of Entomology*, 60, 213–232.

Heil, M., Baumann, B., Krüger, R., and Linsenmair, K. E. (2004). Main nutrient compounds in food bodies of Mexican *Acacia* ant-plants. *Chemoecology*, 14, 45–52.

Heil, M., Delsinne, T., Hilpert, A. et al. (2002). Reduced chemical defence in ant-plants? A critical re-evaluation of a widely accepted hypothesis. *Oikos*, 99, 457–468.

Heil, M., Fiala, B., Baumann, B. and Linsenmair, K. E. (2000). Temporal, spatial and biotic variations in extrafloral nectar secretion by *Macaranga tanarius*. *Functional Ecology*, 14, 749–757.

Heil, M., Fiala, B., Kaiser, W. and Linsenmair, K. E. (1998). Chemical contents of *Macaranga* food bodies: adaptations to their role in ant attraction and nutrition. *Functional Ecology*, 12, 117–122.

Heil, M., Fiala, B., Linsenmair, K. E., Zotz, G. and Menke, P. (1997). Food body production in *Macaranga triloba* (Euphorbiaceae): a plant investment in anti-herbivore defence via symbiotic ant partners. *Journal of Ecology*, 85, 847–861.

Heil, M., Gonzàlez-Teuber, M., Clement, L. W. et al. (2009). Divergent investment strategies of *Acacia* myrmecophytes and the Academy coexistence of mutualists and exploiters. *Proceedings of the National of Sciences*, 106, 18091–18096.

Heil, M., Hilpert, A., Fiala, B. et al. (2002). Nutrient allocation of *Macaranga triloba* ant plants to growth, photosynthesis and indirect defence. *Functional Ecology*, 16, 475–483.

Heil, M., Hilpert, A., Krüger, R. and Linsenmair, K. E. (2004). Competition among visitors to extrafloral nectaries as a source of ecological costs of an indirect defence. *Journal of Tropical Ecology*, 20, 201–208.

Heil, M., Koch, T., Hilpert, A. et al. (2001). Extrafloral nectar production of the ant-associated plant, *Macaranga tanarius*, is an induced, indirect, defensive response elicited by jasmonic acid. *Proceedings of the National Academy of Sciences*, 98, 1083–1088.

Hemming, J. D. and Lindroth, R. L. (1999). Effects of light and nutrient availability on aspen: growth, phytochemistry, and insect performance. *Journal of Chemical Ecology*, 25, 1687–1714.

Herms, D. A. and Mattson, W. J. (1992). The dilemma of plants: to grow or defend. *Quarterly Review of Biology*, 67, 283–335.

Holland, J. N., Chamberlain, S. A. and Horn, K. C. (2010). Temporal variation in extrafloral nectar secretion by reproductive tissues of the senita cactus, *Pachycereus schottii* (Cactaceae), in the Sonoran Desert of Mexico. *Journal of Arid Environments*, 74, 712–714.

Howe, H. F. and Westley, L. C. (1988). *Ecological Relationships of Plants and Animals*. New York: Oxford University Press.

Huang, W., Siemann, E., Wheeler, G. S. et al. (2010). Resource allocation to defence and growth are driven by different responses to generalist and specialist herbivory in an invasive plant. *Journal of Ecology*, 98, 1157–1167.

Izaguirre, M. M., Mazza, C. A., Astigueta, M. S., Ciarla, A. M. and Ballaré, C. L. (2013). No time for candy: passionfruit (*Passiflora edulis*) plants down-regulate damage-induced extra floral nectar production in response to light signals of competition. *Oecologia*, 173, 213–221.

Izzo, T. J. and Vasconcelos, H. L. (2002). Cheating the cheater: domatia loss minimizes the effects of ant castration in an Amazonian ant-plant. *Oecologia*, 133, 200–205.

Itioka, T. (2005). Diversity of anti-herbivore defenses in *Macaranga*. In *Pollination Ecology and the Rain Forest: Sarawak Studies*, D. W. Roubik, S. Sakai and A. A. H. Karim (eds.). Ecological Studies, 174. New York: Springer, pp. 158–171.

Jones, I. M. and Koptur, S. (2015a). Dynamic extrafloral nectar production: The timing of leaf damage affects the defensive response in *Senna mexicana* var. *chapmanii* (Fabaceae). *American Journal of Botany* 102, 58–66.

(2015b). Quantity over quality: light intensity, but not red/far-red ratio, affects extrafloral nectar production in *Senna mexicana* var. *chapmanii*. *Ecology and Evolution*, 5, 4108–4114.

Katayama, N. and Suzuki, N. (2011). Anti-herbivory defense of two *Vicia* species with and without extrafloral nectaries. *Plant Ecology*, 212, 743–752.

Kersch, M. F. and Fonseca, C. R. (2005). Abiotic factors and the conditional outcome of an ant-plant mutualism. *Ecology*, 86, 2117–2126.

Kitajima, K., Llorens, A. M., Stefanescu, C. et al. (2012). How cellulose-based leaf toughness and lamina density contribute to long leaf lifespans of shade-tolerant species. *New Phytologist*, 195, 640–652.

Kobayashi, S., Asai, T., Fujimoto, Y. and Kohshima, S. (2008). Anti-herbivore structures of *Paulownia tomentosa*: morphology, distribution, chemical constituents and changes during shoot and leaf development. *Annals of Botany*, 101, 1035–1047.

Koptur, S. (1985). Alternative defenses against herbivores in *Inga* (Fabaceae: Mimosoideae) over an elevational gradient. *Ecology*, 66, 1639–1650.

(1992). Extrafloral nectary-mediated interactions between insects and plants. *Insect Plant Interactions*, 4, 81–129.

Koricheva, J. and Romero, G. Q. (2012). You get what you pay for: reward-specific trade-offs among direct and ant-mediated defences in plants. *Biology Letters*, 8, 628–630.

Kost, C. and Heil, M. (2005). Increased availability of extrafloral nectar reduces herbivory in Lima bean plants (*Phaseolus lunatus*, Fabaceae). *Basic and Applied Ecology*, 6, 237–248.

(2006). Herbivore-induced plant volatiles induce an indirect defence in neighbouring plants. *Journal of Ecology*, 94, 619–628.

(2008). The defensive role of volatile emission and extrafloral nectar secretion for lima bean in nature. *Journal of Chemical Ecology*, 34, 2–13.

Lach, L., Parr, C. L. and Abott, K. L. (2010). *Ant Ecology*. New York: Oxford University Press.

Letourneau, D. K. and Barbosa, P. (1999). Ants, Stem Borers, and Pubescence in *Endospermum* in Papua New Guinea1. *Biotropica*, 31, 295–302.

LeVan, K. E., Hung, K. L. J., McCann, K. R., Ludka, J. T. and Holway, D. A. (2014). Floral visitation by the Argentine ant reduces pollinator visitation and seed set in the coast barrel cactus, *Ferocactus viridescens*. *Oecologia*, 174, 163–171.

Li, T., Holopainen, J. K., Kokko, H., Tervahauta, A. I. and Blande, J. D. (2012). Herbivore-induced aspen volatiles temporally regulate two different indirect defences in neighbouring plants. *Functional Ecology*, 26, 1176–1185.

Malé, P. J. G., Leroy, C., Dejean, A., Quilichini, A. and Orivel, J. (2012). An ant symbiont directly and indirectly limits its host plant's reproductive success. *Evolutionary Ecology*, 26, 55–63.

Maschwitz U., Fiala B., Davies S. J. and Linsenmair K. E. (1996). A south-east asian myrmecophyte with two alternative inhabitants: *Camponotus* or *Crematogaster* as partners of *Macaranga lamellate*. *Ecotropica* 2, 26–132.

McKey, D. (1974). Ant-plants: selective eating of an unoccupied Barteria by a Colobus monkey. *Biotropica*, 6, 269–270.

Millán-Cañongo, C., Orona-Tamayo, D. and Heil, M. (2014). Phloem sugar flux and jasmonic acid-responsive cell wall invertase control extrafloral nectar secretion in *Ricinus communis*. *Journal of Chemical Ecology*, 40, 760–769.

Miller, T. E. (2014). Plant size and reproductive state affect the quantity and quality of rewards to animal mutualists. *Journal of Ecology*, 102, 496–507.

Mithöfer, A. and Boland, W. (2012). Plant defense against herbivores: chemical aspects. *Annual Review of Plant Biology*, 63, 431–450.

Mody, K. and Linsenmair, K. E. (2004). Plant-attracted ants affect arthropod community structure but not necessarily herbivory. *Ecological Entomology*, 29, 217–225.

Mondor, E. B. and Addicott, J. F. (2003). Conspicuous extrafloral nectaries are inducible in *Vicia faba*. *Ecology Letters*, 6, 495–497.

Murase, K., Itioka, T., Nomura, M. and Yamane, S. (2003). Intraspecific variation in the status of ant symbiosis on a myrmecophyte, *Macaranga bancana*, between primary and secondary forests in Borneo. *Population Ecology*, 45, 221–226.

Ness, J. H. (2003). *Catalpa bignonioides* alters extrafloral nectar production after herbivory and attracts ant bodyguards. *Oecologia*, 134, 210–218.

Ness, J. H. (2006). A mutualism's indirect costs: the most aggressive plant bodyguards also deter pollinators. *Oikos*, 113, 506–514.

Newman, J. R. and Wagner, D. (2013). The influence of water availability and defoliation on extrafloral nectar secretion in quaking aspen (*Populus tremuloides*). *Botany*, 91, 761–767.

Newman, J. R., Wagner, D. and Doak, P. (2015). Impact of extrafloral nectar availability and plant genotype on ant (Hymenoptera: Formicidae) visitation to quaking aspen (Salicaceae). *The Canadian Entomologist*, 148, 1–7.

Nogueira, A., Rey, P. J. and Lohmann, L. G. (2012). Evolution of extrafloral nectaries: adaptive process and selective regime changes from forest to savanna. *Journal of Evolutionary Biology*, 25, 2325–2340.

O'Dowd, D. J. (1979). Foliar nectar production and ant activity on a neotropical tree, *Ochroma pyramidale*. *Oecologia*, 43, 233–248.

   (1982). Pearl bodies as ant food: an ecological role for some leaf emergences of tropical plants. *Biotropica*, 14, 40–49.

Oliveira, K. N., Coley, P. D., Kursar, T. A. et al. (2015). The effect of symbiotic ant colonies on plant growth: a test using an *Azteca-Cecropia* system. *PloS one*, 10, e0120351.

Palmer, T. M., Stanton, M. L., Young, T. P. et al. (2008). Breakdown of an ant-plant mutualism follows the loss of large herbivores from an African savanna. *Science*, 319, 192–195.

Philpott, S. M., Perfecto, I., Armbrecht, I. and Parr, C. L. (2010). Ant diversity and function in disturbed and changing habitats. *Ant Ecology*. New York: Oxford University Press, 137–157.

Pulice, C. E. and Packer, A. A. (2008). Simulated herbivory induces extrafloral nectary production in *Prunus avium*. *Functional Ecology*, 22, 801–807.

Radhika, V., Kost, C., Mithöfer, A. and Boland, W. (2010). Regulation of extrafloral nectar secretion by jasmonates in lima bean is light dependent. *Proceedings of the National Academy of Sciences*, 107, 17228–17233.

Raine, N. E., Willmer, P. and Stone, G. N. (2002). Spatial structuring and floral avoidance behavior prevent ant-pollinator conflict in a Mexican ant-acacia. *Ecology*, 83, 3086–3096.

Rico-Gray, V. and Oliveira, P. S. (2007). *The Ecology and Evolution of Ant-Plant Interactions*. London: University of Chicago Press.

Risch, S. J. and Rickson, F. R. (1981). Mutualism in which ants must be present before plants produce food bodies. *Nature*, 291, 149–150.

Rosumek, F. B., Silveira, F. A., Neves, F. D. S. et al. (2009). Ants on plants: a meta-analysis of the role of ants as plant biotic defenses. *Oecologia*, 160, 537–549.

Rudgers, J. A. and Strauss, S. Y. (2004). A selection mosaic in the facultative mutualism between ants and wild cotton. *Proceedings of the Royal Society of London B: Biological Sciences*, 271, 2481–2488.

Rudgers, J. A., Strauss, S. Y. and Wendel, J. F. (2004). Trade-offs among anti-herbivore resistance traits: insights from *Gossypieae* (Malvaceae). *American Journal of Botany*, 91, 871–880.

Rutter, M. T. and Rausher, M. D. (2004). Natural selection on extrafloral nectar production in *Chamaecrista fasciculata*: the costs and benefits of a mutualism trait. *Evolution*, 58, 2657–2668.

Schupp, E. W. and Feener, D. H. (1991). Phylogeny, lifeform, and habitat dependence of ant-defended plants in a Panamanian forest. In *Ant-Plant Interactions,* Huxley, C. R. and Cutler, D. F. (eds.). Oxford: Oxford University Press, pp. 175–197.

Slaymaker, D. H., Navarre, D. A., Clark, D. et al. (2002). The tobacco salicylic acid-binding protein 3 (SABP3) is the chloroplast carbonic anhydrase, which exhibits antioxidant activity and plays a role in the hypersensitive defense response. *Proceedings of the National Academy of Sciences*, 99, 11640–11645.

Smith, L. L., Lanza, J. and Smith, G. C. (1990). Amino acid concentrations in extrafloral nectar of *Impatiens sultani* increase after simulated herbivory. *Ecology*, 71, 107–115.

Stamp, N. (2003). Out of the quagmire of plant defense hypotheses. *The Quarterly Review of Biology*, 78, 23–55.

Stanton, M. L. and Palmer, T. M. (2011). The high cost of mutualism: effects of four species of East African ant symbionts on their myrmecophyte host tree. *Ecology*, 92, 1073–1082.

Stephenson, A. G. (1982). The role of the extrafloral nectaries of *Catalpa speciosa* in limiting herbivory and increasing fruit production. *Ecology*, 63, 663–669.

Steward, J. L. and Keeler, K. H. (1988). Are there trade-offs among antiherbivore defenses in *Ipomoea* (Convolvulaceae)? *Oikos*, 53, 79–86.

Trager, M. D., Bhotika, S., Hostetler, J. A. et al. (2010). Benefits for plants in ant-plant protective mutualisms: a meta-analysis. *PLoS one*, 5, e14308.

Trager, M. D. and Bruna, E. M. (2006). Effects of plant age, experimental nutrient addition and ant occupancy on herbivory in a neotropical myrmecophyte. *Journal of Ecology*, 94, 1156–1163.

Tripathi, S. N. and Raghubanshi, A. S. (2013). Seedling growth of five tropical dry forest tree species in relation to light and nitrogen gradients. *Journal of Plant Ecology*, 7, 250–263.

Wagner, D. and Fleur Nicklen, E. (2010). Ant nest location, soil nutrients and nutrient uptake by ant associated plants: Does extrafloral nectar attract ant nests and thereby enhance plant nutrition? *Journal of Ecology*, 98, 614–624.

Walters, D. (2011). *Plant Defense: Warding off Attack by Pathogens, Herbivores and Parasitic Plants*. Hoboken, NJ: John Wiley & Sons.

Wang, Y., Carrillo, J., Siemann, E. et al. (2013). Specificity of extrafloral nectar induction by herbivores differs among native and invasive populations of tallow tree. *Annals of Botany*, mct129.

Webber, B. L., Abaloz, B. A. and Woodrow, I. E. (2007). Myrmecophilic food body production in the understorey tree, *Ryparosa kurrangii* (Achariaceae), a rare Australian rainforest taxon. *New Phytologist*, 173, 250–263.

Weber, M. G. and Keeler, K. H. (2013). The phylogenetic distribution of extrafloral nectaries in plants. *Annals of Botany*, 111, 1251–1261.

Wäckers, F. L. and Bezemer, T. M. (2003). Root herbivory induces an above-ground indirect defence. *Ecology Letters*, 6, 9–12.

Wäckers, F. L., Zuber, D., Wunderlin, R. and Keller, F. (2001). The effect of herbivory on temporal and spatial dynamics of foliar nectar production in cotton and castor. *Annals of Botany*, 87, 365–370.

Wildermuth, M. C., Dewdney, J., Wu, G., and Ausubel, F. M. (2001). Isochorismate synthase is required to synthesize salicylic acid for plant defence. *Nature*, 414, 562–565.

Xu, F. F. and Chen, J. (2015). Extrafloral nectar secretion is mainly determined by carbon fixation under herbivore-free condition in the tropical shrub *Clerodendrum philippinum* var. *simplex*. *Flora-Morphology, Distribution, Functional Ecology of Plants*, 217, 10–13.

Yamawo, A. and Hada, Y. (2010). Effects of light on direct and indirect defences against herbivores of young plants of *Mallotus japonicus* demonstrate a trade-off between two indirect defence traits. *Annals of Botany*, 106, 143–148.

Yamawo, A., Hada, Y. and Suzuki, N. (2012a). Variations in direct and indirect defenses against herbivores on young plants of *Mallotus japonicus* in relation to soil moisture conditions. *Journal of Plant Research*, 125, 71–76.

Yamawo, A., Katayama, N., Suzuki, N. and Hada, Y. (2012b). Plasticity in the expression of direct and indirect defence traits of young plants of *Mallotus japonicus* in relation to soil nutritional conditions. *Plant Ecology*, 213, 127–132.

Yamawo, A., Tagawa, J., Hada, Y. and Suzuki, N. (2014). Different combinations of multiple defence traits in an extrafloral nectary bearing plant growing under various habitat conditions. *Journal of Ecology*, 102, 238–247.

Yamawo, A., Tokuda, M., Katayama, N., Yahara, T. and Tagawa, J. (2015). Ant-attendance in extrafloral nectar-bearing plants promotes growth and decreases the expression of traits related to direct defenses. *Evolutionary Biology*, 42, 191–198.

9 Interhabitat Variation in the Ecology of Extrafloral Nectar Production and Associated Ant Assemblages in Mexican Landscapes

Cecilia Díaz-Castelazo, Nathalia Chavarro-Rodríguez, and Victor Rico-Gray[*]

Introduction

Ant-plant mutualisms are model systems for understanding plant defense and the ecology and evolution of interspecific mutualism (Rico-Gray & Oliveira, 2007). There is evidence that ants attracted to the extrafloral nectar secreted by some plant species may act as natural enemies of herbivores, extrafloral nectar being thus an induced biotic defense of plants (congruent with optimal defense) (Heil, 2015). However, a remarkable source of variation in ant-plant mutualisms mediated by extrafloral nectar is provided by interhabitat variation and abiotic variables influencing plant physiology and nectar production (e.g. Rico-Gray et al., 2012). Carbon-rich resources such as extrafloral nectar should be less costly for plants to produce where carbon is unlimited, for example, when excess sunlight is available to favor high photosynthetic activity. This is a favored condition in sunlight-rich habitats such as forest edges (Bentley, 1976), coastal areas (Díaz-Castelazo et al., 2004), cerrado vegetation (Oliveira and Freitas, 2004), soil rich tropical habitats (Dáttilo et al., 2013b), deserts (Pemberton, 1988), forest canopies (Blüthgen et al., 2000) and tropical tree plantations (Chavarro-Rodríguez et al., 2013).

[*] We are especially grateful to Paulo R. Guimarães, Ingrid R. Sánchez-Galván and Rafael L. G. Raimundo, coauthors of the article "Long-term temporal variation in the organization of an ant-plant network," for their help with the analysis and interpretation of ant-plant ecological networks at La Mancha (CICOLMA). We thank Miguel Ángel Hernández-Villanueva, María de Jesús Fernández-Martínez, Elmy Gutierrez Barrera and Vania Ramírez-Flores, former students of CD-C at project number 0106575, financed by the Consejo Nacional de Ciencia y Tecnología (CONACYT), for fieldwork support or data analysis. Fieldwork was partially financed by CONACYT through grant number 46840 to VR-G, scholarship number 224705 to NC-R, and by the Instituto de Ecología, AC (INECOL) project number 2003011143 to CD-C. Gibrán Pérez-Toledo helped CD-C with ant identification. We also thank María de Jesús Peralta Méndez, Arturo Bonet Ceballos and the personnel of CICOLMA for logistic support in the field.

Interactions between plants bearing extrafloral nectaries (EFNs) and foraging ants are also context-dependent regarding the outcome of the interaction. At some sites with high levels of herbivory (or derived negative impacts for plants) and good availability of ant foragers, a higher probability of interaction will exist in settings where they are more necessary and beneficial (Rudgers & Gardener, 2004; Rudgers & Strauss, 2004). Similarly, ant species identity (in the context of their foraging habits and biology, recruitment and competitive behavior), numerical dominance and the intimacy with the plant species upon which ants forage partly determine the interaction outcome. The outcomes of each interaction between ants and EFN-bearing plants vary depending on biotic changes in interacting organisms (internal factors such as phenology and species identity) as well as in their physical environment (external factors such as climatic variation), all of which may affect each interaction (Del-Claro et al., 2016). Compared to other mutualisms (e.g. pollination, Waser & Ollerton, 2006), potential ant-plant facultative mutualisms may be more stable over time within sites, due perhaps to the relative immobility and longevity of ant colonies. Furthermore, EFN-mediated ant-plant mutualism can be established quickly among non-native and invasive species (Chapter 12), indicating that its easy assembly is due to ecological fitting (Heil, 2015).

Many studies on the spatial variation of ant-plant interactions mediated by extrafloral nectar exist (see examples in Rico-Gray & Oliveira, 2007), focusing on many scales of variation (i.e. geographical, altitudinal, interhabitat, among populations, among plant individuals, etc.). The overall results include:

- For natural ecosystems, potentially mutualistic or commensal ant-plant interactions (e.g. see Rico-Gray & Oliveira, 2007) show high spatiotemporal variation in the interaction patterns, as well as in the outcomes. In the context of small-scale geographical variation (regional or interhabitat), plants with ant rewards such as extrafloral nectar or food bodies are more common (more abundant or taxonomically overrepresented) in secondary habitats than in mature forests (Schupp & Feener, 1991; Díaz-Castelazo et al., 2004). In their native habitats, plants with EFNs may be adapted to their ant visitors (and the other way around) through long-term interactions, but also specialized herbivores adapted to their host plants may outsmart the biotic defense system (Koptur & Lawton, 1988; Koptur, 2005).
- For facultative ant-plant mutualisms, the community-level pattern of interactions shows a strong turnover in species composition, but constancy in structural and functional community-level properties (Díaz-Castelazo et al., 2004), emerging as a nested interaction pattern (asymmetric specialization) with the presence of some "supergeneralist" species that maintain most links with their counterpart (Díaz-Castelazo et al., 2010). Furthermore, ants may affect EFN traits across plant populations of certain plant species (Rudgers & Strauss, 2004).
- In the case of obligate ant-plant mutualisms (Guimarães et al., 2007), the resultant ecological networks (community-level interaction pattern and specialization) show a high dependence of interacting partners, higher specialization and community-level compartmentalization, when compared with facultative ant-plant mutualisms.

- Daily turnover of ant species, with distinct ant assemblages for day and night periods, may potentially result in different outcomes of the interaction (Oliveira et al., 1999; Díaz-Castelazo et al., 2004; Dáttilo et al., 2014b).
- For managed or highly disturbed habitats, such as coffee plantations (e.g. Perfecto & Sediles, 1992; Perfecto & Vandermeer, 1994, 2002; Philpott et al., 2004a, 2004b), some ant species at plants with food rewards (shade trees) tend to remove herbivorous arthropods from cultivation, affecting arthropod densities of many different orders (with synergistic effects on the predators of herbivores; Chapter 17). For other cultivars and environments (e.g. Perfecto, 1990, 1991; van Rijn & Sabelis, 2005), the proximity of any plant species to those providing food (i.e. EFN-bearing plants) for natural enemies (such as aggressive ants) may locally increase the predation pressure on target herbivores and favor biological control. Despite these findings, the importance of EFN for the communities of plants and arthropods in invasive and agricultural ecosystems is still limited (Heil, 2015).

Insights gained from simpler agroecosystems suggest directions for investigation in natural situations. For example, tree species diversity in silvicultural systems may influence interactions between a focal plant species (mahogany) and ants (Campos-Navarrete et al., 2015) – where ants had a negative effect on sap feeders in mahogany monocultures, but did not influence these herbivores in polycultures, suggesting that ant effects were influenced by tree species diversity. The process behind these patterns could be higher niche diversity through enriched food sources for the arthropod herbivores, or simply that higher plant cover and complex vegetation structure provide herbivores with refuges from ant predators. Thus, natural systems (i.e. tropical deciduous forests) compared with tree monocultures (such as red cedar plantations) provide contrasting ant-plant-herbivore interaction patterns (Chavarro-Rodriguez et al., 2013); similar contrasts may occur in coffee plantations depending on the agroecological matrix where those crop plants are grown (Perfecto & Vandermeer, 2002).

Interhabitat Variation in the Abundance, Secretory Activity and Morphology of EFNs of Plant Species, and Their Associated Ant Community in a Coastal Tropical Ecosystem

In a coastal tropical seasonal ecosystem in Veracruz (Mexico), Díaz-Castelazo et al. (2004, 2005, 2013) investigated extrafloral nectar production and activity, nectar sugar content, morphology and distribution of EFNs, as well as their correlates with the associated ant community in terms of species composition and visitation frequency, both in different habitats and seasons. The abundance and natural occurrence of plants bearing EFNs at different plant communities vary greatly: plant cover of EFN-bearing plant species, estimated at two distinct time periods (years 2000 and 2010) in representative coastal vegetation types (Díaz-Castelazo et al., 2004; Sánchez-Galván et al., 2012), is higher in the pioneer dune

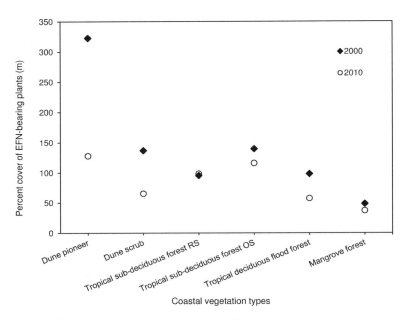

Figure 9.1.   Cover of plants with extrafloral nectaries (EFNs) in different coastal communities in Veracruz, Mexico. Vegetation types or plant communities have different transect lengths (cover expressed length occupied by EFN-bearing plants along transects). Based on data from Díaz-Castelazo et al., (2004); Sánchez-Galván et al. (2012).

vegetation, followed by tropical sub-deciduous forest and coastal dune scrub. The main differences in abundance of plants with EFNs are partly explained by mangrove forest having the lowest cover for these plants (Figure 9.1). A seasonal change in percent cover of EFN-bearing plant species was also recorded due to the occurrence of an extreme weather event, i.e. hurricane Karl, in September 2010 at the coastal ecosystems, which was the case for the 2010 sampling (Sánchez-Galván et al., 2012).

Detailed attributes for EFNs were also studied at the same coastal tropical ecosystem in Veracruz (Díaz-Castelazo et al., 2005), specifically for the 20 most abundant EFN-bearing plant species. Although the 20 selected plant species comprise less than a third of the EFN-bearing plant species richness in the study site, their joint abundance is about 61 percent of the overall abundance of plants with EFNs present (Díaz-Castelazo et al., 2004). Furthermore, the frequency of visits by ants (occurrence of an ant species on a plant with EFNs in each census) in the 20 selected plants constitutes 88 percent of their occurrence in all the EFN-bearing plants at the study site, suggesting the relevance of these plant species for potential facultative mutualism with ants.

Mean volume and sugar concentration of EF-nectar were estimated for those 20 plants species (Table 9.1): *Prestonia mexicana* (Apocynaceae), *Bidens pilosa* (Asteraceae), *Amphilophium paniculatum*, *Mansoa hymenaea*, *Tabebuia rosea* (Bignoniaceae), *Cordia spinescens* (Boraginaceae), *Opuntia stricta* (Cactaceae),

Table 9.1 Distribution of Extrafloral Nectaries (EFNs): Circumscribed (CIRC), Disperse/Discrete (DISCR)

Distribution	Morphology	Mean volume	Mean percentage of sugar	Plant species
CIRC	U	2.12	31.53	*Caesalpinia crista*
CIRC	U	0.78	32	*Crotalaria incana*
DISCR	C	1.6	3.5	*Bidens pilosa*
DISCR	C	2.19	16.4	*Macroptilium atropurpureum*
DISCR	P	0.65	7.25	*Cordia spinescens*
DISCR	P	0.67	13.78	*Prestonia mexicana*
DISCR	S	0.18	74.5	*Mansoa hymaenea*
DISCR	S	0.1	0.83	*Callicarpa acuminata*
DISCR	S	0.15	71	*Tabebuia rosea*
DISCR	S	1.25	12.14	*Amphilophium paniculatum*
DISCR	Ft	0.68	8.18	*Cedrela odorata*
CIRC	H	3.5	6.4	*Conocarpus erectus*
CIRC	H	0.6	11	*Ipomoea pescaprae*
CIRC	Fn	1.4	9.58	*Calopogonium caerulium*
CIRC	Fn	0.21	43.44	*Crotalaria incana*
CIRC	Fn	2.35	12.75	*Canavalia rosea*
CIRC	Fn	3.67	2.67	*Opuntia stricta*
CIRC	Fn	1.34	21.25	*Macroptilium atropurpureum*
CIRC	E	3.68	28.84	*Senna occidentalis*
CIRC	E	3.06	27.53	*Turnera ulmifolia*
CIRC	E	3	77.5	*Acacia cornigera*
CIRC	E	1.45	5	*Chamaecrista chamaecristoides*

EFN morphologies: elevated (E), flattened (Ft), hollow (H) and functional (Fn) EFNs. Non-vascularized EFNs included scale-like (S), capitate (C), peltate (P) and unicellular secretory trichomes (U). Mean volume (°Brix) and sugar concentration obtained from five individuals of each species.

Conocarpus erectus (Combretaceae), *Ipomoea pescaprae* (Convolvulaceae), *Acacia cornigera*, *Caesalpinia crista*, *Calopogonium caeruleum*, *Canavalia rosea*, *Chamaecrista chamaecristoides*, *Crotalaria incana*, *Macroptilium atropurpureum* and *Senna occidentalis* (all Fabaceae), *Cedrela odorata* (Meliaceae), *Turnera ulmifolia* (Turneraceae) and *Callicarpa acuminata* (Verbenaceae).

Extrafloral nectar was collected from a set of glands (the "sets" are described in the pages that follow) from five to ten plant individuals of each species, after 12 h of accumulation (overnight) using standard-size paper bags and tanglefoot (Tanglefoot Co., Grand Rapids, Michigan) to exclude insects. Among plant species, equivalent surfaces or sets of EFNs were bagged to reduce as much as possible the heterogeneity of size or distribution of the glands among plant species: two terminal branches or whorls per individual, whether on leaves, nodes or reproductive structures, depending on the EFN location in each plant species. Extrafloral nectar was cumulatively obtained from these sets of glands within a plant individual (one measure per individual). Nectar volume was estimated using 1- or 5-mL disposable microcapillary tubes, and nectar concentration was measured with 0–32 percent, 28–62 percent and 58–92 percent Bausch and Lomb sugar hand-held refractometers (which measure sugar concentration on a weight/weight basis). When nectar was evident but too viscous to permit measurements, a known amount of distilled water was applied to the set of previously bagged nectaries. The solution was then collected and the proportional volume (V) and sugar concentration (C) was calculated [C1 = (V2C2)/V1].

Detailed attributes for EFNs, including gland morphology, distribution among plant organs and the mean volume secretion and sugar concentration of extrafloral nectar, are shown in Table 9.1. Plant species differ in their EFN attributes. Vascularized glands included the following morphologies: elevated (E), flattened (Ft), hollow (H) and functional (Fn) EFNs. Non-vascularized EFNs included scale-like (S), capitate (C), peltate (P) and unicellular secretory trichomes (U) (Díaz-Castelazo et al., 2005, modified from Zimmerman, 1932). Regarding these morphologies, elevated and functional EFNs produced higher mean nectar volumes. Regarding the distribution of EFNs among plant organs, we used a general characterization (Díaz-Castelazo et al., 2005), differentiating the EFNs which are glands circumscribed to particular plant organs (at specific or modular locations) from the EFNs that are disperse among plant organs (i.e. secretory trichomes on leaf surfaces). Plant species that have circumscribed nectaries (CIRC) produced larger mean nectar volumes (2.06 µl) than those plants with disperse or discrete (DISCR) nectaries (0.53 µl). However, the amount of active glands in a plant individual may be higher for disperse nectaries, since these glands are structurally simpler than those of circumscribed nectaries.

A yearly census for ants foraging on the EFNs of 20 selected plant species at the study site (Díaz-Castelazo et al., 2004) revealed that the frequency of ants foraging on the different EFN morphological types clearly differs ($\chi^2 = 1091.7$). Moreover, the range of total associated ant species visiting plants within each type of nectary is different among EFN distribution types. The range of visits to circumscribed nectaries was between 9 and 17 ant species, while it was between 20 and 23 ant species for disperse/discrete nectaries. Thus, similar to gland morphology, EFN distribution in plant organs may influence visitation rates, but mostly by a distinct composition of associated ant species.

Ant Assemblages at EFN-Bearing Plants and Emergent Networks of Ant-Plant Interactions: Long-Term Spatiotemporal Variation

Frequency and abundance of ants foraging on the extrafloral nectar of all EFN-bearing plant species at the same coastal tropical ecosystem in Veracruz was also estimated but for a long-term period, including and contrasting three censuses in three decades (for two years in each decade: 1990, 2000 and 2010) (Figure 9.2; Díaz-Castelazo et al., 2013). Overall, 54 ant species were recorded foraging at the EFNs of 76 plant species. Richness of ant-plant interactions (i.e. the number of interactions between ant and plant species) was 142, 186 and 215, for the 1990 period, the 2000 period and the 2010 period, respectively, increasing, thus, with time. In contrast, the visitation frequencies (i.e. the summation of the number of times all pairwise ant-plant interactions occurred) decreased with time: 1588, 1352 and 620 for the 1990 period, the 2000 period and the 2010 period, respectively. The noticeably lower visitation frequencies recorded for the 2010 period could be the result of the abiotic disturbance caused by Hurricane Karl during September 2010 (Sánchez-Galván et al., 2012).

The graphic and analytical tool (derived from network theory) of complex networks of plant-animal interactions (or "ecological networks") (Jordano et al., 2003) provided further insight on the structure and specialization of the ant-plant communities (Díaz-Castelazo et al., 2013) mentioned earlier. A common structural pattern in plant-animal mutualistic interaction networks is nestedness (Bascompte et al., 2003), a non-random pattern of structure of the interacting community, with highly heterogeneous distributions of the number of interactions per species, where generalist species interact with most species and specialist species only with generalists; the most generalist plant and animal species interact among them generating a dense core of interactions to which the rest of the community is attached. The ant-plant interactions we studied for the 1990, 2000 and 2010 periods were all significantly nested, but the 1990 network had a slightly stronger nested pattern compared to the two other networks (Figure 9.2).

Another interesting trend is that while network size increases with time (from 142 to 215 pairwise interactions), the pattern was different between the ant and plant guilds. Communities had more plant species in the 1990 period, which decreased through time for the 2010 period (a real decline in the richness of EFN-bearing plants). However, fewer ant species were recorded visiting EFN-bearing plants in the 1990 period, and the number of ant species increased through time to the 2010 period.

Core species within an ecological network (Dáttilo et al., 2014a) are those that sustain the majority of interactions, that is, the species with more links (or higher degree), or those connected with high-degree species. The definition of species components of a central core or periphery within an interaction network is based in the formula proposed by Dáttilo et al. (2013a): $Gc = (ki- kmean) / (\sigma k)$ Where ki is the mean number of links for a given ant species i, $kmean$ is the mean number of links for all ant species in the network and σk is the standard deviation of the number of links for ant species. $Gc > 1$ are ant species present on the central core, and $Gc < 1$ are peripheral ant species. For our ant-plant study system, a more detailed analysis of core species composition for the 2000 period revealed that the core ant species were: *Camponotus planatus*, *Camponotus mucronatus*, and *Camponotus atriceps* (Díaz-Castelazo et al., unpublished data).

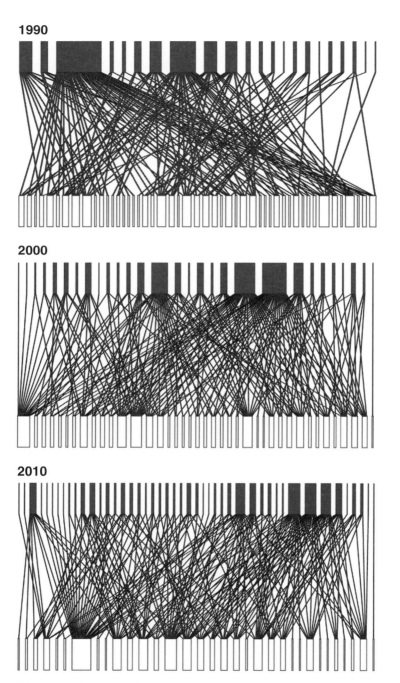

Figure 9.2.   Three-period extrafloral nectary-mediated ant-plant interaction networks (above) and ant degree values (linkage level) for each network (below), in 1990, 2000 and 2010 censuses. Each node is species-specific and identifies an ant (black) or a plant (white). The size of the nodes (squares) is proportional to species' contribution to network structure. Modified from Díaz-Castelazo et al. (2013), with permission from Oxford University Press.

Another network parameter used to assess relevance of species within a community is species strength. Bascompte et al. (2006) explain that for each plant-animal species interacting pair, there are two estimates of mutual dependence (defined in two adjacency matrices P and A): the dependence $dPij$ of plant species i on animal species j (i.e. the fraction of all animal visits coming from this particular animal species), and the dependence $dAji$ of animal species j on plant species i (i.e. the fraction of all visits by this animal species going to this particular plant species). Derived from these dependences at an interaction network, species strength is the sum of dependencies of each species across all its partners (counterpart) (Bascompte et al., 2006). The species strength of an animal species, for example, is defined as the sum of dependences of the plants relying on this animal, a measure then, of the importance of this animal from the perspective of the plant set. Species strength quantifies the importance of a given species for the community it interacts with. It sums to the number of species in the other group. In contrast to the dependence value of a given species that has a maximum of 1 (if it relies only on one partner), the value of species strength is relative to the number of species at the network. Thus, species strength values are more useful for comparisons with other species within the network.

For the ant-plant interactions that we studied for the 1990, 2000 and 2010 periods, some ant species had important strength values (in contrast to plants). Eight ant species had strength values above 1 (relevant) and the core ant species *Camponotus planatus* had strength values over 6; this species was the most relevant or important visitor of EFN-bearing plants.

The community-level ant-plant interactions studied are interhabitat, seasonally and temporally variable, because of high turnover in species composition (Díaz-Castelazo et al. 2010) and preferential attachment of species (imposed either by abiotic restrictions or by the biology of species). Other aspects that remain highly variable in community-level EFN-mediated ant-plant interactions are the ecological relevance of species (evidenced in core species identity or strength values). However, some aspects of the ant-plant interactions studied remained unaltered despite differing circumstances (not even disrupted after a hurricane): (1) the overall structural nested pattern of the assemblages or networks (Rico-Gray & Oliveira, 2007; Díaz-Castelazo et al., 2013); (2) the presence of ant core components, which frequently are highly behaviorally dominant nectar foragers (Dáttilo et al. 2014a), thus being potentially mutualistic or beneficial partners of certain plant species (Oliveira et al., 1999; Cuautle & Rico-Gray, 2003); (3) the differences among habitats (vegetation physiognomies) in the richness of ant-plant interactions, since more interactions occur in open habitats (such as pioneer dune vegetation or coastal dune scrub) and fewer interactions in shaded habitats such as tropical sub-deciduous forests. These interhabitat differences also occur among seasons, and significant habitat-season interaction of factors is common at our tropical ecosystem studied (Figure 9.3).

In the face of these overall findings, we can argue that despite regional scale geographical variation and among-habitat heterogeneity in vegetation structure (as

188   Cecilia Díaz-Castelazo et al.

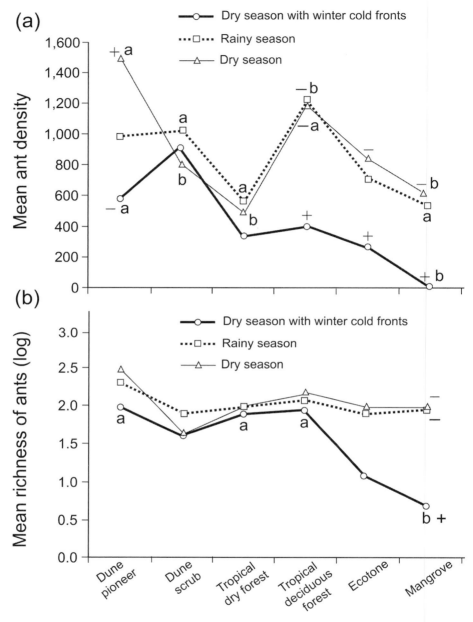

Figure 9.3. Mean values of ant density index (a) and ant species richness (b) for vegetation types and seasons. Significant differences at P < 0.05 (Tukey HSD test) for factor interactions are shown among vegetation types within a season as (a) and (b) and among seasons within vegetation types as (+) and (−). Vegetation types are the same (analogous) as those in Figure 9.1. Modified from Díaz-Castelazo et al. (2004), with permission from Taylor & Francis.

well as seasonality), the nested, asymmetric pattern of community-level ant-plant interactions is consistent in this coastal tropical ecosystem. The overall asymmetric structure (i.e. generalist species having most interactions) may reflect environmental conditions that favor interactions of some species. Increased photosynthetic activity of plants in open habitats may result in higher carbohydrate availability in extrafloral nectar, and thus increased attractiveness to ants, or a higher density of EFN-bearing plant life forms (such as vines). EFN-bearing plants growing in sunlight obtain a measurable benefit from ant visitation, whereas the same plant species growing under shaded conditions have no such benefit (Bentley, 1976; Kersch & Fonseca, 2005). For some plant species, the size of EFNs and nectar secretion are higher under intense light conditions compared to low light conditions (Yamawo & Hada, 2010), and a similar trend is found for the ant abundance foraging on these glands (Rudgers & Gardener, 2004; Yamawo & Hada, 2010; Chapter 8). In our study system, there is evidence that open habitats compared to shaded habitats (vegetation types with relatively dense canopy cover) do sustain different ant and plant species abundances and distinct floristic similarities (Díaz-Castelazo et al., 2004), and render also contrasting network patterns or species assemblages for other types of insect-plant interactions (López-Carretero et al., 2014). To further study the functional ecology of ant-plant interactions mediated by EFNs, an interhabitat approach centered on a single plant species – in order to reduce variation in EFN morphology or distribution and drastic differences in selective pressures – is useful. We will follow with such a case study.

Interhabitat Variation in the Activity of EFNs in Leaves of the Red Cedar *Cedrela odorata*, and Associated Ant Activity

Cedrela odorata (Meliaceae) is a native deciduous tree of tropical America (Pennington & Sarukhán, 2005). Given the characteristics of its wood, it has been considered a very valuable timber species and it has been intensively planted at tropical lowland sites in southeastern Mexico (Fernández-Martínez & Díaz-Castelazo, 2009). In this region, and specifically in the state of Veracruz, it naturally occurs in dry and semidecidous forests (as well as in derived transformed habitats). An important feature of this tree species in the context of the present study is that it has small but numerous EFNs dispersed over the young leaves and branches (Figure 9.4; Díaz-Castelazo et al., 2005; Chavarro-Rodriguez et al., 2013). These secretory structures are minute, somewhat oval (0-5mm long x 0-2mm wide), parenchymatose and closely pressed against the fundamental tissue of stems and branches (Díaz-Castelazo et al., 2005). The glands correspond to the "flattened nectaries" originally described by Zimmerman (1932), in which the glandular surface is scarcely above or just beneath the surface level of the surrounding mesophyll.

Previous studies at these sites (Hernández-Villanueva, 2010) showed that *C. odorata* trees have active EFNs both day and night and is frequently visited by ant foragers, with ant assemblages presenting important variation in species composition,

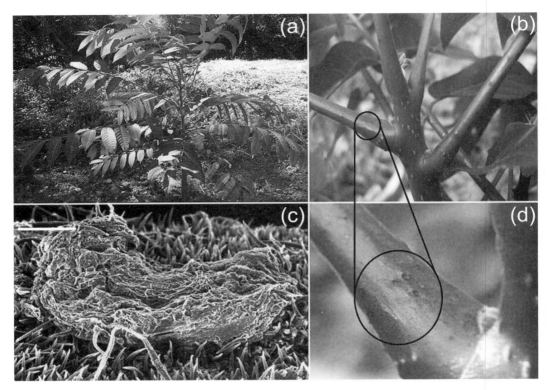

Figure 9.4.   (a) *Cedrela odorata* young plant typically growing in open habitat. (b) Distribution of extrafloral nectaries (EFNs) on the rachis. (c) Scanning electron microscopy (c = scale bar 1mm) showing the flattened morphology of the EFN. (d) Enlargement of a set of active EFNs. Photo credits: C. Díaz-Castelazo (a, c); M.A. Hernández-Villanueva (b, d).

and in the abundance of specific ant foragers. Distribution of the EFNs on plant vegetative tissue and gland morphology is shown in Figure 9.4.

On 40 randomly selected *C. odorata* trees per site, four leaves per individual were bagged with mesh bags to allow accumulation of extrafloral nectar. After 12 hours we estimated the proportion of active EFNs per plant, both day and night (Hernández-Villanueva, 2010). We found that the proportion of active EFNs at each site was generally lower in plantations than forests (Kruskal – Wallis test, $H' = 57.17$, $P < 0.0001$).

The average number of active EFNs per leaf was significantly different among study sites (Kruskal – Wallis test, $H' = 62.68$, $P < 0.005$). Day/night comparison of EFN activity revealed that at sites with natural vegetation such as La Mancha (Figure 9.5a), EFNs were more active at night: average of active glands per leaf was significantly higher at night (1.60 ± 0.33) than in the day (0.83 ± 0.15) (mean ± SE) (Wilcoxon paired test: $Z = -1.9225$, $P = 0.05$). At the site Cerro Gordo (Figure 9.5b), the average active glands per leaf was significantly higher at night (1.92 ± 0.30) than during the day (1.03 ± 0.16) (mean ± SE) (Wilcoxon paired test: $Z = -3.8813$, $P <$

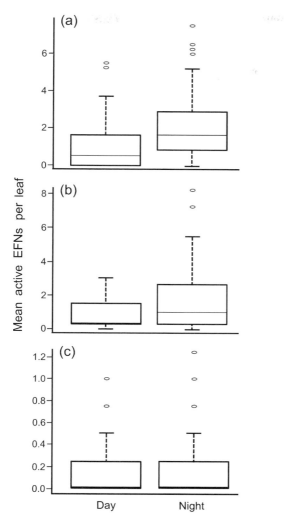

Figure 9.5.   Percentage of active extrafloral nectaries (EFNs) per site (4 leaves per tree, 40 individuals per site, all EFNs count, only glands with secretion were included in the percentage). Boxes show EFN activity during day and night periods (whiskers denote ± standard error). Sites are as follows: (a) La Mancha, (b) Cerro Gordo, (c) San José (respectively, natural forest, silvicultural agroecosystem and cedar plantation). Original data from BSc thesis by Hernández-Villanueva (2010).

0.005). The site with intensive tree management contrasted with both the forest and the agrosilvicultural sites: at San José (Figure 9.5c), the average of active glands per leaf did not differ between day and night (0.16 ± 0.04 and 0.15 ± 0.04) (Z = 0.424, $P > 0.05$).

In addition to the activity of EFNs, we studied interhabitat variation in all EFN secretory attributes of *C. odorata*, including plant physiology and associated ant assemblages, in different habitats that are part of a management gradient (from forest to monoculture plantation). Fieldwork was conducted in the central part of the state of Veracruz, during the rainy and dry seasons of 2010 and 2011, at three sites with different growth conditions (Fernández-Martínez & Díaz-Castelazo, 2009). These conditions depended on the degree of anthropogenic management: the first site is a semi-natural habitat at La Mancha (LM), the same coastal tropical ecosystem where we performed the ant-plant interaction studies described earlier, where *C. odorata* grows in tropical sub-deciduous forest. The second site is a silvicultural agroecosystem in Cerro Gordo (CG), where the six-year-old cedar plantation was not managed, and trees were heterogeneously distributed, surrounded both by a corn field and lowland tropical sub-deciduous and dry forest. The third site is a silvicultural ranch San José (SJ), where a 12-year-old cedar plantation is managed using standard practices for commercial planting, surrounded by introduced grasses and remnants of lowland tropical dry forest.

In all sites studied, *C. odorata* trees presented damage caused by the mahogany shoot borer *Hypsipylla grandella* (Lepidoptera, Pyralidae), which bores into apical shoots, thus producing ramifications (lateral branching). In plantations and forests of central Veracruz, there is a higher incidence of *H. grandella*-infested trees in plantations (92.5 percent) than in forests (42.5 percent), and overall a higher percentage of shoot damage at plantations than forests, where more damage outbreaks (incidence) occur (Fernández-Martínez & Díaz-Castelazo, 2009). Thus, *C. odorata* trees at the intensive plantation site (SJ) experienced noticeably higher herbivory rates compared to the other sites, forest and agroecosystem (Fernández-Martínez & Díaz-Castelazo, 2009).

As for the associated ant community, 20 ant species were found, with forest sites showing higher richness (7–12 species) than plantations (0–7 species). Ant frequency at EFNs was lower for the managed plantation at San José (which has fumigation practices) than at sites with no management ($H = 65.86$, $P < 0.0001$).

Also at San José we found a significant positive correlation between the total amount of EFNs and ant visitation frequency ($rho = 0.34$, $P < 0.05$). Similarly, at this plantation, which was the site with the highest damage levels by *H. grandella*, we found a significant positive correlation between the percentage of active EFNs and shoot borer damage ($rho = 0.36$, $P < 0.05$), suggesting herbivory-induced nectar secretion, presumably to recruit ants as natural enemies. Ants may have different potentials for controlling *H. grandella* in plantations and forests, so we should consider this jointly with abiotic and management-derived variables. Only at La Mancha was there a significant negative correlation found between the proportion of active EFNs and ant visitation frequency ($rho = -0.35$, $P < 0.05$). This is a natural, seasonal environment and factors other than EFN activity may influence ant activity, such as abiotic factors and the abundance of other EFN-bearing plants, or plants supporting arthropods producing honeydew. Given that some ant species foraging on cedar EFNs are opportunists, more attractive extrafloral nectar sources (i.e. plants providing higher volume or richer reward) may decrease ant visits to *C. odorata*.

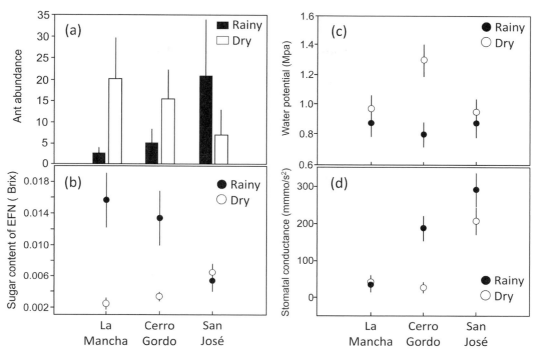

Figure 9.6. Values (±SE) of (a) ant abundance, (b) concentration of dissolved sugars in extrafloral nectar (EFN), (c) water balance (potential) and (d) stomatal conductance in compound leaves of *Cedrela odorata* in each of the growth environments in two seasons (rainy and dry) in central Veracruz, Mexico. (a) and (b) modified from Chavarro-Rodriguez et al. (2013), with permission from the Canadian Science Publishing.

Extrafloral Nectar Sugar Concentration and Sugar Composition, Water Balance and Abundance of Ant Foragers at *Cedrela odorata* EFNs

Estimates of sugar concentration in nectar and characterization of nectar sugars were also made (Chavarro-Rodríguez et al., 2013), but only for daytime hours. The concentration of sugars in the EFNs of individuals of *C. odorata* is generally higher during the rainy season, but this depends also on the site. In general, the mean sugar concentration was lower for the intensively managed site San José than for the more forested ones, but this is still season-dependent, being thus influenced both by the growing environment as well as by water availability (higher in the rainy season; see the results and factor interaction in Figure 9.6). As for the characterization of sugars in the extrafloral nectar of *C. odorata*, four main sugars were identified: glucose, fructose, sucrose and turanose. Glucose was present in the highest proportion, followed by sucrose and turanose, with somewhat different proportions in each habitat.

The interhabitat variation in EFN-mediated ant-plant interactions is noticeable for the functional ecology of *C. odorata* (Chavarro-Rodriguez et al., 2013). We investigated water balance attributes of *C. odorata* trees in different habitats and seasons, and its possible effects on extrafloral nectar attributes. Trees whose water balance was analyzed differed in their stomatal conductance (water loss by transpiration) among habitats, between seasons (rainy and dry season), and a significant habitat by season interaction was found. Individuals of *C. odorata* in habitats with greater degrees of perturbation (i.e. less tree cover at plantations) lost more water by transpiration than those in a more preserved environment (i.e. surrounded by natural vegetation). Thus, in the highly disturbed monoculture tree plantations, cedar trees lost more water, and this trend is reflected in the sugar content of extrafloral nectar. Especially in the dry season, the sugar content of extrafloral nectar is lower for cedar trees in the disturbed habitat than for those in preserved habitats (i.e. subdeciduous forest). However, during the rainy season when conditions are milder, the pattern of sugar content at the extrafloral nectar of cedar trees is the opposite – lower in the disturbed habitat. This could be explained by the fact that water potential, an estimate of the water available within a plant, did not differ among habitats or seasons. This fact implies that trees (Chavarro-Rodriguez et al., 2013) were not suffering much water stress, so in the rainy season (when less water loss takes place) the sugar content of *C. odorata* individuals increases, as appropriate for a biotic defense strategy.

Also, greater availability of water during the rainy season may explain that the highest concentration of extrafloral nectar was found during this season in the three growth environments of *C. odorata*. Plants with EFNs can modulate the volume and concentration of secreted nectar depending on the availability of useful substances in the phloem, given that sugars and amino acids are concentrated in cells under the EFNs where phloem transformation to nectar takes place (Pacini & Nepi, 2007).

The *Cedrela odorata* system allows us to assess the influence of different plant water conditions on the quality of extrafloral nectar. We found site (forest, agroecosystem, plantation) and season (i.e. rainy, dry) effects for both water balance and the quality of extrafloral nectar. Our results show that (1) water has a significant influence on extrafloral nectar quality and (2) environmental disturbance that differed at each site was associated with differences in the loss of water through transpiration. Season also affected the levels of dissolved sugars in extrafloral nectar. This had a direct, site-dependent impact on the abundance of ants foraging on EFNs.

Concomitant inter-habitat variation in the abundance of ants foraging at the EFNs of *C. odorata* was found, but with opposite patterns to nectar sugar concentration (Figure 9.6a, b). Sixteen ant genera were found at *C. odorata* trees considering both seasons in the three environments (Chavarro-Rodriguez et al., 2013), being the more abundant genera *Crematogaster* (419 records/individuals), followed by *Tetramorium* (204), *Cephalotes* (104) and *Pseudomyrmex* (92), but with different dominance among habitats. Significant differences were found between the abundance of ants per site ($x^2 = 72.11$, $P < 0.0001$) and per season ($x^2 = 28.13$, $P <$

0.0001), and the interaction between these factors ($x^2 = 55.40$, $P < 0.0001$). The site with the greatest abundance of ants during the rainy season was the plantation, and the lowest abundance was reported for the preserved forest (Figure 9.6a). This pattern was reversed during the dry season (Figure 9.6a). In a subsequent analysis, significant differences were found between the abundance of ants and the concentration of dissolved sugars ($x^2 = 12.95$, $P < 0.0001$).

The site with greater richness and abundance of plant species and higher land cover was preserved forest (La Mancha), where the original vegetation (tropical deciduous forest) is still dominant in the landscape (Rico-Gray et al. 1998), followed by the agroecosystem (Cerro Gordo) and finally the major disturbance site (commercial planting, San José) which in turn has the lowest abundance and less diversity of plant species (Fernández-Martínez & Díaz-Castelazo, 2009). Surrounding vegetation at each site and the abiotic/microclimatic conditions may influence ant abundance as well. Ant guilds of related genera have different tolerances to disturbances. It should be noted that the dominant ant fauna at *C. odorata* EFNs at the more disturbed site is composed of generalized Myrmicinae and opportunists (sensu Andersen, 2000), while the forested site is controlled by the dominant Dolichoderinae and subdominant Camponotini (sensu Andersen, 2000), the latter being more representative of preserved sites.

Season had a direct, site-dependent impact on the abundance of ants foraging on EFNs: only at the most disturbed site (where noticeably higher damage by the shoot borer herbivore occurred) was the abundance of ants foraging on EFNs higher during the rainy season, when herbivore pressure increased. This pattern was consistent with an induced biotic defense strategy in *C. odorata* (Hernández-Villanueva, 2010; Chavarro-Rodriguez et al., 2013).

This more detailed examination of a single native species growing in different locations has provided novel information on the ecology of multitrophic interactions of an economically important timber species. In the case of *C. odorata*, there is strong evidence of context-dependent results (interhabitat/seasonal variation) in plant defense mediated by EFNs and ant protectors.

Concluding Remarks

The occurrence of liquid rewards on leaves plays a key role in mediating the foraging ecology of foliage-dwelling ants, and facultative ant-plant mutualisms are important in structuring the community of canopy arthropods, susceptible to spatial variation. Facultative ant-plant interactions mediated by EFNs are quite variable with respect to participant species, but surprisingly similar in their community-level interaction pattern (i.e. resultant ecological networks). The outcomes of the interactions for involved species are varied, and it is those outcomes that are a constant source of discovery for those in our field.

Future studies involving multisite comparisons of facultative ant-plant mutualisms, besides evaluating among-site differences in abiotic factors (e.g. rainfall, minimum temperature, light incidence), should take samples at several locations per site and also take into account physiognomic variation (vegetation associations) within sites (Rico-Gray & Oliveira, 2007). These suggested study issues are based in the ecological evidence that the amount and quality of EFN secreted by plants in a community can vary according to local ecosystem features, such as canopy cover (derived from vegetation structure), soil pH (Dáttilo et al., 2013b; Kersch & Fonseca, 2005) and abiotic variables that may influence the trophic interactions between EFN and their visitors.

High inter-habitat variation is evidenced through the following patterns: preserved habitats have higher richness of interacting species or interaction frequencies as compared with strongly disturbed habitats (i.e. anthropogenic or abiotic disturbance). Furthermore, for facultative ant-plant mutualisms, preserved habitats have lower abundances of each interacting species in contrast to disturbed habitats. The latter seems to occur partly because at disturbed habitats the trophic interactions are altered, and the opportunistic or tolerant species (such as invasive ants or ruderal plants) dominate the interaction spectrum. This seems to occur in the *Cedrela odorata* studied system as well as at other partially disturbed habitats (Ness & Bronstein, 2004; Dáttilo et al., 2014a). Thus the context-dependent pattern of EFN-mediated ant-plant interactions is a biotic defense strategy for plants that is possible when particular requirements for resource availability and ant defensive abilities are met.

References

Andersen, A. N. (2000). A global ecology of rainforest ants: functional groups in relation to environmental stress and disturbance. In *Ants: standard methods for measuring and monitoring biodiversity*, ed. D. Agosti, J. D. Majer, L. E. Alonso & T. R. Schultz: Smithsonian Institution Press, pp. 35–44.

Bascompte, J., Aizen, M., Fontaine, C. et al. (2010). Symposium 6: mutualistic networks. *Bulletin of the Ecological Society of America* 91(3), 367–370.

Bascompte, J., Jordano, P., Melián, C. J. & Olesen, J. M. (2003). The nested assembly of plant–animal mutualistic networks. *Proceedings of the National Academy of Sciences* 100, 9383–9387.

Bascompte, J., Jordano, P. & Olesen, J. M. (2006). Asymmetric coevolutionary networks facilitate biodiversity maintenance. *Science* 312, 431–433.

Bentley, B. L. (1976). Plants bearing extrafloral nectaries and the associated ant community: Interhabitat differences in the reduction of herbivore damage. *Ecology* 57, 815–820.

Blüthgen, N., Verhaagh, M., Goitía, W. et al. (2000). How plants shape the ant community in the Amazonian rainforest canopy: the key role of extrafloral nectaries and homopteran honeydew. *Oecologia* 125(2), 229–240.

Campos-Navarrete, M. J., Abdala-Roberts, L., Munguía-Rosas, M. A. & Parra-Tabla, V. (2015). Are tree species diversity and genotypic diversity effects on insect herbivores mediated by ants? *PloS One* 10(8): e0132671.

Chavarro-Rodríguez, N., Díaz-Castelazo, C. & Rico-Gray, V. (2013). Characterization and functional ecology of the extrafloral nectar of *Cedrela odorata* in contrasting growth environments in central Veracruz, Mexico. *Botany* 91, 695–701.

Cuautle M. & Rico-Gray, V. (2003). The effect of wasps and ants on the reproductive success of the extrafloral nectaried plant. *Turnera ulmifolia* Turneraceae. *Functional Ecology*, 17, 417–423.

Dáttilo, W., Díaz-Castelazo, C. & Rico-Gray, V. (2014a). Ant dominance hierarchy determines the nested pattern in ant-plant networks. *Biological Journal of the Linnean Society* 113, 405–414.

Dáttilo, W., Fagundes, R., Gurka, C. A. Q. et al. (2014b). Individual-based ant-plant networks: diurnal-nocturnal structure and species-area relationship. *PLoS One* 9(6), e99838.

Dáttilo, W., Guimarães, P. R., & Izzo, T. J. (2013a). Spatial structure of ant–plant mutualistic networks. *Oikos*, 122, 1643–1648.

Dáttilo, W., Marquitti, F. M. D., Guimarães, P. R. & Izzo, T. J. (2014c). The structure of ant-plant ecological networks: is abundance enough?. *Ecology* 95, 475–485.

Dáttilo, W., Rico-Gray, V., Rodrigues, D. J. & Izzo, T. J. (2013b). Soil and vegetation features determine the nested pattern of ant-plant networks in a tropical rainforest. *Ecological Entomology* 38, 374–380.

Del-Claro, K., Rico-Gray, V., Torezan-Silingardi, H. M. et al. (2016). Loss and gains in ant–plant interactions mediated by extrafloral nectar: fidelity, cheats, and lies. *Insectes Sociaux* 63(2), 207–221.

Díaz-Castelazo, C., Guimarães, P., Jordano, P. et al. (2010). Changes of a mutualistic network over time: reanalysis over a 10-year period. *Ecology* 91(3), 793–801.

Díaz-Castelazo, C., Rico-Gray, V., Oliveira, P. S. & Cuautle, M. (2004). Extrafloral nectary-mediated ant–plant interactions in the coastal vegetation of Veracruz, Mexico: richness, occurrence, seasonality and ant foraging patterns. *Ecoscience* 11, 472–481.

Díaz-Castelazo, C., Rico-Gray, V., Ortega, F. & Ángeles, G. (2005). Morphological and secretory characterization of extrafloral nectaries in plants of coastal Veracruz, Mexico. *Annals of Botany* 96(7), 1175–1189.

Díaz-Castelazo, C., Sánchez-Galván, I. R., Guimarães, P. R., Raimundo, R. L. G. & Rico-Gray, V. (2013). Long–term temporal variation in the organization of an ant-plant network. *Annals of Botany* 111, 1285–1293.

Fernández-Martínez, M. J. & Díaz-Castelazo, C. (2009). Caracterización ecológica de *Cedrela odorata* y patrones de infestación por *Hypsipyla grandella* en selvas y plantaciones de Veracruz. In *Serie memorias científicas 15. XXII Reunión científica Tecnológica Forestal y Agropecuarias*, ed. INIFAP, Veracruz, México, pp. 301–310.

Guimarães P. R. Jr., Rico-Gray, V., Oliveira, P. S. et al. (2007). Interaction intimacy affects structure and coevolutionary dynamics in mutualistic networks. *Current Biology* 17, 1797–1803.

Heil, M. (2015). Extrafloral nectar at the plant-insect interface: a spotlight on chemical ecology, phenotypic plasticity, and food webs. *Annual Review of Entomology* 60, 213–232.

Hernández-Villanueva, M. A. (2010). Interacción insecto planta mediada por nectarios extraflorales del cedro rojo (*Cedrela odorata*, Meliaceae) en selvas y plantaciones del centro de Veracruz. BSc thesis, Benemérita Universidad de Puebla. Puebla, México.

Jordano, P., Bascompte, J., & Olesen, J. M. (2003). Invariant properties in coevolutionary networks of plant–animal interactions. *Ecology Letters*, 6, 69–81.

Kersch, M. F. & Fonseca, C. R. (2005). Abiotic factors and the conditional outcome of an ant–plant mutualism. *Ecology* 86(8), 2117–2126.

Koptur, S. (2005). Nectar as fuel for plant protectors. In *Plant-provided food for carnivorous insects: a protective mutualism and its applications*, ed. F. L. Wäckers, P. C. J. van Rijn & J. Bruin. Cambridge: Cambridge University Press, pp. 75–108.

Koptur, S. & Lawton, J. H. (1988). Interactions among vetches bearing extrafloral nectaries, their biotic protective agents, and herbivores. *Ecology* 69, 278–293.

López-Carretero, A., Díaz-Castelazo, C., Boege, K. & Rico-Gray, V. (2014). Evaluating the spatio-temporal factors that structure network parameters of plant-herbivore interactions. *PLoS One* 9(10), e110430.

Ness, J. H. & Bronstein, J. L. (2004). The effects of invasive ants on prospective ant mutualists. *Biological Invasions* 6, 445–461.

Oliveira, P. S. & Freitas, A. V. L. (2004). Ant-plant-herbivore interactions in the neotropical cerrado savanna. *Naturwissenschaften* 91, 557–570.

Oliveira, P. S., Rico-Gray, V., Díaz-Castelazo, C. & Castillo-Guevara, C. (1999). Interactions between ants, extrafloral nectaries and insect herbivores in neotropical coastal sand dunes: Herbivore deterrence by visiting ants increases fruit set in *Opuntia stricta* (Cactaceae). *Functional Ecology* 13, 623–631.

Pacini, E. & Nepi, M. (2007). Nectar production and presentation. In *Nectaries and nectar*, ed. S. W. Nicolson, M. Nepi & E. Pacini. Dordrecht: Springer, pp. 167–214.

Pemberton, R. W. (1988). The abundance of plants bearing extrafloral nectaries in Colorado and Mojave desert communities of Southern California. *Madroño* 35(3), 238–246.

Pennington, T. D. & Sarukhán, J. (2005). *Árboles tropicales de México: Manual para la identificación de las principales especies*. 3rd ed. México, D. F. Fondo de cultura económica, UNAM.

Perfecto, I. (1990). Indirect and direct effects in a tropical agroecosystem: the maize-pest-ant system in Nicaragua. *Ecology* 71, 2125–2134.

Perfecto, I. (1991). Ants (Hymenoptera: Formicidae) as natural control agents of pests in irrigated maize in Nicaragua. *Journal of Economic Entomology* 84, 65–70.

Perfecto, I. & Sediles, A. (1992). Vegetational diversity, ants (Hymenoptera: Formicidae), and herbivorous pests in a neotropical agroecosystem. *Environmental Entomology* 21, 61–67.

Perfecto, I. & Vandermeer, J. H. (1994). Understanding biodiversity loss in agroecosystems: reduction of ant diversity resulting from transformation of the coffee ecosystem in Costa Rica. *Entomological Trends in Agricultural Science* 2, 7–13.

(2002). Quality of agroecological matrix in a tropical montane landscape: ants in coffee plantations in southern Mexico. *Conservation Biology* 16, 174–182.

Philpott, S. M., Greenberg, R., Bichier, P. & Perfecto, I. (2004a). Impacts of major predators on tropical agroforest arthropods: comparisons within and across taxa. *Oecologia* 140, 140–149.

Philpott, S. M., Maldonado, J., Vandermeer, J. & Perfecto, I. (2004b). Taking trophic cascades up a level: behaviorally-modified effects of phorid flies on ants and ant prey in coffee agroecosystems. *Oikos* 105, 141–147.

Rico-Gray, V., Díaz-Castelazo, C., Ramírez-Hernández, A., Guimarães P. R. Jr. & Holland, J. N. (2012). Abiotic factors shape temporal variation in the structure of a mutualistic ant-plant network. *Arthropod-Plant Interactions* 6, 189–295.

Rico-Gray, V., García-Franco J. G., Palacios-Ríos, M. et al. (1998). Geographical and seasonal variation in the diversity of ant-plant association in Mexico. *Biotropica* 30, 190–200.

Rico-Gray, V. & Oliveira, P. S. (2007). *The ecology and evolution of ant–plant interactions.* Chicago: University of Chicago Press.

Rudgers, J. A. & Gardener, M. (2004). Extrafloral nectar as a resource mediating multispecies interactions. *Ecology* 85, 1495–1502.

Rudgers, J. A. & Strauss, S. Y. (2004). A selection mosaic in the facultative mutualism between ants and wild cotton. *Proceedings of the Royal Society of London B: Biological Sciences* 271(1556), 2481–2488.

Sánchez-Galván I. R., Díaz-Castelazo, C. & Rico-Gray, V. (2012). Effect of hurricane Karl on a plant–ant network occurring in coastal Veracruz, Mexico. *Journal of Tropical Ecology* 28, 603–609.

Schupp, E. W. & Feener, D. H. (1991). Phylogeny, lifeform and habitat dependence of ant-defended plants in a Panamanian forest. In *Ant-plant interactions*, ed. C. R. Huxley & D. F. Cutler, eds. Oxford: Oxford University Press, pp. 175–197.

van Rijn, P. C., & Sabelis, M. W. (2005). Impact of plant-provided food on herbivore-carnivore dynamics. In *Plant-provided food for carnivorous insects: a protective mutualism and its applications*, ed. F. L. Wäckers, P. C. J. van Rijn & J. Bruin. Cambridge: Cambridge University Press, pp. 223–266.

Waser, N. M. & Ollerton, J. (2006*). Plant-pollinator interactions: from specialization to generalization.* Chicago: University of Chicago Press.

Yamawo, A. & Hada, Y. (2010). Effects of light on direct and indirect defences against herbivores of young plants of *Mallotus japonicus* demonstrate a trade-off between two indirect defence traits. *Annals of Botany*, 106(1), 143–148.

Zimmerman, J. G. (1932). Uber die extrafloralen Nektarien der Angiospermen. *Beihefte zum Botanischen Zentralblatt* 49, 99–196.

10 Integrating Ecological Complexity into Our Understanding of Ant-Plant Mutualism: Ant-Acacia Interactions in African Savannas

Todd M. Palmer and Truman P. Young[*]

Introduction

Ant-plant protection mutualisms, and mutualisms more broadly, are typically defined as "+ / +" interactions, indicating that species on each side of the interaction have a positive effect on the per capita growth rate of their partner. But that simple representation belies potentially great complexity: each "+" sign represents the net effects of the interaction's costs and benefits on the lifetime fitness of participants, and these costs and benefits may shift both spatially and temporally with environmental conditions, with variation in the guild of mutualist partners, and with variation in the community of interacting species outside of the mutualism. As these conditions change, so too may the strength of mutualism, with consequences that can strongly influence the communities in which these interactions are embedded. To understand mutualism, therefore, requires moving beyond traditional pairwise approaches (Stanton, 2003), and accounting for both the complexity inherent in mutualist networks and the ways in which the surrounding environment affects these interactions. Doing this accounting correctly is important: our entire understanding of mutualism, from the evolution of traits and behaviors to the stability of mutualism itself, is predicated on understanding how these interactions integrate to influence the lifetime fitness of the participants.

To illustrate the complexity of mutualism, and to provide examples of how this complexity may affect its ecological and evolutionary dynamics, in this chapter we review ant-acacia interactions within the savannas and bushlands of East Africa. In particular, we focus on research investigating the well-studied mutualism between *Acacia* (*Vachellia*) *drepanolobium*, a widespread and abundant myrmecophyte

[*] We thank Mpala Research Centre for logistical support over the course of our research on *Acacia drepanolobium* over the past two decades, John Lemboi, Simon Akwam, James Lengingiro, Alfred Inauzuri and Jackson Ekadeli for their extremely capable field assistance, and the National Science Foundation for their support during the writing of this chapter (NSF DEB-1149980 and NSF DEB-1556905 to TMP and NSF DEB 12-56034 to TPY).

(ant-plant), and its suite of symbiotic ant associates. Our approach highlights the "community ecology" of this ant-plant association, shedding light on the spatio-temporal variability of the ant-plant interaction, and examining both how the broader community and environment impacts the mutualism, and how the mutualism in turn affects the broader community. We focus on *A. drepanolobium* because it has been more intensively studied than other ant-plant associations in East Africa. There are many other extrafloral nectar-bearing species in the region, many of which are visited by suites of obligate or facultative ant associates, still awaiting research addressing the nature of these ant-plant interactions.

Natural History of the *Acacia drepanolobium* – Ant Symbiosis

Large areas of eastern and southern Africa are underlain with high-clay vertisols called "black cotton" soils. Over much of these, a wooded grassland community has developed dominated by a single tree species, *Acacia drepanolobium*. This tree is an ant acacia, providing food and housing to ant colonies, and receiving protection from herbivores from at least some of its ant associates (Madden and Young, 1992; Stanton and Palmer, 2011). The tree serves as an important source of forage for a variety of mammals, including giraffe, rhino, and other large browsing herbivores (Madden and Young, 1992; Martins, 2010).

In Laikipia and elsewhere (e.g., the Athi Plains in Kenya, many areas within the Great Rift Valley, and the Grumeti area of the Serengeti in Tanzania), *A. drepanolobium* is a foundation species (sensu Dayton, 1972), forming a virtual canopy monoculture (>90 percent of the woody cover), and capable of achieving high densities (> 1500 stems (> 0.5 m tall)/ha, Young et al., 1998). The tree is defended by pairs of straight spines (modified stipules) at each branch node, which effectively reduce herbivory by large mammals, and whose length can be induced by large mammal herbivory. In addition, approximately one pair of spines in four is swollen at the base to produce a large (2–5 cm) hollow structure, which serve as domatia which several (but not all) symbiont ant species use for living space and to rear brood. In addition, the leaves of *A. drepanolobium* are characterized by extrafloral nectaries along the lower petiole, on which some (but not all) symbiotic ant species feed. This ant-plant symbiosis was first described in depth by Hocking (1970), and has since become an iconic example of the richness and complexities of mutualism (Young et al., 1997; Palmer et al., 2000; Martins, 2010).

Throughout its range, *A. drepanolobium* associates with a guild of ant species that vary strongly in their interactions with the host plant (Table 10.1). Here we discuss the four best-studied ant associates, and what is currently known about each unique association between these ant species and their host plant. The first three species, *Tetraponera penzigi*, *Crematogaster mimosae*, and *C. nigriceps*, are obligate associates of the host plant that appear to depend solely on the swollen spine domatia of the acacia for brood rearing. The fourth species, *C. sjostedti*, is a more generalized twig and cavity nester (a condition thought to be ancestral to residing within

Table 10.1 Ant Species Vary in Their Relationships with *Acacia drepanolobium*

ANT SPECIES	Dominance rank of mature colonies[1]	Avg # (± s.e.) trees per colony[2]	Dominance rank of queens[3]	Anti-herbivore defense[4]	Use of nectaries	Scale tending?	Domatia density[6]	Fruiting of host plants[7]
Crematogaster sjostedti	1	13.2 (2.4)	–	LOW	LOW	YES	MED	HIGH
C. mimosae	2	4.8 (1.1)	3	HIGH	HIGH	YES	MED	MED
C. nigriceps	3	2.8 (0.7)	2	HIGH	HIGH	NO	HIGH	–[8]
Tetraponera penzigi	4	1.2 (0.1)	1	MED	N/A[5]	NO	LOW	MED

[1] Dominance ranks taken from Palmer et al. (2000); [2] average colony size differs significantly among ant species; ANOVA, $F_{3,159} = 78.07$, $p < 0.0001$; [3] in contests among queens for establishment sites in swollen spine domatia (*Crematogaster sjostedti* colonies appear to spread primarily by fission, and independent foundress queens have not been observed), Stanton et al. (2002); [4] Palmer and Brody (2007); [5] *Tetraponera penzigi* destroys host tree leaf nectaries, Palmer et al. (2002); [6] for size-matched trees, see Palmer (2004); [7] from Brody et al. (2010); [8] *C. nigriceps*-occupied trees rarely flower because this ant species castrates host plants Stanton et al. (1999).

Note: The most competitively dominant ant species (*C. sjostedti*) appears to be a relatively ineffective host-tree defender.

specialized domatia, Davidson and McKey, 1993; Chomicki et al., 2015), and can be found in cavities within both live *A. drepanolobium* and dead snags, as well as hollows within the trunks of other woody plant species in these habitats.

These four ant species compete strongly both inter- and intra-specifically for *A. drepanolobium* trees; each tree hosts only a single ant species, and virtually all trees are occupied. Because of the high host plant densities in many areas, inter-specific and intraspecific conflicts among neighboring colonies for the possession of host plants are common, and the outcome of competition is determined largely by colony size (Palmer, 2004). There is a hierarchy in dominance of mature colonies of the four ant species (from low to high): *Tetraponera penzigi* < *C. nigriceps* < *C. mimosae* < *C. sjostedti*. This hierarchy is paralleled by a gradient in the mean average number of trees occupied by each ant species (and therefore colony size, Table 10.1). It also sets up a successional series, with the least dominant ant, *T. penzigi*, occupying the smallest trees on average, and *C. sjostedti* the tallest. In the sections that follow we discuss in turn each ant species, and its unique relationship with *A. drepanolobium*.

The Four Acacia Ant Species

Tetraponera penzigi, one of the most widespread associates of *A. drepanolobium* (Hocking, 1970), is a narrow-bodied, monogynous (one queen per colony) species with colonies that typically span only 1–2 canopies (Table 10.1). This species is a fungal farmer (Visiticao, 2011) which also gleans small food items (e.g., pollen, fungal spores) from surfaces of the host plant (Palmer, 2003). *Tetraponera penzigi* modifies host plants in two ways that decrease the frequency of takeovers by aggressive neighboring colonies. First, this narrow-bodied species maintains very small entry holes on swollen spine domatia, which are too small to permit passage of the larger-bodied *Crematogaster* species. Consequently, when taking over *T. penzigi*-occupied host plants, *Crematogaster* species must enlarge holes by chewing (Palmer et al., 2002). Second, *T. penzigi* workers chew and destroy virtually all leaf nectaries on their host trees. These "priority effects" increase the costs (hole enlargement) and decrease the benefits (nectary availability) of host plant takeover, reducing the probability of competitive displacement by aggressive *Crematogaster* neighbors (Palmer et al., 2002).

Crematogaster nigriceps is also a competitively subordinate, monogynous species, with colonies that typically span 2–3 host plants. This species also engages in behaviors that might be related to its position low in the dominance hierarchy, both forms of meristem pruning. First, they remove many (Stapley, 1998; Martins, 2013) or most (Stanton et al., 1999) of the axillary meristem at the nodes along the branches, leaving only those at swollen spines, and greatly reducing the number of leaves on each tree. The behavior also strongly reduces flower production, in some areas effectively sterilizing host trees (but see Martins, 2013; Tarnita et al., 2014). While the reasons for this sterilization behavior have not been identified experimentally,

C. nigriceps   C. mimosae

Figure 10.1.   Contrasting effects of different resident ant species on *Acacia drepanolobium* architecture.
Crematogaster nigriceps prunes host plant canopies, and their host trees (left panel) can
be identified from a distance by their compact canopies, high level of branching, restricted
lateral growth, and high densities of swollen spines. Contrastingly, host plants occupied by
C. mimosae (right panel) have greater lateral growth, lower levels of branching, and lower
swollen spine densities. Photo credit: Todd M. Palmer.

it may increase host plant allocation to domatia production to resident ants, as
noted for other systems (Frederickson, 2009). Consistent with this hypothesis,
C. nigriceps-occupied trees produce domatia in much greater abundance than
trees occupied by the other three ant species (Figure 10.1). Second, *C. nigriceps*
engages in selective pruning of apical meristems, reducing the lateral growth of
their host trees, but largely in the direction of trees occupied by competitively
dominant species, in particular *C. mimosae*, slowing or even preventing the physical contact between branches that serve as a corridor for ant takeovers (Stanton
et al., 1999).

Crematogaster mimosae is often the most abundant ant species in the system, in
some areas occupying ca. 50 percent of host trees (Hocking, 1970; Young et al.,
1997; Palmer, 2004; Martins, 2010). Colonies of this species can be either monogynous (single-queen) or polygynous (multiple-queen) (Stanton et al., 2002, 2005), and
typically occupy an average of 4–5 host plants. *Crematogaster mimosae* is among
the most mutualistic of the four ant species, supporting high rates of plant growth
and low host-plant mortality (Figure 10.2) (Palmer et al., 2008a). Nonetheless, in
addition to harvesting nectar from the nectaries and foraging off-tree (as do all

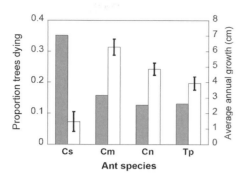

Figure 10.2. Contrasting effects of different ant species on *Acacia drepanolobium* vital rates. Average annual growth rate (white bars ± SEM) and cumulative mortality (gray bars) for host trees occupied by the four acacia ant species over an eight-year observation period. Tp = *T. penzigi*, Cn = *C. nigriceps*, Cm = *C. mimosae*, and Cs = *C. sjostedti*. Host plants occupied by *C. sjostedti* suffer much higher mortality than those occupied by the other acacia ant species. *Crematogaster mimosae* promotes the highest rates of plant growth, followed by *C. nigriceps* and *T. penzigi* (adapted from Palmer et al., 2008a).

three *Crematogaster* species in this system) (Palmer, 2003, 2004), *C. mimosae* tends scale insects (Coccidae) on its host trees (*Hockiana insolitus* and *Ceroplastes* spp., Baker, 2015).

  Crematogaster sjostedti is a non-obligate symbiont on *A. drepanolobium*, and can be found both on other acacia species (*A. seyal* var *fistula*, Young et al., 1997, and *A. zanzibarica*, Cochard and Edwards, 2011) as well as in hollow cavities within other woody plants and occasionally dead snags. This species occupies the greatest number of trees per (multi-queen) colony (Table 10.1), is competitively dominant, and is often found on the oldest and largest trees. Unlike the other three species, *C. sjostedti* does not live in swollen spine domatia, but instead occupies and raises brood in cavities in the stems of their host trees. (In some locations, *C. sjostedti* has been observed to construct and reside within papery nests on the outside of branches, Cochard and Edwards, 2011). Many of these cavities are excavated by the larvae of large cerambycid beetles, which *C. sjostedti* appears to facilitate (Palmer et al., 2008b), and which likely decrease the rates of plant growth and increase mortality (Figure 10.2). *Crematogaster sjostedti* is the least active against herbivores (Palmer et al., 2008b), and least mutualistic of the four ant species (Table 10.1) *Crematogaster sjostedti* also tends scale insects (*Ceroplastes* spp.), although at lower frequency than *C. mimosae*.

  Species coexistence among the acacia ants – intense competition among colonies for host trees, coupled with a linear dominance hierarchy, motivates the question of how these four ant species coexist. Prior work has demonstrated that a number of mechanisms operate to support species coexistence in this guild, including competition-colonization tradeoffs, niche partitioning and priority effects (Stanton et al., 1999; Palmer, 2001; Palmer et al., 2002; Stanton et al., 2002; Palmer, 2003,

2004; Stanton et al., 2005). Tradeoffs in colonization versus competitive ability are particularly prominent in this system; *Tetraponera penzigi* and *C. nigriceps* are strong colonists, both investing reproductive effort heavily in the production of queens, while *C. mimosae* and *C. sjostedti* are dominant competitors, investing disproportionately in the production of workers (Stanton et al., 2002). Intriguingly, the dominance hierarchy in conflicts among queens for establishment sites (swollen spine domatia) is exactly the *opposite* of that for mature colonies (Table 10.1)! This ontogenetic reversal in competitive ability further reinforces the strong colonization and establishment success of *T. penzigi* and *C. nigriceps*. And for the dominant *C. mimosae* and *C. sjostedti*, rapid rates of colony growth are reinforced by these species' dominance in foraging for protein sources off of host plants (Palmer, 2004), and may be further supported by their tending of scale insects (K. M. Prior and T. M. Palmer, unpublished manuscript).

The four acacia ant species differ strongly, not only in their natural histories but also in the costs imposed and benefits provided to host plants. In the next section, we discuss this variation within the mutualism, and consider its ecological consequences.

Complexity within the *Acacia drepanolobium*-Ant Symbiosis: Variation in Benefits and Costs

The net benefits of mutualism are a balance between the goods or services obtained, and the costs that accrue in obtaining those benefits (Chapter 11). There may be several axes of benefit to plants in an ant-plant interaction, including protection from herbivory (Janzen, 1966), fertilization (Huxley, 1978; Sagers et al., 2000), and the pruning of neighboring plants (Janzen, 1969; Davidson et al., 1988; Fiala et al., 1989; Morawetz et al., 1992). So too can there be a variety of costs, including direct (e.g., metabolic) costs such as the production of extrafloral nectar, food bodies,and domatia, and other costs ("ecological costs," Frederickson et al., 2012) such as the collateral deterrence of beneficial insects such as pollinators (Ness, 2006), the tending by ants of scale insects or aphids (Styrsky and Eubanks, 2007, but see Pringle et al., 2011), and even sterilization by particular ant associates (Stanton et al., 1999; Yu et al., 2004). Predicting how the "balance of trade" in mutualism may shift across spatial or temporal environmental variation requires that we decompose mutualism into its constituent costs and benefits.

Variation in the Benefit of Defensive Symbiotic Ant Species

Acacia ants rapidly respond to host plant disturbances, emitting pungent alarm pheromones to which nestmates immediately respond and recruit (Wood and Chong, 1975). These workers swarm onto the muzzle of larger herbivores (or onto the bodies of invertebrate herbivores) and locate softer tissue (e.g., mucus membranes) and bite down and/or sting. This defense can be quite formidable, as tens to hundreds of workers may swarm onto herbivores as they feed. The four acacia ant species vary strongly in their levels of aggressive response to disturbance (Table 10.1), as well as the extent of herbivore damage to their host plants. Following both real and simulated browsing, *C. mimosae* and *C. nigriceps* display the strongest recruitment of workers to the site of the disturbance, while *T. penzigi*'s response is relatively weak, and *C. sjostedti*'s response is almost non-existent (Palmer and Brody, 2007; Martins, 2010). Correspondingly, levels of both vertebrate and invertebrate herbivory tend to be higher on *T. penzigi-* and *C. sjostedti*-occupied trees, and lower on acacias occupied by *C. mimosae* and *C. nigriceps* (Palmer and Brody, 2007). These differences are clearly driven by differences in aggressive defense by symbiotic ants; in a five-year ant removal experiment, the overall levels of attack (by vertebrate browsers, cerambycid beetles, and stem- and leaf-galling insects) increased on ant-removal trees for *C. mimosae* and *C. nigriceps*, while attack rates on trees occupied by *C. sjostedti* and *T. penzigi* did not differ from their paired trees from which these species had been removed. Patterns of browsing by black are congruent with these findings. Martins (2010) found that rhinos fed on acacias occupied by the aggressively defensive *C. mimosae* and *C. nigriceps* significantly less often than expected based on their relative abundances, and fed more often than expected on acacias occupied by *T. penzigi*.

Yet even the most aggressive acacia ants do not appear to effectively deter all African browsers. For example, evidence for giraffe deterrence is equivocal. In one study conducted in the Kajiado district of Kenya, Martins (2010) found that Maasai giraffes (*Giraffa camelopardalis tippelskirchi*) fed on acacias occupied by *C. mimosae* and *C. nigriceps* significantly less often than expected by chance, while feeding on trees occupied by *T. penzigi* more frequently than expected. In the Athi Kapiti plains of Kenya, Madden and Young (1992) found that while *Crematogaster* ants shortened the duration of feeding bouts by giraffe calves, they did not reduce feeding bout durations for adult giraffes. In Laikipia, experimental density reductions of *C. mimosae* had no effect on giraffe damage to host plants over one year (Palmer and Brody, 2013), while in a 4.5-year ant removal experiment (Stanton and Palmer, 2011), damage consistent with giraffe feeding was significantly higher on paired ant-removal trees than ant-occupied trees for both *C. mimosae* and *C. nigriceps*.

Tolerance to herbivory appears to play a large role in *Acacia drepanolobium*'s defensive repertoire. In a longer-term ant removal experiment, acacias without resident ants grew faster and reproduced more than ant-occupied trees, despite much higher levels of herbivory on ant removal plants (Stanton and Palmer, 2011). These data suggest that acacia ants may impose large metabolic costs to host plants which may not be offset, at least over the shorter term, by protection from chronic but non-lethal herbivory. It is important to emphasize, however, that even this 4.5-year experiment is fairly short-term, relative to the long lifespan of this tree species (> 150 yrs). Over longer time scales, chronic herbivory, especially by the wood boring larvae of cerambycid beetles, is likely to have a large impact on plant growth and

Figure 10.3.   An African elephant (*Loxodonta africana*) sniffs an *Acacia drepanolobium*, assessing how well-defended the plant is by symbiotic ants. The four acacia ant species produce pungent and different volatile alarm pheromones (Wood et al., 2002) when they detect herbivores, which elephants may use as cues that indicate both the identity of the ant species in residence, as well as the density of the resident colony. Elephants tend to attack *Acacia drepanolobium* with very low densities of acacia ants, or those occupied by *Crematogaster sjostedti*, the least aggressive ant species. Photo: Kathleen Rudolph. (A black-and-white version of this figure will appear in some formats. For the color version, please refer to the plate section.)

survival (Palmer et al., 2008a). In general, tolerance of herbivory may have evolved in *A. drepanolobium* in part due to the variable nature of protection conferred by different ant species. Because ant species turn over frequently on host plants, tolerance of herbivory may enable plants occupied by poor defenders to continue to survive until they are replaced by more aggressive ant defenders. We discuss this possibility in greater detail later in this chapter.

Elephants: Key Drivers of the Acacia drepanolobium-*Ant Mutualism*

If ants impose such high metabolic costs to host plants, yet do not appear to provide sufficient benefits in the form of protection to offset these costs, then why associate with ants at all? The key to this puzzle, emerging from two decades of study of this mutualism, appears to be elephants (Figure 10.3). Elephants are a distinctive herbivore in that they are capable of imposing catastrophic herbivory to trees, stripping branches, knocking down and destroying the main stem, and girdling trees by feeding on bark and phloem/cambium. Several studies have demonstrated that ants are highly effective defenders of host plants from catastrophic herbivory by elephants (Goheen and Palmer, 2010; Palmer and Brody, 2013). In the most comprehensive of these, Goheen and Palmer (2010) used a combination of experiments, feeding trials with captive elephants and remote sensing to show that acacia ants regulate woody plant cover in *A. drepanolobium* savannas by strongly deterring elephant herbivory. The efficacy of ant defense likely results from a combination of high ant densities on host plants (up to 90,000 workers on some trees, Hocking, 1970) and their tendency to attack areas of thin skin and mucous membranes by biting down and holding fast with their mandibles. Despite the massive size and eponymous thick skin of these pachyderms, the "Achilles heel" of these herbivores appears to be their trunks, whose inside surfaces are highly innervated to allow for fine motor coordination and sensitivity. While giraffes use their long, prehensile tongues to swipe away ants from their muzzles (Palmer and Goheen, unpublished data), elephants must use their trunks to feed, essentially inserting their "noses" directly into a canopy swarming with aggressive ants. Elephants also avoid attacks by bees (Vollrath and Douglas-Hamilton, 2002), further suggesting that swarming Hymenoptera may be a potent defense against these enormous and powerful herbivores.

Although catastrophic herbivory by elephants can occur, the annual risk to any individual plant is likely fairly low, owing to the relatively low abundance of elephants and high density of trees in many savanna and bushland systems. As a consequence, ant removal experiments over shorter time periods (relative to the long lifespan of the tree) may (Palmer and Brody, 2013) or may not (Stanton and Palmer, 2011) reveal these risks. Nonetheless, because intense herbivory by elephants can greatly reduce fitness, this rare form of herbivory likely generates potent selection on acacias for strong defense against these herbivores (Goheen and Palmer, 2010). Maintaining ant colonies, despite their high metabolic costs, appears to act as a critical "insurance policy" against rare but potentially lethal elephant herbivory. In this and other ant-plant systems (Pringle et al., 2013), rare events may be easily missed in short-term experiments, highlighting the need to conduct long-term studies of ant-plant associations, particularly when host plants are long-lived.

Collateral Damage: Deterrence of Pollinators

Although the aggressive ant guards clearly benefit plants by attacking herbivores, these same ants could be a liability when it comes to beneficial visitors such as pollinators (Chapter 13). However, both *A. drepanolobium* and its closely related congener *A. zanzibarica* appear to have evolved mechanisms to deal with this conflict of interest. In a population of *A. zanzibarica* in Mkomazi, Tanzania, Willmer and Stone (1997) demonstrated that resident ants are deterred from flowers during the peak of pollen dehiscence, probably via a chemical in the pollen itself. In a later study that included both *A. drepanolobium* and *A. seyal* var. *fistula*, a related species with an ecotype that produces swollen spines, Willmer et al. (2009) similarly showed that floral volatiles produced by these species exerted a deterrent effect on ants, although the effects were weaker for these species than for *A. zanzibarica*.

Overall, the benefits provided by different acacia ants clearly vary substantially. But benefits are only one side of the "coinage" in the economy of mutualisms. In the next section, we turn our attention to the costs imposed by different ant species, again revealing strong variation and highlighting the strikingly high costs to host plants of maintaining ant associates.

Variation in Costs Imposed by Different Acacia Ant Species

Much of the research on ant-plant protective mutualisms has centered on iden-tifying and quantifying the benefits of these interactions to participants, and in particular to host plants (reviewed in Heil and McKey, 2003). Interestingly, far less research has focused on the costs of mutualism to the participants, but research on the *Acacia drepanolobium* ant mutualism suggests that these direct costs can in some cases be quite high. For example, in a 4.5-yr study, removing the three *Crematogaster* species from host plants resulted in much higher rates of growth and higher rates of reproduction for host trees (Stanton and Palmer, 2011).

In contrast, host plants from which *Tetraponera penzigi* was removed did not grow or reproduce more than control trees where colonies were left in residence. *Tetraponera penzigi* destroys host plant nectaries and does not tend scale insects, suggesting that the high costs borne by trees occupied by the *Crematogaster* spe-cies largely result from photosynthate consumed by these three ant associates, both directly in the form of extrafloral nectar, and indirectly through honeydew exuded by scale insects, which are tended by both *C. mimosae* and *C. sjostedti*.

Extrafloral nectar and scale exudates are two very different pathways by which plant photosynthate is supplied to acacia ants. Rates of extrafloral nectar produc-tion appear to be under the plant's control, as EFNs are a plastic trait that responds to variation in herbivory, with trees producing fewer active nectaries when herbi-vores are experimentally excluded (Huntzinger et al., 2004; Palmer et al., 2008a). At the level of individual nectaries, it is not known how extrafloral nectar secretion is induced, but observations suggest it may be mediated by the acacia ants them-selves. When resident colonies are experimentally removed from host plants, nectar-ies become inactive (Stanton and Palmer, 2011), and other non-acacia ant species (e.g., *Pheidole megacephala*) do not appear to be able to induce nectar production on these trees (Riginos et al., 2015).

When host plants change patterns of EFN production, shifts in host plant occupancy can occur, mediated through changes in the dynamics of competition between neighboring ant colonies. A reduction in active nectaries in response to an experimental reduction in herbivore pressure on host plants caused reductions in the size of *C. mimosae* colonies, making them vulnerable to takeover by neighboring

C. sjostedti colonies, which are less mutualistic (Palmer et al., 2008a). These results raise the intriguing possibility that *A. drepanolobium* may exert some control over the identity of its resident colonies through "partner screening" (Archetti et al., 2011), where the conditions established by the host plant select for the best adapted mutualist (Heil, 2013). When host plants are small (i.e., saplings) and have not yet emerged from the grass layer, competition or other forms of environmental stress constrain EFN production, which in turn favors the establishment and persistence of the moderately defensive and less metabolically demanding *T. penzigi* (Palmer, unpublished data). As acacias grow to larger sizes, nectar production may be less costly relative to the plant's overall carbon budget, allowing hosts to produce more nectar favoring the aggressive and metabolically demanding *C. mimosae* and *C. nigriceps* (Palmer et al., 2010).

As contrasted with EFN tending, colonies that tend scale insects have greater control of resource provisioning by their hosts. Tending scale insects may augment colony growth and activity level and buffer colonies from times when plants retrench their provisioning of extrafloral nectar (e.g., during drought or after exclusion of large herbivores, Palmer et al., 2008a). Recent experiments (K. M. Prior and T. M. Palmer, unpublished) point to the important role that scale insect associates play in the growth and energetics of *C. mimosae* colonies, an ecologically dominant species in many areas where *A. drepanolobium* occurs (Hocking, 1970; Young et al., 1997; Martins, 2010). Colonies whose scale insects were removed had lower activity levels and lower recruitment to simulated herbivory, and were more likely to be taken over by neighboring colonies than colonies whose scale insects were left intact. Scale-removal trees also produced more domatia (owing to increased branch growth) and more fruit, suggesting that scale tending by ants is costly to host plants. Yet despite these costs, the net effects of harboring scale insects may be positive for *A. drepanolobium*; acacias from which scale insects were removed were significantly more likely to be destroyed by elephants than trees where scale insects were present, suggesting that the increased colony activity and/or size driven by excess carbohydrates from scale insects feeds back to acacias through stronger ant defense.

The Costs of Ant Wars

Wars between neighboring colonies can have costly consequences for both ants and host plants. Following battles, one or both warring colonies are at increased risk for hostile takeovers from yet *other* neighboring colonies (Rudolph and McEntee, 2016; Ruiz-Guajardo et al., 2017), suggesting that these conflicts make colonies highly vulnerable to takeover by other neighbors. But *C. mimosae* may employ strategies to deal with those consequences, at least in intraspecific battles: in an elegant set of experiments, Rudolph and McIntee (2016) showed that for more than half of the intraspecific battles they experimentally induced, relatedness within the victorious colony declined, possibly as a result of the adoption of non-kin brood, or by queen-right fusion of the warring colonies. These results point to the primacy of maintaining high worker densities within this intensely competitive community. Yet even with the possibility of colony fusion, post war colonies are typically at reduced densities, and as a result often suffer increased mammalian herbivory (Rudolph and McEntee, 2016), including catastrophic herbivory by elephants (Palmer and Brody, 2013).

The High Cost of Mutualism Constrains Host Plant Distribution

The high cost of hosting ant associates also appears to play a prominent role in restricting the distribution of *A. drepanolobium* to nutrient-rich black cotton soils. Many areas in which this tree species occurs are bordered by nutrient-poor sandy clay loam soils, which are dominated by less well-defended acacia species, support higher densities of herbivores, and from which *A. drepanolobium* is largely absent. This sharp disjunction appears to result from the interaction between herbivory and the costs of maintaining the defensive mutualism: in a set of transplant experiments, Pringle et al. (2016) demonstrated that resource addition and herbivore exclusion increased the performance of *A. drepanolobium* saplings transplanted to nutrient-poor red soils, while having no effect on *A. drepanolobium* establishment in nutrient-rich black cotton soils. Saplings exposed to herbivory on red soils had much lower survival than their black soil counterparts, and the overall higher level of herbivory on red soils constrained the number of domatia and EFNs produced by host plants. The authors concluded that resource limitation on nutrient-poor red soils constrains defensive investment for *A. drepanolobium*, increasing the vulnerability of this myrmecophyte to high levels of herbivory on red soils, and thereby restricting its distribution to more fertile black cotton soils. These results add to a growing body of literature demonstrating that mutualism can play important roles in constraining the realized niches of participant species (reviewed in Palmer et al., 2015).

With such variation in the costs and benefits of partnering with different ant species, it is perhaps unsurprising that each ant species differentially affects host plant vital rates. In the sections that follow, we review these differential impacts and how they integrate to affect the lifetime fitness of *A. drepanolobium*.

Ant Species Exert Strong and Differential Influences on Host Plant Vital Rates

Experiments and long-term observations have demonstrated that the four acacia ants exert very different impacts on host plant vital rates (Palmer et al., 2010); plants occupied by *C. mimosae*, *T. penzigi* and *C. nigriceps* have the highest growth and survival, while plants occupied by *C. sjostedti* have both the highest fruit production and the greatest mortality (Table 10.1, Figure 10.2). Intriguingly, the most "optimal" ant associate (from the host plant perspective) may change as trees progress through their ontogeny; at early ontogenetic stages, plants occupied by

T. penzigi and *C. nigriceps* have the highest survivorship, probably owing to the fact that these species seldom abandon small host plants. By contrast, *C. sjostedti* and *C. mimosae*, with colonies that often extend across > 5 host plant canopies, often abandon small "satellite" trees during periods of environmental stress (e.g., drought) – and abandoned trees have considerably lower survival rates than trees with any species of ant (Palmer et al.; 2010). As plants grow larger and gain more value to ant colonies, *C. mimosae* is less likely to abandon them, and plants occupied by this species have higher rates of growth and reproduction than those occupied by *T. penzigi* and *C. nigriceps*.

A Lifetime of Partnerships: Integrating Costs and Benefits to Host Plants across the Guild of Acacia Ants

With such strong variation in the costs and benefits of associating with different ant partners, including antagonists such as *Crematogaster sjostedti*, and sterilizing partners like *C. nigriceps*, it is reasonable to ask whether associating with these ants yield net benefits to host plants in the first place. Addressing this question requires evaluating the effects of partnering with different ant species on the lifetime fitness of host plants, a challenging task given the long (>150-year) lifespan of these trees. In a study that combined long-term (eight-year) monitoring of >1700 trees, demographic modeling and experiments, Palmer et al. (2010) showed that the fitness benefits to *Acacia drepanolobium* increased as trees interacted with more symbiotic ant species, including the sterilizing *C. nigriceps* and the antagonistic *C. sjostedti*. This seemingly paradoxical result emerged from the order in which different ant species typically occupy host plants (Figure 10.4). During early life stages, when survivorship is far more important than reproduction, *A. drepanolobium* is frequently occupied by *T. penzigi* and *C. nigriceps*, two species which promote high host plant survival due to their low rates of host plant abandonment. Later in life, *C. mimosae* becomes the most common occupant, and this species both protects host plants and is associated with moderate levels of reproduction. At very late life stages (Young et al., 1997), host plants become increasingly occupied by the less mutualistic *C. sjostedti*, investing less in ant rewards, and shunting resources toward reproduction (Table 10.1). Hence, host plants are able to trade-off survivorship and reproduction at different life stages, such that the full suite of ant occupants results in the highest rates of *A. drepanolobium* population growth (Palmer et al., 2010). Across a tree's lifetime, it is modeled to undergo many transitions in occupancy (> 15, Palmer et al., 2010) and to partner with each of the four acacia ant species.

Much about the dynamics of the *A. drepanolobium*-ant mutualism has been revealed from the study of host plants and ants, and it is found that no mutualism functions in isolation of the community that surrounds it. In the next section we turn our attention to the broader community in which this interaction occurs, examining both biotic and abiotic drivers that shape the dynamics of the mutualistic association.

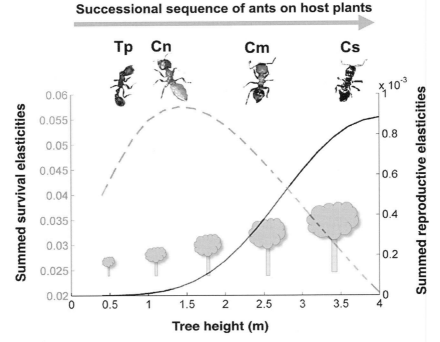

Figure 10.4. Sensitivity of *A. drepanolobium* population growth (🌱) to host plant survival (dashed line) and reproduction (solid line) across the ontogeny of the host plant. The successional sequence of acacia ants is shown at the top, Tp = *Tetraponera penzigi*, Cn = *Crematogaster nigriceps*, Cm = *C. mimosae* and Cs = *C. sjostedti*. Early in plant ontogeny, occupation by the strongly colonizing *T. penzigi* and *C. nigriceps* is most likely, and both species confer high survivorship benefits, while *C. nigriceps* temporarily sterilizes host plants while in residence. Later in life, occupation by *C. mimosae* becomes more likely; this species confers both strong survivorship and moderate reproduction. At late life stages, host plants are more likely to be colonized by the antagonistic *C. sjostedti* and invest heavily in reproduction while suffering higher rates of mortality. Host plants are able to trade off survival and reproduction at different life stages, maximizing fitness in the face of strong variation in the costs and benefits of associating with different symbiotic ant species (adapted from Palmer et al., 2010).

Effects of the Broader Community on the Acacia-Ant Mutualism: Biotic, Abiotic and Anthropogenic Effects

Termite Mounds Structure Competition and Patterns of Host Plant Occupancy among Acacia Ants

In some habitats where *Acacia drepanolobium* is found, strong spatial heterogeneity in soils is generated through the subterranean action of termites from the genus *Odontotermes*. Through their mound-building activities, these termites generate "islands of fertility," with increased water infiltration (Arshad, 1981; Palmer,

2003; Fox-Dobbs et al., 2010), higher levels of organic carbon, nitrogen and phosphorus (Palmer, unpublished data), higher rates of *A. drepanolobium* growth and more abundant invertebrates, an important and N-rich prey source for the three *Crematogaster* species (Palmer, 2003; Fox-Dobbs et al., 2010; Pringle et al., 2010). This increase in the resource base favors more the more aggressively foraging *C. sjostedti* and *C. mimosae*, resulting in larger colonies of both of these species near mounds. The underlying heterogeneity in resources generated by the action of termites "cascades upward" to shape the spatial distribution of ants on host plants; in productive termite mound microhabitats, the competitively dominant *C. sjostedti* and *C. mimosae* are disproportionately successful in displacing the subordinate *C. nigriceps* and *T. penzigi* from large host plants, while these subordinate species are more successful in less productive inter-mound areas (Palmer, 2003).

Large Herbivores

Several lines of evidence suggest that large browsing herbivores – in particular elephants – strongly influence both the ecology and evolution of this ant-plant mutualism. In a long-term, large-scale exclosure experiment, we found that *A. drepanolobium* reduces its investment in both extrafloral nectar and domatia production in the absence of vertebrate herbivory. This reduction in investment in ant associates shifts the balance of competition from the nectar-dependent and strongly mutualistic *C. mimosae* in favor of the antagonistic *C. sjostedti*, which does not depend on host plant rewards (Palmer et al., 2008a). As a consequence, *C. sjostedti* becomes the most abundant ant occupant where herbivores have been eliminated, with negative consequences for host plants (Figure 10.1). Consequently, browsing by large mammals serves to reinforce the protective mutualism between acacias and *C. mimosae*, by inducing reward production that allows this strongly mutualistic species to retain its competitive edge. These are the very herbivores most threatened by human activities, including habitat conversion, competition with livestock and direct killing.

Fire

Africa has been called the fire continent, and much of this has been anthropogenic fire, which may date back to hundreds of thousands of years (Archibald et al., 2012; Archibald, 2016). Natural and anthropogenic fire are likely drivers of the evolution in *A. drepanolobium* of thick-bark (Midgley et al., 2016) and ready coppicing after the loss of aboveground biomass (Okello et al., 2001; Okello et al., 2008). However, perhaps even more striking evidence of fire as an evolutionary force in *A. drepanolobium* ecosystems is illustrated by the behavior of acacia ants. When both *C. mimosae* and *C. nigriceps* detect smoke, they quickly initiate an evacuation of host plants, rapidly moving brood and alates down the stem of the tree and into cracks within the heavy clay soil surrounding the tree's base (Palmer et al., 2008b). This evacuation behavior is highly effective;

survival rates after fires of colonies of *C. mimosae* and *C. nigriceps* are 85 percent and 70 percent, respectively (Sensenig et al. 2017). In contrast, fewer than 10 percent of the colonies of *T. penzigi* survived controlled burns. It appears that the behavior of *C. sjostedti* of living in stem cavities makes them less vulnerable to fires than living in swollen spines; approximately 50 percent of their colonies survive fires. Re-colonization of trees after fire occurs by two mechanisms. First, colonies that successfully evacuate domatia to safe havens belowground may simply emerge after the fire and return to their host tree. Second, trees that have lost their colonies can be recolonized by ants from nearby trees with surviving colonies. As a consequence, burned areas have disproportionately more *C. nigriceps* colonies and disproportionately fewer *T. penzigi* colonies than unburned areas (Sensenig et al. 2017).

Fuelwood Collection

Acacia drepanolobium is widely collected for fuelwood and charcoal production (Okello and Young, 2000). Because *A. drepanolobium* readily coppices after the loss of aboveground tissue, this harvesting usually does not kill the trees. However, when harvesting intervals are short (as they often are in fuel-limited environments), these coppicing trees are kept in small-statured populations (authors' personal observation, and Andrews and Bamford, 2008). We have visited two sites in different parts of Kenya (Athi Plains and Naivasha) characterized by intense repeat harvesting of *A. drepanolobium*. At both sites, the ant community was essentially limited to the early successional ant species *T. penzigi* and *C. nigriceps* (T. M. Palmer and T. P. Young, personal observation)

How Does the Ant-Plant Mutualism Affect the Broader Community?

Because *Acacia drepanolobium* is a foundation species (sensu Dayton, 1972), the effects of acacia ants on plant vital rates have the potential to reverberate through the entire community. For example, Goheen and Palmer (2010) showed that by effectively protecting host plants from elephants, acacia ants play a central role in regulating woody plant cover in black cotton habitats. This is especially impressive when one considers that the average ant weighs about 5 mg, while a 5000-kg elephant (the size of a large male) is literally a billion times more massive! Because woody plant cover regulates a host of ecosystem properties in savannas, including carbon storage, fire-return intervals, predation-risk, food web dynamics, nutrient cycling and soil-water relations (Belsky et al., 1989; Pringle and Fox-Dobbs, 2008; Holdo et al., 2009; Riginos et al., 2009; Ford et al., 2014; Riginos, 2015), these minute yet pugnacious bodyguards can exert powerful indirect effects across entire landscapes.

The key role of the ant-acacia mutualism in shaping black cotton savannas is becoming increasingly evident with the advent of a recent invasion in some parts of

Figure 10.5. Elephant damage to *Acacia drepanolobium* increases strongly in areas where the native acacia-ant mutualism has been disrupted by invasion by *Pheidole megacephala*. The invasive ant displaces native acacia ants, but does not protect host trees from mammalian herbivores. Thus, invaded areas (bottom panel) have a much higher frequency of moderate to severe elephant damage to host plants, relative to non-invaded savannas (top panel). Photo credits: Todd M. Palmer.

East Africa by *Pheidole megacephala*, the "big-headed ant" (Chapter 15). *Pheidole megacephala* forms massive super-colonies and aggressively preys on arthropods while displaying little or no aggression toward larger animals. This species is a recent invader of the Laikipia region (ca. within the last decade), and in numerous areas is disrupting the *A. drepanolobium*-ant mutualism by nearly or completely extermin-ating native acacia ants on host plants. In contrast to native acacia ants, *P. mega-cephala* does not protect trees from vertebrate herbivores. As a result, in areas where it has invaded, browsing by elephants has increased substantially, resulting in higher levels of damage and mortality to host plants (Figure 10.5, Riginos et al., 2015). Over the longer-term, disruption of the ant-acacia mutualism by *P. megacephala* may strongly alter the extent of woody plant cover in these savannas, with cascading consequences for the entire community.

Summary and Future Directions

More than two decades of research on *A. drepanolobium* and its ant associates has revealed much about the complex and variable nature of ant-plant symbiosis. The system has also served as a model for integrating the study of mutualism into a community context, revealing how both biotic and abiotic environmental variation can shape ant-plant interactions, and demonstrating how a foundational mutualism can structure its surrounding community at large spatial scales. Yet much remains to be learned about this widespread association: can host plants exert control over the identity of their ant occupants? What are the physiological mechanisms underlying host plant allocation to ant rewards versus other carbon demands? Does the multi-species nature of the ant-plant association buffer *A. drepanolobium* from environmental variation? What are the landscape-scale consequences of the mutualisms' disruption by *P. megacephala*? The list of questions is long, and research has only begun to scratch the surface of this intriguing study system.

Study of this symbiosis may also help to illuminate more general principles that underlie many of the worlds' mutualisms of conservation concern. Like corals and their dinoflagellate associates, and tropical trees and their pollinators or dispersers, *A. drepanolobium* is a long-lived species which interacts over its ontogeny with multiple shorter-lived partner species. A thorough understanding of the drivers of contingency in costs and benefits within mutualist networks may lead to a more predictive framework for understanding context-dependent outcomes in these widespread interactions. The broad distribution and highly tractable nature of the *A. drepanolobium*-ant interaction makes it an ideal candidate for these investigations.

References

Andrews, P., and M. Bamford. (2008). Past and present vegetation ecology of Laetoli, Tanzania. *Journal of Human Evolution* 54,78–98.

Archetti, M., F. Ubeda, D. Fudenberg, J. Green, N. E. Pierce, and D. W. Yu. (2011). Let the right one in: a microeconomic approach to partner choice in mutualisms. *American Naturalist* 177,75–85.

Archibald, S. (2016). Managing the human component of fire regimes: Lessons from Africa. *Philosophical Transactions of The Royal Society B* 371:20150346. http://dx.doi.org/10.1098/rstb.2015.0346

Archibald, S., A.C. Staver, and S.A. Levin. (2012). Evolution of human-driven fire regimes in Africa. *Proceedings of the National Academy of Sciences of the United States of America* 109, 847–852.

Arshad, M. A. (1981). Physical and chemical properties of termite mounds of two species of Macrotermes (Isoptera, Termitidae) and the surrounding soils of the semi-arid savanna of Kenya. *Soil Science* 132,161–174.

Baker, C. (2015). Complexity in Mutualisms: Indirect Interactions with Multiple Parties. Doctoral Dissertation, Harvard University, Cambridge, MA.

Belsky, A. J., R. G. Amundson, J. M. Duxbury, S. J. Riha, A. R. Ali, and S. M. Mwonga. (1989). The effects of trees on their physical, chemical, and biological environments in a semi-arid savanna in Kenya. *Journal of Applied Ecology* 26,1005–1024.

Brody, A. K., T. M. Palmer, K. Fox-Dobbs, and D. F. Doak. (2010). Termites, vertebrate herbivores and the fruiting success of Acacia drepanolobium. *Ecology* 91,399–407.

Chomicki, G., P. S. Ward, and S. S. Renner. (2015). Macroevolutionary assembly of ant/plant symbioses. Pseudomyrmex ants and their ant-housing plants in the Neotropics. *Proceedings of the Royal Society B-Biological Sciences* 282,20152200.

Cochard, R., and P. J. Edwards. (2011). Structure and biomass along an Acacia zanzibarica woodland-savanna gradient in a former ranching area in coastal Tanzania. *Journal of Vegetation Science* 22,475–489.

Davidson, D. W., J. T. Longino, and R. R. Snelling. (1988). Pruning of host plant neighbors by ants: An experimental approach. *Ecology* 69,801–808.

Davidson, D. W., and D. McKey. (1993). The evolutionary ecology of symbiotic ant – plant relationships. *Journal of Hymenoptera Research* 2,13–83.

Dayton, P. K. (1972). Toward an understanding of community resilience and the potential effects of enrichments to the benthos at McMurdo Sound, Antarctica. Pages 81–96. In *Proceedings of the Colloquium on Conservation Problems*. Allen Press, Lawrence, Kansas.

Fiala, B., U. Maschwitz, T. Y. Pong, and A. J. Helbig. (1989). Studies of a South East Asian ant-plant association: protection of *Macaranga* trees by *Crematogaster borneensis*. *Oecologia* 79,463–470.

Ford, A. T., J. R. Goheen, T. O. Otieno, P. Arcese, T. M. Palmer, R. Woodroffe, D. Ward, and R. M. Pringle. (2014). Large carnivores make savanna tree communities less thorny. *Science* 346,346–349.

Fox-Dobbs, K., D. F. Doak, A. K. Brody, and T. M. Palmer. (2010). Termites create spatial structure and govern ecosystem function by affecting N-2 fixation in an East African savanna. *Ecology* 91,1296–1307.

Frederickson, M. E. (2009). Conflict over reproduction in an ant-plant symbiosis: why *Allomerus octoarticulatus* ants sterilize *Cordia nodosa* trees. *American Naturalist* 173,675–681.

Frederickson, M. E., A. Ravenscraft, G. A. Miller, L. M. A. Hernandez, G. Booth, and N. E. Pierce. (2012). The direct and ecological costs of an ant-plant symbiosis. *American Naturalist* 179,768–778.

Goheen, J. R., and T. M. Palmer. (2010). Defensive plant-ants stabilize megaherbivore-driven landscape change in an African savanna. *Current Biology* 20,1768–1772.

Heil, M. (2013. Let the best one stay: screening of ant defenders by Acacia host plants functions independently of partner choice or host sanctions. *Journal of Ecology* 101,684–688.

Heil, M., and D. McKey. (2003). Protective ant-plant interactions as model systems in ecological and evolutionary research. *Annual Review of Ecology Evolution and Systematics* 34,425–453.

Hocking, B. (1970). Insect associations with the swollen thorn acacias. *Transactions of the Royal Entomological Society of London* 122,211–255.

Holdo, R. M., A. R. E. Sinclair, A. P. Dobson, K. L. Metzger, B. M. Bolker, M. E. Ritchie, and R. D. Holt. (2009). A disease-mediated trophic cascade in the Serengeti and its implications for ecosystem C. *Plos Biology* 7, e1000210.

Huntzinger, M., R. Karban, T. P. Young, and T. M. Palmer. (2004). Relaxation of induced indirect defenses of acacias following exclusion of mammalian herbivores. *Ecology* 85,609–614.

Huxley, C. R. (1978). The ant-plants *Myrmecodia* and *Hydnophytum* (Rubiaceae), and the relationships between their morphology, ant occupants, physiology and ecology. *New Phytologist* 80,231–268.

Janzen, D. H. (1966). Coevolution of mutualism between ants and acacias in Central America. *Evolution* 20,249–275.

  (1969). Allelopathy by myrmecophytes: the ant *Azteca* as an allelopathic agent of *Cecropia*. *Ecology* 50,147–153.

Madden, D., and T. P. Young. (1992). Symbiotic ants as an alternative defense against giraffe herbivory in spinescent *Acacia drepanolobium*. *Oecologia* 91,235–238.

Martins, D. J. (2010). Not all ants are equal: obligate acacia ants provide different levels of protection against mega-herbivores. *African Journal of Ecology* 48,1115–1122.

  (2013). Effect of parasitoids, seed-predators and ant-mutualists on fruiting success and germination of *Acacia drepanolobium* in Kenya. *African Journal of Ecology* 51,562–570.

Midgley, J. J., T. Sawe, P. Abanyam, K. Hintsa, and P. Gacheru. (2016). Spinescent East African savannah acacias also have thick bark, suggesting they evolved under both an intense fire and herbivory regime. *African Journal of Ecology* 54,118–120.

Morawetz, W., M. henzl, and B. Wallnöfer. (1992). Tree killing by herbicide producing ants for the establishment of pure *Tococa occidentalis* populations in the Peruvian Amazon. *Biodiversity and Conservation* 1,19–33.

Ness, J. H. (2006). A mutualism's indirect costs: the most aggressive plant bodyguards also deter pollinators. *Oikos* 113,506–514.

Okello, B. D., T. G. O'Connor, and T. P. Young. (2001). Growth, biomass estimates, and charcoal production of *Acacia drepanolobium* in Laikipia, Kenya. *Forest Ecology and Management* 142,143–153.

Okello, B. D., and T. P. Young. (2000). Effects of fire, bruchid beetles and soil type on the germination and seedling establishment of *Acacia drepanolobium*. *African Journal of Range and Forage Science* 17,45–61.

Okello, B. D., T. P. Young, C. Riginos, D. Kelly, and T. G. O'Connor. (2008). Short-term survival and long-term mortality of Acacia drepanolobium after a controlled burn. *African Journal of Ecology* 46,395–401.

Palmer, T. M. (2001). Competition and Coexistence in a Guild of African Acacia-Ants. PhD dissertation. University of California Davis, Davis, CA.

  (2003). Spatial habitat heterogeneity influences competition and coexistence in an African acacia ant guild. *Ecology* 84,2843–2855.

  (2004). Wars of attrition: colony size determines competitive outcomes in a guild of African acacia-ants. *Animal Behaviour* 68,993–1004.

Palmer, T. M., and A. K. Brody. (2007). Mutualism as reciprocal exploitation: ant guards defend foliar but not reproductive structures of an African ant-plant. *Ecology* 88,3004–3011.

(2013). Enough is enough: the effects of symbiotic ant abundance on herbivory, growth and reproduction in an African acacia. *Ecology* 94,683–691.

Palmer, T. M., D. F. Doak, M. L. Stanton, J. L. Bronstein, E. T. Kiers, T. P. Young, J. R. Goheen, and R. M. Pringle. (2010). Synergy of multiple partners, including freeloaders, increases host fitness in a multispecies mutualism. *Proceedings of the National Academy of Sciences of the United States of America* 107,17234–17239.

Palmer, T. M., E. G. Pringle, A. C. Stier, and R. D. Holt. (2015). Mutualism in a community context. Pages 159–180. In J. L. Bronstein, editor. *Mutualism*. Oxford University Press, Oxford.

Palmer, T. M., M. L. Stanton, T. P. Young, J. R. Goheen, R. M. Pringle, and R. Karban. (2008a). Breakdown of an ant-plant mutualism follows the loss of large herbivores from an African Savanna. *Science* 319,192–195.

(2008b). Putting ant-Acacia mutualisms to the fire – response. *Science* 319,1760–1761.

Palmer, T. M., T. P. Young, and M. L. Stanton. (2002). Burning bridges: Priority effects and the persistence of a competitively subordinate acacia-ant in Laikipia, Kenya. *Oecologia* 133,372–379.

Palmer, T. M., T. P. Young, M. L. Stanton, and E. Wenk. (2000). Short-term dynamics of an acacia ant community in Laikipia, Kenya. *Oecologia* 123,425–435.

Pringle, E. G., E. Akcay, T. K. Raab, R. Dirzo, and D. M. Gordon. (2013). Water stress strengthens mutualism among ants, trees, and scale insects. *Plos Biology* 11, e1001705.

Pringle, E. G., R. Dirzo, and D. M. Gordon. (2011). Indirect benefits of symbiotic coccoids for an ant-defended myrmecophytic tree. *Ecology* 92,37–46.

Pringle, R. M., D. F. Doak, A. K. Brody, R. Jocque, and T. M. Palmer. (2010). Spatial pattern enhances ecosystem functioning in an African savanna. *Plos Biology* 8, e1000377.

Pringle, R. M., and K. Fox-Dobbs. (2008). Coupling of canopy and understory food webs by ground-dwelling predators. *Ecology Letters* 11,1328–1337.

Pringle, R. M., K. M. Prior, T. M. Palmer, T. P. Young, and J. R. Goheen. (2016). Large herbivores promote habitat specialization and beta diversity of African savanna trees. *Ecology* 97,2640–2657.

Riginos, C. (2015). Climate and the landscape of fear in an African savanna. *Journal of Animal Ecology* 84,124–133.

Riginos, C., J. B. Grace, D. J. Augustine, and T. P. Young. (2009). Local versus landscape-scale effects of savanna trees on grasses. *Journal of Ecology* 97,1337–1345.

Riginos, C., M. A. Karande, D. I. Rubenstein, and T. M. Palmer. (2015). Disruption of a protective ant-plant mutualism by an invasive ant increases elephant damage to savanna trees. *Ecology* 96,654–661.

Rudolph, K. P., and J. P. McEntee. (2016). Spoils of war and peace: enemy adoption and queen-right colony fusion follow costly intraspecific conflict in acacia ants. *Behavioral Ecology* 27,793–802.

Ruiz-Guajardo, J. C., D. Grossenbacher, R. K. Grosberg, T. M. Palmer, and M. L. Stanton. (2017). Impacts of worker density in colony-level aggression, expansion, and survival of the acacia-ant *Crematogaster mimosae*. *Ecological Monographs* 87: 246–259.

Sagers, C. L., S. M. Ginger, and R. D. Evans. (2000). Carbon and nitrogen isotopes trace nutrient exchange in an ant-plant mutualism. *Oecologia* 123,582–586.

Sensenig, R. L., D. K. Kimuyu, J. C. Ruiz-Guajardo, C. Riginos, and T. P. Young. (2017). Fire disturbance disrupts an acacia ant-plant mutualism in favor of a subordinate ant species. *Ecology* 98,1455–1464.

Stanton, M. L. (2003). Interacting guilds: moving beyond the pairwise perspective on mutualisms. *American Naturalist* 162,S10-S23.

Stanton, M. L., and T. M. Palmer. (2011). The high cost of mutualism: effects of four species of East African ant symbionts on their myrmecophyte host tree. *Ecology* 92,1073–1082.

Stanton, M. L., T. M. Palmer, and T. P. Young. (2002). Competition-colonization trade-offs in a guild of African acacia-ants. *Ecological Monographs* 72,347–363.

(2005). Ecological barriers to early colony establishment in three coexisting acacia-ant species in Kenya. *Insectes Sociaux* 52,393–401.

Stanton, M. L., T. M. Palmer, T. P. Young, A. Evans, and M. L. Turner. (1999). Sterilization and canopy modification of a swollen thorn acacia tree by a plant-ant. *Nature* 401,578–581.

Stapley, L. (1998). The interaction of thorns and symbiotic ants as an effective defence mechanism of swollen-thorn acacias. *Oecologia* 115,401–405.

Styrsky, J. D., and M. D. Eubanks. (2007). Ecological consequences of interactions between ants and honeydew-producing insects. *Proceedings of the Royal Society B-Biological Sciences* 274,151–164.

Tarnita, C. E., T. M. Palmer, and R. M. Pringle. (2014). Colonisation and competition dynamics can explain incomplete sterilisation parasitism in ant-plant symbioses. *Ecology Letters* 17,1290–1298.

Visiticao, J. M. (2011). Multi-Species Interactions in African Ant-Acacias. PhD Dissertation, Harvard University, Cambridge, MA.

Vollrath, F., and I. Douglas-Hamilton. (2002). African bees to control African elephants. *Naturwissenschaften* 89,508–511.

Willmer, P. G., C. V. Nuttman, N. E. Raine, G. N. Stone, J. G. Pattrick, K. Henson, P. Stillman, L. McIlroy, S. G. Potts, and J. T. Knudsen. (2009). Floral volatiles controlling ant behaviour. *Functional Ecology* 23,888–900.

Willmer, P. G., and G. N. Stone. (1997). How aggressive ant-guards assist seed-set in *Acacia* flowers. *Nature* 388,165–167.

Wood, W. F., and B. Chong. (1975). Alarm pheremones of the east African acacia symbionts: *Crematogaster mimosae* and *C. negriceps*. *Journal of the Georgia Entomological Society* 10,332–334.

Young, T. P., B. D. Okello, D. Kinyua, and T. M. Palmer. (1998). KLEE: a long-term multi-species herbivore exclusion experiment in Laikipia, Kenya. *African Journal of Range Forage Science* 14,94–102.

Young, T. P., C. H. Stubblefield, and L. A. Isbell. (1997). Ants on swollen-thorn acacias: species coexistence in a simple system. *Oecologia* 109,98–107.

Yu, D. W., H. B. Wilson, M. E. Frederickson, W. Palomino, R. De la Colina, D. P. Edwards, and A. A. Balareso. (2004). Experimental demonstration of species coexistence enabled by dispersal limitation. *Journal of Animal Ecology* 73,1102–1114.

11 Ecological and Evolutionary Responses of Protective Ant-Plant Mutualisms to Environmental Changes

Doyle McKey and Rumsaïs Blatrix[*]

Species Interactions and Environmental Change

Environmental Change as a Threat to Biotic Interactions, Particularly Mutualisms

As habitats become increasingly fragmented and degraded, and as climates change, species linked by tight interdependencies may suffer extinction or coextinction (Dunn et al., 2009). Long before the last individuals of such tightly linked species die, their interactions are extinct, having ceased to play whatever role they may once have played in the dynamics of communities and the functioning of ecosystems. The extinction of interactions is thus an even more fundamental threat to biodiversity and ecosystem functioning than is the extinction of individual species themselves (Janzen, 1974; Bascompte & Jordano, 2013).

Tight interdependencies may result from indirect effects in food webs, as when the disappearance of a top predator or of megaherbivores sets in motion a cascade of events that drive some species to extinction (Estes et al., 2011). They may also result from direct positive interactions between two or more species. The benefits of these interactions may be one-sided (species A has a positive effect on species B, but species B has no effect on species A) or reciprocal. Systems characterized by reciprocal positive interactions – mutualisms – have been a particularly important focus of research on how environmental change affects species interactions (Dunn et al., 2009). Among the open questions are the following: Can environmental change lead to disassembly of mutualisms? When mutualisms are disassembled, what are the fates of the partner species? Can disruption of mutualisms lead to extinction cascades?

[*] We thank Tom Fayle for stimulating discussion and for useful insights and comments which improved the manuscript. Our work on ant-plant interactions has been supported by grants from two programs of the French Agence Nationale de la Recherche, "Young scientists" (research agreement no. ANR-06-JCJC-0127), "Biodiversity" (IFORA project) and "Sixth extinction" (C3A project).

The Diversity of Relationships between Environmental Change and the Dynamics of Mutualisms

Although much emphasis has been placed on environmental change as a threat to mutualisms, the relationships between the two are actually quite diverse. First of all, mutualisms are not just passive receivers of environmental change; their constituent partners also adapt to change. Environmental change drives coevolution, whether of mutualism or of other interactions, and times or places where environmental change is rapid may be hotspots in the coevolutionary mosaic (Thompson, 2005). Second, positive interactions within species networks can buffer the effects of environmental change (Tylianakis et al., 2008). The following example of facilitation by nurse-plants illustrates this point. When moist subtropical Tertiary woodlands of California, Chile and the Mediterranean Basin became drier mediterranean shrublands, species already adapted to drier climates joined the initial subtropical flora. Some components of this initial flora disappeared as climates dried, but others persisted and adapted. Their persistence – for periods at least long enough to evolve adaptations to the changed environment – was facilitated by drought-adapted new arrivals that acted as "nurse plants" for relict species from a wetter time, providing shady and relatively moist microhabitats on which these holdovers came to depend (Valiente-Banuet et al., 2006; Lortie, 2007). Third, some kinds of environmental change may even strengthen mutualisms, including some ant-plant mutualisms (Pringle et al., 2013). Despite the existence of such examples, it remains true that few studies have addressed how mutualisms adapt to environmental change, or how mutualisms could affect the consequences of environmental change. A final scenario that must not be discounted is that some mutualistic interactions may not be influenced by environmental change, at least not of the kinds and at the level of intensity observed today. The proportion of mutualistic interactions unaffected by current changes is unknown.

Finally, in one aspect of global environmental change, the increased rate of biotic exchange among regions, mutualisms are a big part of the problem. Invasive mutualisms can transform ecosystems and lead to loss of biodiversity. The most pervasive example is provided by the mutualisms with domesticated plants and animals that have allowed *Homo sapiens*, the most mutualistic animal of them all, to dominate the planet (Diamond, 2002). Other examples abound. The actinorhizal symbiosis between *Myrica gale* and its nitrogen-fixing symbiotic *Frankia* bacteria drove the invasion of the Hawaiian archipelago by this plant, greatly increasing soil stocks of nitrogen and transforming ecosystem functioning (Vitousek & Walker, 1989). Likewise, seed dispersal mutualisms with frugivorous vertebrates (or seed-transporting ants) have enabled many fleshy-fruited (or elaiosome-bearing) exotic plants to invade large areas rapidly (Richardson et al., 2000).

Rapid Environmental Change and the Evolutionary Ecology of Mutualisms: Open Questions

In the context of the rapid environmental changes we are currently experiencing, several important questions arise: How does change affect the ecology of each actor in the mutualism? How do all changes, taken together, affect the net outcomes of interactions? These questions are complex. Many kinds of change are occurring simultaneously and different actors in the mutualism may have dramatically different life histories and therefore different responses to the same change. How does environmental change affect the longer-term evolutionary dynamics of mutualisms? When changing ecological dynamics result in reduced benefits or increased costs to one or more partners, does selection lead to dissolution of mutualisms? When mutualisms are disassembled, what are the fates of the former partners? When ecological dynamics lead to the strengthening of mutualism, or to altered patterns of associations between species, how do changed selective pressures affect the evolution of mutualistic traits?

Ant-Plant Protection Mutualisms as Model Systems for Studying the Consequences for Biodiversity of Global Environmental Change

Most research on the fate of mutualisms under environmental change has focused on three kinds of mutualisms: pollination mutualisms, seed-dispersal mutualisms between frugivorous animals and fleshy-fruited plants, and ant-plant mutualisms (Ings et al., 2009). Of these three, ant-plant mutualisms are arguably the model systems from which we stand to learn the most about the ecological and evolutionary consequences of environmental change. This is because the ecological and physiological diversity of actors in ant-plant mutualisms far outstrips that in the other two kinds of mutualisms. Ant-plant mutualisms can be seen as microcosms of biotic communities (Blatrix et al., 2009), complex but still tractable systems that provide insight into general questions.

First, ant-plant mutualisms comprise a great diversity of interactions (Rico-Gray & Oliveira, 2007), ranging from opportunistic associations in which extrafloral nectaries and other food rewards attract diverse ants, each of which may visit many plant species, to longer-term interactions where ants reside in more or less specialized structures (termed "domatia") provided by one or a few kinds of host plants (termed "myrmecophytes" or "ant-plants"). Even within these symbiotic interactions, there is great variation in the degree of specialization and specificity of the partner ant and plant species. This diversity allows great scope for studying the effects of the structure of interaction networks on the responses of biodiversity to environmental change.

Second, "ant-plant" mutualisms are in fact complex food webs. Many involve hemipteran trophobionts that play an essential role in making plant resources such as phloem sap available to ants (McKey et al., 2005). These trophobionts in turn depend on intracellular bacteria to manufacture essential amino acids and other nutrients lacking in phloem sap. Ants themselves host other bacteria (Russell et al., 2009), the roles of which are just beginning to be elucidated. Most plant-resident ants also cultivate fungi in domatia of their host (Mayer et al., 2014). These fungi contribute to the ants' nutrition and appear to play a key role in recycling nitrogenous waste products (Defossez et al., 2011; Blatrix et al., 2012). Ants deposit their excrement, the cadavers of dead nestmates and debris collected from the surfaces of the host plant on these fungal patches (Defossez et al., 2009), supporting dense populations of nematodes and bacteria, of whose role in these food webs nothing is known. Furthermore, via these food webs ants also "feed" the host plant, directly or indirectly. Most symbiotic ant-plant "protection" mutualisms are thus also nutritional mutualisms (Mayer et al., 2014).

Third, the ecological basis of ant-plant protection mutualisms, their raison d'être, is the web of interactions between plants and their herbivores and pathogens. The ecological and evolutionary dynamics of interactions between plants and their ant mutualists under environmental change are thus dependent on how these same changes affect the interactions between plants and their enemies. Understanding the responses of ant-plant protection mutualisms to environmental change thus requires integrating the vast literature on how environmental change affects plant-herbivore and plant-pathogen interactions.

"Ant-plant" protection mutualisms englobe interactions among organisms from all the kingdoms of life. Owing to the physiological and ecological diversity of all these actors, ant-plant mutualisms are affected by virtually every driver of global change, often simultaneously. Studying interaction networks around ants and plants can help prepare us for the even more complex task of understanding the response of entire ecosystems to environmental change, at both ecological and evolutionary timescales.

The Scope of This Chapter

In this chapter, we will consider how each of the drivers of global change might affect the ecological dynamics of each of the various actors in ant-plant mutualisms and of the networks in which they are embedded. We also examine how these effects are integrated, over longer time scales, in the evolutionary outcome of the entire set of interactions.

The impacts of several drivers of environmental change on ant-plant protection mutualisms have been the subject of substantial research, and are treated elsewhere in this book. Land-use change (Chapter 3) and other disturbances fragment habitats and create mosaics of primary and secondary forests (Chapter 2). The cascading effects of the loss of the megaherbivores that helped drive the evolution of a protection mutualism are analyzed in Chapter 10. The effects of biotic invasions on protection mutualisms are extensively explored in Chapters 12–15.

However, gaps remain in currently conducted research, even for these most frequently studied aspects of environmental change. Although ecological responses to fragmentation, habitat change and biological invasions have been characterized, few studies examine the potential consequences for the evolutionary dynamics of ant-plant interactions.

Furthermore, several important dimensions of environmental change – notably climate change and CO_2 fertilization – have received little attention. Only a few studies of thermal tolerances of ants, and of a few associated organisms, offer some insights into the potential impact of warming. Most importantly, few studies attempt to integrate (1) the effects of multiple factors acting together, or (2) the interactions among the multiple trophic levels that together determine the eco-evolutionary dynamics of "ant-plant" interactions. Our chapter focuses on these neglected aspects. Our emphasis is on framing the questions, not on offering conclusions. Given the present state of research, firm conclusions would be premature.

Diversity in the Structure of Ant-Plant Mutualistic Networks

The increasing application of network theory in community ecology has opened new questions, and has provided tools to examine old, but still unsettled, questions. Ant-plant mutualisms exhibit a diversity of network structures. In some specialized symbiotic mutualisms, partners are host-specific. In extreme cases, a single mutualist ant species is associated with a single host plant species. These associations constitute highly compartmentalized ant/plant networks (Dáttilo et al., 2013). In other symbiotic mutualisms, partners are less specific: each of a number of plant-ant species can be associated with a range of ant-plant hosts. Non-symbiotic mutualisms are even less specific: many ant species, and many plant species, can each be associated with a broad range of partners. Networks in these "facultative" ant-plant mutualisms often have a "nested" structure (reciprocal generalist species plus specialists that interact mostly with generalists) (Guimarães et al., 2006).

These differences in network structure, as well as the physiological differences between the constituent species of the networks, condition the effects of environmental change on species interactions. Ants and plants can be expected to respond differently to many kinds of environmental change. For instance, ants might be more affected by elevated temperature, but less affected by elevated CO_2, than plants. These differences in ant and plant responses to change can reduce the size of the area that is suitable to both partners, leading to the disassembly of associations in many sites. Change is much more likely to threaten the persistence of mutualistic interactions when these interactions are obligate and specific. In contrast, interactions (whether symbiotic or not) between generalists should be less affected by environmental change, because even under altered conditions each partner has a higher probability of finding a suitable partner in its range. For example, the non-specific mutualism between an epiphytic fern species and various protective ants that dwell in its root system has been shown to be resilient to the conversion of rain forest to oil palm plantation (Fayle et al., 2015). The more specialized the interaction, the less resilient it should be to disassembly.

Environmental changes are likely to decrease the proportion of specialized mutualisms in general (Kiers et al., 2010; Brodie et al., 2014; Fontúrbel & Murúa, 2014). One of the reasons for this is that specialized obligate mutualists may be "locked in," unable to adapt to changed conditions. In many cases, coevolution of partners in specialized mutualisms has led to the loss of traits (Ellers et al., 2012). Although loss of these traits is adaptive in the context of the interaction, trait loss, by forcing dependence on the partner, may lead into an evolutionary dead-end. For example, invertase activity in the foliar nectaries of Mexican *Acacia* species that host symbiotic ants acts as a partner-choice filter, by making their nectar unattractive to most ants other than the species of *Pseudomyrmex* that live in them (Kautz et al., 2009). Because their hosts produce invertase, these specialist symbionts have lost this capability, now redundant. This makes them dependent on the host, and escaping this dependence when environmental change disrupts the association may be difficult. Re-evolution of an enzymatic function is a much more improbable evolutionary step than the initial loss. In networks of generalist plants and ants, each capable of association with a diversity of partners, each partner has a higher probability of continuing to encounter a suitable partner when environmental conditions are altered, compared to mutualisms involving more specialized partners.

Ants and plants are not the sole participants in these networks. Ant-plant mutualisms, both EFN-mediated opportunistic protection mutualisms and ant-plant symbioses, involve numerous actors. These include the herbivores and pathogens against which ants protect plants, as well as additional mutualist partners that are often essential to the functioning of the network. Bacteria, fungi, hemipteran trophobionts and nematodes have all been shown to be linked through trophic interactions with ants and plants (Gaume et al., 1998; Defossez et al., 2011; Leroy et al., 2011; Blatrix et al., 2012; Mayer et al., 2014; Maschwitz et al., 2016) and to be key partners in ant-plant mutualisms (Mayer et al., 2014). For example, some ants that visit extrafloral nectaries, providing protection to the plants that bear them, host bacteria that help them cope with diets particularly rich in carbohydrates and lacking in essential amino acids (Russell et al., 2009). Phloem sap-sucking hemipterans, including those that are ant trophobionts, host endocellular symbiotic bacteria that enable them to cope with the same adaptive problem (Douglas, 2009). Some of these actors – for example, *Buchnera* and other endocellular symbionts of aphids, scale insects and other phloem-sucking hemipterans (Moran & Baumann, 2000) – are strongly buffered against environmental variation and environmental change by their ancient, constant, vertically transmitted symbioses with their hosts. Others may be sensitive to environmental change. Just as for ants and plants, their ranges of tolerance for abiotic conditions such as temperature, and their degree of host specificity, can condition the network's response to environmental change. For example, some microbial partners of ants may be affected by climate change. Arboreal ants of the formicine tribe Camponotini host endosymbiotic *Blochmannia* bacteria that contribute to their nutrition. Subjecting the ant hosts of *Blochmannia* to four days of increased temperature reduces populations of the symbiont by 99 percent (Fan & Wernegreen, 2013). This example, reminiscent of coral bleaching and of bacterial symbiont elimination in insects in general (Wernegreen, 2012), shows that the response of ant-plant mutualisms to climate change may be conditioned by effects on one or more of the other actors involved in these systems.

Ecological and Evolutionary Responses of Ant-Plant Mutualisms to Environmental Change

The short-term ecological responses of ants, plants and their partners and adversaries to environmental change will alter the ecological setting in which selection acts, conditioning the evolutionary responses of actors in these mutualisms. However, it may be difficult to predict which actors will evolve, what traits will change, how rapidly change will occur, whether interactions persist and, if so, how the altered interactions will function.

In practice, ecological and evolutionary responses are difficult to disentangle (Merilä & Hendry, 2014). However, it is important to consider them separately, at least from a theoretical point of view, because they may diverge. In the following sections, we discuss the effects of different dimensions of global environmental change on ant-plant protection mutualisms. In each section, we treat ecological responses and speculate about evolutionary responses over the long term. Throughout, however, the two kinds of response, driven by the same processes and acting on overlapping time scales, are necessarily interwoven.

Climate Change and Ant-Plant Protection Mutualisms

Some studies suggest that climate change can weaken ant-plant mutualisms, reduce the range of environments in which mutualisms can occur, or even lead to their disassembly. Ants and plants are expected to respond differently to climatic variables. Whereas ants may be affected by elevated temperature, plants are likely to be affected by a combination of variables, tolerating higher temperatures if water availability is sufficiently high to maintain transpirational cooling of leaves. Climate change could thus reduce the range within which plant and ant can co-occur.

Two environmental factors, temperature and water availability, play predominant roles in explaining the distribution of terrestrial biodiversity and the functioning of terrestrial ecosystems. Both are affected by climate change, and both can be expected to strongly influence ant-plant mutualisms. Higher temperatures lead to higher metabolic rates in insects (Bale et al., 2002) and in some insects can lead to an increase in the number of generations per year (DeLucia et al., 2012). Higher herbivory pressure under increasing average temperature has been documented in the fossil record at the geological scale (Wilf & Labandeira, 1999). Warming could thus potentially increase the benefits to plants of ant protection. However, at least one instance suggests that plant-visiting or plant-dwelling ants might have lower tolerance to higher temperatures than many phytophagous insects, so that higher temperatures could decrease the protective efficiency of ants (Fitzpatrick et al., 2013).

Droughts can increase the frequency of insect outbreaks (Coley, 1998). Also, when plants are weakened by drought or other stresses, their tolerance of herbivory is decreased. Thus under drought stress the same amount of herbivory could impose a higher cost. Under drought conditions the benefits to plants of ant protection could thus be increased. As with warming, however, the effect of droughts on ant-plant mutualisms depends not only on how droughts affect interactions between plants and herbivores, but also partly on how they affect plant-visiting or plant-resident ants, as well as other actors in the mutualistic networks.

The initial, rapid response of populations to environmental change is determined by their phenotypic plasticity. An important aspect of plasticity is the range of tolerance for key abiotic environmental conditions. As in other organisms (Sunday et al., 2011), ants of temperate zones have evolved a much broader temperature-tolerance range than tropical ants. Thus, as in other organisms, tropical ant species may be less well equipped (less "pre-adapted") to survive rapid climate change than temperate-zone ants (Diamond et al., 2012). Ant-plant mutualisms are most frequent and diverse in the tropics, and symbiotic ant-plant mutualisms are completely restricted to tropical regions. These mutualisms could thus be particularly affected by climate change.

The climatic tolerance of other actors in these networks can also condition the response of ant-plant mutualisms to climate change. As seen in a previous section, some microbial symbionts of ants have restricted temperature tolerances that could affect the response of networks. Survival, growth and reproduction of hemipteran trophobionts of ants are also influenced by climate (Bale et al., 2002). The cost to the plant of herbivory inflicted by ant-tended hemipterans is often counterbalanced by the benefits of the protection the tending ants provide against other herbivores (Zhang et al., 2012). However, the outcome of ant/plant/hemipteran interactions is context dependent (Barton & Ives, 2014; Marquis et al., 2014). Thus, if climate change affects hemipterans, the effects could potentially cascade through the mutualistic network.

Higher temperatures can change where and when ants forage (Stuble et al., 2013). Could such changes affect ant visitation at extrafloral nectaries or the activities of ants on myrmecophytes, thereby affecting the protection they confer against herbivores and (for myrmecophytes) other benefits? Fitzpatrick et al. (2014) showed that there was a succession over the daily cycle in the presence of different ant species at extrafloral nectaries of *Ferocactus wislizeni*. The species of ants that were most efficient at protecting the plant were also those least tolerant of high temperatures. During the warmest hours of the day, the plant was thus less efficiently protected from the plant's principal herbivore, which had a higher thermal tolerance than the ants. In such a system, the duration of the period of each day suitable for the most protective ants will be shortened with global warming. This could lead to less efficient protection, or even to dissolution of the mutualism.

The only other example sufficiently well-developed to be discussed here does not involve warming, but rather drought, the frequency of which is expected to increase under climate change. Plants under water stress are expected to conserve water and energy. Do ant-plants under water stress reduce the water- and energy-rich rewards they offer to ants? Pringle et al. (2013) showed that in the symbiosis between the tree *Cordia alliodora* and *Azteca pittieri*, carbohydrate rewards to ants were actually greater in plants exposed to stronger drought. They argued that the maintenance of large resident ant colonies under recurrent water stress was driven by the risk of extreme events of herbivory. Level of herbivory is not constant but varies among years. Trees lacking a sufficient force of *Azteca* workers run the risk of occasional complete defoliation. Water stress reduces the tree's probability of surviving such an event. Thus the maintenance of a large worker force is particularly advantageous where drought conditions are frequent. The hypothesis that the evolutionary dynamics of symbiotic ant-plant mutualisms could be driven by herbivory during extreme events that stress host plants, rather than "average" levels of herbivory, is of particular interest in the context of global climate change. The frequency of extreme climatic events is predicted to increase, and this could affect the cost/benefit ratio of maintaining ants, potentially leading to shifts in how ant-plants allocate resources.

Interpreting the ecological and evolutionary dynamics of the *Cordia/Azteca* interaction is complicated by the fact that the rewards for ants are not provided directly by the tree, but via honeydew produced by trophobiotic hemipterans tended by ants inside the tree's domatia. The rate of provisioning of ant rewards is determined by the number of these trophobionts. Whether this number is regulated by the tending ants or by traits of the host plant is not known. Thus, "who is in the driver's seat" in steering the system's dynamics is unclear. If the evolutionary interests of ants, plants and hemipterans are entirely convergent, the answer to this question makes little difference. However, if partners have partially conflicting interests, then predictions about the system's dynamics could diverge, particularly if one partner is more likely to "drive" the interaction than others, owing to greater phenotypic plasticity or behavioral flexibility, or faster evolutionary rates.

Climate change alters the spatial distribution of environmental conditions. A shift in the distribution of the climatic niche of a species (or of a mutualism) will lead to shifts in its geographical range. At the leading edge, where populations are colonizing new sites, selection often favors variants with better dispersal capacity. The resulting "colonization front syndrome" (Thomas et al., 2001; Phillips et al., 2006; Léotard et al., 2009) involves a whole set of correlated traits. Increased dispersal capacity often comes at the cost of decreased competitive ability. In ants, for example, dispersal capacity can be increased by the onset of sexual maturity at small colony size and by greater investment in dispersing foundresses; but these traits are accompanied by reduced investment in workers, potentially reducing the colony's chances of survival. Such "colonization/competition trade-offs" can

Figure 11.1. The myrmecophytic plant *Leonardoxa africana* subsp. *africana* has hollow internodes in which nests the mutualistic ant *Petalomyrmex phylax*. This symbiosis is endemic to a narrow strip of coastal rainforest in Cameroon and is threatened by anthropogenic pressure, in particular land-use change. Photo credit: Rumsais Blatrix. (A black-and-white version of this figure will appear in some formats. For the color version, please refer to the plate section.)

potentially weaken symbiotic ant-plant mutualisms. The benefits ants confer on plants are provided by the worker force of the plant's resident ant colony. Because protection increases with the size of the worker force (Duarte Rocha & Godoy Bergallo, 1992; Gaume et al., 1998; Heil et al., 2001), traits of ants that reduce the rate of colony growth can reduce defense and thereby the benefits the ant colony confers to the plant.

The consequences of range shift and colonization-front dynamics have been studied in only one ant-plant symbiotic mutualism, that between the rainforest understory tree *Leonardoxa africana africana* (Fabaceae) and its host-specific, mutualistic ant symbiont, *Petalomyrmex phylax* (Formicinae) (Figure 11.1; Léotard et al., 2009; Blatrix et al., 2013). Colonies of this ant at the colonization front possess several traits that increase dispersability: larger queens, queens with longer wings and higher sexuals/workers investment ratio (Dalecky et al., 2007; Léotard et al., 2009). Moreover, colonies at the colonization front are monogynous, whereas colonies in the center of the system's range are secondarily polygynous: when the colony reaches sexual maturity, some of the alate females it produces do not disperse but remain in the colony as additional queens (Dalecky et al., 2005), providing a larger worker force and a lower risk of colony death. All these shifts in the allocation of resources between colony growth/survival and dispersal can affect the benefits that ants confer on plants. In addition, recruitment of workers induced by damage to leaves was lower at the colonization front than in a population in the central, stable part of the system's range (Vittecoq et al., 2012). Thus, both the reduced investment in growth relative to reproduction and the lower behavioral investment in defense per individual worker result in a strategy that is less mutualistic as a whole at the colonization front.

These shifts in allocation at the colonization front, although dramatic, appear to be transient. Behind the colonization front, the advantage of greater dispersability quickly disappears, and more competitive phenotypes of *Petalomyrmex*, with multiple-queen colonies, smaller alates and larger worker forces, quickly replace the original colonists. In contrast to its resident ants, the host tree *Leonardoxa* showed no indication of a shift to a more dispersive or a less mutualistic phenotype at the colonization front (Léotard et al., 2009). This tree grows very slowly and has a very long lifespan. With a generation time much longer than that of its mutualistic ant, it lacked the time to evolve such dramatic trait shifts. Thus the tree retained its mutualistic phenotype unaltered, and when the more mutualistic phenotype of *Petalomyrmex* arrived just behind the colonization front, the mutualism was essentially restored, with no lasting alteration of its dynamics.

CO_2 Enrichment

Increased atmospheric concentration of CO_2 could have complex effects on ant-plant mutualisms. First, increased CO_2 concentrations affect the chemical composition of plants, altering the nutritional quality of their leaves and other tissues for herbivores. These changes can affect herbivory rate, altering the numbers and kinds of herbivores against which plants must be defended. Indeed, plants grown under elevated $[CO_2]_{atm}$ often have higher C/N ratios in their tissues (Robinson et al., 2012; Niziolek et al., 2013). This in turn usually leads to higher rates of consumption by herbivores (Coley, 1998; Lindroth, 2010; DeLucia et al., 2012; Couture et al., 2015). In addition, hemipterans seem to benefit from better performance under elevated CO_2 (Robinson et al., 2012; Ryalls et al., 2016). Second, CO_2 enrichment can reduce the effectiveness of plant chemical defenses (e.g. Ballhorn et al., 2011). Carbon-based chemical defense compounds may be augmented (Stiling & Cornelissen, 2007; Robinson et al., 2012), but overall the diversity of chemical defenses is decreased. Furthermore, elevated CO_2 can alter chemical defenses by modifying hormone signaling (DeLucia et al., 2012). Data so far suggest that CO_2 enrichment, along with warming, will result in increased rates of herbivory. Finally, as carbon availability to plants increases, and as plant rewards to ant mutualists are mostly carbon-based (Pringle, 2016), it should cost plants less to sustain mutualistic ants under CO_2 enrichment. With the predicted concomitant increase in herbivory and reduction in the effectiveness of plant chemical defenses, selection could favor the increased production of rewards for ants, in both opportunistic and symbiotic ant-plant associations. This evolutionary trend could be facilitated by standing variation. Indeed, increased herbivory is known in some plants to induce increased investment in ant rewards. For example, the myrmecophyte *Cordia nodosa* produces larger caulinary domatia when herbivory pressure is higher (Frederickson et al., 2013).

Does CO_2 enrichment affect the chemical composition of rewards for ants? If so, does this affect the functioning of ant-plant mutualisms? Tree-dwelling tropical ants, which depend heavily on plant-provided carbon-rich food sources, have long since adapted to exploit a diet that is heavy on carbohydrates and light on protein, compared to the ancestral diets of predatory ants (Davidson et al., 2003; Davidson, 2005; Cook & Davidson, 2006). Will the diets of tropical tree-dwelling ants become even more unbalanced? CO_2 enrichment increases the C/N ratio of many plant tissues; phytophagous insects thus encounter carbohydrate-enriched food. Whether plant-produced rewards for ants such as extrafloral nectar and honeydew also have higher C/N ratios under CO_2 enrichment has been studied in only a few cases, with few notable effects so far discovered (Sun et al., 2009). Whether an increase in C/N ratio of plant-produced rewards would have an effect on plant-associated ants is open to question. At least some phytophagous insects faced with increased C/N content of their food have rapidly evolved adaptations that negate the initial negative effects of this alteration in food quality (Warbrick-Smith et al., 2006). Furthermore, if herbivory rate increases with warming and CO_2 enrichment, as projected, plant-associated ants may be able to obtain adequate protein to counter any stoichiometric imbalance in the plant-derived part of their diet.

The global phenomenon of CO_2 enrichment is accompanied in many regions by anthropogenic inputs of plant-available nitrogen in the atmosphere and in soils (Vitousek et al., 1997). Increased N availability could also affect ant-plant symbioses, both directly and through interactions between the carbon and nitrogen cycles. Studies have shown the transfer of nitrogen from ants to plants in many ant-plant symbioses, not only in the ant-epiphytes upon which attention to this aspect was initially focused, but also in many soil-rooted trees, shrubs and vines. Many of these symbioses can thus be viewed as both nutritional and protection mutualisms (Mayer et al., 2014). Where the availability of nitrogen to plants is increased by anthropogenic inputs in the atmosphere, in soils, or both, plants may need less help from ants to acquire this important nutrient. How such a change might affect ant-plant mutualisms is difficult to predict. On the one hand, increased nitrogen availability could decrease the benefits that ants confer on plants, thereby weakening the mutualism. Effects could be similar to those observed with mycorrhizae, where plants invest less in root-associated mutualists when soil fertility is higher (Kiers & Denison, 2008; Johnson, 2010; P, rather than N, is often the most important nutrient transferred in mycorrhizal mutualisms), and fungal species are favored that are less efficient in nutrient scavenging and transfer to plants. Furthermore, increased nitrogen availability could negate any competitive advantage ant-plants have relative to other plant species without special adaptations for acquiring nitrogen. On the other hand, increased nitrogen availability, by removing nitrogen limitation, could enable ant-plant symbioses to respond to increased atmospheric concentrations of CO_2 by increased growth.

Habitat Fragmentation/Disturbance and Species Range Dynamics

A first dimension to take into account for understanding the effect of habitat fragmentation and disturbance on ant-plant mutualisms is the plant's ecological niche relative to disturbance. In some ant-plant symbioses, the plant is adapted to early successional stages after disturbance. For these, fragmentation of mature forest and disturbance increases the area of suitable habitat, and these positive effects may outweigh potential negative effects on their interaction with ants. For instance, pioneer ant-plant symbioses, such as those involving the plant genera *Barteria* (Africa), *Cecropia* (America) and *Macaranga* (Asia), thrive along roads and clearings opened through mature rain forest. In contrast, ant-plant symbioses adapted to the understory of mature forest, such as *Leonardoxa* in Africa (Gaume et al., 1997) and *Hirtella* (Orivel et al., 2011) in the Neotropics, are expected to be particularly sensitive to habitat fragmentation and disturbance.

Land-use changes leading to habitat fragmentation could weaken horizontally transmitted ant-plant mutualisms, which depend on dispersal for the establishment of the association with each new generation. Failure of ant mutualists to disperse to unoccupied plants, either juveniles that have not yet acquired a colony or larger plants that have lost their colony, could strongly reduce benefits to the plant. In systems where plants may be occupied by a diversity of ant occupants, habitat fragmentation could further weaken mutualisms by favoring ant species with high dispersal ability, because theory predicts they should often be less effective mutualists for plants (see the end of the paragraph on climate change). Over evolutionary timescales, competition/colonization trade-offs can alter selection pressures acting on the allocation of resources to dispersal and to worker production (and hence in mutualism with the plant). Examples illustrating this theme (e.g. Yu et al., 2004; Dalecky et al., 2007; Léotard et al., 2009) show the importance of taking into account conflicts of interest between partners in understanding the evolutionary dynamics of mutualisms.

For example, among populations of the myrmecophytic tree *Leonardoxa africana* subsp. *africana*, the proportion of trees occupied by the dispersive parasite of the mutualism, the ant *Cataulacus mckeyi*, rather than by the mutualistic and less dispersive ant, *Petalomyrmex phylax*, was positively correlated with disturbance rate (Debout et al., 2009). Disturbance – primarily branches falling from the canopy – can kill *Petalomyrmex* colonies, creating opportunities for the colonization of the damaged, unoccupied host tree by *Cataulacus*. Disturbance of many types could affect the dynamics of many plant-ant metacommunities. Similarly, in habitats fragmented by deforestation, degradation and dams, poorly dispersing plant-ant species may decrease in frequency or even be selectively removed. The example of *Leonardoxa* and its ants suggests that these may often be the most effective mutualists. Dynamics at the intra-specific level could have similar effects, favoring genotypes with greater dispersability and thus lower protective efficacy. Finally, if conditions such as disturbance and fragmentation that favor high dispersability are no longer transient, as in the *Leonardoxa* colonization-front example presented above, but become permanent features of degraded tropical-forest vegetation, the more slowly evolving plant hosts of these symbioses will have time to forge evolutionary responses to the reduced benefits conferred by their ant associates. These responses could hasten the disassembly of symbiotic mutualisms.

However, the relationship between the dispersability of plant-ants and their capacity to provide mutualistic benefits to the host plant may not always be negative. The South American understory treelet *Cordia nodosa* (Boraginaceae) hosts colonies of resident plant-ants in swollen twig-domatia. Its principal associates, *Allomerus octoarticulatus* (Myrmicinae) and *Azteca* spp. (Dolichoderinae), both protect their hosts (Frederickson, 2005). However, *Allomerus* ants also "castrate" the trees they occupy, removing floral buds (Yu & Pierce, 1998). Trees occupied by *Allomerus* produce few fruits, and with the resources thereby "spared" the tree produces more domatia and other resources for ants (Frederickson, 2009). Although the evolutionary dynamics of this system are more complex than originally thought (Frederickson, 2009; Frederickson et al., 2012), one factor that tends to stabilize the system against such conflicts between mutualists is a difference in dispersal capacity between *Allomerus* and *Azteca*. Although *Allomerus* colonies have higher fecundity, *Azteca* spp. have a greater ability to colonize isolated juvenile *Cordia nodosa* (Yu et al., 2004). This example shows that high dispersal capacity does not necessarily depend on high production of alate sexuals. There may thus not always be a trade-off between dispersal capacity and mutualistic benefits conferred by workers.

Like climate change, habitat fragmentation should also affect specialists more than generalists. However, studies of the effects of fragmentation on networks of symbiotic ant-plants and plant-ants have so far yielded varying results. Bruna et al. (2005) and Passmore et al. (2012) found no strong effects of fragmentation on the characteristics of networks of ant-plant symbioses in the Biological Dynamics of Forest Fragments Project. However, the vegetation between the "islands" of habitat created in this experiment grows back quickly (Laurance et al., 2011). The extent to which fragmentation really imposes a barrier to dispersal to plant-ants (or to ant-plants) in these sites is thus open to question. In contrast, habitat fragmentation caused by the construction of Balbina dam in central Amazonia in Brazil in 1989 led to marked effects on ant-plant symbiotic networks and biodiversity (Emer et al., 2013), including the local extinction of specialist ant species and an increase in the proportion of unoccupied plants. The networks that persist in fragments are nested subsets of the larger networks present in undisturbed landscape. Fragmentation can thus lead to disassembly of specialized mutualisms. Studies such as these are "snapshots" at a single point in time. More detailed studies of the dynamics of extinction (and of eventual colonization and recolonization) are required to determine whether the levels of species richness and habitat occupancy represent values close to an eventual equilibrium, or whether further extinctions will follow given enough time ("extinction debt," Kuussaari et al., 2009).

Defaunation and the Disappearance of Large Herbivores

The symbiosis between *Acacia drepanolobium* and its ant associates in East Africa has shown the profound effects, at a relatively short time scale, of the removal of large mammalian herbivores on the outcome of the mutualism. The spectacular effects are mediated through indirect interactions: in the absence of large herbivores, and under increased competition from other vegetation, *A. drepanolobium* trees invest less in mutualistic traits, i.e. they produce fewer domatia and extrafloral nectaries. This, in turn, changed the composition of the associated ant community and reduced the proportion of trees occupied by the most protecting ant (Palmer et al., 2008). Other African systems are suitable for investigating more subtle, but also more direct, effects of the removal of large herbivores on ant-plant symbioses. Plant-ants in some symbioses are efficient in removing small phytophagous arthropods, but do not attempt to attack larger intruders (Gaume et al., 1997). Typically timid ants that remove eggs and kill or remove small insects, these ants hide inside domatia when the host plant is disturbed by a large mammal (such as a human observer). Other plant-ants storm out of the domatia as soon as they perceive physical disturbance of the host plant and attack intruders, regardless of their size. While these ants are often effective against insect herbivores, they also possess powerful deterrents to vertebrates, such as a powerful sting and venom, or defense chemicals that are sprayed and can irritate vertebrate respiratory tracts. This second group includes large pseudomyrmecines and the myrmicine genus *Crematogaster*. Typical of this second type is the symbiosis between *Tetraponera* ants (*T. aethiops* and *T. latifrons*; Pseudomyrmecinae) and *Barteria* trees (*B. fistulosa* and *B. dewevrei*; Passifloraceae) in rainforests of central Africa. These ant species live exclusively on these two species of *Barteria*. They pour out of the domatia as soon as the plant is disturbed. Interestingly, these species are the largest of their genus and their sting is very potent to humans. They have also been shown to drive monkeys away from their host tree (McKey, 1974). Their protective traits appear to have been driven primarily by selective pressures imposed by large mammals such as elephants and primates. The *Tetraponera/Barteria* symbiosis is widespread and abundant throughout lowland tropical rainforests of Central Africa. Hunting pressure within this area is heterogeneous; in some sites the large mammal fauna is largely intact, whereas in others, large mammalian herbivores were wiped out decades ago. Comparative studies in such sites could reveal the ecological and/or evolutionary responses of *Barteria* and *Tetraponera* to defaunation. Does the removal of large herbivores reduce the benefits to trees of harboring ants and, if so, do plants reduce their investment in plant rewards? Hosting protective ants in the absence of herbivores is costly to plants (e.g. through resources taken by hemipterans tended by ants within domatia; Frederickson et al., 2012). Supporting *Tetraponera* ants and their hemipteran trophobionts is probably particularly costly, owing to the large size of individual workers. Defaunation might lead not to the loss of mutualism, but to evolutionary adjustments to the altered coterie of herbivores, now dominated by insects. Are there shifts in traits of ants (worker size, volume and composition of venom) that could reflect such an adjustment? An intriguing possibility is that the plant may "switch partners." *Barteria* trees are sometimes occupied by small *Crematogaster* spp. ants. In some populations of *B. dewevrei*, many trees bear domatia of smaller-than-usual diameter (4-mm internal diameter, compared to 6 mm, as is typical of most populations of both *B. dewevrei* and *B. fistulosa*) (Kokolo et al., 2016). These trees host only small *Crematogaster* spp. Are these smaller ants more cost-effective protective mutualists in the absence of large mammalian herbivores? Many other differences between the two ants could explain variation in their frequency among tree populations, and this question presently has no answer.

Defaunation of the American continent in pre-historic times may have led to processes similar to those observed in Africa today. However, the longer time elapsed since defaunation of America may well have given rise to evolutionary responses that are perceptible in communities of symbiotic plants and ants today. This theme has never been addressed. A comparative analysis of traits of these communities in America and Africa could generate useful hypotheses about the responses of ant-plant symbioses to defaunation that could be tested in the defaunated areas of Africa today.

Asymmetry in the Fates of Plant-Ants and Ant-Plants When Specific "Obligate" Mutualisms Are Disassembled

What happens when a specific, symbiotic interaction is dissolved? Do both partners go extinct? Or does one or the other persist, in a different range of environments, in altered form, or both? Several examples suggest that symbiotic plant-ants are more vulnerable to extinction than are myrmecophytes, the domatia-bearing ant-plants that host them.

First, there are cases where myrmecophytes continue to persist, even thrive, in the absence of ant associates. Several populations of *Tococa guianensis* (Melastomataceae) in the Brazilian cerrado persist despite the absence of associated ants. The low herbivore pressure in these populations is a likely hypothesis for explaining why this plant survives without any mutualistic partner (Moraes & Vasconcelos, 2009).

Second, analysis of ant-plant phylogenies suggests that adaptations for housing and feeding ants can be lost (Davies et al., 2001; Michelangeli, 2005; Peccoud et al., 2013) – without extinction of the "obligate" ant-plant host. In contrast, we know of no instances where an ant phylogeny clearly shows the persistence of an ant lineage that has lost a specific association with a myrmecophyte host. Studying these historical cases of disassembly could give insights on how environmental change might lead to disassembly of mutualisms today. *Barteria solida* (Passifloraceae), a congener of *B. fistulosa* and *B. dewevrei*, completely lacks domatia. This small tree is restricted to patches of submontane rainforest above 900 m, scattered from Nigeria to Gabon. Molecular evidence shows it clearly to be derived from a common ancestor with the specialized myrmecophyte *B. fistulosa*. It appears to have lost its myrmecophytic traits upon colonization of high-elevation environments (Peccoud et al., 2013). Submontane forests are outside the niche of *Tetraponera*, the most common ant symbiont of *Barteria*, probably because these ants are thermally limited at high elevation. Loss of the ancestral ant symbiont, lower herbivore pressure compared to lowland forests and lower diversity at high elevations of other potential protective ant symbionts could all have contributed to the loss of ant-related traits. This example is reminiscent of the loss of ant-caterpillar mutualism when butterfly species adapted to high-elevation environments (Pellissier et al., 2012). Another *Barteria* species, *B. nigritana*, clearly derived from the other specialized myrmecophyte, *B. dewevrei*, has highly reduced stem domatia and is found in association with non-specialist tree-dwelling ants (Djiéto-Lordon et al., 2004). This species is restricted to scrub vegetation on coastal sand dunes. Its apparent loss of symbiosis with *Tetraponera* may be related to its colonization of this distinctive hot, open habitat.

Third, we know cases of ant-plants that have colonized environments in which their ant associates failed to colonize. For instance, myrmecophytic *Cecropia* have colonized islands and have been introduced to many countries in the tropics. Although their mutualistic *Azteca* ant partners did not follow the trees during these events, the trees (which have lost or reduced ant-attractive traits) thrive in the new environments and are now considered invasive throughout the tropics (Janzen, 1973; Putz & Holbrook, 1988).

Environmental change may lead to the extinction of a mutualistic partner by highly indirect mechanisms. For example, an ant partner might still survive in the changed environment, but be less efficient as a mutualist. This could produce selective pressure on the plant to evolve traits that favor other partners, or to lose ant-attractive traits. Switching of partner ant species is likely to have occurred already in the course of evolution. For instance, the *Pseudomyrmex* ant species that are obligate symbionts of myrmecophytic *Triplaris* are younger than the myrmecophytic state in *Triplaris*, suggesting the complete replacement of the original ant partners (Chomicki et al., 2015).

What Kinds of Studies Are Needed?

Although the details of nutrient fluxes in ant-plant-herbivore multitrophic interactions are of particular interest for understanding how these interactions respond to increased CO_2 and N availability, experimental CO_2 and N fertilization is a more direct way to predict the effect of these changes. Unfortunately, there are still too few studies that have attempted this approach with ant-plant-herbivore interactions.

The thermal tolerances of individual ant species have been shown to predict the response of ant communities to higher temperatures under experimental conditions, at least in environments close to species' physiological limits (Diamond et al., 2012). Investigating the thermal limits of ants involved in opportunistic or symbiotic mutualisms with plants would thus be of great interest. These thermal limits will condition whether ants will actually be able to respond to the changes that are predicted to increase the benefits of ant-mediated protection in a context of global change.

Disassembly of ant-plant mutualisms at the evolutionary scale has been inferred from the phylogenetic approach. In addition, current cases of mutualism breakdown can be related to variation in the ecological and environmental context. Linking these two approaches may help us reach a better understanding of the fate of ant-plant mutualisms under ongoing global changes.

Conclusion

Ant-plant protection mutualisms are often complex mutualistic networks involving organisms with a great diversity of life histories. Each of these may respond differently to environmental change. In addition, different types of simultaneously occurring changes can evoke different, sometimes opposite, responses from the same organism (Tylianakis et al., 2008). Understanding the response of a particular interaction to global change requires integrating the type of change, the reaction norms of each species with respect to each variable subject to change, and the ecological context that conditions selective forces acting on each partner in the network. An important component of this context dependency is the potential for synergies among changes in their effects on network dynamics (e.g. Zvereva & Kozlov, 2006). How a particular mutualistic network responds to global environmental change results from the combination of the responses of each actor to each kind of change (and their interactions), and consequently may be very difficult to predict.

Global change will affect the structure of mutualistic networks, leading to selective loss of specialists, reducing the diversity both of species and of interactions. Precisely how network characteristics will be affected by the removal of specialists depends on the network itself and no general predictions can be drawn. However, non-symbiotic protection mutualisms mediated by the attraction of ants to food sources such as extrafloral nectar and honeydew are neither obligatory nor specific, and thus are expected to show greater resilience to environmental change. In symbiotic protection mutualisms, the degree of specificity and dependency is variable. Some ant-plant symbioses will be more resilient than others.

Ants, plants, herbivores and ant-associated bacteria are all affected by high temperatures and drought, and these will increase in severity and frequency with climate change. In some cases, these changes could weaken mutualisms, but some mutualisms might be strengthened.

Thus, several kinds of environmental change – warming, drought stress and enriched CO_2 – are predicted to increase herbivore pressure on plants. Ant partners may thus help to mitigate the effects of all these kinds of change on the plants they visit or permanently inhabit. In addition, as carbon becomes less limiting under

CO_2 enrichment, plants may experience increasing N limitation. Just as plants associated with nitrogen-fixing root symbionts escape N limitation and maintain lower C/N ratios in their tissues than other plants (Robinson et al., 2012), plants whose ant associates contribute to their N nutrition could also be favored. As ant-plant symbioses may well be nutritional symbioses in most cases (Mayer et al., 2014), the presence of ants (and associated microorganisms) may help to maintain the C/N stoichiometry of these plants.

References

Bale, J. S., Masters, G. J., Hodkinson, I. D. et al. (2002). Herbivory in global climate change research: direct effects of rising temperature on insect herbivores. *Global Change Biology*, 8, 1–16.

Ballhorn, D. J., Schmitt, I., Fankhauser, J. D., Katagiri, F. and Pfanz, H. (2011). CO2-mediated changes of plant traits and their effects on herbivores are determined by leaf age. *Ecological Entomology*, 36, 1–13.

Barton, B. T. and Ives, A. R. (2014). Direct and indirect effects of warming on aphids, their predators, and ant mutualists. *Ecology*, 95, 1479–84.

Bascompte, J. and Jordano, P. (2013). *Mutualistic networks*. Princeton: Princeton University Press.

Blatrix, R., Bouamer, S., Morand, S. and Selosse, M. A. (2009). Ant-plant mutualisms should be viewed as symbiotic communities. *Plant Signaling & Behavior*, 4, 554–6. doi:10.1111/j.1469-8137.2009.02793.x

Blatrix, R., Djiéto-Lordon, C., Mondolot, L. et al. (2012). Plant-ants use symbiotic fungi as a food source: new insight into the nutritional ecology of ant-plant interactions. *Proceedings of the Royal Society B: Biological Sciences*, 279, 3940–7. doi:10.1098/rspb.2012.1403.

Blatrix, R., McKey, D. and Born, C. (2013). Consequences of past climate change for species engaged in obligatory interactions. *Comptes Rendus Geoscience*, 345, 306–15.

Brodie, J. F., Aslan, C. E., Rogers, H. S. et al. (2014). Secondary extinctions of biodiversity. *Trends in Ecology & Evolution*, 29, 664–72.

Bruna, E. M., Vasconcelos, H. L. and Heredia, S. (2005). The effect of habitat fragmentation on communities of mutualists: Amazonian ants and their host plants. *Biological Conservation*, 124, 209–16. doi:10.1016/j.biocon.2005.01.026.

Chomicki, G., Ward, P. S. and Renner, S. S. (2015). Macroevolutionary assembly of ant/plant symbioses: *Pseudomyrmex* ants and their ant-housing plants in the Neotropics. *Proceedings of the Royal Society B-Biological Sciences*, 282, 20152200. doi:10.1098/rspb.2015.2200.

Coley, P. D. (1998). Possible effects of climate change on plant/herbivore interactions in moist tropical forests. *Climatic Change*, 39, 455–72.

Cook, S. C. and Davidson, D. W. (2006). Nutritional and functional biology of exudate-feeding ants. *Entomologia Experimentalis et Applicata*, 118, 1–10.

Couture, J. J., Meehan, T. D., Kruger, E. L. and Lindroth, R. L. (2015). Insect herbivory alters impact of atmospheric change on northern temperate forests. *Nature Plants*, 1, 15016.

Dalecky, A., Debout, G., Estoup, A., McKey, D. B. and Kjellberg, F. (2007). Changes in mating system and social structure of the ant *Petalomyrmex phylax* are associated with range expansion in Cameroon. *Evolution*, 61, 579–95. doi:10.1111/j.1558-5646.2007.00044.x.

Dalecky, A., Gaume, L., Schatz, B., McKey, D. and Kjellberg, F. (2005). Facultative polygyny in the plant-ant *Petalomyrmex phylax* (Hymenoptera: Formicinae): sociogenetic and ecological determinants of queen number. *Biological Journal of the Linnean Society*, 86, 133–51.

Dáttilo, W., Izzo, T. J., Vasconcelos, H. L. and Rico-Gray, V. (2013). Strength of the modular pattern in Amazonian symbiotic ant–plant networks. *Arthropod-Plant Interactions*, 7, 455–61.

Davidson, D. W. (2005). Ecological stoichiometry of ants in a New World rain forest. *Oecologia*, 142, 221–31.

Davidson, D. W., Cook, S. C., Snelling, R. R. and Chua, T. H. (2003). Explaining the abundance of ants in lowland tropical rainforest canopies. *Science*, 300, 969–72. doi:10.1126/science.1082074.

Davies, S. J., Lum, S. K. Y., Chan, R. and Wang, L. K. (2001). Evolution of myrmecophytism in western Malesian *Macaranga* (Euphorbiaceae). *Evolution*, 55, 1542–59.

Debout, G., Dalecky, A., Ngomi, A. and McKey, D. (2009). Dynamics of species coexistence: maintenance of a plant-ant competitive metacommunity. *Oikos*, 118, 873–84. doi:10.1111/j.1600-0706.2009.16317.x.

Defossez, E., Djiéto-Lordon, C., McKey, D., Selosse, M. A. and Blatrix, R. (2011). Plant-ants feed their host plant, but above all a fungal symbiont to recycle nitrogen. *Proceedings of the Royal Society B: Biological Sciences*, 278, 1419–26. doi:10.1098/rspb.2010.1884.

Defossez, E., Selosse, M. A., Dubois, M. P. et al. (2009). Ant-plants and fungi: a new three-way symbiosis. *New Phytologist*, 182, 942–9. doi:10.1111/j.1469-8137.2009.02793.x.

DeLucia, E. H., Nabity, P. D., Zavala, J. A. and Berenbaum, M. R. (2012). Climate change: resetting plant-insect interactions. *Plant Physiology*, 160, 1677–85.

Diamond, J. (2002). Evolution, consequences and future of plant and animal domestication. *Nature*, 418, 700–7.

Diamond, S. E., Sorger, D. M., Hulcr, J. et al. (2012). Who likes it hot? A global analysis of the climatic, ecological, and evolutionary determinants of warming tolerance in ants. *Global Change Biology*, 18, 448–56. doi:10.1111/j.1365-2486.2011.02542.x.

Djiéto-Lordon, C., Dejean, A., Gibernau, M., Hossaert-McKey, M. and McKey, D. (2004). Symbiotic mutualism with a community of opportunistic ants: protection, competition, and ant occupancy of the myrmecophyte *Barteria nigritana* (Passifloraceae). *Acta Oecologica*, 26, 109–16.

Douglas, A. E. (2009). The microbial dimension in insect nutritional ecology. *Functional Ecology*, 23, 38–47. doi:10.1111/j.1365-2435.2008.01442.x.

Duarte Rocha, C. F. and Godoy Bergallo, H. (1992). Bigger ant colonies reduce herbivory and herbivore residence time on leaves of an ant-plant: *Azteca muelleri* vs. *Coelomera ruficornis* on *Cecropia pachystachya*. *Oecologia*, 91, 249–52.

Dunn, R. R., Harris, N. C., Colwell, R. K., Koh, L. P. and Sodhi, N. S. (2009). The sixth mass coextinction: are most endangered species parasites and mutualists? *Proceedings of the Royal Society B: Biological Sciences*, 276, 3037–45. doi:10.1098/rspb.2009.0413.

Ellers, J., Kiers, T. E., Currie, C. R., McDonald, B. R. and Visser, B. (2012). Ecological interactions drive evolutionary loss of traits. *Ecology Letters*, 15, 1071–82.

Emer, C., Venticinque, E. M. and Fonseca, C. R. (2013). Effects of dam-induced landscape fragmentation on Amazonian ant-plant mutualistic networks. *Conservation Biology*, 27, 763–73. doi:10.1111/cobi.12045.

Estes, J. A., Terborgh, J., Brashares, J. S. et al. (2011). Trophic downgrading of planet Earth. *Science*, 333, 301–6.

Fan, Y. and Wernegreen, J. J. (2013). Can't take the heat: high temperature depletes bacterial endosymbionts of ants. *Microbial Ecology*, 66, 727–33.

Fayle, T. M., Edwards, D. P., Foster, W. A., Yusah, K. M. and Turner, E. C. (2015). An ant-plant by-product mutualism is robust to selective logging of rain forest and conversion to oil palm plantation. *Oecologia*, 178, 441–50.

Fitzpatrick, G., Davidowitz, G. and Bronstein, J. L. (2013). An herbivore's thermal tolerance is higher than that of the ant defenders in a desert protection mutualism. *Sociobiology*, 60, 252–8.

Fitzpatrick, G., Lanan, M. C. and Bronstein, J. L. (2014). Thermal tolerance affects mutualist attendance in an ant–plant protection mutualism. *Oecologia*, 176, 129–38.

Fontúrbel, F. E. and Murúa, M. M. (2014). Microevolutionary effects of habitat fragmentation on plant-animal interactions. *Advances in Ecology*, 2014, 379267.

Frederickson, M. E. (2005). Ant species confer different partner benefits on two neotropical myrmecophytes. *Oecologia*, 143, 387–95. doi:10.1007/s00442-004-1817-7.

(2009). Conflict over reproduction in an ant-plant symbiosis: why *Allomerus octoarticulatus* ants sterilize *Cordia nodosa* trees. *The American Naturalist*, 173, 675–81. doi:10.1086/597608

Frederickson, M. E., Ravenscraft, A., Hernandez, L. M. A. et al. (2013). What happens when ants fail at plant defence? *Cordia nodosa* dynamically adjusts its investment in both direct and indirect resistance traits in response to herbivore damage. *Journal of Ecology*, 101, 400–9.

Frederickson, M. E., Ravenscraft, A., Miller, G. A. et al. (2012). The direct and ecological costs of an ant-plant symbiosis. *The American Naturalist*, 179, 768–78. doi:10.1086/665654.

Gaume, L., McKey, D. and Anstett, M. C. (1997). Benefits conferred by "timid" ants: active anti-herbivore protection of the rainforest tree *Leonardoxa africana* by the minute ant *Petalomyrmex phylax*. *Oecologia*, 112, 209–16.

Gaume, L., McKey, D. and Terrin, S. (1998). Ant-plant-homopteran mutualism: how the third partner affects the interaction between a plant-specialist ant and its myrmecophyte host. *Proceedings of the Royal Society of London, Series B*, 265, 569–75.

Guimarães, P. R., Rico-Gray, V., dos Reis, S. F. and Thompson, J. N. (2006). Asymmetries in specialization in ant-plant mutualistic networks. *Proceedings of the Royal Society B: Biological Sciences*, 273, 2041–7.

Heil, M., Hilpert, A., Fiala, B. and Linsenmair, K. E. (2001). Nutrient availability and indirect (biotic) defence in a Malaysian ant-plant. *Oecologia*, 126, 404–8.

Ings, T. C., Montoya, J. M., Bascompte, J. et al. (2009). Ecological networks – beyond food webs. *Journal of Animal Ecology*, 78, 253–69.

Janzen, D. H. (1973). Dissolution of mutualism between *Cecropia* and its *Azteca* ants. *Biotropica*, 5, 15–28.

(1974). The deflowering of Central America. *Natural History*, 83, 48.

Johnson, N. C. (2010). Resource stoichiometry elucidates the structure and function of arbuscular mycorrhizas across scales. *New Phytologist*, 185, 631–47. doi:10.1111/j.1469-8137.2009.03110.x.

Kautz, S., Lumbsch, H. T., Ward, P. S. and Heil, M. (2009). How to prevent cheating: a digestive specialization ties mutualistic plant-ants to their ant-plant partners. *Evolution*, 63, 839–53. doi:10.1111/j.1558-5646.2008.00594.x.

Kiers, E. T. and Denison, R. F. (2008). Sanctions, cooperation, and the stability of plant-rhizosphere mutualisms. *Annual Review of Ecology, Evolution and Systematics*, 39, 215–36. doi:10.1146/annurev.ecolsys.39.110707.173423.

Kiers, E. T., Palmer, T. M., Ives, A. R., Bruno, J. F. and Bronstein, J. L. (2010). Mutualisms in a changing world: an evolutionary perspective. *Ecology Letters*, 13, 1459–74. doi:10.1111/j.1461-0248.2010.01538.x.

Kokolo, B., Atteke, C., Ibrahim, B. and Blatrix, R. (2016). Pattern of specificity in the tripartite symbiosis between *Barteria* plants, ants and Chaetothyriales fungi. *Symbiosis*, 69, 169–74. doi:10.1007/s13199-016-0402-2.

Kuussaari, M., Bommarco, R., Heikkinen, R. K. et al. (2009). Extinction debt: a challenge for biodiversity conservation. *Trends in Ecology & Evolution*, 24, 564–71.

Laurance, W. F., Camargo, J. L. C., Luizao, R. C. C. et al. (2011). The fate of Amazonian forest fragments: a 32-year investigation. *Biological Conservation*, 144, 56–67. doi:10.1016/j.biocon.2010.09.021.

Léotard, G., Debout, G., Dalecky, A. et al. (2009). Range expansion drives dispersal evolution in an equatorial three-species symbiosis. *Plos One*, 4, e5377.

Leroy, C., Sejalon-Delmas, N., Jauneau, A. et al. (2011). Trophic mediation by a fungus in an ant-plant mutualism. *Journal of Ecology*, 99, 583–90. doi:10.1111/j.1365-2745.2010.01763.x.

Lindroth, R. L. (2010). Impacts of elevated atmospheric CO_2 and O_3 on forests: phytochemistry, trophic interactions, and ecosystem dynamics. *Journal of Chemical Ecology*, 36, 2–21.

Lortie, C. J. (2007). An ecological tardis: the implications of facilitation through evolutionary time. *Trends in Ecology & Evolution*, 22, 627–30.

Marquis, M., Del Toro, I. and Pelini, S. L. (2014). Insect mutualisms buffer warming effects on multiple trophic levels. *Ecology*, 95, 9–13.

Maschwitz, U., Fiala, B., Dumpert, K., bin Hashim, R. and Sudhaus, W. (2016). Nematode associates and bacteria in ant-tree symbioses. *Symbiosis*, 69, 1–7. doi:10.1007/s13199-015-0367-6.

Mayer, V. E., Frederickson, M. E., McKey, D. and Blatrix, R. (2014). Current issues in the evolutionary ecology of ant-plant symbioses. *New Phytologist*, 202, 749–64. doi:10.1111/nph.12690.

McKey, D. (1974). Ant-plants: selective eating of an unoccupied *Barteria* by a *Colobus* monkey. *Biotropica*, 6, 269–70.

McKey, D., Gaume, L., Brouat et al. (2005). The trophic structure of tropical ant-plant-herbivore interactions: community consequences and coevolutionary dynamics. In *Biotic interactions in the tropics: Their role in the maintenance of species diversity,* Burselm, D., Pinard, M., Hartley, S., eds. Cambridge: Cambridge University Press, pp. 386–413.

Merilä, J. and Hendry, A. P. (2014). Climate change, adaptation, and phenotypic plasticity: the problem and the evidence. *Evolutionary Applications*, 7, 1–14.

Michelangeli, F. A. (2005). *Tococa (Melastomataceae)*. New York: New York Botanical Garden Press.

Moraes, S. C. and Vasconcelos, H. L. (2009). Long-term persistence of a neotropical ant-plant population in the absence of obligate plant-ants. *Ecology*, 90, 2375–83.

Moran, N. A. and Baumann, P. (2000). Bacterial endosymbionts in animals. *Current Opinion in Microbiology*, 3, 270–5.

Niziolek, O. K., Berenbaum, M. R. and DeLucia, E. H. (2013). Impact of elevated CO_2 and increased temperature on Japanese beetle herbivory. *Insect Science*, 20, 513–23.

Orivel, J., Lambs, L., Malé, P. J. G., Leroy, C., Grangier, J., Otto, T., Quilichini, A. and Dejean, A. (2011). Dynamics of the association between a long-lived understory myrmecophyte and its specific associated ants. *Oecologia*, 165, 369–76.

Palmer, T. M., Stanton, M. L., Young, T. P. et al. (2008). Breakdown of an ant-plant mutualism follows the loss of large herbivores from an African savanna. *Science*, 319, 192–5.

Passmore, H. A., Bruna, E. M., Heredia, S. M. and Vasconcelos, H. L. (2012). Resilient networks of ant-plant mutualists in Amazonian forest fragments. *Plos One*, 7, e40803. doi:10.1371/journal.pone.0040803.

Peccoud, J., Piatscheck, F., Yockteng, R. et al. (2013). Multi-locus phylogenies of the genus *Barteria* (Passifloraceae) portray complex patterns in the evolution of myrmecophytism. *Molecular Phylogenetics and Evolution*, 66, 824–32.

Pellissier, L., Litsios, G., Fiedler, K. et al. (2012). Loss of interactions with ants under cold climate in a regional myrmecophilous butterfly fauna. *Journal of Biogeography*, 39, 1782–90.

Phillips, B. L., Brown, G. P., Webb, J. K. and Shine, R. (2006). Invasion and the evolution of speed in toads. *Nature*, 439, 803. doi:10.1038/439803a.

Pringle, E. G. (2016). Integrating plant carbon dynamics with mutualism ecology. *New Phytologist*, 210, 71–5.

Pringle, E. G., Akçay, E., Raab, T. K., Dirzo, R. and Gordon, D. M. (2013). Water stress strengthens mutualism among ants, trees and scale insects. *PLoS Biology*, 11, e1001705.

Putz, F. E. and Holbrook, N. M. (1988). Further observations on the dissolution of mutualism between *Cecropia* and its ants: the Malaysian case. *Oikos*, 53, 121–5. doi:10.2307/3565671.

Richardson, D. M., Allsopp, N., D'Antonio, C. M., Milton, S. J. and Rejmanek, M. (2000). Plant invasions – the role of mutualisms. *Biological Reviews*, 75, 65–93.

Rico-Gray, V. and Oliveira, P. S. (2007). *The ecology and evolution of ant-plant interactions*. Chicago and London: University of Chicago Press.

Robinson, E. A., Ryan, G. D. and Newman, J. A. (2012). A meta-analytical review of the effects of elevated CO_2 on plant–arthropod interactions highlights the importance of interacting environmental and biological variables. *New Phytologist*, 194, 321–36.

Russell, J. A., Moreau, C. S., Goldman-Huertas, B. et al. (2009). Bacterial gut symbionts are tightly linked with the evolution of herbivory in ants. *Proceedings of the National Academy of Sciences of the United States of America*, 106, 21236–41. doi:10.1073/pnas.0907926106.

Ryalls, J. M., Moore, B. D., Riegler, M. et al. (2016). Climate and atmospheric change impacts on sap-feeding herbivores: a mechanistic explanation based on functional groups of primary metabolites. *Functional Ecology*, 41, 161–171. doi:10.1111/1365–2435.12715.

Stiling, P. and Cornelissen, T. (2007). How does elevated carbon dioxide (CO_2) affect plant–herbivore interactions? A field experiment and meta-analysis of CO_2-mediated changes on plant chemistry and herbivore performance. *Global Change Biology*, 13, 1823–42.

Stuble, K. L., Pelini, S. L., Diamond, S. E. et al. (2013). Foraging by forest ants under experimental climatic warming: a test at two sites. *Ecology and Evolution*, 3, 482–91.

Sun, Y. C., Jing, B. B. and Ge, F. (2009). Response of amino acid changes in *Aphis gossypii* (Glover) to elevated CO_2 levels. *Journal of Applied Entomology*, 133, 189–97.

Sunday, J. M., Bates, A. E. and Dulvy, N. K. (2011). Global analysis of thermal tolerance and latitude in ectotherms. *Proceedings of the Royal Society of London B: Biological Sciences*, 278, 1823–30.

Thomas, C. D., Bodsworth, E. J., Wilson, R. J. et al. (2001). Ecological and evolutionary processes at expanding range margins. *Nature*, 411, 577–81.

Thompson, J. A. (2005). *The geographic mosaic of coevolution*. Chicago: University of Chicago Press.

Tylianakis, J. M., Didham, R. K., Bascompte, J. and Wardle, D. A. (2008). Global change and species interactions in terrestrial ecosystems. *Ecology Letters*, 11, 1351–63. doi:10.1111/j.1461-0248.2008.01250.x.

Valiente-Banuet, A., Rumebe, A. V., Verdú, M. and Callaway, R. M. (2006). Modern Quaternary plant lineages promote diversity through facilitation of ancient Tertiary lineages. *Proceedings of the National Academy of Sciences of the United States of America*, 103, 16812–7.

Vitousek, P. M., Aber, J. D., Howarth, R. W. et al. (1997). Human alteration of the global nitrogen cycle: sources and consequences. *Ecological Applications*, 7, 737–50. doi:10.2307/2269431.

Vitousek, P. M. and Walker, L. R. (1989). Biological invasion by *Myrica faya* in Hawai'i: plant demography, nitrogen fixation, ecosystem effects. *Ecological Monographs*, 59, 247–65.

Vittecoq, M., Djiéto-Lordon, C., McKey, D. and Blatrix, R. (2012). Range expansion induces variation in a behavioural trait in an ant-plant mutualism. *Acta Oecologica*, 38, 84–8.

Warbrick-Smith, J., Behmer, S. T., Lee, K. P., Raubenheimer, D. and Simpson, S. J. (2006). Evolving resistance to obesity in an insect. *Proceedings of the National Academy of Sciences of the United States of America*, 103, 14045–9.

Wernegreen, J. J. (2012). Mutualism meltdown in insects: bacteria constrain thermal adaptation. *Current Opinion in Microbiology*, 15, 255–62.

Wilf, P. and Labandeira, C. C. (1999). Response of plant-insect associations to Paleocene-Eocene warming. *Science*, 284, 2153–6.

Yu, D. W. and Pierce, N. E. (1998). A castration parasite of an ant-plant mutualism. *Proceedings of the Royal Society of London, Series B*, 265, 375–82.

Yu, D. W., Wilson, H. B., Frederickson, M. E. et al. (2004). Experimental demonstration of species coexistence enabled by dispersal limitation. *Journal of Animal Ecology*, 73, 1102–14.

Zhang, S., Zhang, Y. and Ma, K. (2012). The ecological effects of the ant–hemipteran mutualism: a meta-analysis. *Basic and Applied Ecology*, 13, 116–24.

Zvereva, E. L. and Kozlov, M. V. (2006). Consequences of simultaneous elevation of carbon dioxide and temperature for plant-herbivore interactions: a metaanalysis. *Global Change Biology*, 12, 27–41.

Part IV

Effect of Invasive Ants on Plants and Their Mutualists

12 Playing the System: The Impacts of Invasive Ants and Plants on Facultative Ant-Plant Interactions

Suzanne Koptur, Ian M. Jones, Hong Liu, and Cecilia Díaz-Castelazo[*]

Introduction

Extrafloral nectaries (EFNs) are sugar-secreting glands located outside of flowers; they are structurally diverse, and may be found on almost any vegetative or reproductive plant structure (Bentley, 1977a; Koptur, 1992). Although a wide range of ecological functions have been suggested for EFNs (Baker et al., 1978; Becerra & Venable, 1989; Wagner & Kay, 2002; Gonzalez-Teuber & Heil, 2009; Heil, 2011), they are most noted for providing indirect defense against herbivory by attracting natural enemies (Janzen, 1966; Inouye & Taylor, 1979; Koptur 1984; Heil et al., 2001; Heil, 2015). Ants represent the most common visitors to EFNs, and have regularly been observed to benefit host plant fitness (Bentley, 1977b; Koptur, 1992; Rosumek et al., 2009; Heil, 2015).

Myrmecophytes are plants that provide domatia, and food bodies and/or EFN, and engage in obligate interactions with ants (Chapters 10 and 11). A far greater number of plants, however, known as myrmecophiles, provide only EFN and engage in facultative interactions with ants. Because of the non-specialized nature of their interactions, the EFN that these plants provide is open to exploitation by any number of ant species, some of which may provide no benefits, or even negatively affect plant fitness (Koptur & Lawton, 1988; Torres-Hernandez et al., 2000; Ness et al., 2006). This variation in partner quality represents an important ecological cost of EFN production for plants. In this chapter, however, we focus not on the costs for individual plants, but on the costs for native species and ecosystems. We address the question: Does EFN in disturbed environments support and facilitate species invasions?

[*] We thank Robin Currey, Phil Gonsiska, Chad Husby, Maria Cristina Rodriguez, Carl Weekley, and Hipolito Paulino Neto for collections and observations that contributed to our data; Jaeson Clayborn for advice on ant-trapping, help with ant species determination (along with Mark Deyrup) and constructive comments on the manuscript, which were also provided by Brittany Harris, Adel Peña, and Maria Cleopatra Pimienta, as well as editor Paulo Oliveira and two anonymous reviewers. This is Publication Number 335 of the Tropical Biology Program at Florida International University.

The Role of Plant-Based Resources in Supporting Invasive Ants

In the southern United States, no invasive ant species is more ubiquitous than the red imported fire ant, *Solenopsis invicta*. The uncontrolled spread of this highly invasive species is, in part, a result of its ability to infiltrate mutualistic networks. Wilder et al. (2011) showed that a lack of interspecific competition in its invasive range has allowed *S. invicta* to dominate plant-based carbohydrate resources, both EFN and hemipteran honeydew. Indeed, stable isotope analyses have shown that *S. invicta* occupies a lower trophic position in the United States than in its native Argentina, where other arboreal foraging ants can exclude it from mutualist-derived resources (ibid.).

This kind of behavioral plasticity is a common feature of highly invasive species. Savage and Whitney (2011) manipulated EFN availability on a native shrub, *Morinda citrifolia* (Rubiaceae), on the Samoan islands. The invasive ant, *Anoplolepis gracilipes*, responded more strongly to increased EFN availability, in terms of recruitment activity and aggressive behavior, than did native ant species. The invasion of *A. gracilipes* on the Samoan islands has progressed over recent decades, and its distribution is highly correlated with the presence of EFN-producing plants. Not surprisingly, in areas where *A. gracilipes* is present, the abundance and diversity of native ants has been reduced (Savage et al., 2009).

Other studies have also made the link between ant invasions and plant-derived resources. Eubanks (2001) described the patchy distribution of *S. invicta* in agricultural habitats in the southern United States and determined that much of the pattern could be attributed to the presence of ant-tended aphids. The Argentine ant, *Linepithema humile*, became dominant in South African vineyards only after the introduction of honeydew-excreting insects (Addison & Samways, 2000). The presence of EFN-producing plants may influence ant invasions in a similar manner (e.g., Lach, 2003; Ness & Bronstein, 2004).

Many highly invasive ant species share a suite of traits that allow them to dominate plant carbohydrate resources. Several of the most successful species, including both *S. invicta* and *L. humile,* produce multi-queened, multi-nested supercolonies that lack intraspecific aggression (Holway et al., 2002). Individual nests can also be highly movable in response to available resources. Colonies of *S. invicta*, for example, will often produce satellite nests at the base of plants when harvesting EFN or tending aphids (Kaakeh & Dutcher, 1992; Koptur et al., 2015). A high level of aggression is a character shared by almost all invasive ants (Lach, 2003) and some, including *L. humile*, even have modified crops that allow them to take in more liquid food (Davidson, 1998).

Although invasive ants are often well equipped to dominate plant-derived resources, it is less clear whether they make effective mutualistic partners for plants. On an EFN-producing tree, *Acacia lamprocarpa* (Fabaceae), invasive *A. gracilipes* ants display greater recruitment behavior and aggression toward herbivores than the native Weaver ants, *Oecophylla smaragdina*. Herbivore damage to the leaves of *A. lamprocarpa*, however, is greater in the presence of invasive ants than when native ants are resident (Lach & Hoffman, 2011). In South Africa, *L. humile* ants displace native ants on *Protea nitida* (Proteaceae), where they tend membracid planthoppers. Unlike the native ants, however, *L. humile* are often found in the inflorescences, and have been shown to deter pollinators (Lach, 2007). These examples highlight that invasive ants can negatively impact their plant partners; however, the majority of studies have shown that invasive ants do benefit plants, whether it be through a reduction in herbivory (Koptur, 1979; de la Fuente & Marquis, 1999; Oliveira et al., 1999; Fleet & Young, 2000; Ness, 2003), an increase in fruit or seed production (Koptur, 1979; Horvitz & Schemske, 1984; Oliveira et al., 1999; McLain, 1983; Fleet & Young, 2000) or an increase in plant growth rate (de la Fuente & Marquis, 1999).

Mutualisms play a key role in the functioning of ecosystems. In disturbed habitats, however, generalist interactions between ants and plants often involve introduced species. In these cases, such interactions can enhance invasion success, and further disrupt ecosystem integrity. The most commonly documented impact of ant invasions is, not surprisingly, the displacement of native ant species. Since its arrival in the southern United States, *S. invicta* has substantially reduced the range of its congener, *Solenopsis geminata*, along with numerous other ant species (Gotelli & Arnett, 2000). In the Galapagos Islands, the little fire ant, *Wasmannia auropunctata*, has had a similar impact on several native ant species (Lubin, 1984). The effects of invasive ants are, however, not limited to other ant species. In Hawaii, the loss of numerous and diverse native insects has been attributed to the invasion of *Pheidole megacephala* (Zimmerman, 1970). On Christmas Island, the arrival of *A. gracilipes* ants has impacted the populations of the red land crab, a keystone species on the island, with cascading effects on the entire ecosystem (Green et al., 1999).

The Role of Ants in Supporting Invasive Plants

While it appears that plant-based resources have facilitated invasions by several ant species, the exploitation of facultative ant-plant interactions is a two-way street. Of the approximately 4,000 plant species that bear EFNs, the majority are pioneer plants capable of adapting and thriving in changeable abiotic and biotic conditions (Weber & Keeler, 2013). Here we explore how this characteristic has contributed to the spread of EFN plants into new environments, particularly those that have been heavily impacted by humans. We address the question: Do generalist ants, either native or invasive, facilitate plant invasions?

The evolution of increased competitive ability hypothesis (EICA) predicts that, in the absence of their coevolved natural enemies, plants should decrease their investment in indirect defenses and, instead, focus their resources toward growth and reproduction (Blossey & Notzold, 1995). Indeed, populations of the Chinese tallow tree, *Triadica sebifera* (Euphorbiaceae), in their native range have been shown to produce more EFN than their invasive conspecifics in the United States (Carrillo et al., 2012). As an extension to this theory, one would predict that investment in

EFN would be reduced in urban or highly disturbed environments. Indeed, Rios et al. (2008) collected seeds of *Chamaecrista fasciculata* (Fabaceae) from populations in natural and urban environments, and reared them in controlled greenhouse conditions. Plants derived from urban populations had smaller EFNs and produced less EFN. Conversely, EFN-producing plants in their native range are better equipped to exploit local ant populations than their invasive competitors. In Chinese tallow, induced EFN production is significantly greater in response to damage by specialist herbivores found only in their native range, than to damage by generalist herbivores (Carrillo et al., 2012b). Also, in China, a native passion vine, *Passiflora siamica* (Passifloraceae), produces significantly more EFN per leaf than its invasive congener, *P. coccinea* (Xu & Chen, 2009).

One factor that appears to contribute to the success of non-native EFN plants, however, is the arrival or presence of invasive ants. The invasive Argentine ant, *Linepithema humile* (*Iridomyrmex humilis*, previously), for example, may have facilitated the naturalization of two non-native vetch species, *Vicia sativa* and *V. augustifolia* (Fabaceae), across northern California (Koptur, 1979) by reducing damage to leaves from surface-feeding herbivores. In Mauritius an invasive ant, *Technomyrmex albipes*, has been shown to benefit an invasive tree, *Leucaena leucocephala* (Fabaceae), by removing herbivores. In contrast, the same invasive ant negatively impacts a native tree, *Scaevola taccada* (Goodeniaceae), by tending sap-sucking hemipterans (Lach et al., 2010). In Puerto Rico, the population of a non-native orchid, *Spathoglottis plicata* (Orchidaceae), had been kept in check by a native weevil seed predator. The arrival of fire ants, *S. invicta*, on the island, however, has led to the deterrence of these weevils, and the elevation of *S. plicata* to invasive status (Ackerman et al., 2014). Invasive ants may also outcompete native ants, especially in altered habitats: fire ants and invasive *Pheidole* ants colonized clear-cut forest areas, and numbers of native ants were significantly reduced (Zettler et al., 2004).

The Hawaiian Islands represent an ideal system in which to study the impacts of invasive ants, as it is generally accepted that ants were absent from the island prior to their human introduction (Keeler, 1985; Krushelnycky et al., 2005). As one would expect, very few plant species that are endemic to these islands possess EFNs; however, many invasive species bearing EFNs have proliferated on the islands since the arrival of ants (Junker et al., 2011). In addition to defending invasive plants, ants have been shown to act as nectar robbers on many native plants that lack the floral defenses exhibited by many of their invasive counterparts (Bleil et al., 2011; Junker et al., 2011).

Ant-Plant Interactions in South Florida

In south Florida the native flora contains a high proportion of plants that bear EFNs, many of which have been shown to facilitate mutualistic interactions with ants and other beneficial insects (Koptur, 1992; Koptur et al., 2015). The human

Figure 12.1. Legume plants with extrafloral nectaries in the urban environment – clockwise from upper left: (a) Native plant landscape with *Senna ligustrina* and *S. chapmannii* (foreground), *Lysiloma latisiliquum* canopy; (b) *Senna surattensis* in landscape in front of South Miami city hall; (c) *Senna chapmannii* hedge in front of sculpture by Metrorail station; (d) *L. latisiliquum* as parking lot tree. Photo credits: Suzanne Koptur.

population in southern Florida has grown dramatically over the past century (Barrios et al., 2011), and development in Miami-Dade and Monroe Counties has progressed rapidly from very little to almost complete urbanization.

Many native EFN-producing plants are utilized in the urban landscape as garden plants and shade trees (Figure 12.1a, c, d). Non-native congeneric species, also bearing EFNs, have been introduced as ornamentals in the same area (Figure 12.1b), and some have become invasive. The proliferation of human activity in south Florida has also seen the introduction of several non-native ant species, most notably the red imported fire ant *S. invicta* and the now naturalized *Pseudomyrmex gracilis*. In this section we describe ant-plant interactions in several native legume plants, in natural areas where they have been studied. We describe patterns of interactions among native plants and their exotic relatives with ants, both native and introduced species. We consider the effects of invasive ants on extrafloral nectary-mediated mutualisms in south Florida, the impact of these interactions on populations of native herbivores, and the resulting fitness benefits to plants (both native and non-native) bearing EFNs.

Wild tamarind, *Lysiloma latisiliquum* (Fabaceae), is an EFN-producing tree native to south Florida. Ant exclusion experiments, conducted in Everglades National Park, showed that ants provide *L. latisiliquum* with protection against leaf-feeding herbivores, particularly during leaf expansion and development (Koptur, unpublished data). Extrafloral nectaries on the leaves attracted four species of ants (*Pheidole dentata*, *Pseudomyrmex elongata*, *P. gracilis* and *Solenopsis geminata*), two of which are non-native. The most abundant ant on the plants was *P. gracilis*, the elongate twig ant. This species was introduced to the Miami area in around 1960 (Whitcomb et al., 1972) and now is present worldwide (Wetterer, 2010). This solitary forager nests in twigs and is an important predator on caterpillars and other arthropods, including the cloudless sulfur, *Phoebis sennae*, a native pierid butterfly that utilizes *L. latisiliquum* as a hostplant, preferentially ovipositing on the new foliage. Future work should consider how the introduction of *P. gracilis* (and other aggressive ants) has affected populations of these native butterflies.

Most *Pseudomyrmex* ants that form mutualisms with plants prefer hexose-rich nectar, as they lack invertase, the enzyme that cleaves sucrose. As a result, many myrmecophytic plants produce hexose-rich nectar as a way to discourage nectar robbing by non-mutualistic ants (Kautz et al., 2009). *Pseudomyrmex gracilis*, however, provides an exception to this rule, as it does produce invertase. This species is, therefore, well placed to exploit generalist ant-plants, such as *Lysiloma latisiliquum*, that produce largely sucrose-based nectar. A congener of wild tamarind, *Lysiloma sabicu*, has been widely used in landscaping in south Florida, and non-native generalist ant-plants like this may well have facilitated the spread of invasive ants. As a close relative of a native plant, however, *L. sabicu* may also provide a service for native herbivores, creating connections between remaining natural landscape fragments. Indeed, *L. sabicu* has recently been shown to host the rare pink spot sulfur butterfly, *Aphrissa neleis* (Warren, 2011).

Senna is a species-rich genus of caesalpinioid legumes, the diversification of which has been attributed to the evolution of EFNs (Marazzi et al., 2013). Many species of the genus, both native (Figure 12.2) and non-native (Figure 12.1 b), are abundant in south Florida, and represent important host plants for sulfur caterpillars. *Senna mexicana* var. *chapmanii* (henceforth, *S. chapmanii*) is native to pine rockland habitats, and the presence of ants has been shown to reduce herbivory and increase plant reproductive fitness in this species (Koptur et al., 2015; Jones et al., 2016). Nine ant species were observed foraging on *S. chapmanii*, including both *P. gracilis* and *S. invicta* (Koptur et al., 2015); similar studies, conducted a decade

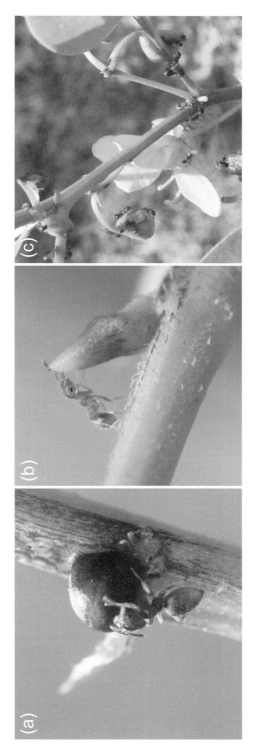

Figure 12.2. *Senna* extrafloral nectaries with ant visitors – left to right: (a) *Senna chapmannii* with *Wasmannia auropunctata*; (b) *Senna ligustrina* with same ant species; (c) *Solenopsis invicta* on *Senna chapmannii*. Photos by Maria Cleopatra Pimienta (a & b) and Ian Jones (c). (A black-and-white version of this figure will appear in some formats. For the color version, please refer to the plate section.)

later, also observed foraging by the little fire ant, *Wasmannia auropunctata*, a more recent arrival in south Florida (Figure 12.2; see also Wetterer & Porter, 2003; Jones et al., 2016). *Senna occidentalis* is a close relative of *S. chapmanii*, and is invasive in south Florida where it serves as a host plant for the sulfur butterflies. In Texas, another part of its invasive range, *S. occidentalis,* has been shown to benefit specifically from the presence of the red imported fire ant, *S. invicta*; these aggressive ants dominate the EFNs, and remove sulfur caterpillars (*Phoebis sennae* and *Abaeis nicippe*) from the plants (Fleet & Young, 2000).

The proliferation of *S. invicta* in south Florida has undoubtedly had a negative effect on native ant species, and may pose a threat to a host of other native organisms as diverse as butterflies and sea turtles (Allen et al., 2001). Although controlling the spread of these invasive ants has proven difficult, it may be possible to utilize them as part of biological control programs that protect other native organisms. Two native *Opuntia* species (Cactaceae) – *O. stricta* and *O. humifusa* – host the highly invasive moth, *Cactoblastis cactorum*, in south Florida. Aggressive ants, in particular *S. invicta*, have been shown to reduce *C. cactorum* numbers by attacking eggs and larvae. The numbers of defensive ants on *Opuntia* species can be enhanced by planting the native EFN-producing legume *Chamaecrista fasciculata* beneath the cacti (Jezorek et al., 2011).

In recent times, there has been a move to increase the numbers of native plant species in human-dominated environments, for many reasons: to enhance the surroundings for wildlife, including butterflies, bees, and birds (Minno & Minno, 1999; Mathew & Anto, 2007; Koi & Daniels, 2015); to educate the public about native plants in landscaping, making connections with the natural heritage of the region (Wild Ones®, 2004), promoting place-based learning so that connections can be made with the larger environment of the planet (Billick & Price, 2010); and to provide connections between remaining natural landscape fragments (Haddad et al., 2003; Maschinski & Wright, 2006). It is interesting to compare the ant-plant associations of native plant species in natural areas versus urban areas, and compare their interactions with those of non-native congeneric or closely related plants as well. In this study we defined natural areas as places that have been relatively undisturbed by humans, where plants occur naturally; urban areas are those where vegetation has been removed by human development, and then replanted with plants in landscaping and gardens.

We sought to document associations of ants and plants; so to augment our ongoing observations from various studies in our plant ecology lab, we used pit-fall traps. We attached vials of soapy water upright to smaller branches of plants, with their openings at the surface of the branch, so that wandering insects fall in and do not leave; these traps were placed on individual plants in urban and natural areas to collect ants and other arthropods on different plants in different situations. Each plant monitored received three vials, which were collected after one week; these were refrigerated until contents were examined in the lab and specimens pointed for determination and vouchers. We aimed to monitor ten individuals of each plant species in each situation, but for several species in the urban areas there were fewer

(a)

NATURAL AREAS plants/ants

Ant species columns:
1 - Brachymyrmex obscurior*; 3 - Camponotus emeryodicatus*; 4 - Camponotus floridanus; 5 - Camponotus inaequalis*; 6 - Camponotus planatus*; 7 - Camponotus rasilis*; 9 - Camponotus tortuganus; 8 - Camponotus sexguttatus*; 10 - Cardiocondyla emeryi*; 11 - Odontomachus brunneus; 12 - Odontomachus ruginodis; 13 - Paratrechina longicornis*; 14 - Pheidole dentata; 16 - Pheidole moerens*; 17 - Pseudomyrmex ejectus; 18 - Pseudomyrmex elongatus; 19 - Pseudomyrmex gracilis*; 20 - Pseudomyrmex simplex; 21 - Solenopsis geminata; 22 - Solenopsis invicta*; 25 - Wasmannia auropunctata*

Plant	1	3	4	5	6	7	9	8	10	11	12	13	14	16	17	18	19	20	21	22	25
B - Acacia pinetorum	1	0	0	0	0	0	0	0	0	0	0	0	0	0	0	0	0	1	0	0	0
D - Erythrina herbacea	0	0	0	0	0	0	0	0	0	0	0	1	0	0	0	0	0	0	1	0	0
F - *Leucaena leucocephala	1	0	1	0	0	0	0	0	0	0	0	0	0	0	0	0	0	1	0	0	0
G - Lysiloma latisiliquum	0	0	1	0	0	0	0	0	0	0	0	1	1	0	0	1	1	1	1	0	0
J - Pithecellobium guadalupense	0	0	1	0	0	0	0	0	0	0	0	0	0	0	0	0	0	0	1	0	0
K - Pithecellobium unguis-cati	0	0	0	0	0	0	0	0	1	0	0	0	0	0	0	0	0	0	0	0	0
M - *Senna bicapsularis	1	0	0	0	0	0	0	0	0	0	0	0	0	0	0	0	0	0	1	0	0
N - Senna ligustrina	0	0	1	0	0	0	0	1	0	0	0	1	0	0	0	1	0	1	1	0	0
O - Senna chapmannii	1	1	1	1	1	1	1	1	1	1	0	1	0	1	0	1	1	0	1	1	1

(b)

URBAN plants/ants

Ant species columns:
1 - Brachymyrmex obscurior*; 2 - Camponotus castaneus; 4 - Camponotus floridanus; 5 - Camponotus inaequalis*; 8 - Camponotus sexguttatus*; 9 - Camponotus tortuganus; 10 - Cardiocondyla emeryi*; 13 - Paratrechina longicornis*; 15 - Pheidole megacephala*; 16 - Pheidole moerens*; 17 - Pseudomyrmex ejectus; 18 - Pseudomyrmex elongatus; 19 - Pseudomyrmex gracilis*; 20 - Pseudomyrmex simplex; 22 - Solenopsis invicta*; 23 - Tapinoma melanocephalum*; 24 - Technomyrmex difficilis*; 25 - Wasmannia auropunctatus*

Plant	1	2	4	5	8	9	10	13	15	16	17	18	19	20	22	23	24	25
A - *Acacia chundra	0	0	0	0	0	0	1	0	1	0	0	1	0	0	1	0	0	0
B - Acacia pinetorum	1	0	0	0	0	0	1	0	0	0	0	0	0	0	0	0	0	0
C - *Cassia bakeriana	0	0	0	0	0	1	0	0	0	0	0	0	0	0	0	0	0	0
D - Erythrina herbacea	1	0	0	0	0	0	0	0	0	0	0	0	0	0	0	1	0	0
E - *Erythrina variegata	1	0	0	0	0	0	1	0	1	0	0	0	0	0	0	0	1	0
F - *Leucaena leucocephala	0	0	0	0	0	0	1	0	0	0	0	0	0	0	0	0	0	1
G - Lysiloma latisiliquum	0	0	1	0	0	0	0	0	0	0	0	0	0	0	0	1	0	0
H - *Lysiloma sabicu	0	0	0	0	0	0	0	0	0	0	0	0	0	0	0	1	0	0
I - *Pithecellobium arboreum	0	0	0	0	0	0	0	0	0	0	0	0	0	0	0	1	0	0
J - Pithecellobium guadalupense	0	0	1	0	0	0	1	0	0	0	0	0	0	0	0	1	0	0
L - *Senna alata	1	1	0	0	1	0	1	0	0	1	1	1	1	1	0	0	0	0
M - *Senna bicapsularis	0	0	0	0	0	0	0	1	0	0	0	0	1	0	0	0	0	0
N - Senna ligustrina	1	0	0	0	1	1	0	0	0	0	0	0	0	0	0	1	0	0
O - Senna chapmannii	1	0	0	1	1	1	1	0	1	1	0	0	1	0	1	1	1	0
P - *Senna polyphylla	1	0	0	0	0	0	0	0	0	0	0	0	0	0	0	0	0	0
Q - *Senna surattensis	1	0	0	0	0	0	0	0	0	0	0	0	0	0	0	0	1	0

Figure 12.3. Legume plants with extrafloral nectaries and ants in (a) natural and (b) urban habitats in south Florida – species not native are highlighted and indicated with an asterisk*.

individuals available (*C. fistula* – 4; *S. bicapsularis* – 5; *L. sabicu* – 4). Combining our past observations with the results of the deliberate samples, and including only the plant species and ant species for which we had documented an association, we can create a matrix of interactions for plants in natural and urban environments (Figure 12.3).

We can see that there is a greater variety of legume plants with EFNs in urban areas (16 species vs. 9 in the natural areas for the genera monitored in our study), as many of the urban plants are non-native (10 of the 16 species monitored) and only 2 non-native plants were encountered in the natural areas. There are more species of ants observed to be associated with the nectary-bearing legume plants in natural areas (21 species vs. 18 observed in urban areas). In both urban and natural areas, more than half of the ant species associated with EFN-producing legumes are not native to south Florida (13/21 in natural areas; 12/18 in urban areas).

The more common a plant species is in natural areas, the larger number of associated ant species it had: *Senna chapmannii* was observed in association with 17 of the 21 ant species in natural areas. This species is also one of the most popular and heavily promoted plants for butterfly gardening (Minno & Minno, 1999) and is widely planted in both home landscapes and city beautification projects (Figure 12.1 a, c); in urban sampling we found that it was associated with 12 of the 18 ants encountered in urban samples.

It appears that the more common a plant species is, the greater the proportion of its ant associates are non-native: of the ants associated with *S. chapmannii*, in natural areas all of the non-native ant species were its associates, and only five out of eight native ant species were its associates. In urban sampling, all but one of the non-native ant species were its associates, and only one of the six native ant species encountered associated with *S. chapmannii*. Less common in natural areas, *Senna ligustrina* is associated with a total of six species of ants in natural areas, and four of those are native ants; in urban areas, where it is also widely planted for butterfly gardens, three of the four associated ant species are non-native. Also infrequent in natural areas, but very characteristic of the pine rockland habitat, *Acacia pinetorum* was found associated with only two ant species in natural areas, one native, the other non-native; in urban areas, where it is utilized as a hostplant in butterfly gardening, both of its ant associates were non-native.

Lysiloma latisiliquum, a native pioneer tree species in pine rocklands that grows larger as succession proceeds to hardwood hammocks, is associated with seven ant species in natural areas, five of which are native ants; it is widely planted in native plant landscaping (Figure 12.1 a), and in urban area sampling we found only two ant associates, one native, one non-native. This apparent sparsity of urban ant associations contrasts with some very attractive non-native woody species which, in urban areas, had many more associated ant species than some of the native hostplants (e.g., *Acacia chundra* with four species, three of which were non-native; *Senna alata* with nine species, four of which were non-native). The invasive *Leucaena leucocephala* and *Senna bicapsularis* were each associated with only two ant species, and all of those ant associates are non-native.

Network nestedness analysis using ANINHADO software (Aninhado 3.0.2) (Guimarães & Guimarães, 2006), and network-level analysis using BIPARTITE package (Bipartite 2.05) for R software (R software v. 3.1.2, R Development Core Team, 2014) (Dormann et al., 2009), allows us to compare the structure of these different communities in a preliminary analysis. Such analyses can elucidate

Table 12.1 Network Metrics for Interactions between Selected Legume Plants and Ants in Natural versus Urban Habitats of Southern Florida

Network metrics	Natural habitat	Urban habitat
Number of plant species	9	16
Number of ant species	21	18
Number of associations	42	52
Mean number for plant species	1.00	1.49
Mean number for ant species	1.94	1.28
Network connectance	0.245	0.180
Nestedness value (NODF)	46.78 ($P < 0.01$)	40.13 ($P < 0.01$)
Robustness/resilience to random extinction of partners for plants	0.750	0.676
Robustness/resilience to random extinction of partners for ants	0.674	0.695

Values of specialization or dependence asymmetry cannot be calculated for qualitative data (binary matrices).

the general patterns of ant-plant interactions, as well as the extent of nestedness (Lewinsohn & Inacio Prado, 2006) and specialization of interactions in these different communities across the south Florida landscape, as others have done for ant-plant associations in other parts of the world (Diaz-Castelazo et al., 2010; Dattilo et al., 2013). Our sampling was not as extensive or as quantitative as some recent studies have been (Ivens et al., 2016; Sendoya et al., 2016), but we present our findings as they allow another basis for comparison.

Using the qualitative data from the simple matrices of associations shown in Figure 12.3, we estimated the nestedness value (NODF estimator) compared to the nestedness value for each one of the 1,000 network replicates for each interaction matrix, using ANINHADO software. We found that both networks (natural and urban areas) were significantly nested (Table 12.1): both the network for natural areas, with its NODF value of 46.78, and the network for urban areas with its NODF value of 40.13, were higher ($P < 0.01$) than the NODF values of 1,000 random networks. For the same qualitative interaction matrices, we estimated with BIPARTITE package of R software (Dormann et al., 2009) using the "Networklevel" function, the Shannon diversity of interactions, the Niche overlap of each trophic level and the Robustness for each trophic level (against secondary extinctions); network-level specialization and dependence asymmetry were not estimated given that our interaction data were binary matrices.

For natural areas (Figure 12.4a; Table 12.1), the network has low connectance (0.245, 1 being the highest connectance possible). With more species at the higher trophic level (ants), 5 was the highest degree of association for ant species (the most connected ant species in the network had interactions with five plant species). It is a non-modular network since it has only one compartment or module. This network

(a) Natural areas

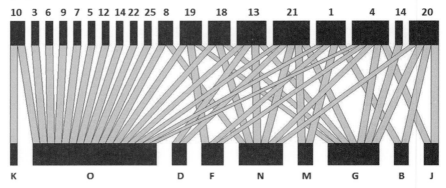

(b) Urban areas

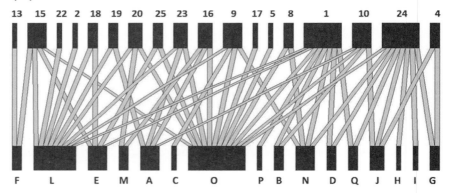

Figure 12.4.   Graphs of interactions for ants and plants in (a) natural and (b) urban areas from network analyses. Plant species are: A, *Acacia chundra*; B, *Acacia pinetorum;* C, *Cassia bakeriana*; D, *Erythrina herbacea*; E, *Erythrina variegata;* F, *Leucaena leucocephala*; G, *Lysiloma latisiliquum*; H, *Lysiloma sabicu*; I, *Pithecellobium arboreum*; J, *Pithecellobium guadalupense*; K, *Pithecellobium unguis-cati*; L, *Senna alata*; M, *Senna bicapsularis*; N, *Senna ligustrina*; O, *Senna mexicana* var. *chapmannii*; P, *Senna polyphylla*; Q, *Senna surattensis*. Ant species are: 1, *Brachymyrmex obscurior*; 2, *Camponotus castaneus*; 3, *Camponotus emeryodicatus*; 4, *Camponotus floridanus*; 5, *Camponotus inaequalis*; 6, *Camponotus planatus*; 7, *Camponotus rasilis*; 8, *Camponotus sexguttatus*; 9, *Camponotus tortuganus*; 10, *Cardiocondyla emeryi*; 11, *Odontomachus brunneus*; 12, *Odontomachus ruginodis*; 13, *Paratrechina longicornis*; 14, *Pheidole dentata*; 15, *Pheidole megacephala*; 16, *Pheidole moerens*; 17, *Pseudomyrmex ejectus*; 18, *Pseudomyrmex elongatus*; 19, *Pseudomyrmex gracilis*; 20, *Pseudomyrmex simplex*; 21, *Solenopsis geminata*; 22, *Solenopsis invicta*; 23, *Tapinoma melanocephalum*; 24, *Technomyrmex difficilis*; 25, *Wasmannia auropunctata*.

has a Shannon diversity (of interactions) value of 3.73, in which plants have higher diversity of interacting partners (1.94), but ants have higher niche overlap than plants (0.53, vs. 0.25, of a possible maximum value of 1). In this natural areas network, plants are more robust or resilient to random extinctions of partners (i.e. ants) than the opposite.

For urban areas (Figure 12.4b; Table 12.1), the network has a very low connectance (0.180, lower than the natural areas network). This urban network also has more species at the higher trophic level (ants), as does the natural areas network, 8 being the highest degree of association for ant species (the most connected ant species in the network had interactions with eight plant species). It is also a non-modular network since it has only one compartment or module. Its Shannon diversity (of interactions) value is 3.95, in which plants have a slightly higher diversity of interacting partners (1.49), and ants have a slightly higher niche overlap than plants (0.282, vs. 0.218, out of a maximum value of 1), both less dramatic than the differences in the natural areas network. In the urban areas network, plants and ants are similarly robust or resilient to random extinctions of partners.

Like other mutualistic networks, ant-plant networks have been shown to be nested (Guimarães et al., 2006), with a core of reciprocal generalists plus specialist species that interact with generalists. Symbiotic interactions (such as those found with myrmecophytes) are species-poor and compartmentalized, compared with non-symbiotic interactions (such as those with myrmecophylic ant-plant interactions, like those in this study) that are species-rich and nested (Guimarães et al., 2007). In south Florida natural areas, the generalist ant species (with the highest number of plant associates) are *Camponotus floridanus* and *Solenopsis geminata*; the generalist plants are *S. chapmanni* and *L. latisiliquum* (as discussed earlier), all native species. In urban areas, the ant species with highest number of plant associates are *Brachymyrmex obscurior* and *Tapinoma melanocephalum*; the most generalist plants are *S. chapmanni* and *S. alata*, more exotic species on each side of the interaction.

In a study comparing networks in the same location sampled ten years later, both networks had similar nested topology; even with the presence of new species a decade later, the contributions of each species to nestedness stayed the same (Diaz-Castelazo et al., 2010). Our comparison of disturbed urban environments and relatively undisturbed natural areas represents the most extreme case of how ant-plant networks have changed in south Florida over recent decades. The severity of these changes is such that species in the urban network show lower connectance and more resilience to random extinction of their partners. The pervasive use of pesticides to control arthropods in urban areas is common, and Miami gardens and city landscapes are no exception. This may be why there were fewer ant species and a lower number of associations found in urban areas. Future sampling should distinguish between "green" areas, where pesticide use is limited, and those maintained by conventional means.

Our first look at these comparisons provides suggestions of some interesting patterns in associations between legume plants with extrafloral nectaries and ants in natural and urban areas of south Florida. It seems that in many situations, non-native ants are a perfect fit with legume plants with extrafloral nectaries in both natural and urban areas. These facultative mutualisms may facilitate the naturalization and colonization of horticultural species escaping from cultivation, especially if their ant partners provide them with some protection against generalist herbivores. Since introduced plants do not usually have to contend with coevolved, specialized herbivores, they may gain a competitive advantage. Future studies on the details of interactions of various plant species with the ants utilizing their nectar may allow us to see which ants are mutualists, and which are simply *"aprovechados"* (opportunists, sensu Soberon & Martinez del Rio, 1985), taking advantage of the reward but providing no services, or interfering with benefits from mutualist ant partners. Through this work we hope to better understand the ecology of ant-plant interactions in disturbed environments and assess their potential to facilitate invasive species.

References

Ackerman, J. D., Falcon, W., Molinari, J., Vega, C., Espino, I., & Cuevas, A. A. (2014). Biotic resistance and invasional meltdown: consequences of acquired interspecific interactions for an invasive orchid, *Spathoglottis plicata* in Puerto Rico. *Biological Invasions*, 16, 2435–2447.

Addison, P., & Samways, M. J. (2000). A survey of ants (Hymenoptera: Formicidae) that forage in vineyards in the Western Cape Province, South Africa. *African Entomology*, 8, 251–260.

Allen, C., Forys, E., Rice, K., & Wojcik, D. (2001). Effects of fire ants (Hymenoptera: Formicidae) on hatching turtles and prevalence of fire ants on sea turtle nesting beaches in Florida. *Florida Entomologist*, 84, 250–253.

Baker, H. G., Opler, P. A., & Baker, I. (1978). A comparison of the amino acid complements of floral and extrafloral nectars. *Botanical Gazette,* 139, 322–332.

Barrios, B., Arellano, G., & Koptur, S. (2011). The effects of fire and fragmentation on occurrence and flowering of a rare perennial plant. *Plant Ecology*, 212, 1057–1067.

Becerra, J. X. I., & Venable, D. L. (1989). Extrafloral nectaries a defense against ant-homoptera mutualism. *Oikos,* 55, 276–280.

Bentley, B. L. (1977a). Extrafloral nectaries and protection by pugnacious bodyguards. *Annual Review of Ecology and Systematics,* 88, 407–427.

(1977b). The protective function of ants visiting the extrafloral nectaries of *Bixa orellana* (Bixaceae). *Journal of Ecology,* 65, 27–38.

Billick, I., & Price, M. V. (eds.). (2010). *The Ecology of Place: Contributions of Place-Based Research to Ecological Understanding*. Chicago: University of Chicago Press.

Bleil, R., Blüthgen, N., & Junker, R. R. (2011). Ant-plant mutualism in Hawaii? Invasive ants reduce flower parasitism but also exploit floral nectar of the endemic shrub *Vaccinium reticulatum* (Ericaceae). *Pacific Science*, 65, 291–300.

Blossey, B., & Notzold, R. (1995). Evolution of increased competitive ability in invasive nonindigenous plants: a hypothesis. *Journal of Ecology*, 83, 887–889.

Carrillo, J., Wang, Y., Ding, J., Klootwyk, K., & Siemann, E. (2012a). Decreased indirect defense in the invasive tree, *Triadica sebifera*. *Plant Ecology*, 213, 945–954.

Carrillo, J., Wang, Y., Ding, J., & Siemann, E. (2012b). Induction of extrafloral nectar depends on herbivore type in invasive and native Chinese tallow seedlings. *Basic and Applied Ecology*, 13, 449–457.

Dattilo, W., Rico-Gray, V., Rodrigues, D. J., & Izzo, T. J. (2013). Soil and vegetation features determine the nested pattern of ant–plant networks in a tropical rainforest. *Ecological Entomology*, 38(4), 374–380.

Davidson, D. W. (1998). Resource discovery versus resource domination in ants: a functional mechanism for breaking the trade-off. *Ecological Entomology*, 23, 484–490.

de la Fuente, M. A. S., & Marquis, R. J. (1999). The role of ant-tended extrafloral nectaries in the protection and benefit of a neotropical rainforest tree. *Oecologia*, 118, 192–202.

Diaz-Castelazo, C., Guimarães, P. R., Jordano, P., Thompson, J. N., Marquis, R. J., & Rico-Gray, V. (2010). Changes of a mutualistic network over time: reanalysis over a 10-year period. *Ecology*, 91, 793–801.

Dormann, C. F., Frund, J., Blüthgen, N., & Gruber, B. (2009). Indices, graphs, and null models: analyzing bipartite ecological networks. *The Open Ecology Journal*, 2, 7–24.

Eubanks, M. D. (2001). Estimates of the direct and indirect effects of red imported fire ants on biological control in field crops. *Biological Control*, 21, 35–43.

Fleet, R. R., & Young, B. L. (2000). Facultative mutualism between imported fire ants (*Solenopsis invicta*) and a legume (*Senna occidentalis*). *Southwestern Naturalist*, 45, 289–298.

Gonzalez-Teuber, M., & Heil, M. (2009). The role of extrafloral nectar amino acids for the preferences of facultative and obligate ant mutualists. *Journal of Chemical Ecology*, 35, 459–468.

Gotelli, N. J., & Arnett, A. E. (2000). Biogeographic effects of red fire ant invasion. *Ecological Letters*, 3, 257–261.

Green, P. T., O'Dowd, D. J., & Lake, P. S. (1999). Alien ant invasion and ecosystem collapse on Christmas Island, Indian Ocean. *Aliens*, 9, 2–4.

Guimarães, P. R., & Guimarães, P. (2006). Improving the analyses of nestedness for large sets of matrices. *Environmental Modelling and Software*, 21, 1512–1513.

Guimarães, P. R., Rico-Gray, V., dos Reis, S. F., & Thompson, J. N. (2006). Asymmetries in specialization in ant-plant mutualistic networks. *Proceedings of the Royal Society of London, Series B, Biological Sciences*, 273, 2041–2047.

Guimarães, P. R., Rico-Gray, V., Oliveira, P. S., Izzo, T. J., dos Reis, S. F., & Thompson, J. N. (2007). Interaction intimacy affects structure and coevolutionary dynamics in mutualistic networks. *Current Biology*, 17(20), 1797–1803.

Haddad, N. M., Bowne, D. R., Cunningham, A., Danielson, B. J., Levey, D. J., Sargent, S., & Spira, T. (2003). Corridor use by diverse taxa. *Ecology*, 84, 609–615.

Heil, M. (2011). Nectar: generation, regulation and ecological functions. *Trends in Plant Science*, 16, 191–200.

   (2015). Extrafloral nectar at the plant-insect interface: a spotlight on chemical ecology, phenotypic plasticity, and food webs. *Annual Review of Entomology*, 60, 213–232.

Heil, M., Koch, T., Hilpert, A., Fiala, B., Boland, W., & Linsenmair, K. E. (2001). Extrafloral nectar production of the ant-associated plant, *Macaranga tanarius*, is an induced, indirect, defensive response elicited by jasmonic acid. *Proceedings of the National Academy of Sciences of the United States of America*, 98, 1083–1088.

Holway, D. A., Lach, L., Suarez, A. V., Tsutsui, N. D., & Case, T. J. (2002). The causes and consequences of ant invasions. *Annual Review of Ecology and Systematics*, 33, 181–233.

Horvitz, C. C., & Schemske, D. W. (1984). Effects of ants and an ant-tended herbivore on seed production of a neotropical herb. *Ecology*, 65, 1369–1378.

Inouye, D. W., & Taylor, O. R. (1979). A temperate region plant-ant-seed predator system: consequences of extrafloral nectar secretion by *Helianthella quinquenervis*. *Ecology*, 60, 1–8.

Ivens, A. B. F., Beeren, C. V., Blüthgen, N., & Kronauer, D. J. C. (2016). Studying the complex communities of ants and their symbionts using ecological network analysis. *Annual Review of Entomology*, 61, 353–371.

Janzen, D. H. (1966). Coevolution between ants and acacias in Central America. *Evolution*, 20, 249–275.

Jezorek, H., Stiling, P., & Carpenter, J. (2011). Ant predation on an invasive herbivore: can an extrafloral nectar-producing plant provide associational resistance to *Opuntia* individuals? *Biological Invasions*, 13, 2261–2273.

Jones, I. M., Koptur, S., Gallegos, H. R., Tardanico, J. P, Trainer, P. A., & Peña, J. (2016). Changing light conditions in pine rockland habitats affect the intensity and outcome of ant-plant interactions. *Biotropica*, 49, 83–91.

Junker, R. R., Daehler, C. C., Doetterl, S., Keller, A., & Blüthgen, N. (2011). Hawaiian ant-flower networks: nectar-thieving ants prefer undefended native over introduced plants with floral defenses. *Ecological Monographs*, 81, 295–311.

Kaakeh, W., & Dutcher, J. D. (1992). Foraging preference of red imported fire ants (Hymenoptera: Formicidae) among three species of summer cover crops and their extracts. *Journal of Economic Entomology*, 85, 389–394.

Kautz, S., Lumbsch, H. T., Ward, P. S., & Heil, M. (2009). How to prevent cheating: a digestive specialization ties mutualistic plant-ants to their ant-plant partners. *Evolution*, 6, 839–853.

Keeler, K. H. (1985). Extrafloral nectaries on plants in communities without ants: Hawaii. *Oikos*, 44, 407–414.

Koi, S., & Daniel, J. (2015). New and revised life history of the Florida hairstreak *Eumaeus atala* (Lepidoptera: Lycaenidae) with notes on its current conservation status. *Florida Entomologist*, 98(4), 1134–1147.

Koptur, S. (1979). Facultative mutualism between weedy vetches bearing extrafloral nectaries and weedy ants in California. *American Journal of Botany*, 66, 1016–1020.

(1984). Experimental evidence for defense of *Inga* (Mimosoideae) saplings by ants. *Ecology*, 65, 1787–1793.

(1992). Plants with extrafloral nectaries and ants in Everglades habitats. *The Florida Entomologist*, 75(1), 38–50.

Koptur, S., Jones, I. M., & Peña, J. E. (2015). The influence of host plant extrafloral nectaries on multitrophic interactions: An experimental investigation. *PLOSone*, 10(9), e0138157.

Koptur, S., & Lawton, J.H. (1988). Interactions among vetches bearing extrafloral nectaries, their biotic protective agents, and herbivores. *Ecology*, 69, 278–293.

Krushelnycky, P. D., Loope, L. L., & Reimer, N. J. (2005). The ecology, policy, and management of ants in Hawaii. *Proceedings of the Hawaiian Entomological Society*, 37, 1–25.

Lach, L. (2003). Invasive ants: unwanted partners in ant-plant interactions? *Annals of the Missouri Botanical Garden*, 90, 91–108.

Lach, L. (2007). A mutualism with a native membracid facilitates pollinator displacement by Argentine ants. *Ecology*, 88, 1994–2004.

Lach, L., & Hoffmann, B. D. (2011). Are invasive ants better plant-defense mutualists? A comparison of foliage patrolling and herbivory in sites with invasive yellow crazy ants and native weaver ants. *Oikos*, 120, 9–16.

Lach, L., Tillberg, C. V., & Suarez, A. V. (2010). Contrasting effects of an invasive ant on a native and an invasive plant. *Biological Invasions*, 12, 3123–3133.

Lewinsohn, T. M., & Inacio Prado, P. (2006). Structure in plant/animal interaction assemblages. *Oikos*, 113(1), 174–184.

Lubin, Y. D. (1984). Changes in the native fauna of the Galapagos Islands following invasions by the little red fire ant, *Wasmannia auropunctata*. *Biological Journal of the Linnean Society*, 21, 229–242.

Marazzi, B., Conti, E., Sanderson, M. J., McMahon, M. M., & Bronstein, J. L. (2013). Diversity and evolution of a trait mediating ant-plant interactions: Insights from extrafloral nectaries in *Senna* (Leguminosae). *Annals of Botany*, 111, 1263–1275.

Maschinski, J., & Wright, S. (2006). Using ecological theory to plan restorations of the endangered Beach jacquemontia (Convolvulaceae) in fragmented habitats. *Journal for Nature Conservation*, 14, 180–189.

Mathew, G., & Anto, M. (2007). In situ conservation of butterflies through establishment of butterfly gardens: A case study at Peechi, Kerala, India. *Current Science*, 93(3), 337–347.

McLain, D. K. (1983). Ants, extrafloral nectaries, and herbivory on the passion vine *Passiflora incarnata*. *American Midland Naturalist*, 110, 433–439.

Minno, M., & Minno, M. (1999). *Florida Butterfly Gardening: A Complete Guide to Attracting, Identifying, and Enjoying Butterflies*. Gainesville, FL: University Press of Florida.

Ness, J. H. (2003). Contrasting exotic *Solenopsis invicta* and native *Forelius pruinosus* ants as mutualists with *Catalpa bignonioides*, a native plant. *Ecological Entomology*, 28, 247–251.

Ness, J. H., & Bronstein, I. L. (2004). The effects of invasive ants on prospective ant mutualists. *Biological Invasions*, 6, 445–461.

Ness, J. H., Morris, W. F., & Bronstein, J. L. (2006). Integrating quality and quantity of mutualistic service to contrast ant species protecting *Ferocactus wislizeni*. *Ecology*, 87, 912–921.

Oliveira, P. S., Rico-Gray, V., Diaz-Castelazo, C., & Castillo-Guevara, C. (1999). Interactions between ants, extrafloral nectaries, and insect herbivores in Neotropical sand dunes: herbivore deterrence by visiting ants increases fruit set in *Opuntia stricta* (Cactaceae). *Functional Ecology*, 13, 623–631.

Rios, R. S., Marquis, R. J., & Flunker, J. C. (2008). Population variation in plant traits associated with ant attraction and herbivory in *Chamaecrista fasciculata* (Fabaceae). *Oecologia*, 156(3), 577–588.

Rosumek, F. B., Silveira, F. A. O., Neves, F. d. S., Barbosa, N. P. d. U., Diniz, L., Oki, Y., Pezzini, F., Fernandes, G. W., & Cornelissen, T. (2009). Ants on plants: a meta-analysis of the role of ants as plant biotic defenses. *Oecologia*, 160, 537–549.

Savage, A. M., Rudgers, J. A., & Whitney, K. D. (2009). Elevated dominance of extrafloral nectary-bearing plants is associated with increased abundances of an invasive ant and reduced native ant richness. *Diversity and Distributions*, 15, 751–761.

Savage, A. M., & Whitney, K.D. (2011). Trait-mediated indirect interactions in invasions: unique behavioral responses of an invasive ant to plant nectar. *Ecosphere*, 2, 106.

Sendoya, S. F., Blüthgen, N., Tamashiro, J. Y., Fernandez, F., & Oliveira, P. S. (2016). Foliage-dwelling ants in a neotropical savanna: effects of plant and insect exudates on ant communities. *Arthropod-Plant Interactions*, 10, 183–195.

Soberon Mainero, J., & Martinez del Rio, C. (1985). Cheating and taking advantage in mutualistic associations. In *The Biology of Mutualism*, ed. D. A. Boucher. New York: Oxford University Press, pp. 192–216.

Torres-Hernandez, L., Rico-Gray, V., Castillo-Guevara, C., & Vergara, J. A. (2000). Effect of nectar-foraging ants and wasps on the reproductive fitness of *Turnera ulmifolia* (Turneraceae) in a coastal sand dune in Mexico. *Acta Zoologica Mexicana*, 81, 13–21.

Wagner, D., & Kay, A. (2002). Do extrafloral nectaries distract ants from visiting flowers? An experimental test of an overlooked hypothesis. *Evolutionary Ecology Research*, 4, 293–305.

Warren, A. D., & Calhoun, J. V. (2011). Notes on the historical occurrence of *Aphrissa neleis* in Southern Florida, USA (Lepidoptera: Pieridae: Coliadinae). *News of the Lepidopterists' Society*, 53(1), 3–7.

Weber, M. G., & Keeler, K. H. (2013). The phylogenetic distribution of extrafloral nectaries in plants. *Annals of Botany*, 111(6), 1251–1261.

Wetterer, J. K. (2010). Worldwide spread of the graceful twig ant, *Pseudomyrmex gracilis* (Hymenoptera: Formicidae). *Florida Entomologist*, 93, 535–540.

Whitcomb, W. H., Denmark, H. A., Buren, W. F., & Carroll, J. F. (1972). Habits and present distribution in Florida of the exotic ant, *Pseudomyrmex mexicanus* (Hymenoptera: Formicidae). *Florida Entomologist*, 55, 31–33.

Wild Ones® Natural Landscapers Ltd. (2004). *Wild Ones: Native Plants, Natural Landscapes – Landscaping with Native Plants*, 4th edition. Downloaded from US Environmental Protection Agency website (https://archive.epa.gov/greenacres/web/pdf/wo_2004b.pdf).

Wilder, S. M., Holway, D. A., Suarez, A. V., LeBrun, E. G., & Eubanks, M. D. (2011). Intercontinental differences in resource use reveal the importance of mutualisms in fire ant invasions. *Proceedings of the National Academy of Sciences of the USA*, 108, 20639–20644.

Xu, F. F., & Chen, J. (2009). Comparison of the differences in response to the change of the extrafloral nectar-ant–herbivore interaction system between a native and an introduced *Passiflora* species. *Acta Botanica Yunnanica*, 31, 543–550.

Zettler, J. A., Taylor, M. D., Allen, C. R., & Spira, T. P. (2004). Consequences of forest clear-cuts for native and non-indigenous ants (Hymenoptera: Formicidae). *Annals of the Entomological Society of America*, 97, 513–518.

Zimmerman, E. C. (1970). Adaptive radiation in Hawaii with special reference to insects. *Biotropica*, 2, 32–38.

13 Biological Invasions and Ant-Flower Networks on Islands

Nico Blüthgen, Christopher Kaiser-Bunbury, and Robert R. Junker

Oceanic island biotas are characterised by high levels of endemism, yet there are general rules that drive the ecology and evolution of island species. Mac Arthur and Wilson's seminal work on island biogeography (Mac-Arthur and Wilson 1967) continues to provide a framework to predict how island size and distance to a mainland shape species assemblages on islands. Interestingly, the authors' equilibrium theory was inspired by Wilson's earlier work on ant communities in Melanesia (Wilson 1961). Islands also provided the conceptual foundation for Darwin's work on how isolation leads to pronounced speciation, resulting in high endemism, shifts in life histories and unique life forms. While the remoteness of islands helped humans to understand and study fundamental principles in ecology and evolution, the same conditions render island biotas particularly vulnerable to the dispersal of competitive mainland species by humans. Island species and ecosystems suffer severe impacts from the spread of invasive alien species (IAS), which place islands yet again in the spotlight of scientific advancement – this time in the fields of invasion biology (Hansen 2015), climate change (Karnauskas et al. 2016) and conservation and restoration ecology (Kaiser-Bunbury and Blüthgen 2015). Extinctions of endemic vertebrates and severe habitat transformation are among the best-documented consequences of IAS. Island ecosystems are threatened because their unique life forms, which evolved in relative isolation and without natural enemies, lost costly defences or mobility and are now confronted with novel predators, parasites and competitors. Human-assisted colonisation by rats and cats, for example, have been a major driver of extinctions of endemic birds and other island taxa (Medina et al. 2011; Harper and Bunbury 2015), many of which performed critical ecosystem functions (e.g., Boyer and Jetz 2014). Island endemics are thus a major target for conservation actions and require specific management interventions.

In this chapter we particularly focus on the potential vulnerability of the islands' endemic flowering plants – particularly their interactions with pollinators – to invasive ants.

Setting the Scene: Ants on Islands

Ants are prominent players on many islands, more often in the role of invaders rather than as threatened native species. This perspective may be biased given that rare and endemic ant species are less intensively studied, whereas the biology and impact of ant invaders are better known due to their prominence, ecological relevance and wide distribution. However, the ants' current prominent role in biological invasions on islands is beyond any doubt, and the rather insignificant ecological role of ants in the evolutionary history of many islands contributes to the severity of their impacts on native ecosystems today.

On very remote archipelagos such as Hawaii, native ants are scant or even entirely absent (Moller 1996; Krushelnycky et al. 2005) (Box 13.1). The same holds true for other social insects (Gillespie and Roderick 2002). As expected from the theory of island biogeography, large islands close to the mainland have more native ant species than small and isolated islands (Box 13.1). Very large islands such as Madagascar thus have a particularly diverse native ant fauna (Fisher 1997; Brühl et al. 1998). Some remote archipelagos, however, also have an unexpected high diversity and endemism, e.g., among 187 ant species recorded in Fiji over 70 per cent are endemic (Sarnat and Economo 2012). On islands with high ant endemism, native ant communities may be particularly vulnerable to invasions as they are not only victims of predatory invasive ants but also competitively inferior concerning resource acquisition and adaptation to anthropogenic habitats (Holway et al. 2002; Lach and Hooper-Bui 2010).

While islands have been relatively poorly populated by ants in the past, these 'gaps' were rapidly filled since human colonisation. Remote Pacific islands today contain the largest number of introduced ants compared to any other biogeographic region (McGlynn 1999). Dozens of exotic ant species now live in the Galapagos, Hawaii and other remote islands (Lubin 1984; Krushelnycky et al. 2005). Ants dispersed across different island habitats in very short time spans, often facilitated by habitat disturbance by humans. Consequences for native arthropods can be severe. On the Galapagos, native ant species and other insects were reduced in sites invaded by the little fire ant, *Wasmannia auropunctata* (Lubin 1984). A strong reduction in arthropod diversity was also recorded in Hawaii shortly after the yellow crazy ant *Anoplolepis gracilipes* appeared (Cole et al. 1992). Yellow crazy ants effectively preyed upon a key herbivore on the Christmas Islands, a land crab, which extirpated local populations and caused cascading effects on the forest understorey vegetation that grew to high densities in the absence of the crab (O'Dowd et al. 2003). Most negative effects can be assigned to predation by ants of native animals and interference competition with native species (Holway et al. 2002; Lach and Hooper-Bui 2010). Ground-dwelling arthropods are particularly vulnerable and represent the majority of cases published so far. Pitfall traps provide a simple method of quantifying the impact of ant invasions, which may explain the bias towards reports on declining ground-dwelling arthropod fauna. In addition, exposed tree-dwelling

Box 13.1 Island Biogeography of Native and Introduced Ants

We analysed the variation in a number of ant species per island or archipelago (group of islands) based on well-documented species lists (including subspecies) in the AntWeb and AntMaps online databases, which are partly complementary in terms of ant species listed and spatial units used (AntWeb: www.antweb.org; AntMaps: www.antmaps.org Janicki et al. 2016), both accessed 1 March 2016). In AntWeb, 161 species were globally listed as 'introduced ants' (native in one region, introduced elsewhere). We also added all species listed for Hawaii to this incomplete list of introduced ants, since it is generally assumed that this archipelago has no (or only a single) native ant species (Krushelnycky et al. 2005); the extended introduced ant list in our analysis thus contains 172 species. We split each island species list into potentially introduced species (irrespective of their origin) and all other species, which are called 'native' in the context of this analysis; morpho-species were omitted and only valid species were used. We used 66 islands from AntWeb and additional 19 from AntMaps, which is restricted to native ants (not used for computing introduced ants). While both databases are considered non-exhaustive, the data from 23 islands on the number of ant species per island are highly correlated (log-transformed; $r = 0.95$, $p < 10^{-12}$), supporting that they can be pooled in the context of comparative analyses. The area of the island (km^2) and their degree of isolation were taken from an online database on islands provided by the United Nations Environment Programme (UNEP, http://islands.unep.ch/) as used in Caujape-Castells et al. (2010). The isolation index in the UNEP database is defined as the sum of the square roots of the following three minimum distances: (1) to the nearest island of similar size or larger; (2) the nearest island group or archipelago and (3) the nearest continent. Index values are reported only for remote islands; for the remaining islands we roughly estimated the index largely based on (2) and (3) for the purpose of our comparative analysis. Data from archipelagos are summed (area) or averaged (isolation) from the single islands.

Consistent with the theory of island biogeography, the number of native ant species increased with island size and decreased with its isolation across 85 islands (Figure 13.1a). Species richness increased with latitude, and about half of the variation in species richness was explained by island area (Table 13.1). After accounting for latitude and area, the remaining variation was only marginally significantly explained by the islands' isolation index. The number of exotic ants increased slightly with island area and isolation, i.e., more exotic ants were found on isolated islands compared to islands closer to mainland areas. Interestingly, the proportion of exotic to native ants (here termed 'ant alienism') decreased with island area (Figure 13.1b) and increased with isolation (Figure 13.1c). Overall, small and remote islands seem to be particularly dominated by exotic ants, possibly due to intensive human traffic as well as a lower resistance of native communities.

Table 13.1 Effects of Island Biogeography (Absolute Latitude, Log Area and Degree of Isolation) for the Number of Native Ant Species Recorded on Each Island and the Proportion of Introduced Ants (Ant Alienism); Linear Multiple Regression

Predictor	Richness of native ants				Ant alienism (% introduced)			
	β	F	df	p	β	F	df	p
Latitude	−0.043	17.2	1	$< 10^{-4}$	−0.002	3.4	1	0.07
log(area)	0.410	72.8	1	$< 10^{-12}$	−0.039	12.1	1	< 0.001
Isolation	−0.008	3.7	1	0.058	0.002	7.0	1	0.011
Full model	$n = 83, r^2 = 0.60, p < 10^{-14}$				$n = 65, r^2 = 0.27, p < 0.001$			

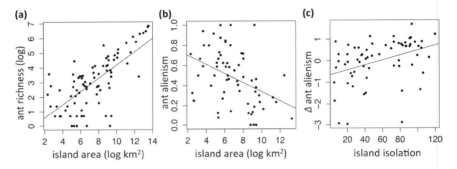

Figure 13.1. (a) Native ant richness increases with island area ($r^2 = 0.50, p < 0.001$). (b) The relative proportion of introduced ants ('ant alienism') decreases with island area ($r^2 = 0.16, p = 0.001$). (c) The residuals of (b) increase with island isolation ('Δ ant alienism'; $r^2 = 0.12, p = 0.005$).

species are also reduced by invasive ants (Hill et al. 2003; Kaiser-Bunbury et al. 2014), whereas highly mobile, flying insects such as bees and flies are less likely to be directly affected by ants during their adult life stage. It was reported, however, that invasive ants reduce the densities of larvae or nests of, e.g., ground-nesting *Hylaeus* bees (Colletidae) in Hawaii (Cole et al. 1992).

Negative effects via exploitation competition are generally more difficult to detect than effects by predation or interference competition. Many of the impacts by exotic ants may therefore be subtle, indirect and complex, yet they can be as severe as the direct loss of native arthropod fauna by ant predation (Lach and Hooper-Bui 2010). These effects may be particularly important in animal-plant mutualisms. For example, endemic birds were interrupted in their handling of fruits by yellow crazy ants on Christmas Islands, with possible consequences for frugivory and fruit dispersal (Davis et al. 2009). Whereas some invasive ants also contribute to myrmecochorous seed dispersal, others such as small invasive red fire ants (*Solenopsis invicta*) trigger a reduction in seed-dispersal services by competitively excluding larger native ants (Ness and Bronstein 2004). Variation in the net effects of novel species interactions also depends on the plant species or seed type. As an example from the mainland, myrmecochorous seed dispersal of South African *Protea* were found to be disrupted by Argentine ants *Linepithema humile* (Bond and Slingsby 1984). In contrast, the same invasive ant species may contribute in the dispersal of elaiosome-bearing seeds of endemic *Anchusa crispa* (Boraginaceae) on Corsica (Quilichini and Debussche 2000). Effects on trophobiotic interactions with hemipterans are similarly ambiguous (Ness and Bronstein 2004). Certain throphobionts such as scale insects benefit from invasive ants on islands – and reciprocally facilitate the ants' success in their new habitat (Hill et al. 2003; Abbott and Green 2007). One such key impact – the disruption of plant-pollinator associations on islands – is the topic of this chapter.

Unprepared Hosts: Endemic Plants on Islands

Most islands have a high degree of plant endemism (Carlquist 1974; Whittaker and Fernández-Palacios 2007). The world's islands, while only about 4 per cent of the Earth's land surface, hold about a quarter of all plant species, and these species are 9.5 times more likely to be endemic compared to those in mainland regions (Kier et al. 2009). Plant endemism also varies considerably across islands. New Caledonia, islands in Polynesia-Micronesia, the Eastern Pacific region, the Atlantic and the Western Indian Ocean (WIO) harbour substantially more endemic plant species compared to islands, for example, in South East Asia (all but Borneo), Japan or New Zealand (Kier et al. 2009). The main determinants behind plant species richness on islands are a combination of geological origin, island size, degree of isolation, elevation and abiotic parameters such as temperature and precipitation. The strongest single predictor of endemism is island size, accounting for two-thirds of plant species richness, followed by elevation, which predicts 40 per cent of the species richness pattern (Kreft et al. 2008). Thus, large mountainous islands have a richer flora than small flat islands. Isolation has the opposite effect; plant species richness declines with distance to the nearest mainland, as predicted by MacArthur and Wilson's theory of island biogeography. Interestingly, temperature and precipitation play only a marginal role in determining plant species richness on islands (but see Chown et al. 1998 for the influence of sea surface temperature on plant species richness – Kreft et al. 2008). Species richness and degree of endemism on islands are positively related, as the rate of speciation increases with species diversity (Emerson and Kolm 2005). Another global analysis of patterns of plant endemism in search of biodiversity hotspots provides partial support for the relationship between species richness and endemism in plants (Myers et al. 2000). The authors identified Madagascar and the WIO Islands, the Philippines and Polynesia-Micronesia as areas with relatively high degrees of plant endemism, yet New Caledonia, Wallacea and New Zealand have considerably fewer unique species than reported by Kier et al. (2009). These differences derive partly from variations in the methodology used by the two studies, and it also hints towards a more fundamental discrepancy in the definition of endemic species in terms or rarity and geographic spread (Kruckeberg and Rabinowitz 1985). Despite the inconsistencies in the described distribution of island plant endemism, both studies identify tropical islands as plant endemism hotspots (Myers et al. 2000; Kier et al. 2009).

The same processes that direct the evolution of island endemics – rare and isolated colonisation events paired with subsequent radiation of lineages – determine the reproductive strategies of island floras. Moreover, a general shortage of pollinators and a disharmonic set of pollinator guilds are characteristic for many islands, which coerce island plants into relying on generalised interactions with pollinators to secure pollination services (Kaiser-Bunbury et al. 2010). Alternatively, island plants may depend on wind for pollination, or display autogamy (self-fertilisation). Baker's rule (1967) postulates that self-compatible hermaphrodites should have an advantage in colonising and establishing on isolated islands. Yet, strong support for Baker's rule in woody island plants is still lacking (e.g., in New Zealand; Newstrom and Robertson 2005), albeit successful plant invasion was linked to high levels of self-fertilisation (Rambuda and Johnson 2004). Wind pollination (anemophily) as a breeding strategy for pollinator-deprived island endemics may be restricted to subantarctic islands, while temperate and tropical islands harbour more species with floral traits suggestive of biotic pollination (Lord 2015).

Even if island floras in milder climates are more dependent on animals for pollination, many groups of pollinators, which are prominent across continents, are largely missing from isolated oceanic islands. For example, native social bees, much like ants and other social insects, are rarely found on oceanic islands. To attract a wide range of available island pollinators, mostly flies, solitary bees, beetles, birds and reptiles, island plants adapted their breeding strategies (Barrett 1996) and evolved simple and open flowers, usually devoid of any bright colours, accessible to many potential mutualists (Carlquist 1974). The floras of New Zealand (Webb and Kelly 1993) and Juan Fernandez Islands (Anderson et al. 2001) are representative examples of typical reproductive strategies of island plants. Particularly common are hermaphroditic flowers with comparatively small, bowl-shaped, short corollas, which are usually dull in colour. To avoid inbreeding as a consequence of sharing pollinators among many endemics, dioecy and the temporal (dichogamy) and spatial (herkogamy) separation of the sexes in hermaphrodite flowers are disproportionately widespread in such floras (Bawa 1982; Sakai et al. 1995; Barrett 1996).

By shedding costly signalling colours and complex floral architecture and defence mechanisms to exclude floral larcenists, endemic plants not only increased the attractiveness of their flowers to a broad range of native pollinators, but it also rendered them more vulnerable to the exploitation by non-native species. With human colonisation, social insects, many of which are specialised on floral nectar for food, e.g., honeybees and bumblebees, arrived on many remote islands. As endemics on oceanic islands are intrinsically rare (i.e., true rarities; Kruckeberg and Rabinowitz 1985), the effects of introduced social insects on the long-term viability of endemic plants and their mutualist pollinators are of major conservation concern (Kaiser-Bunbury et al. 2010). Both honeybees and bumblebees are useful and effective pollinators of many wild and crop species in their local range. Once introduced to islands beyond their natural range, their role as pollinators of native species is seen more critically. Honeybees (*Apis mellifera*), for example, are known to provide pollination services to many island endemics (Dupont et al. 2003; Kato and Kawakita 2004; Philipp et al. 2006; Kaiser-Bunbury et al. 2011), yet their overall impact on native pollinators is generally considered detrimental (Kato et al. 1999; Hansen et al. 2002; Dupont et al. 2004; Traveset and Richardson 2006). The same is true for introduced bumblebees on islands; *Bombus terrestris*, for example, was introduced to Tasmania where it outcompetes two native megachilid bees from the genus *Chalicodoma* on a scarce floral resource (Hingston and McQuillan 1999). While the negative effects of exotic pollinators on endemic plants are buffered by the pollination services provided by the exotic species, island endemics have little to gain from the introduction and spread of another group of social insects that frequently invade islands: ants.

Ants on Flowers – Consequences for Pollination

Pollination by ants is a rare exception, largely confined to some plant species that flower close to the ground in dense stands in hot and dry habitats (Hickman 1974; Beattie 2006). Moreover, most ants do not feed on pollen (exceptions: Czechowski et al. 2011). Virtually all ants, however, readily feed on sugar solutions (extrafloral nectar, honeydew), including floral nectar when experimentally presented outside a flower (Junker and Blüthgen 2008). Given their appetite for nectar, it is surprising that ants are absent from flowers of many, if not most, plant species. This apparent lack of flower-visiting (nectar-thieving) ants may be explained by floral morphology that prevents ants from visiting flowers, or by secondary metabolites which are either dissolved in nectar or emitted by flowers. The first documented observations that ants do not exploit available floral nectar originate from tropical latitudes (Pijl 1955; Janzen 1977) where ants are known to rapidly find and consume sugary resources that are rather limited in these habitats. Although we are not aware of a study that systematically mapped the proportion of flowering plants visited by ants onto different biomes, we detected a trend towards a higher proportion of plants with ants as floral visitors on oceanic islands compared to continental areas (Junker et al. 2011a) (Figure 13.2). For example, on the Seychelles island of Mahé, the invasive white-footed ant (*Technomyrmex albipes*) and the yellow crazy ant (*A. gracilipes*) heavily exploited 97 per cent and 58 per cent, respectively, of endemic flowering species in inselberg plant communities (Kaiser-Bunbury unpublished data; for

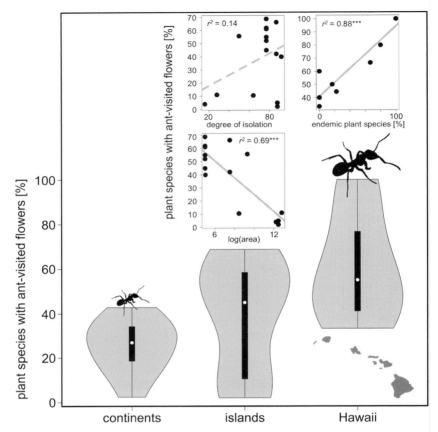

Figure 13.2. Percentage of plant species with ant-visited flowers in habitats on continents, oceanic islands and in Hawaii. Note that many ant-flower interactions, especially those on Hawaii, involve invasive ants. Violin plots show range, interquartile range and median of data combined with a kernel density plot. The percentage of plant species with ant-visited flowers differs between Hawaii and the other two biogeographic regions (ANOVA: $F_{2,32} = 6.9**$). In addition, the percentage of plant species with ant-visited flowers as functions of island area and isolation (for islands) or the percentage of endemic plant species within the Hawaiian habitats is shown. For methods and data see Junker et al. (2011a). Map of Hawaii was taken from freevectormaps.com.

methods see Kaiser-Bunbury et al. 2011). On the Juan Fernández Islands, introduced Argentine ants (*Linepithema humile*) were reported to be among the very few flower visitors of endemic *Wahlenbergia* species (Campanulaceae) that offer nectar, but are most likely self- or wind-pollinated (Anderson et al. 2001). While these observations occurred near the only human settlement on the protected island, the spread of the interactions across uninhabited parts of the islands may only be a matter of time.

Communities on the Hawaiian islands comprise a particularly high proportion of flowering plant species that are exploited by invasive ants (Junker et al. 2011a) (Figure 13.2). These ants most heavily attend flowers of several endemic plant species that evolved in the absence of ants. For instance, the endemic *Metrosideros polymorpha* (Myrtaceae) and *Vaccinium reticulatum* (Ericaceae) are strongly exploited by nectar-seeking ants (Lach 2008b; Junker et al. 2010; Bleil et al. 2011). Thus, there is a high potential of interference by ants with pollination on isolated oceanic islands which often harbour more invasive than native ants (Box 13.1).

The literature on flower-ant interactions predominantly shows negative effects of both native and non-native ants on pollination and plant reproduction. Even if they rob nectar and thus do not transfer pollen, the net effect of floral ant visitation may not necessarily be negative. In a few reported cases, ants indirectly benefit pollination and plant reproduction. For instance, ants of the genus *Ectatomma* do not pollinate the flowers of the shrub *Psychotria limonensis* (Rubiaceae) directly but aggressively force pollinators to leave flowers earlier, which facilitates pollen dispersal (Altshuler 1999). Since ants defend floral resources against competitors or prey on insects on flowers, ants may protect flowers against parasites. *Oecophylla smaragdina* ants on the flowers of the shrub *Melastoma malabathricum* (Melastomataceae) sit and wait for prey and preferentially attack less efficient pollinators but avoid large carpenter bees (*Xylocopa* spp.; Apidae), the main pollinator of the plant. In this case, the pollinator filter effect resulted in higher fruit set in *M. malabathricum* (Gonzalvez et al. 2013), but it is unclear whether such filtering decreases the plant species' resilience in the long term. An unexpected mutualistic interaction between invasive ants and the endemic shrub *Vaccinium reticulatum* has been observed in Hawaii. Invasive ants strongly reduced flower parasitism by introduced plume moths *Stenoptilodes littoralis* (Pterophoridae), which feed on flowers and developing fruits (Bleil et al. 2011).

These 'positive' examples of ant-flower interactions with direct or indirect benefits for native plant reproductive success are in strong contrast to numerous studies demonstrating negative effects of native and exotic ants (Lach and Hooper-Bui 2010). Negative effects often involve the depletion of nectar by ants (exploitation competition) that abundantly recruit workers to nectar sources and thereby deplete the nectar of entire plants (Herrera et al. 1984; Junker et al. 2010; Bleil et al. 2011). Reduced quantities of nectar by nectar-robbing ants may then trigger a reduction in visitation frequency of pollinating birds, geckos or insects and, thus, fruit set of plants (Kalinganire et al. 2001; Hansen and Müller 2009; Junker et al. 2010). Besides exploitation competition, interference competition is commonly observed when ants and pollinators forage on the same nectar source (Rodriguez-Girones et al. 2013). Rewarding resources are aggressively defended by ants against competitors, including invertebrate and vertebrate pollinators that are approaching to collect floral nectar (Tsuji et al. 2004; Ness 2006; Bleil et al. 2011). The effectiveness of the defensive behaviour of ants is dependent on ant density and identity (Gaume et al. 2005; Ness 2006). Invasive ant species are known to achieve higher abundances and to be more aggressive than naturalised but non-invasive exotic species

(Holway et al. 2002), therefore imposing a stronger negative impact on invaded habitats *via* exploitation and interference competition on plant pollination compared to non-invasive exotic ant species. Examples from Hawaii where invasive ants have severe negative effects on native ecosystems (Krushelnycky et al. 2005) support this notion by clearly showing that invasive ants such as the yellow crazy ant, Argentine ant and big-headed ant (*Pheidole megacephala*) commonly displace pollinators from flowers (Lach 2008b; Junker et al. 2010; Bleil et al. 2011).

Despite increasing evidence on the effects of ant-flower-pollinator interactions and the (negative) consequences for both native plants and pollinators, such work is primarily based on single-species systems and short-term studies. There is increasing concern that exotic ants may have community-wide disruptive effects and long-term consequences for native plant-pollinator communities. On the island of Praslin in the Seychelles, for example, the invasive yellow crazy ant displaced an endemic community of arboreal vertebrate and invertebrate palm specialists from its main food source, male flowers of the coco de mer palm *Lodoicea maldivica* (Kaiser-Bunbury et al. 2014). It was suggested that the displaced fauna plays a key role in the maintenance of the ancient coco de mer forest (Edwards et al. 2015). Invasive ants in Hawaii were reported to locally eliminate groundnesting bees *Nesoprosopis* sp. that were abundant in ant-free sites (Medeiros et al. 1986). The total displacement of native arthropods by exotic ants including a number of important Hawaiian pollinators has been documented in the Haleakala National Park, Maui, Hawaii (Cole et al. 1992), which indicates that exotic ants have the potential to alter local community compositions of arthropods. A detailed analysis of arthropod community composition as a function of the presence/absence of ants revealed that overall species richness declined in ant-invaded sites and that endemic species were particularly affected (Krushelnycky and Gillespie 2008). Population declines in important pollinators of endemic plant species such as *Nesoprosopis* sp. may eventually result in reductions in plant abundance, too. Likewise, flower-visiting invasive ants strongly altered the composition of the arthropod community usually associated with ant-free flowers of *Leucospermum conocarpodendron* (Proteaceae) in South Africa as an example for the mainland (Lach 2008a). Given these alterations, species communities and abundances by ants, it is not surprising that flower-visitor networks in ant-invaded sites markedly differ in their topology from networks in ant-free sites despite identical plant and pollinator species assemblages (Hanna et al. 2015). At an early stage of invasion, however, successful invasive species management may restore plant-pollinator interactions. In a field study in Hawaii, Hanna et al. (2013) reduced *Vespula pensylvanica* (Vespidae) populations that outcompete pollinators of *M. polymorpha* for floral resources and recorded an increase in visitation rates of pollinators. Large-scale control or even eradications of invasive ant populations in restoration projects, however, are mostly unsuccessful or of limited duration (Krushelnycky et al. 2005). Generally, eradication strategies may have a higher chance of long-term success when applied to small islands (Gaigher and Samways 2013).

Avoidance Strategies of Plants to Reduce Negative Effects by Ants

The negative effects associated with ants (native or non-native) consuming floral nectar suggest that plants would benefit from traits that prevent ants from accessing floral resources. Indeed, a number of studies discussed and/or demonstrated that unpalatable nectar, morphological barriers and repellent floral scents are effective in reducing ant visitation or completely excluding ants from flowers. Janzen (1977) postulated the hypothesis that flowers offer nectar unpalatable to ants to explain the general absence of ants from flowers. In the following years the hypothesis was extensively tested generating inconsistent results (Haber et al. 1981). Screenings across multiple species from a large selection of plant families revealed that unpalatable nectar is an exception rather than the rule (Junker and Blüthgen 2008; Junker et al. 2011a).

Flower morphology, which is assumed to be an adaptation to efficient pollen deposition (Armbruster et al. 2004), also has important functions as filter in selectively permitting visitors to access floral rewards (Herrera et al. 1984). Although a restricting morphology was mostly discussed and investigated in the context of proboscis length of insects (Stang et al. 2006), ants are also efficiently excluded from flowers by morphology. For example, in cases where head capsules of ants are wider than the corolla tube of flowers the nectar remains inaccessible for ants (Galen and Cuba 2001). In addition to such structural properties of flowers, other morphological features of plants may effectively hinder ants or other crawling arthropods from entering the inflorescences. Most prominently, the 'greasy pole syndrome' proposed by Harley (1991) states that flower stalks covered with a slippery wax layer are impassable for ants. Actually, it has been shown that ants are able to pass such barriers but prefer not to do so if they have alternatives (Gorb and Gorb 2011). It has also been proposed that trichomes may serve similar functions (Junker et al. 2011a).

The ecological function of floral scents exceeds the purpose of attracting pollinators by providing a defence mechanism against floral antagonists such as ants (Junker 2016). In behavioural bioassays using various methodological approaches it was repeatedly shown that ants are repelled by floral scents of plants from a broad taxonomic spectrum (Junker and Blüthgen 2008; Willmer et al. 2009; Galen et al. 2011). These results indicate that ants exert selective pressure on floral scent emissions, particularly in plant species that have flowers with no morphological barriers (Willmer et al. 2009; Galen et al. 2011; Junker et al. 2011a). Olfactometer tests suggest that common floral monoterpenes (Knudsen et al. 2006) are often responsible for ant repellence (Junker and Blüthgen 2008; Junker and Blüthgen 2010b; Junker et al. 2011b). Similarly, compounds originating from other biochemical pathways may be ant repellents (Junker and Blüthgen 2010a).

Island plants that evolved in the absence of ants usually lack morphological barriers, unpalatable nectar and repellent floral scents. They are thus particularly vulnerable to invasive ants that frequently deplete all available nectar (Junker et al. 2011a). In Hawaii, introduced plant species usually either have morphological barriers or repellent floral scent resulting in very high visitation rates to endemic and native plant species compared to introduced flowering plants (Junker et al. 2011a) (Figure 13.2). Interestingly, plant species endemic to the Hawaiian islands emit less monoterpenes on average than invasive plants species (Llusia et al. 2010), which could explain the lower potential of endemic plants to repel ants.

A Timeline of Ant Invasions with Reference to Plant-Pollinator Interactions

Oceanic islands often have disharmonic and species-poor assemblages of plants and animals (MacArthur and Wilson 1967; Denslow 2003) rendering island biotas susceptible to the establishment and spread of exotic species. Human activities, such as the global transportation system, allow many tramp species to reach remote areas where they may establish and eventually become invasive (Mooney et al. 2005). The first few weeks after arrival on a new island are crucial for successful establishment. Ant species reach islands either as founding queens or as small colonies, whereas the latter is more likely to establish successfully in a new habitat (Hee et al. 2000). In both cases the propagules must find a suitable nesting site to found a first colony. Later growth and spread of the naturalised population is then initiated by this founder colony. Ant species known to be invasive in many parts of the world usually have generalised nesting habits, allowing them to establish in a variety of habitats (Holway et al. 2002). Furthermore, invasive ant species have the strong ability to discover and recruit new food sources and to dominate these occupied resources over a long period of time (Krushelnycky et al. 2010). Their ability in discovering *and* dominating resources contributes to the success of invasive ants over native ant species that are typically more specialised in either discovering *or* dominating resources (Holway 1999). The competitive dominance of invasive ants facilitates their monopolisation of food resources after successful colony establishment and growth (Drescher et al. 2011). At an early stage, however, the impact of introduced ants is locally restricted within the range of action of a single colony (Haines and Haines 1978a). As a result of heavy exploitation of floral nectar and imposing strong interference competition on plant mutualists (e.g., native pollinators; Figure 13.3), invasive ants may then begin to affect the reproductive success of plant individuals in close vicinity to the founding colony.

Ants feature independent or dependent colony foundation, which determines the speed of spread of species within landscapes (Hölldobler and Wilson 1990). For the latter, expansion may occur *via* 'budding', i.e., mated queens leave the natal nest with a number of workers and establish a new but dependent nest nearby their natal nest. Individuals from the natal and the new nest usually tolerate each other, and together these unicolonial assemblages of mutually tolerant nests form supercolonies (Giraud et al. 2002; Tsutsui and Suarez 2003). While independent colony foundation is more commonly found across the Formicidae, ant species are more likely to become invasive if they spread through dependent colony foundation

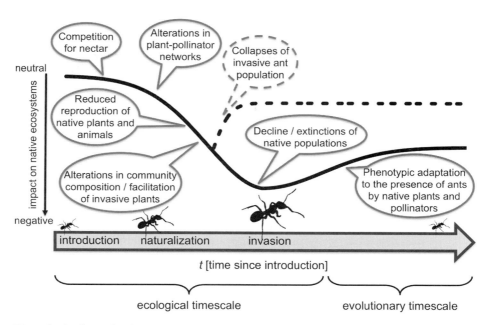

Figure 13.3. Hypothesis about the chronosequence of ant invasions with reference to plant-pollinator interactions in ecological and evolutionary timescales. The solid line represents the impact of introduced ants on native ecosystems as a function of the establishment of the invasive ant species. The dashed line illustrates a case of potential declines of invasive ant populations.

(Krushelnycky et al. 2010). Some invasive species such as yellow crazy ants even disperse in both modes, which allows local dominance by multi-nest colonies and fast long-distance dispersal (Krushelnycky et al. 2010). The polygynous supercolonies thus reach abundances that allow them to exploit, monopolise and deplete all resources available in habitat with consequences for native flora and fauna (Figure 13.3).

Although dependent dispersal of invasive ants is rather slow with a maximum of some hundreds of metres per year, invaded areas are promptly and severely affected by ants that occur in high densities (Holway et al. 2002; Krushelnycky et al. 2010). Dependent dispersal of ants through landscapes results in a clear invasion front, demarcating a well-defined border between invaded and ant-free areas. Consequently, the impact that invasive ants inflict on native communities can also vary greatly at a relatively short distance, depending on whether certain parts of the vegetation are in the invaded or the ant-free zone. In Hawaii, for example, the activity by endemic *Hylaeus* bees on native flowers was close to zero within sites that were heavily invaded by ants and increased at the periphery and towards the non-invaded area (Junker and Blüthgen, unpublished data). It is likely that the bees were locally extinct in the centre of the invaded area and foraged in the periphery, while nesting outside the invaded range. As *Hylaeus* bees nest in pre-existing cavities in the ground or in plants (Daly and Magnacca 2003), their brood is easily access-ible to ants supporting the notion that ants displaced these bees in invaded areas (although their impact is debated; see Magnacca (2007)). At an invasion front (close to the Hilina Pali road within the Hawaii Volcanoes National Park) of Argentine ants, a species with dependent colony dispersal, immediate effects were observed of ant invasion on visits of *Hylaeus* bees to flowers of *M. polymorpha* (Junker, unpublished data). We checked 41 flowering trees in front of and behind the invasion of Argentine ants for the presence of ants and *Hylaeus* bees on flow-ers. On 6 trees we observed neither ants nor bees; on the remaining 35 trees we mostly observed either ants foraging for nectar (11 trees) or *Hylaeus* bees collect-ing pollen (13 trees). On 11 trees both ants and bees collected floral resources, yet usually not on the same inflorescences (in 2 cases only). These observations suggest strong avoidance of ants by bees, with potential negative consequences for both the endemic bee population and the long-term viability of *M. polymorpha* (Figure 13.3). Invasion fronts offer a unique opportunity to study localised effects of ant invasion on the native ecosystem in a natural experimental setup with invaded and non-invaded sites.

Native ecosystems may recover from, or at least mitigate, the negative effects of invasive ants on an either ecological or evolutionary timescale (Figure 13.3). In several cases, colonies of invasive ants have disappeared locally after a few dec-ades of high population densities, a population dynamic observed also from other invasive invertebrates (Simberloff and Gibbons 2004). Although the causes of such collapses have not been clearly identified, a few explanations were brought forward, including inbreeding depression and Allee effects (Taylor and Hastings 2005; Vogel et al. 2010), perhaps associated with prey depletion (Haines and Haines 1978b) and parasites and pathogens (Vogel et al. 2009; Morrison 2012). Alternatively, they are outcompeted by other, more recently invading ant species (LeBrun et al. 2014). Populations of invasive ants in new regions are typically characterised by very nar-row genetic bottlenecks, often based on a single foundress or nest as a starting point of the population (Cooling and Hoffmann 2015). Notably, islands – where such bottlenecks may be particularly severe – represent the best-documented examples of such invasive ant declines. Yellow crazy ants have declined strongly on Mahé island, Seychelles, within five years (Haines and Haines 1978b). One haplotype of the same species has declined after seven years in the Tokelau archipelago (Gruber et al. 2013). In New Zealand, where high densities of Argentine ants have established from a single introduced nest, colonies have collapsed within two decades and dis-appeared from many invaded regions (Cooling et al. 2012). Some prey populations may recover after temporary declines in invasive ants, e.g., land crabs on Christmas Islands which had suffered from yellow crazy ants (Abbott et al. 2014). For other victims, crashes of biological invasions may be too late for recovery (Simberloff and Gibbons 2004). Depending on the islands' ecological degradation and the climatic niche of the invasive species, climate warming may either promote or inhibit the success of invasive ants. Whereas the survival of Argentine ant nests was higher in warmer regions in New Zealand (Cooling et al. 2012), modelling suggests a future decline of big-headed ants particularly in Oceania (Bertelsmeier et al. 2013) – a fate shared with many native species.

The rise and fall of invasive ant populations and eco-evolutionary causes and consequences may be explained by the 'taxon cycle hypothesis' (Wilson 1961). This hypothesis states that introduced ants are initially well adapted to spread in non-native habitats. Subsequently, the populations decline in abundance due to ecological specialisation and associated increased inter-specific competition with either native or freshly introduced species that are now about to invade the area (Economo and Sarnat 2012). Alternatively, counter-adaptations of native plant and animal species responding to the dominance of the invaders may cause the decline in invasive ant populations (Figure 13.3). We are not aware of a study that specifically tested this hypothesis in relation to pollination ecology, but the strong selective pressure of invasive ants on native organisms (Langkilde 2009) suggests the potential for evolutionary adaptations of flowering plants and pollinators. In theory, native organisms may evolve mechanisms to avoid, tolerate or even benefit from invasive ants. Selection may favour plant individuals that either have floral morphological barriers or emit ant-repellent floral scents to avoid encounters with ants (compare to Junker et al. 2011a). Alternatively, a larger spatial separation of nectar and the reproductive organs anthers and stigmas could help the plants to tolerate ant visits on flowers if ants and pollinators partition floral resources (compare to Junker et al. 2010). Pollinator species may likewise adapt to the presence of ants by avoidance or toleration strategies. Finally, plants may even benefit from ants with increased reproductive success either because ants reduce the impact of herbivores or floral parasites (compare to Bleil et al. 2011) or plants may utilise ants as pollen vectors. An example beyond plant-pollinator interactions reveals the potential of native organisms to respond with phenotypic adaptations towards ant invasions. Fence lizards (*Sceloporus undulatus*; Phrynosomatidae) native to the United States adapted behaviourally and morphologically to the presence of invasive fire ants (*Solenopsis invicta*). The likelihood that adult lizard fled after encounters with *S. invicta* increased with the time since the habitat was invaded by the ants and the relative hind limb length also positively correlated with time since invasion (Langkilde 2009). Most interestingly, these adaptations occurred within less than 40 generations (70 years) suggesting both a strong selection by ants and a strong potential of adaptations to invaders. Future studies may investigate phenotypic adaptations in plants and pollinators that may provide a chance to cope with the negative impacts of invasive ants.

The impact of invasive ants on pollination may be twofold: direct effects on diversity and community composition of pollinator species through predation, and indirect interference and exploitation competition and cascading effects on plants mediated by declining pollinator activities. Where both endemism and invasions are pronounced, endemic communities may be particularly susceptible, prone to declines or even extinctions. Impacts of ant invasions are thus likely to increase with the degree of plant endemism (per cent species endemic to that island or endemism richness), corresponding to the degree of isolation over evolutionary

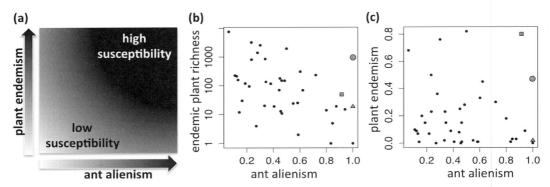

Figure 13.4. (a) A schematic hypothesis about how islands vary in their susceptibility of native plant-pollinator communities against invasive ants, depending on plant endemism and in ant alienism. Islands differ in both parameters, i.e., plant endemism and ant alienism: (b) number of endemic plant species plotted against relative number of introduced ant species across 40 islands, and (c) proportion of endemic plants versus proportion of introduced ants. Large grey circle: Hawaii, triangle: Turks and Caicos, square: Virgin Islands. Note that these three islands change their relative 'susceptibility' positions depending on the choice of criteria such as absolute versus relative number of endemic plant species (b vs. c).

timescales and associated with potential losses of defences. The impact of ants may be relatively low on islands where native or endemic ants have played a significant role in the evolutionary past, and will increase with the relative proportion or absolute number of ant species that have recently invaded the island (Figure 13.4). We thus hypothesise that both plant endemism and ant alienism are among the most important determinants of the (partly irreversible) impact on plant-pollinator interactions.

In conclusion, island ecosystems are particularly susceptible to negative impacts of biological invasions due to their evolutionary and biogeographic history and the lack of important functional groups, such as predatory mammals, prior to recent introductions by humans. Similarly, ants and other social insects have often been absent or relatively species-poor on remote oceanic islands before human colonisation. Today, ants are present on virtually all oceanic islands and represent a serious threat to native island faunas, floras and their associated ecosystem functions. Ants are frequent flower visitors, which are generally considered harmful for pollination. In mainland biotas flowers have evolved multiple defence mechanisms, such as narrow tubes or other structures to obstruct access to floral resources and floral scents to repel ants. Invasive ants may thus disrupt pollination of endemic island plants that lack defences against ants – unlike their mainland counterparts. Ant interactions with pollinators, however, are complex and do not always translate into a net negative effect on pollination. Interactions between invasive ants, native flowers and their visitors are widespread on oceanic islands, but generalisations require more systematic comparisons in the future.

Oceanic island ecosystems provide the largest global 'experiment' with spatially independent replicates to test the direct and indirect effects of ant invasions on ecosystem functions. Our increased knowledge of the ants' impacts and their dynamics will hopefully contribute to effectively control the species and to restore invaded ecosystems.

References

Abbott, K. L. & P. T. Green. (2007). Collapse of an ant-scale mutualism in a rainforest on Christmas Island. *Oikos* 116: 1238–1246.

Abbott, K. L., P. T. Green & D. O'Dowd. (2014). Seasonal shifts in macronutrient preferences in supercolonies of the invasive yellow crazy ant *Anoplolepis gracilipes* (Smith, 1857) (Hymenoptera: Formicidae) on Christmas Island, Indian Ocean. *Austal Entomology* 53: 337–346.

Altshuler, D. L. (1999). Novel interactions of non-pollinating ants with pollinators and fruit consumers in a tropical forest. *Oecologia* 119: 600–606.

Anderson, G. J., G. Bernardello, T. F. Stuessy & D. J. Crawford. (2001). Breeding system and pollination of selected plants endemic to Juan Fernández Islands. *American Journal of Botany* 88: 220–233.

Armbruster, W., C. Pelabon, T. Hansen & C. Mulder. (2004). Floral integration and modularity: distinguishing complex adaptations from genetic constraints. In M. Pigliucci & K. Preston, eds. *Phenotypic integration: studying the ecology and evolution of complex phenotypes*. Oxford: Oxford University Press, pp. 23–49.

Baker, H. G. (1967). Support for Baker's law – as a rule. *Evolution* 21: 853–856.

Barrett, S. C. H. (1996). The reproductive biology and genetics of island plants. *Philosophical Transactions of the Royal Society of London Series B-Biological Sciences* 351: 725–733.

Bawa, K. S. (1982). Outcrossing and the incidence of dioecism in island floras. *American Naturalist* 119: 866–871.

Beattie, A. J. (2006). The evolution of ant pollination systems. *Botanische Jahrbücher Systematik* 127: 43–55.

Bertelsmeier, C., G. M. Luque & F. Courcham. (2013). Global warming may freeze the invasion of big-headed ants. *Biological Invasions* 15: 1561–1572.

Bleil, R., N. Blüthgen & R. R. Junker. (2011). Ant-plant mutualism in Hawaii? Invasive ants reduce flower parasitism but also exploit floral nectar of the endemic shrub *Vaccinium reticulatum* (Ericaceae). *Pacific Science* 65: 291–300.

Bond, W. & P. Slingsby. (1984). Collapse of an ant-plant mutualism: the Argentine ant (*Iridomyrmex humilis*) and myrmeco-chorous Proteaceae. *Ecology* 65: 1031–1037.

Boyer, A. G. & W. Jetz. (2014). Extinctions and the loss of ecological function in island bird communities. *Global Ecology and Biogeography* 23: 679–688.

Brühl, C. A., G. Gunsalam & K. E. Linsenmair. (1998). Stratification of ants (Hymenoptera, Formicidae) in a primary rain forest in Sabah, Borneo. *Journal of Tropical Ecology* 14: 285–297.

Carlquist, S. (1974). *Island biology*. New York: Columbia University Press.

Caujape-Castells, J., A. Tye, D. J. Crawford, A. Santos-Guerra et al. (2010). Conservation of oceanic island floras: present and future global challenges. *Perspectives in Plant Ecology, Evolution and Systematics* 12: 107–129.

Chown, S. L., N. J. M. Gremmen & K. J. Gaston. (1998). Ecological biogeography of Southern Ocean Islands: species-area relationships, human impacts, and conservation. *The American Naturalist* 152: 562–575.

Cole, F. R., A. C. Medeiros, L. L. Loope & W. W. Zuehlke. (1992). Effects of the Argentine ant on arthropod fauna of Hawaiian high-elevation shrubland. *Ecology* 73: 1313–1322.

Cooling, M., S. Hartley, D. A. Sim & P. J. Lester. (2012). The widespread collapse of an invasive species: Argentine ants (*Linepithema humile*) in New Zealand. *Biology Letters* 8: 430–433.

Cooling, M. & B. D. Hoffmann. (2015). Here today, gone tomorrow: declines and local extinctions of invasive ant populations in the absence of intervention. *Biological Invasions* 17: 3351–3357.

Czechowski, W., B. Marko, K. Erős & E. Csata. (2011). Pollenivory in ants (Hymenoptera: Formicidae) seems to be much more common than it was thought. *Annales Zoologici* 61: 519–525.

Daly, H. V. & K. N. Magnacca. (2003). *Insects of Hawaii*. Honolulu, HI: University of Hawaii Press.

Davis, N. E., D. J. O'Dowd, R. Mac Nally & P. T. Green. (2009). Invasive ants disrupt frugivory by endemic island birds. *Biology Letters*: rsbl20090655.

Denslow, J. S. (2003). Weeds in paradise: thoughts on the invasibility of tropical islands. *Annals of the Missouri Botanical Garden* 90: 119–127.

Drescher, J., H. Feldhaar & N. Blüthgen. (2011). Interspecific aggression, resource monopolization and ecological dominance of *Anoplolepis gracilipes* within an ant community in Malaysian Borneo. *Biotropica* 43: 93–99.

Dupont, Y. L., D. M. Hansen & J. M. Olesen. (2003). Structure of a plant-flower-visitor network in the high-altitude sub-alpine desert of Tenerife, Canary Islands. *Ecography* 26: 301–310.

Dupont, Y. L., D. M. Hansen, A. Valido & J. M. Olesen. (2004). Impact of introduced honey bees on native pollination interactions of the endemic *Echium wildpretii* (Boraginaceae) on Tenerife, Canary Islands. *Biological Conservation* 118: 301–311.

Economo, E. P. & E. M. Sarnat. (2012). Revisiting the ants of Melanesia and the taxon cycle: historical and human-mediated invasions of a tropical archipelago. *The American Naturalist* 180: E1–E16.

Edwards, P. J., F. Fleischer-Dogley & C. N. Kaiser-Bunbury. (2015). The nutrient economy of *Lodoicea maldivica*, a monodominant palm producing the world's largest seed. *New Phytologist* 206: 990–999.

Emerson, B. C. & N. Kolm. (2005). Species diversity can drive speciation. *Nature* 434: 1015–1017.

Fisher, B. L. (1997). Biogeography and ecology of the ant fauna of Madagascar (Hymenoptera: Formicidae). *Journal of Natural History* 31: 269–302.

Gaigher, R. & M. J. Samways. (2013). Strategic management of an invasive ant-scale mutualism enables recovery of a threatened tropical tree species. *Biotropica* 45: 128–134.

Galen, C. & J. Cuba. (2001). Down the tube: pollinators, predators, and the evolution of flower shape in the Alpine Skypilot, *Polemonium viscosum*. *Evolution* 55: 1963–1971.

Galen, C., R. Kaczorowski, S. L. Todd, J. Geib, et al. (2011). Dosage-dependent impacts of a floral volatile compound on pollinators, larcenists, and the potential for floral evolution in the Alpine Skypilot *Polemonium viscosum*. *The American Naturalist* 177: 258–272.

Gaume, L., M. Zacharias & R. M. Borges. (2005). Ant-plant conflicts and a novel case of castration parasitism in a myrmecophyt. *Evolutionary Ecology Research* 7: 435–452.

Gillespie, R. G. & G. K. Roderick. (2002). Arthropods on islands: colonization, speciation, and conservation. *Annual Review of Entomology* 47: 595–632.

Giraud, T., J. S. Pedersen & L. Keller. (2002). Evolution of supercolonies: the Argentine ants of southern Europe. *Proceedings of the National Academy of Sciences of the United States of America* 99: 6075–6079.

Gonzalvez, F. G., L. Santamaria, R. T. Corlett & M. A. Rodriguez-Girones. (2013). Flowers attract weaver ants that deter less effective pollinators. *Journal of Ecology* 101: 78–85.

Gorb, E. & S. Gorb. (2011). How a lack of choice can force ants to climb up waxy plant stems. *Arthropod-Plant Interactions* 5: 297–306.

Gruber, M. A. M., A. R. Burne, K. L. Abbott, R. J. Pierce et al. (2013). Population decline but increased distribution of an invasive ant genotype on a Pacific atoll. *Biological Invasions* 15: 599–612.

Haber, W. A., G. W. Frankie, H. G. Baker, I. Baker et al. (1981). Ants like flower nectar. *Biotropica* 13: 211–214.

Haines, I. H. & J. B. Haines. (1978a). Colony structure, seasonality and food-requirements of crazy ant. *Ecological Entomology* 3: 109–118.

   (1978b). Pest status of the crazy ant, *Anoplolepis longipes* (Jerdon) (Hymenoptera: Formicidae), in the Seychelles. *Bulletin of Entomological Research* 68: 627–638.

Hanna, C., D. Foote & C. Kremen. (2013). Invasive species management restores a plant-pollinator mutualism in Hawaii. *Journal of Applied Ecology* 50: 147–155.

Hanna, C., I. Naughton, C. Boser, R. Alarcon et al. (2015). Floral visitation by the Argentine ant reduces bee visitation and plant seed set. *Ecology* 96: 222–230.

Hansen, D. M. (2015). Non-native megaherbivores: the case for novel function to manage plant invasions on islands. *AoB Plants* 7: plv085; doi:10.1093/aobpla/plv085.

Hansen, D. M. & C. B. Müller. (2009). Invasive ants disrupt gecko pollination and seed dispersal of the endangered plant *Roussea simplex* in Mauritius. *Biotropica* 41: 202–208.

Hansen, D. M., J. M. Olesen & C. G. Jones. (2002). Trees, birds and bees in Mauritius: exploitative competition between introduced honey bees and endemic nectarivorous birds? *Journal of Biogeography* 29: 721–734.

Harley, R. (1991). The greasy pole syndrome. In C. R. Huxley & D. F. Cutler, eds. *Ant-plant interactions*. Oxford: Oxford University Press, pp. 430–433.

Harper, G. A. & N. Bunbury. (2015). Invasive rats on tropical islands: their population biology and impacts on native species. *Global Ecology and Conservation* 3: 607–627.

Hee, J. J., D. A. Holway, A. V. Suarez & T. J. Case. (2000). Role of propagule size in the success of incipient colonies of the invasive Argentine ant. *Conservation Biology* 14: 559–563.

Herrera, C. M., J. Herrera & X. Espadaler. (1984). Nectar thievery by ants from southern Spanish insect-pollinated flowers. *Insectes Sociaux* 31: 142–154.

Hickman, J. C. (1974). Pollination by ants – low-energy system. *Science* 184: 1290–1292.

Hill, M., K. Holm, T. Vel, N. J. Shah et al. (2003). Impact of the introduced yellow crazy ant *Anoplolepis gracilipes* on Bird Island, Seychelles. *Biodiversity and Conservation* 12: 1969–1984.

Hingston, A. B. & P. B. McQuillan. (1999). Displacement of Tasmanian native megachilid bees by the recently introduced bumblebee *Bombus terrestris* (Linnaeus, 1758) (Hymenoptera: Apidae). *Australian Journal of Zoology* 47: 59–65.

Hölldobler, B. & E. O. Wilson. (1990). *The ants*. Cambridge, MA: Harvard University Press.

Holway, D. A. 1999. Competitive mechanisms underlying the displacement of native ants by the invasive Argentine ant. *Ecology* 80: 238–251.

Holway, D. A., L. Lach, A. V. Suarez, N. D. Tsutsui et al. (2002). The causes and consequences of ant invasions. *Annual Review of Ecology and Systematics* 33: 181–233.

Janicki, J., N. Narula, M. Ziegler, B. Guénard et al. (2016). Visualizing and interacting with large-volume biodiversity data using client-server web-mapping applications: the design and implementation of antmaps.org. *Ecological Informatics* 32: 185–193.

Janzen, D. H. (1977). Why don't ants visit flowers? *Biotropica* 9: 252.

Junker, R. R. (2016). Multifunctional and diverse floral scents mediate biotic interactions embedded in communities. In J. D. Blande & R. T. Glinwood, eds. *Deciphering chemical language of plant communication*. Heidelberg: Springer, pp. 257–282.

Junker, R. R., R. Bleil, C. C. Daehler & N. Blüthgen. (2010). Intra-floral resource partitioning between endemic and invasive flower visitors: consequences for pollinator effectiveness. *Ecological Entomology* 35: 760–767.

Junker, R. R. & N. Blüthgen. (2008). Floral scents repel potentially nectar-thieving ants. *Evolutionary Ecology Research* 10: 295–308.

(2010a). Dependency on floral resources determines the animals' responses to floral scents. *Plant Signaling and Behavior* 5: 1014–1016.

(2010b). Floral scents repel facultative flower visitors, but attract obligate ones. *Annals of Botany* 105: 777–782.

Junker, R. R., C. C. Daehler, S. Dötterl, A. Keller et al. (2011a). Hawaiian ant-flower networks: nectar-thieving ants prefer undefended native over introduced plants with floral defenses. *Ecological Monographs* 81: 295–311.

Junker, R. R., J. Gershenzon & S. B. Unsicker. (2011b). Floral odour bouquet loses its ant repellent properties after inhibition of terpene biosynthesis. *Journal of Chemical Ecology* 37: 1323–1331.

Kaiser-Bunbury, C. & N. Blüthgen. (2015). Integrating network ecology with applied conservation: a synthesis and guide to implementation. *AoB Plants* 7: plv076.

Kaiser-Bunbury, C. N., H. Cuthbert, R. Fox, D. Birch et al. (2014). Invasion of yellow crazy ant *Anoplolepis gracilipes* in a Seychelles UNESCO palm forest. *Neobiota* 22: 43–57.

Kaiser-Bunbury, C. N., A. Traveset & D. M. Hansen. (2010). Conservation and restoration of plant-animal mutualisms on oceanic islands. *Perspectives in Plant Ecology Evolution and Systematics* 12: 131–143.

Kaiser-Bunbury, C. N., T. Valentin, J. Mougal, D. Matatiken et al. (2011). The tolerance of island plant-pollinator networks to alien plants. *Journal of Ecology* 99: 202–213.

Kalinganire, A., C. E. Harwood, M. U. Slee & A. J. Simons. (2001). Pollination and fruit-set of *Grevillea robusta* in western Kenya. *Austral Ecology* 26: 637–648.

Karnauskas, K. B., J. P. Donnelly & K. J. Anchukaitis. (2016). Future freshwater stress for island populations. *Nature Climate Change*: doi:10.1038/nclimate2987.

Kato, M. & A. Kawakita. (2004). Plant-pollinator interactions in New Caledonia influenced by introduced honey bees. *American Journal of Botany* 91: 1814–1827.

Kato, M., A. Shibata & T. Yasui. (1999). Impact of introduced honeybees, *Apis mellifera*, upon native bee communities in the Bonin (Ogasawara) Islands. *Research of Population Ecology* 41: 217–228.

Kier, G., H. Kreft, T. M. Lee, W. Jetz et al. (2009). A global assessment of endemism and species richness across island and mainland regions. *Proceedings of the National Academy of Sciences* 106: 9322–9327.

Knudsen, J. T., R. Eriksson, J. Gershenzon & B. Stahl. (2006). Diversity and distribution of floral scent. *Botanical Review* 72: 1–120.

Kreft, H., W. Jetz, J. Mutke, G. Kier et al. (2008). Global diversity of island floras from a macroecological perspective. *Ecology Letters* 11: 116–127.

Kruckeberg, A. R. & D. Rabinowitz. (1985). Biological aspects of endemism in higher plants. *Annual Review of Ecology and Systematics* 16: 447–479.

Krushelnycky, P. D. & R. G. Gillespie. (2008). Compositional and functional stability of arthropod communities in the face of ant invasions. *Ecological Applications* 18: 1547–1562.

Krushelnycky, P. D., D. Holway & E. G. LeBrun. (2010). Invasion process and causes of success. In L. Lach, C. L. Parr & K. L. Abbott, eds. *Ant ecology.* Oxford: Oxford University Press, pp. 115–136.

Krushelnycky, P. D., L. L. Loope & N. J. Reimer. (2005). The ecology, policy, and management of ants in Hawaii. *Proceedings of the Hawaiian Entomological Society* 37: 1–25.

Lach, L. (2008a). Argentine ants displace floral arthropods in a biodiversity hotspot. *Diversity and Distributions* 14: 281–290.

(2008b). Floral visitation patterns of two invasive ant species and their effects on other hymenopteran visitors. *Ecological Entomology* 33: 155–160.

Lach, L & L. M. Hooper-Bui. (2010). Consequences of ant invasions. In L. Lach, C. L. Parr & K. L. Abbott, eds. *Ant ecology.* Oxford: Oxford University Press, pp. 261–286.

Langkilde, T. (2009). Invasive fire ants alter behavior and morphology of native lizards. *Ecology* 90: 208–217.

LeBrun, E. G., N. T. Jones & L. E. Gilbert. (2014). Chemical warfare among invaders: a detoxification interaction facilitates an ant invasion. *Science* 343: 1014–1017.

Llusia, J., J. Penuelas, J. Sardans, S. M. Owen et al. (2010). Measurement of volatile terpene emissions in 70 dominant vascular plant species in Hawaii: aliens emit more than natives. *Global Ecology and Biogeography* 19: 863–874.

Lord, J. M. (2015). Patterns in floral traits and plant breeding systems on Southern Ocean Islands. *AoB Plants* 7.

Lubin, Y. D. (1984). Changes in the native fauna of the Galapagos Islands following invasion by the little red fire ant, *Wasmannia auropunctata. Biological Journal of the Linnean Society* 21: 229–242.

MacArthur, R. H. & E. O. Wilson. (1967). *The theory of island biogeography.* Princeton: Princeton University Press.

Magnacca, K. N. (2007). Conservation status of the endemic bees of Hawai'i, *Hylaeus* (*Nesoprosopis*) (Hymenoptera: Colletidae). *Pacific Science* 61: 173–190.

McGlynn, T. P. (1999). The worldwide transfer of ants: geographical distribution and ecological invasions. *Journal of Biogeography* 26: 535–548.

Medeiros, A. C., L. L. Loope & F. R. Cole. (1986). Distribution of ants and their effects on endemic biota of Haleakala and Hawaii Volcanoes National Park: a preliminary assessment. In C. W. Smith & C. P. Stone, eds. *Proceedings 6th conference in natural sciences.* Hawaii Volcanoes National Park. Honolulu. Honolulu, HI: University of Hawaii, pp. 39–52.

Medina, F. M., E. Bonnaud, E. Vidal, B. R. Tershy et al. (2011). A global review of the impacts of invasive cats on island endangered vertebrates. *Global Change Biology* 17: 3503–3510.

Moller, H. (1996). Lessons for invasion theory from social insects. *Biological Conservation* 78: 125–142.

Mooney, H. A., R. N. Mack, J. A. McNeely, L. E. Neville et al. (eds.) (2005). *Invasive alien species.* Washington, Covelo, London: Island Press.

Morrison, L. W. 2012. Biological control of *Solenopsis* fire ants by *Pseudacteon* parasitoids: theory and practice. *Psyche: A Journal of Entomology* : 2012: Article ID 424817, http://dx.doi.org/10.1155/2012/424817.

Myers, N., R. A. Mittermeier, C. G. Mittermeier, G. A. B. da Fonseca et al. (2000). Biodiversity hotspots for conservation priorities. *Nature* 403: 853–858.

Ness, J. H. (2006). A mutualism's indirect costs: the most aggressive plant bodyguards also deter pollinators. *Oikos* 113: 506–514.

Ness, J. H. & J. L. Bronstein. (2004). The effects of invasive ants on prospective ant mutualists. *Biological Invasions* 6: 445–461.

Newstrom, L. & A. Robertson. (2005). Progress in understanding pollination systems in New Zealand. *New Zealand Journal of Botany* 43: 1–59.

O'Dowd, D. J., P. T. Green & P. S. Lake. (2003). Invasional 'meltdown' on an oceanic island. *Ecology Letters* 6: 812–817.

Philipp, M., J. Bocher, H. R. Siegismund & L. R. Nielsen. (2006). Structure of a plant-pollinator network on a pahoehoe lava desert of the Galapagos Islands. *Ecography* 29: 531–540.

Pijl, L. v. d. (1955). Some remarks on myrmecophytes. *Phytomorphology* 5: 190–200.

Quilichini, A. & M. Debussche. (2000). Seed dispersal and germination patternsin a rare Mediterranean island endemic (*Anchusa crispa* Viv., Boraginaceae). *Acta Oecologica* 21: 303–313.

Rambuda, T. D. & S. D. Johnson. (2004). Breeding systems of invasive alien plants in South Africa: does Baker's rule apply? *Diversity & Distributions* 10: 409–416.

Rodriguez-Girones, M. A., F. G. Gonzalvez, A. L. Llandres, R. T. Corlett et al. (2013). Possible role of weaver ants, Oecophylla smaragdina, in shaping plant-pollinator interactions in South-East Asia. *Journal of Ecology* 101: 1000–1006.

Sakai, A. K., W. L. Wagner, D. M. Ferguson & D. R. Herbst. (1995). Biogeographical and ecological correlates of dioecy in the Hawaiian flora. *Ecology* 76: 2530–2543.

Sarnat, E. M. & E. P. Economo. (2012). *The ants of Fiji.* Berkeley: University of California Press.

Simberloff, D. & L. Gibbons. (2004). Now you see them, now you don't! – population crashes of established introduced species *Biological Invasions* 6: 161–172.

Stang, M., P. G. Klinkhamer & E. Van Der Meijden. (2006). Size constraints and flower abundance determine the number of interactions in a plant-flower visitor web. *Oikos* 112: 111–121.

Taylor, C. M. & A. Hastings. (2005). Allee effects in biological invasions. *Ecology Letters* 8: 895–908.

Traveset, A. & D. M. Richardson. (2006). Biological invasions as disruptors of plant reproductive mutualisms. *Trends in Ecology & Evolution* 21: 208–216.

Tsuji, K., A. Hasyim, Harlion & K. Nakamura. (2004). Asian weaver ants, *Oecophylla smaragdina*, and their repelling of pollinators. *Ecological Research* 19: 669–673.

Tsutsui, N. D. & A. V. Suarez. 2003. The colony structure and population biology of invasive ants. *Conservation Biology* 17: 48–58.

Vogel, V., J. S. Pederson, P. d'Ettorre, L. Lehmann et al. (2009). Dynamics and genetic structure of Argentine ant supercolonies in their native range. *Evolution* 63: 1627–1639.

Vogel, V., J. S. Pederson, T. Giraud, M. Krieger et al. (2010). The worldwide expansion of the Argentine ant. *Diversity and Distributions* 16: 170–186.

Webb, C. J. & D. Kelly. (1993). The reproductive biology of the New Zealand flora. *Trends in Ecology & Evolution* 8: 442–447.

Whittaker, R. J. & J. M. Fernández-Palacios. (2007). *Island biogeography: ecology, evolution, and conservation.* Oxford: Oxford University Press.

Willmer, P. G., C. V. Nuttman, N. E. Raine, G. N. Stone et al. (2009). Floral volatiles controlling ant behaviour. *Functional Ecology* 23: 888–900.

Wilson, E. O. (1961). The nature of the taxon cycle in the Malanesian ant fauna. *The American Naturalist* 95: 169–193.

14 Mutualisms and the Reciprocal Benefits of Comparing Systems with Native and Introduced Ants

Joshua H. Ness and David A. Holway[*]

Introduction

Ant invasions corrupted the dissertation of one of the authors of this chapter. The proposed work focused on the interacting ecologies of nectary-bearing plants, ants, herbivores, and parasitic wasps in the southeastern United States. However, red imported fire ants (*Solenopsis invicta*) consumed some of the participants (larvae and pupae of both the herbivores and parasitoids), competitively excluded others (the native ant community) and displayed relative indifference to the rewards being provided by the plants. An era of profanity ensued as it became apparent that no plan survives contact with a swarm of stinging ants. The research begged the question of whether the invaded incarnations of the system provided any useful information about the functioning of more intact/native incarnations of that or other systems. The second author had a more productive experience with non-natives; introduced ants were a model system to explore competitive dynamics, susceptibility to invasion, and dispersal. In the intervening time, it has become increasingly apparent that species introductions into novel habitats provide excellent opportunities to understand fundamental phenomena and processes in ecology, evolution, and biogeography (e.g., Sax et al., 2005). Here, we focus on mutualisms – reciprocally beneficial interspecific interactions – and explore the ways that non-native ants have contributed to our understanding of mutualisms, the ways that an appreciation of the functioning of these relationships has informed our understanding of the processes that influence the distribution and consequence of invasions, and the application of insights from systems dominated by introduced species to systems with native ants.

The domination of certain locations by introduced ant species has provided some of the most powerful demonstrations of the importance of mutualism as a force that structures communities. For example, the invasion of the Argentine ant *Linepithema humile* into the fynbos ecosystems of South Africa has profound effects on the abundance,

[*] This synthesis was made possible by the contributions of W. F. Morris, J. L. Bronstein, M. C. Lanan, G. Fitzpatrick, J. Hung, K. Le Van, J. Ludka, K. McCann, and S. Menke. Further support was provided by the Faculty Development Committee and the Office of the Dean of Faculty at Skidmore College. The work was further improved by the comments of P. Oliveira, S. Koptur, and an anonymous reviewer.

spatial distribution, and species composition of seedlings (Bond and Slingsby, 1984; Christian, 2001) – a consequence of the diverging responses exhibited by Argentine ants and the native ant fauna when encountering fynbos seeds that are adapted for dispersal by ants. The introduction of these ants into the fynbos system demonstrated the role of the ant-seed mutualism in shaping the interactions of the seeds and seedlings with natural enemies and fire in a global biodiversity hotspot. The loss of function in that mutualism resulting from invasion made it clear that different species of ants were not interchangeable, even within the context of a generalised, facultative mutualism.

The disruption of seed dispersal mutualisms by Argentine ants encourages one to associate invasion with rapid qualitative changes in system functioning (see also O'Dowd et al., 2003). Although that interpretation is not inaccurate and certainly true of settings (e.g., Hawaii) that lacked ants beforehand, the effects of introduced ants can also be linked to traits that are found in natives and/or less disruptive ant species. Smaller ants are often relatively ineffectual seed dispersers, irrespective of origin vis-á-vis the local flora and, to the extent that invasion decreases the size of the median seed-dispersing ant in the local community, dispersal distances may decrease subsequent to invasion (Ness et al., 2004). There is a suite of traits shared by many of the most consequential introduced ants: omnivory, polymorphic workers, a tendency to form expansive supercolonies, the maintenance of multiple, ephemeral nests that can be relocated in response to changing conditions (threats or opportunities), and a real appetite for carbohydrates (Holway et al., 2002). These traits are overrepresented in the most ecologically disruptive of introduced ant species but certainly are not unique to them. Invasions help us to appreciate the consequences of a truncated community, wherein particular niches that may be predisposed towards effective participation in certain types of mutualisms (and ill-suited for others) become well represented and others are lost.

The invasion of islands in the south Pacific and Indian Oceans by *Anoplolepis gracilipes* provides perhaps the clearest demonstration of how a super-abundant, non-native ant, allied with carbohydrate-providing mutualists (Abbott and Green, 2007), can transform plant and animal communities (Hill et al., 2003; O'Dowd et al., 2003; Lester and Tavite, 2004; Davis et al., 2008; Chapter 13). This system represents the textbook case of invasional meltdown, wherein direct and indirect interactions between introduced species ultimately dissemble a community. *Anoplolepis gracilipes* acts as a mutualist of the non-native honeydew-producing hemipterans and facilitates a secondary invasion by giant African land snails by extirpating land crabs (Green et al., 2011). More generally, these studies illustrate the importance of facilitation as an ecological force. For example, some bird species, including the native Christmas Island White Eye (*Zosterops natalis*), experience increased foraging success in the environments transformed by the actions of *A. gracilipes* (Davis et al., 2008), and the exclusion of land crabs can increase seedling recruitment for some species (Green et al., 2008; see also Abbott and Green, 2007).

Invasions have provided opportunities to understand relatively fixed interactions in new ways. A defining feature of the ant-plant symbioses of Africa, southeastern Asia, and Central and South America is the tight link between host plant and a small coterie of specialised ants that occupy the plants in each region. For example, the whistling thorn acacia (*Acacia drepanolobium*) typically associates with a quartet of ant species that can be arrayed along a linear dominance hierarchy and whose short-term effects on the plant range from mutualistic to parasitic (Palmer et al., 2000; Palmer et al., 2010). As a result of the invasion of *Acacia*-dominated habitats by the big-headed ant, *Pheidole megacephala,* this obligate mutualism has a new participant, one that is simultaneously aggressive, competitively dominant, relatively insensitive to plant-produced resources (and perhaps less capable of stimulating reward production), and that indirectly fosters an increased fraction of the plants being occupied by *Tetraponera penzigi*, the least competitive, least aggressive ant in this symbiosis (Riginos et al., 2015). Although ant-plant symbioses can also be re-assembled in the opposite direction, as can occur when ant-plants are introduced into new locations (Wetterer and Wetterer, 2003), the invasion of the *Acacia drepanolobium* system by *Pheidole megacephala* is a particularly powerful example because it allows scientists to evaluate the consequences of the invasion on a greater variety of phenomena directly and indirectly associated with the prospective ant-plant mutualism (e.g., ant transitions within the system, economies of the food-for-protection trade, plant susceptibility to particular natural enemies, and the production of woody cover, forage, and coarse woody debris within the habitat).

There is a growing number of works that critically review the participation and effects of non-native ants in mutualisms (see Holway et al., 2002; Lach, 2003; Ness and Bronstein, 2004; Lach and Hooper-Bùi, 2010; Helms, 2013; Chapter 15). Given the pace and distribution of new and ongoing invasions, as well as the new scientific contributions derived from them, new syntheses will always be necessary and timely. However, our goal here is not to pursue that particular line of work. Rather, we explore how the study of ant invasions contributes to our understanding of prospective mutualisms and, more generally, ant-plant interactions. We know that an understanding of mutualism as a phenomenon puts researchers in a stronger position to describe and predict the distribution and consequence of invasions. We believe the corollary is also true; introduced ants can be models for understanding the ecology of interactions between native ants and prospective partners. Interspecific variation in the behaviour of ant bodyguards, for example, represents a spectrum with native and non-native ant species interspersed along axes relating to tending effectiveness, participation/disruption of other mutualisms, and broader ecological effects resulting from participation in the mutualism. We explore this premise using the interaction webs associated with extrafloral nectary-bearing cacti (*Ferocactus* spp.) and associated ant species as a model system. We start by identifying characteristics shared by the behaviourally dominant, opportunistic ants – introduced and native – that have joined two *Ferocactus*-centred systems, pivot to exploring ecological outcomes associated with the changes in ant community composition, and conclude by assessing phenomena that alter the stability of ant-plant interactions over long time frames.

Figure 14.1. *Linepithema humile* workers foraging at the extrafloral nectaries of *Ferocactus viridescens*. Plants produce carbohydrate-rich extrafloral nectar from modified spines at the crown on the plant, often in locations associated with the production of reproductive structures. The white depression at top centre indicates that a fruit was previously attached at that location. Photo credit: David A. Holway. (A black-and-white version of this figure will appear in some formats. For the color version, please refer to the plate section.)

The *Ferocactus* Systems

The model systems for the forthcoming discussion are *Ferocactus wislizeni* and *F. viridescens* cacti and the interactions of those two plant species with an assemblage of ant species that includes an ecologically opportunistic, behaviourally dominant ant (native southern fire ant *Solenopsis xyloni* and introduced Argentine ant *Linepithema humile*, respectively) as well as two *Crematogaster* species (*C. opuntia* and *C. californica*), which act as common and effective plant mutualists and that can be replaced by *S. xyloni* or *L. humile*. *Ferocactus* species are widespread throughout the arid regions of the southwestern United States and northern Mexico; *Ferocactus wislizeni* and *F. viridescens* are found in the Sonoran Desert and Mediterranean-climate shrub environments (respectively), and much of the work described in the sections that follow occurred in the vicinities of Tucson, Arizona, and San Diego, California.

Upon reaching sufficient size, both cactus species produce carbohydrate-rich extrafloral nectar from modified spines at the crown on the plant (Figure 14.1).

This resource can be sufficient to attract ants to some plants throughout the entire year – a degree of continuity that is unusual for facultative ant-plant mutualism. That *Ferocactus* can benefit from interactions with ants is evidenced by decreases in the presence and/or abundance of herbivorous coreid bugs in the presence of ants in both the *F. wislizeni* (Ness et al., 2006) and *F. viridescens* (Ludka et al., 2015) systems, as well as positive relationships between seed mass and/or fruit production with the protection conferred by the visiting ants (Ness et al., 2006; Ludka et al., 2015). Although the plant populations in both systems are collectively visited by a variety of ants, individual plants are usually visited only by one ant species at a time and the identity of that attendant species can remain consistent for months and sometimes even years (Morris et al., 2005; Ludka et al., 2015).

Research in both systems has focused on evaluating whether and how the consequences for plants are mediated by the identity of the ant species that visit the plant. As a result of the combined work of several research groups, the interactions between individual plants, ants, and other animals – including natural enemies as well as in some cases pollinators and frugivores – has been monitored in both *Ferocactus* systems with regular inspections for multiple years. These inspection intervals ranged from approximately weekly (*F. viridescens* from March 2007 to August 2008, biweekly for *F. wislizeni* from 2003 to 2005) to only once per season (approximately five times per year for *F. wislizeni* in 2006–2009 and 2011–2016). This chapter utilises observations from 2003 to 2016 for *F. wislizeni* and 2007–2011 for *F. viridescens*. Monitoring in the *F. viridescens* system occurred at six sites along the coastline north and south of La Jolla, California, and was organised to include plants tended by *L. humile* or native ants at each site (see LeVan et al., 2014; Ludka et al., 2015). The *F. wislizeni* system described here is a longitudinal study of all *F. wislizeni* individuals in three plots within 1.5 km of one another (see descriptions in Lanan and Bronstein, 2013). All three *F. wislizeni* plots are within the larger protected landscape of the Desert Laboratory in the Sonoran Desert, Arizona, and the western side of site 1 (where *S. xyloni* is most common) is at the edge of the protected landscape and adjacent to suburban development associated with the city of Tucson.

Although *S. xyloni* is native and *L. humile* is introduced to North America, the two species are similar in several respects. First, both can be synanthropic. *Linepithema humile* is the most commonly reported ant pest in or around human structures in regional surveys of the greater San Francisco, Los Angeles, and San Diego areas in California, and *S. xyloni* is the most commonly reported ant pest in these same habitats in the Central Valley of California and Phoenix, Arizona, and the second-most commonly reported in the Los Angeles area (Knight and Rust, 1990; Field et al., 2007; see also Smith, 1936). *Solenopsis xyloni* was absent or rare to the point of being undetectable in 2003 at the Desert Laboratory, and quickly gained control over most plants in one of three long-term study sites during an era that coincided with the suburbanisation of the adjacent landscape (Morris et al., 2005). Second, both species are adept at exploiting opportunities associated with environmental disturbances, such as elevated soil moisture levels

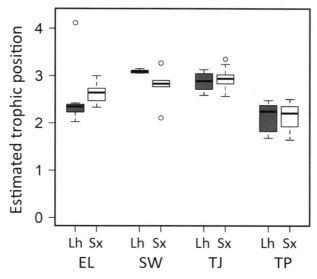

Figure 14.2. Box plots showing estimates of trophic position (1 – primary producer, 2 – herbivore, 3 – primary predator, 4 – secondary predator) for *Linepithema humile* (Lh; gray boxes) and *Solenopsis xyloni* (Sx; white boxes) from four different sites: University of California Elliott Chaparral Reserve (EL), Otay-Sweetwater Unit of the San Diego National Wildlife Refuge (SW), Tijuana River National Estuarine Research Reserve (TJ), and Torrey Pines State Reserve Extension (TP). Box plots show median (bold line), first and third quartiles, and minimum and maximum; unfilled circles represent outliers. Data are re-plotted from Menke et al. (2010).

(e.g., resulting from irrigation) in the case of the Argentine ant (Menke et al., 2007) and grazing and microsite disturbances caused by kangaroo rat excavations in the case of *S. xyloni* (Kerley and Whitford, 2000; Schooley et al., 2000). Third, both species are scavenging predators and can occupy similar trophic positions. Stable isotope comparisons, for example, at paired, invaded, and uninvaded sites revealed striking similarities in the trophic positions of both species (Figure 14.2); even seasonal variation in trophic position appeared somewhat correlated among months for the two species of ants (Menke et al., 2010). Finally, *S. xyloni* and *L. humile* share the distinction of having an abiotic niche that is dissimilar to most of the resident native species that visit *Ferocactus*. For example, the area-independent rate of water loss for *L. humile* is almost three times greater than that of *C. californica* (Schilman et al., 2007). The critical thermal maxima (temperature at which ants have a 0.5 probability to dying when confined) is several degrees lower for *S. xyloni* than *C. opuntia* (47.9 and 50.1°C, respectively), and the two ants have comparable odds of abandoning tended *F. wislizeni* when surface temperatures reach 40.9 versus 45.3°C in the field (Fitzpatrick et al., 2014). One consequence is that the aggregate niche of the ant communities becomes truncated in contexts when *S. xyloni* and *L. humile* competitively exclude other species such as *C. californica* and *C. opuntia*.

Invasion Outcomes: Foraging on Reward-Producing Plants

The spatiotemporal distribution of ant invasions is often positively associated with the presence and/or activities of reward-producing partners (e.g., *Solenopsis invicta* and reward-producing insects (Helms & Vinson, 2002; Styrsky and Eubanks, 2010); *Pheidole megacephala* with EFN-bearing plants (Hoffman et al., 1999) and reward-producing hemipterans (Gaigher et al., 2011); *Anoplolepis gracilepes* with reward-producing hemipterans (Tanaka et al., 2011), EFN-bearing plants (Savage et al., 2009; Lach and Hoffman, 2011), and artificial nectaries (Savage et al., 2011); *Myrmica rubra* with EFN-bearing *Fallopia japonica* (Ness et al., 2012); *Linepithema humile* with honeydew producers (Lach, 2007; Brightwell & Silverman, 2011), and many authors have reported changes in aggregate on-plant foraging by ants as an important outcome of ant invasions.

Here, we explore whether the invasion by native *S. xyloni* changes the phenology of ant-plant interactions in the community. Foraging by *S. xyloni* is strongly influenced by season, and epigeaic foragers can be rare for 3–5 months even in the relatively consistent and mild climates of coastal southern California (Knight and Rust, 1990; Hooper and Rust, 1997). In our Sonoran Desert study site, *S. syloni* is rarely detected at baits or on *F. wislizeni* plants between late winter and the dry summer (~January–June). To the extent that the other ant species fail to recover from the domination of the ant community by *S. xyloni* that occurs during the monsoon and autumn, we expect that plants that benefit from ant-provided protection in the monsoon and autumn (seasons aligned with the relatively fixed foraging dynamics of *S. xyloni*) will do well in *S. xyloni*-dominated settings, and those plants that require services in other seasons have the potential for diminished access to ant mutualists. These plants could include species that attract ants in the Sonoran Desert's dry summer (e.g., *Cylindropuntia versicolor, Opuntia engelmanni, Carnegiea gigantea, Stenocereus thurberi*) as well as species such as *F. wislizeni* that potentially interact with ants throughout the year.

Using the biweekly surveys of *F. wislizeni* from 2003 to 2005, we asked whether the incidence of tending by the trio of *C. opuntia, S. aurea* and *S. xyloni* differed between 2003 and 2004–2005 (the year of the invasion and subsequent years, respectively). All three of these species typically provide greater protection than do other ants in the community (see Ness et al., 2006). To the extent that the *S. xyloni* invasion only changes the relative proportion of tending of *F. wislizeni* among the three ants (and it clearly does; see Figure 14.4), tended plants may still be well protected from natural enemies insofar as the tending ants occur in sufficient abundance on the plants and the overall incidence of tending remains similar (Ness et al., 2006). However, we find that the phenology of on-plant tending changes rather significantly. Based on paired pre- and post-invasion comparisons of 80 *F. wislizeni* plants at the invaded site (site 1; Figure 14.3), the proportion of inspections that included the trio was lower in 2004–2005 than in 2003 in March–April (late spring; mean difference ± SE = –0.51 ± 0.03) was unchanged in May–July (dry summer until onset

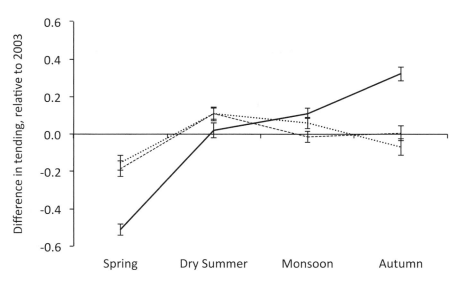

Figure 14.3.   Changes in phenology of plant tending by ants after invasion by the native southern fire ant, *Solenopsis xyloni*. Difference in proportions of plant inspections that documented tending of *Ferocactus wislizeni* by the trio of *S. xyloni*, *S. aurea* and/or *Crematogaster opuntiae* in 2003 versus 2004 and 2005 is shown. Solid, dotted, and dashed lines indicate sites 1, 2, and 3, respectively; site 1 was invaded by *S. xyloni* in 2003. Bars indicate standard errors, and are calculated for change in tending proportions of individual plants over time (n = 80, 67, and 54 plants, respectively).

of monsoon; mean difference = 0.02 ± 0.04), modestly increased in July–September (monsoon; mean difference = 0.11 ± 0.03) and substantially increased in October–December (autumn and early winter; mean difference = 0.32 ± 0.04). Although some of these differences may be the result of climatic differences between 2003 and the two successive years, any inter-era differences are much less pronounced in sites 2 and 3 – adjacent areas that experienced the same climate but were not invaded by *S. xyloni* (inter-era tending differences described for 67 and 54 plants, respectively; Figure 14.3).

Although other workers have demonstrated changes in aggregate on-plant foraging by ants as a consequence of ant invasions, the outcome usually entailed an increase in aggregate foraging (but see Ness, 2003). We also observed changes in the phenology of the ant-plant interaction with the *S. xyloni* invasion; on-plant foraging by the trio of relatively effective ants became rarer in the late spring in the era after the community was invaded, and the incidence increased in the monsoon season and autumn. In this context, the phenology of plant tending by ants is powerfully influenced by the composition of the ant community. Whether diminished tending in the spring has consequences on *F. wislizeni* performance is unclear. Some of the insect herbivores of *F. wislizeni* are bivoltine, with generations in the spring and monsoon (e.g., Vessels et al., 2013), and the first generation may be less likely to encounter protective ants in the spring in *S. xyloni*-dominated communities. The

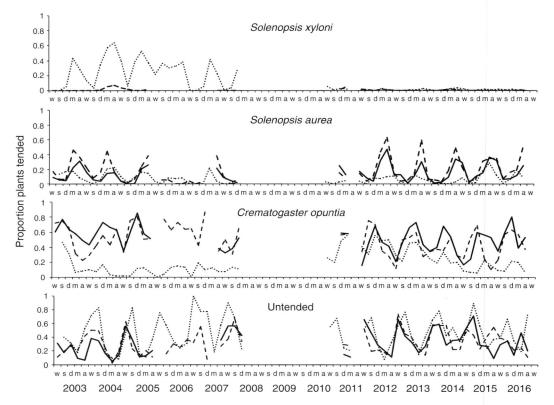

Figure 14.4.   Tending dynamics of *Ferocactus wislizeni*, an extrafloral nectary-bearing plant, by three ant species at three study sites at Tumamoc Hill Desert Laboratory, Tucson Arizona from 2003 to 2015. Proportion of inspected plants that were not tended by ants is also shown. Sites 1, 2, and 3 are shown in dotted lines, solid lines, and dashed lines, and include tending records for 268, 104, and 133 plants, respectively. W, S, D, M, and A refer to winter, spring, dry summer, monsoon, and autumn. Surveys were not performed between consecutive seasons in portions of 2008, 2009, and 2010.

monsoon and autumn – seasons with elevated visitation in the *S. xyloni*-dominated site – coincide with the periods of flowering and fruit production by *F. wislzeni*, and interactions between ants and visiting pollinators and frugivores are the focus of our next section.

Invasion Outcomes: Effects on Pollination and Frugivory

Exotic ants demonstrate that trophobiotic interactions can have ancillary effects on flower visitation by ants and other taxa (e.g., Lach, 2007; Hansen and Muller, 2009; LeVan and Holway, 2015). Because nectaries (floral and extrafloral) and honeydew producers can put ants in the proximity of flowers as well as encourage protective behaviours by the ants, visitation to the flowers by the ants can elicit a variety of cascading effects. Deterrence of pollinators by flower-visiting exotic ants, and an increase in the magnitude of this phenomenon in invaded sites relative to sites lacking the exotic ants, has been demonstrated in a variety of systems. The phenomenon is particularly well documented in the context of *Linepithema humile* invasions (e.g., Lach, 2007; LeVan and Holway, 2015) but is not limited to those settings. For example, Hansen and Muller (2009) demonstrated a sixfold increase in seed production per fruit when mealybug-tending *Technomyrmex albipes* ants were excluded from flowers, and attributed that result to the pronounced decrease in visitation rate and visit duration by geckos that pollinate the plants. Although ant-pollinator interactions certainly do not require an adjacent trophobiotic interaction (e.g., Blancafort and Gomez, 2005; Hanna et al., 2015), these examples certainly support the argument that a fuller reckoning of the consequences of the ant-trophobiont interaction can be ascertained by taking these accompanying effects into account.

Interactions between pollinators and flower-visiting ants in the *Ferocactus* systems demonstrate that nuisance ants can be influential in settings where the ants provide plants with protection from natural enemies. To the extent that analyzing the effects of protective ants on plant reproductive success requires an approach that incorporates effects on flower production, pollination, fruit development, and fruit removal, these systems represent particularly clear examples of how the addition of novel ants can contribute to an understanding of the range of outcomes possible in the study systems, as well as reveal indirect effects that are oblique to the ant-herbivore interaction that are the focus of most investigations of ant-plant protection mutualisms. In the *F. viridescens* system, flowers are more likely to include ants when the plant is tended by exotic *Linepithema humile* rather than the native *Crematogaster californica*. Plants tended by *L. humile* also experience shorter visits by pollinating cactus bees (*Diadasia* spp.) and, perhaps as consequence, produce fewer and smaller seeds per fruit (Levan et al., 2014). In the *F. wislizeni* system, pollinator visits decreased in frequency and duration as the potential for attack by ants in the flowers increased; this aversion was most pronounced in the plants controlled by the native invader, *S. xyloni* and weakest in plants controlled by *C. opuntia* (Ness, 2006). Further, pollinator visitation rates increased when individual plants transitioned from being tended by an aggressive ant to a species less likely to pose a threat in the flowers (Ness, 2006).

Invasions have also demonstrated the potential for ants to deter frugivorous vertebrates. Fruit removal, handling, and pecking by avian frugivores decreases in sites invaded by *Anoplolepis gracilipes* and can be increased via ant-exclusion treatments (Davis et al., 2010), and comparable patterns and experimental effects are observed with *Technomyrmex albipes* and seed-dispersing geckos (Hansen and Muller, 2009). These two studies are powerful because they demonstrate the loss of function within the invaded habitats with a pairing of descriptive data and manipulative experiments (here, ant exclusions). However, neither study compares the responses of the vertebrates to other ants, including those ant species that are presumably replaced or dominated by the exotics. As a result, the studies demonstrate the influence of the ants on the vertebrate-plant interactions but are not well suited for resolving whether the phenomenon is particular to exotic ants, particular to the exotic ant species that are the focus of these works, or evident across a continuum of plant visitation by ants.

The effects of *A. gracilipes* and *T. albipes* invasions inspired us to ask whether changes in the identity of plant-visiting ants in the *Ferocactus* system are associated with changes in the fruit removal rates experienced by individual plants. We hypothesised that fruit removal rates increase when individual plants transition to being tended by a less aggressive species (or visitation by ants lapses) and/or decrease when plants transition to being tended by a more aggressive species. We addressed this hypothesis using the records of ant attendance and fruit removal in the *F. wislizeni* system. Fruits mature in early autumn (September–October), do not passively fall off plants, and leave recognisable scars when removed from the maternal plant. The most common frugivores of *F. wislizeni* are large mammals (mule deer and collared peccary), and delays in fruit removal have the potential to detrimentally affect the viability of resulting seedlings (Eddy, 1961; Ness et al., 2016). We monitored the fruits anchored to individual plants across the 2004 fruiting season at the three sites (site 1 inspected on October 3, October 26, November 16, December 13, and December 27; site 2 inspected on September 28, 20 October, 19 November, and December 18, and site 3 inspected on September 27, November 21, December 18, and December 23) and identified ants visiting the plants during each inspection for fruits. The expected level of deterrence of visitors by the ants was ranked based on the abundance and aggressiveness of the tending species (*Solenopsis xyloni* > *S. aurea* > *Crematogaster opuntia* > *Forelius pruinosus* and rare 'other' ants (*Myrmecocystus* spp., *Camponotus* spp.) > no ants present at the time of inspection; see Ness et al. 2006). Because the incidence of ant visitation decreases from autumn into winter, as does the presence of the most aggressive species (*Solenopsis* spp.), we have few observations where plants become progressively better defended over September–December but many observations where the plants become less well defended (e.g., transition from *Solenopsis* to *Crematogaster*, from *Crematogaster* to unoccupied, etc.). In addition, aggregate fruit availability decreases over the course of the season. As a result, the hypothesis we ultimately tested was *fruit removal rates increase over time when plants transition to being tended by a less aggressive species, relative to the fruit removal rates of those plants where ant identity is consistent over the same interval of time.*

Daily fruit removal rates for an interval of time were calculated as $RR = (F_o - F_{s1})/d$, where F_o and F_{s1} are the number of fruits on the original and first successive inspection dates and d is the number of intervening days. Because the calculation of this rate may be distorted on plants that had all the fruits removed during the interval (e.g., d would then include days where further removal was impossible), calculations were limited to plants that retained at least one fruit at the second inspection date. Removal rates for a second interval were calculated as $(F_{s1} - F_{s2})/d$, where F_{s1} and F_{s2} are the second and third inspections, and so on. Because each time interval also includes two inspections of the ants visiting the plants (at the time of the fruit counts), we can distinguish intervals that include the same ant attendants at the beginning and end of the interval as well as intervals that include changes in ant attendance. For example, a plant that had a trio of successive inspections that always included *Crematogaster* would be described as having a consistent identity across the two fruit removal intervals, and the differences in daily fruit removal rates between the two intervals would be attributed to changes in the foraging environment or motivation of foragers rather than changes in plant defence provided by the ants. A plant with *Crematogaster, Crematogaster, Solenopsis aurea* during the three inspections (i.e., two intervals) would be described as being tended by a more aggressive ant defence during the second interval (a duration that included tending by *Crematogaster* and *Solenopsis*, with *Crematogaster* transitioning to *Solenopsis*) relative to the first interval (*Crematogaster* remaining *Crematogaster*), and a neighboring plant with *S. xyloni, S. xyloni, Crematogaster* would be described as being tended by a less aggressive ant defence during the second interval relative to the first interval, and both these plants also experience the environmental changes described for the consistent plant. Our hypothesis predicts that the inter-interval difference in fruit removal rates will be greater when the plant's ant defences are lessened in the second interval relative to the inter-interval differences that occur for plants with consistent ant identities. We tested this hypothesis with comparisons of the average inter-interval differences between consistent plants versus plants that lost defenders (or transitioned to less aggressive defenders), and paired these two plants types by combinations of site and time such that any spatio-temporal heterogeneity manifested across the combination of three sites and four months should be relatively consistent within a pair. Using a one-tailed, paired t-test, we find significant support for the hypothesis that fruit removal rates increase when plants transition to a less defended state, relative to any changes in fruit removal rates that occur over the same interval for plants with a consistent ant identity ($t = 2.4$, df = 6, p = 0.027).

The effects demonstrated for exotic *Anoplolepis gracilipes* and *Technomyrmex albipes* involve ant presence eliciting pronounced changes in foraging by frugivorous birds and geckos, respectively, whereas the *Ferocactus* system demonstrates some potential for native ants to influence foraging by large frugivorous mammals (as they do with pollinators). Although the *A. gracilipes* and *T. albipes* studies have the benefit of a strong manipulative experiment, the complementary evidence from the *F. wislizeni* system demonstrates that effects are also detectable in the less discrete scenario where the interactions of a fruit-bearing plant with ants changes over time. Interestingly, that turnover in the identity of partner ants can influence interactions between plants and large mammals was presaged by work in systems invaded by non-native ants (e.g., see Riginos et al., 2015 regarding elephants as plant antagonists in the *Acacia* system invaded by *Pheidole megacephala*).

Invasive Ants and Interaction Chronosequences

Invasion fronts are one of the few settings in which ant communities have been monitored and/or resampled over long time scales (~ decades and greater; see also Palmer et al., 2010). These repeated surveys have demonstrated continued invasion/consolidation in some instances (*Pheidole megacephala:* Riginos et al., 2015, Hoffman and Parr, 2006, *Linepithema humile* in California: Holway, 1995), as well as evidence that introduced populations can be susceptible to apparent declines (*Anoplolepis gracilipes*: Haines and Haines, 1978; Cooling and Hoffman, 2015, *Linepithema humile*: Cooling et al., 2012, *Solenopsis invicta*: Morrison, 2002, *Pheidole megacephala*: Majer and De Kock, 1992). For example, *Linepithema humile* infestations from urban environments in New Zealand appear to have a mean 'lifespan' of 10–18 years, with longevity decreasing with increasing latitude (Cooling et al., 2012). In some instances, native ant communities reassemble following the decline of the invader (e.g., Morrison, 2002; Cooling et al., 2012). Population-level declines in invader abundance are not limited to ants (Simberloff and Gibbons, 2004), and whether common, mechanistic explanations exist for these declines remains unclear (Strayer, 2012). For some introduced ants, the genetic uniformity common to supercolonies may have associated costs that engender vulnerability. Colonisation of supercolonies by parasites, and progressive increases in infection rates over time, have been identified as potentially influential drivers (Tragust et al., 2015). Fluctuations in the populations of certain non-native species may have been detected for several reasons, including intrinsic and extrinsic characteristics (e.g., genetic uniformity, enemy accumulation in new habitats, modifications to the resource base that supports focal populations). The visual 'obviousness' of many introduced ant species, as well as the duration and attention given to the monitoring of certain populations, also increase the opportunities to detect these fluctuations, and this begs the question of whether these fluctuations are particularly common and/or profound in invaders or merely particularly detectable.

Chronosequences of this type have the potential to reveal a great deal about the dynamics of interspecific interactions between ants and prospective partners. We see some evidence of the aforementioned phenomena in the *F. wislizeni* system between 2003 and 2016, and highlight three examples related to the stability of the ant-plant interactions over that duration. First, *Solenopsis xyloni* invaded site 1 in 2003 (see Morris et al., 2005) and subsequently dominated *F. wislizeni* tending from 2003 to 2008. Peak tending by *S. xyloni* occurs in the late summer monsoon and autumn (Figure 14.4), and the maximum observed proportion of *F. wislizeni* plants tended by *S. xyloni* at site 1 was 0.48 ± 0.11 (mean $\pm$ SD) in 2003–2007 (data omitted from 2008 because of lack of monsoon and autumn data). Subsequently, it largely disappeared from the site; the maximum proportion of *F. wislizeni* plants tended by *S. xyloni* in 2009–2015 was 0.044 ± 0.038 – a tenfold reduction (Figure 14.4). We recognise that descriptions of visitation to a nectary-bearing plant such as *F. wislizeni* (and other nectary-bearing plant species referenced here) provide an indirect measure of the fluctuations in the *S. xyloni* population. However, *S. xyloni* was also not detected in baiting surveys performed in the 2011 and 2013 summers. Based on those multiple lines of evidence, we conclude that the population crashes that have been reported for several introduced ants species can also be evident in native nuisance ant populations, and that these fluctuations can influence tending dynamics for reward-producing plants.

Second, tending by the other three most common ant species in the *F. wislizeni* systems has fluctuated, and in some cases recovered, at the invaded site in a manner consistent with strong competitive effects of *S. xyloni* (Figure 14.4). The maximum proportion of *F. wislizeni* plants tended by *C. opuntia* at site 1 was 0.46 in 2003 (the year of the *S. xyloni* invasion), 0.15 ± 0.03 (mean $\pm$ SD) in 2004–2008, and 0.43 ± 0.13 in 2009–2015. The maximum proportion of *F. wislizeni* plants tended by *F. pruinosus* at the site was 0.15 in 2003 (the year of the *S. xyloni* invasion), 0.21 ± 0.07 (mean $\pm$ SD) in 2004–2008, and 0.29 ± 0.13 in 2009–2015. Peak tending by *S. aurea* occurs in the late summer monsoon and autumn (as with *S. xyloni*), and the maximum observed proportion of *F. wislizeni* plants tended was 0.17 ± 0.06 (mean $\pm$ SD) in 2003–2007 (data omitted from 2008 because of lack of monsoon and autumn data), and 0.12 ± 0.09 in 2009–2015. In short, we witnessed pronounced decreases followed by a recovery by *C. opuntia* in the periods that coincided with the invasion and loss of *S. xlyoni* from the site, modest differences between those two eras for *S. aurea,* and a trend for increased plant visitation across years for *F. pruinosus.*

Third, we took the opportunity to test whether decreases in colony abundance and density can be linked with changes in the services provided by the ant population to the plant population. We hypothesised that higher densities of discrete colonies provide an element of redundancy to the interactions between the ant and plant populations, as might be expected if discrete colonies have somewhat idiosyncratic dynamics (e.g., inter-colony variation in phenology of dietary preferences, reproductive allocation, interactions with parasites and other natural enemies) and compete enough that a resource is 'captured' as it becomes available (as it might in the waning days of a colony that historically defended it). We used *Crematogaster opuntia* as the model system for this analysis. *Crematogaster opuntia* colonies are polydomous, exchange workers and resources between spatially segregated nests, and can utilise large territories (up to 18,100 m^2; Lanan and Bronstein, 2013). Colony boundaries were mapped in 2007, 2008, and 2009, and densities were 0.5, 2.7, and 5.0 colonies per hectare in sites 1, 2, and 3 (see methodology in Lanan and Bronstein, 2013). Although the areas controlled by particular colonies shifted modestly in space during that interval, the plants within the plots were always within the boundaries of a *C. opuntia* colony, even as the identity of the colony that controlled access to a particular plant in some cases changed over time and other ant species may also visit (and control access to) plants within the boundaries of a *C. opuntia* colony (Lanan and Bronstein, 2013). We tested the prediction that the inter-survey variation for the proportion of plants tended by *C. opuntia* would decrease as the densities of colonies increased. For each site, we calculated the coefficient of variation for the proportion of plants tended by *C. opuntia* during the seasonal inspections, treating each year by season combination as a replicate. The CV for the proportion of plants at the three sites was 0.68, 0.37, and 0.36, respectively, when limited to the seasonal surveys that occurred in 2007–2009, the period during which the colonies were mapped. The relative levels of variation among the sites were similar for seasonal surveys that occurred in 2003–2006 (1.00, 0.32, and 0.21) and 2010–2015 (0.55, 0.43, and 0.37), time periods before and after the colony mapping. As a result, the among-site differences in tending consistency can be predicted based on the among-site differences in the density of colonies ($r^2 = 0.75$, 0.84, and 0.95 for 2007–2009, 2003–2006, and 2010–2016, respectively). We conclude that the presence of discrete colonies – with presumably some elements of independence and difference among them – has the potential to engender more regular interactions between the larger ant population and local plant populations. Whether interactions become more inconsistent as sites are dominated by sprawling supercolonial populations, a characteristic particularly common to many exotic ant populations, remains untested.

Closing

The incorporation of novel participants into interaction webs is an increasingly common feature of global change. As a reviewer of this chapter noted, the 'exotic' in exotic species can encourage the impression of these populations, and their interactions with the native biota as being something markedly different from the norm. Introduced ants certainly provide opportunities to better understand the phenomenon of mutualism as well as the range of consequences that can be associated with variation in partner identity. In particular, invasions put us in a strong position to isolate traits of ant participants that have a large impact, positive or negative, on mutualism function. Except for locations lacking ants, native and introduced species do seem to form a continuum in terms of how their participation influences the outcome of mutualisms. Influential traits seem closely tied to the type of service provided by the ant (e.g., protection, dispersal), and the phenological congruence between the needs of the two participants is an important criterion. A greater focus on these characteristics would have predictive value, as well as provide a balance to the native versus introduced dichotomy that can frame expectations in ways that can be counter-productive. For example, we might assume – wrongly – that an introduced species will be disruptive simply because it is non-native, or that native species are not responsible for the truncation of mutualist communities and the associated ecological reverberations.

The *Ferocactus* ant-plant system is a useful model for exploring ways in which the inclusion of opportunistic, behaviourally dominant ants – exotic and native – into the system alters our understanding of several important dynamics within the interaction web centred around the plants. These include the fundamental components of the ant-plant interaction (e.g., tending phenologies and deterrence of natural enemies), the effects of partner identity (and, more generally, indirect plant defences) on pollination and frugivory, and the phenomena that encourage spatiotemporal variation versus continuity in plant-insect interactions. In each of these instances, we find work – models, predictions, and conclusions – with native and introduced species to be inspiring models for one another.

References

Abbott, K. L. and Green, P. T. (2007). Collapse of an ant-scale mutualism in a rainforest on Christmas Island. *Oikos*, 116, 1238–1246.

Blancafort, X. and Gomez, C. (2005). Consequences of the Argentine ant, *Linepithema humile* (Mayr), invasion on pollination of *Euphorbia characias* (L.) (Euphorbiaceae). *Acta Oecologia*, 28, 49–55.

Bleil, R., Bluthgen, N. and Junker R. R. (2011). Ant-plant mutualism in Hawaii? Invasive ants reduce flower parasitism but also exploit floral nectar of the endemic shrub *Vaccinium reticulatum*. *Pacific Science*, 65, 291–300.

Bond, W and Slingsby, P. (1984). Collapse of an ant-plant mutualism: the Argentine ant (*Iridomyrmex humilis*) and myrmecochorous Proteaceae. *Ecology*, 65, 1031–1037.

Brightwell, R. J. and Silverman, J. (2011). The Argentine ant persists through unfavorable winters via a mutualism facilitated by a native tree. *Environmental Entomology*, 40, 1019–1026.

Christian, C. E. (2001). Consequences of a biological invasion reveal the importance of mutualism for plant communities. *Nature*, 413, 635–639.

Cooling, M., Hartley, S. and Lester, P. J. (2012). The widespread collapse of an invasive ant species: Argentine ants (*Linepithema humile*) in New Zealand. *Biology Letters*, 8, 430–433.

Cooling, M. and Hoffman, B. D. (2015). Here today, gone tomorrow: declines and local extinctions of invasive ant populations in the absence of intervention. *Biological Invasions*, 17, 3351–3357.

Davis, N. E., O'Dowd, D. J., Green, P. T. et al. (2008). Effects of an alien ant invasion on abundance, behavior, and reproductive success of endemic island birds. *Conservation Biology*, 22, 1165–1176.

Davis, N. E., O'Dowd, D. J., MacNally, R. et al. (2010). Invasive ants disrupt frugivory by endemic island birds. *Biology Letters*, 6, 85–88.

Eddy, T. A. (1961). Foods and feeding patterns of the collared peccary in southern Arizona. *Journal of Wildlife Management*, 25, 248–257.

Field, H. C., Evans, W. E., Hartley, R. et al. (2007). A survey of structural ant pests in the Southwestern U.S.A (Hymenoptera: Formicidae). *Sociobiology*, 49, 1–14.

Fitzpatrick, G., Lanan, M. C. and Bronstein, J. L. (2014). Thermal tolerance affects mutualist attendance in an ant-plant mutualism. *Oecologia*, 176, 129–138.

Gaigher, R., Samways, M. J., Henwood, J. et al. (2011). Impact of a mutualism between an invasive ant and honeydew-producing insects on a functionally important tree on a tropical island. *Biological Invasions*, 13, 1717–1721.

Green, P. T., O'Dowd, D. J., Abbott, K. L. et al. (2011). Invasional meltdown: Invader-invader mutualism facilitates secondary invasion. *Ecology*, 92, 1758–1768.

Green, P. T., O'Dowd, D. J. and Lake, P. S. (2008). Recruitment dynamics in a rainforest seedling community: context- independent impact of a keystone consumer. *Oecologia*, 156, 373–385.

Haines, I. H. and Haines, J. B. (1978). Pest status of the crazy ant, *Anoplepis longipes* (Jerdon) (Hymenoptera: Formicidae), in the Seychelles. *Bulletin of Entomological Research*, 68, 627–638.

Hanna, C., Naughton I., Boser, C., Alarcón, R., Hung, K.-L. J. and Holway, D. (2015). Floral visitation by the Argentine ant reduces bee visitation and plant seed set. *Ecology*, 96: 222–230.

Hansen, D. M. and Müller, C. B. (2009). Invasive ants disrupt gecko pollination and seed dispersal of the endangered plant *Roussea simplex* in Mauritius. *Biotropica*, 41, 202–208.

Helms, K. R. (2013). Mutualisms between ants (Hymenoptera: Formicidae) and honeydew-producing insects: Are they important to invasions? *Myrmecological News*, 8, 61–71.

Helms, K. R. and Vinson, S. B. (2002). Widespread association of the invasive *Solenopsis invicta* with an invasive mealybug. *Ecology*, 83, 2425–2438.

Hill, M., Holm, K., Vel, T. et al. (2003). Impact of the introduced yellow crazy ant *Anoplolepis gracilipes* on Bird Island, Seychelles. *Biodiversity and Conservation*, 12, 1969–1984.

Hoffman, B. D., Andersen, A. N. and Hill, G. E. (1999). Impact of an introduced ant on native rain forest invertebrates: *Pheidole megacephala* in monsoonal Australia. *Oecologia*, 120, 595–604.

Hoffman, B. D. and Parr, C. L. (2006). An invasion revisited: the African big-headed ant (*Pheidole megacephala*) in northern Australia. *Biological Invasions*, 10, 1171–1181.

Holway, D. A. (1995). Distribution of the Argentine ant (*Linepithema humile*) in northern California. *Conservation Biology*, 9, 1634–1637.

Holway, D. A., Lach, L., Suarez, A. V. et al. (2002). The causes and consequences of ant invasions. *Annual Review of Ecology and Systematics*, 33, 181–233.

Hooper, L. M. and Rust, M. K. (1997). Food preference and patterns of foraging activity of the southern fire ant (Hymenoptera: Formicidae). *Annals of the Entomological Society*, 90, 246–253.

Kerley, G. I. H. and Whitford, W. G. (2000). Impact of grazing and desertification in the Chihuahuan Desert: plant communities, granivores and granivory. *American Midland Naturalist*, 144, 78–91.

Knight, R. L. and Rust, M. K. (1990). The urban ants of California with distribution notes of imported species. *Southwestern Entomologist*, 15, 167–178.

Lach, L. (2003). Invasive ants: unwanted partners in ant-plant interactions? *Annals of the Missouri Botanical Garden*, 90, 91–108.

(2005). Interference and exploitation competition of three nectar-thieving invasive ant species. *Insectes Sociaux*, 52, 257–262.

(2007). A mutualism with a native membracid facilitates pollinator displacement by Argentine ants. *Ecology*, 88, 1994–2004.

(2008a). Argentine ants displace floral arthropods in a biodiversity hotspot. *Diversity and Distributions*, 14, 281–290.

(2008b). Floral visitation patterns of two invasive ant species and their effects on other hymenopteran visitors. *Ecological Entomology*, 33, 155–160.

Lach, L. and Hoffmann, B. D. (2011). Are invasive ants better plant-defense mutualists? A comparison of foliage patrolling and herbivory in sites with invasive yellow crazy ants and native weaver ants. *Oikos*, 120, 9–16.

Lach, L. and Hooper-Bui, L. M. (2010). Consequences of ant invasions. In Lach L., Parr, C. L. and Abbott, K. (eds.). *Ant ecology*. Oxford: Oxford University Press, pp. 261–286.

Lanan, M. C. and Bronstein, J. L. (2013). An ant's-eye view of an ant-plant protection mutualism. *Oecologia*, 172, 779–790.

Lester, P. J. and Tavite, A. (2004). Long-legged ants, *Anoplolepis gracilipes* (Hymenoptera: Formicidae), have invaded Tokelau, changing composition and dynamics of ant and invertebrate communities. *Pacific Science*, 58, 391–401.

LeVan K. E. and Holway, D. A. (2015). Ant-aphid interactions increase ant floral visitation and reduce plant reproduction via decreased pollinator visitation. *Ecology*, 96, 1620–1630.

LeVan K. E., Hung, K. J., McCann, K. R. et al. (2014). Floral visitation by the Argentine ant reduces pollinator visitation and seed set in the coast barrel cactus, *Ferocactus viridescens*. *Oecologia*, 174, 163–171.

Ludka J., LeVan, K. E. and Holway, D. A. (2015). Infiltration of a facultative ant–plant mutualism by the introduced Argentine ant: effects on mutualist diversity and mutualism benefits. *Ecological Entomology*, 174, 163–171.

Majer, J. D. and De Kock, A. E. (1992). Ant recolonization of sand mines near Richards Bay, South Africa: an evaluation of progress with rehabilitation. *South African Journal of Science*, 88, 31–36.

Menke S. B., Suarez, A. V., Tillberg, C. V. et al. (2010). Trophic ecology of the invasive argentine ant: spatio-temporal variation in resource assimilation and isotopic enrichment. *Oecologia*, 164, 763–771.

Morris, W. F., Wilson, W. G., Bronstein, J. L. et al. (2005). Environmental forcing and the competitive dynamics of a guild of cactus-tending ant mutualists. *Ecology*, 86, 3190–3199.

Morrison, L. W. (2002). Long-term impacts of an arthropod-community invasion by the imported fire ant, *Solenopsis invicta*. *Ecology*, 83, 2337–2345.

Ness, J. H. (2003). Contrasting exotic *Solenopsis invicta* and native *Forelius pruinosus* ants as mutualists with *Catalpa bignoniodes*, a native plant. *Ecological Entomology*, 28, 247–251.
  (2006). A mutualism's indirect costs: the most aggressive plant bodyguards also deter pollinators. *Oikos*, 113, 506–514.

Ness, J. H. and J. L. Bronstein (2004). The effects of invasive ants on prospective ant mutualists. *Biological Invasions*, 6, 445–461.

Ness, J. H., Bronstein, J. L., Andersen, A. N. et al. (2004). Ant body size predicts dispersal distance of ant-adapted seeds: implications of small-ant invasions. *Ecology*, 85, 1244–1250.

Ness, J. H., Morales, M. A., Kenison, E. et al. (2012). Reciprocally beneficial interactions between introduced plants & ants induced by the presence of a third introduced species. *Oikos*, 122, 695–704.

Ness J.H., Morris, W. F. and Bronstein, J. L. (2006). Integrating quality and quantity of mutualistic service to contrast ant species protecting *Ferocactus wislizeni*. *Ecology*, 87, 912–921.

Ness J. H., Pfeffer, M., Stark, J., Guest, A., Combs, L. J. and Nathan, E. (2016). In an arid urban matrix, fragment size predicts access to frugivory and rain necessary for plant population persistence. *Ecosphere*, 7: 1–19 (e01284).

O'Dowd, D. J., Green, P. T. and Lake, P. S. (2003). Invasional 'meltdown' on an oceanic island. *Ecology Letters*, 9, 812–817.

Palmer T. M., Doak, D. F., Stanton, M. L. et al. (2010). Synergy of multiple partners, including freeloaders, increases host fitness in a multispecies mutualism. *Proceedings of the National Academy of Sciences, USA*, 107, 17234–17239.

Palmer T. M., Young, T. P., Stanton, M. L. et al. (2000). Short-term dynamics of an acacia ant community in Laikipia, Kenya. *Oecologia*, 123, 425–435.

Riginos, C., Karande, M. A., Rubenstein, D. I. et al. (2015). Disruption of a protective ant-plant mutualism by an invasive ant increases elephant damage to savanna trees. *Ecology*, 96, 654–661.

Savage, A. M., Johnson, S. D., Whitney, K. D. et al. (2011). Do invasive ants respond more strongly to carbohydrate availability than co-occurring non-invasive ants? A test along an active *Anoplolepis gracilipes* invasion front. *Austral Ecology*, 36, 310–319.

Savage, A. M., Rudgers, J. A. and Whitney, K. D. (2009). Elevated dominance of extrafloral nectary-bearing plants is associated with increased abundances of an invasive ant and reduced native ant richness. *Diversity and Distributions*, 15, 751–761.

Savage, A. M. and Whitney, K. D. (2011). Trait-mediated indirect interactions in invasions: unique behavioral responses of an invasive ant to plant nectar. *Ecosphere*, 2, article 106.

Sax, D. F., Stachowicz, J. J. and Gaines, S. D. (2005). *Species invasions: insights into ecology, evolution, and biogeography*. Sunderland, MA: Sinauer.

Schilman, P. E., Lighton, J. B. and Holway, D. A. (2007). Water balance in the Argentine ant (*Linepithema humile*) compared with five common native ant species from southern California. *Physiological Entomology*, 32, 1–7.

Schooley, R. L., Bestelymeyer, B. T. and Kelly, J. F. (2000). Influence of small-scale disturbances by kangaroo rats on Chihuahuan Desert ants. *Oecologia*, 125, 142–149.

Simberloff, D. and Gibbons, L. (2004). Now you see them, now you don't! – population crashes of established introduced species. *Biological Invasions*, 6, 161–172.

Smith, M. R. (1936). Consideration of the fire ant *Solenopsis xyloni* as an important southern pest. *Journal of Economic Entomology*, 29, 120–122.

Strayer, D. L. (2012). Eight questions about invasions and ecosystem functioning. *Ecology Letters*, 15, 1199–1210.

Styrsky, J. D. and Eubanks, M. D. (2010). A facultative mutualism between aphids and an invasive ant increases plant reproduction. *Ecological Entomology*, 35, 190–199.

Tanaka, H., Ohnishi, H., Tatsuta, H. et al. (2011). An analysis of mutualistic interactions between exotic ants and honeydew producers in the Yanbaru district of Okinawa Island, Japan. *Ecological Research*, 26, 931–941.

Tragust, S., Feldhaar, H., Espadaler, X. et al. (2015). Rapid increase of the parasitic fungus *Laboulbenia formicarum* in supercolonies of the invasive garden ant *Lasius neglectus*. *Biological Invasions*, 17, 2795–2801.

Vessels, H. K., Bundy, C. S. and McPherson, J. E. (2013). Life history and laboratory rearing of *Narnia femorata* (Hemiptera: Coreidae) with descriptions of immature stages. *Annals of the Entomological Society of America*, 106, 575–585.

Wetterer, J. K. and Wetterer, A. L. (2003). Ants (Hymenoptera: Formicidae) on non-native neotropical ant-acacias (Fabales: Fabaceae) in Florida. *Florida Entomologist*, 85, 460–463.

15 Invasion Biology and Ant-Plant Systems in Australia

Lori Lach

Introduction

Australia is a continent known for its ecological idiosyncracies. It goes well beyond the abundance of furry animals with pouches and diversity of creatures that can kill you. Australia's geologically long period of geographic isolation has translated to remarkable floral and faunal endemicity; some 92 per cent of its vascular plants and over 80 per cent of its frogs, reptiles, and mammals are found nowhere else (Chapman, 2009). It spans 35 degrees in latitude, and although much of it is desert and xeric shrublands, it also comprises seven other ecoregions including Mediterranean forests and woodlands, temperate grasslands, and tropical and sub-tropical broadleaf forests (Environmental Resources Information Network, 2012) that also contribute to its plant and invertebrate diversity. It is the flattest and driest of the continents, and its old soils are notably nutrient poor (Orians & Milewski, 2007).

What does any of this have to do with ant-plant interactions? These nutrient-poor soils have at least two consequences that affect ant-plant interactions. Low nutrient soils are more likely to have shrubby plants that depend on ants for seed dispersal (Westoby et al., 1991a; Berg, 1975). Australia is one of the three hotspots for myrmecochory (Lengyel et al., 2010) harbouring an estimated 32 per cent of myrmecochore genera (Warren & Giladi, 2014; Chapters 5 and 6). It is common for 30–50 per cent of a site's flora to be ant-dispersed (Westoby et al., 1991b). Second, plants growing in nutrient-poor soils are limited in nitrogen and phosphorus and have an abundance of carbon. Plants can expend this excess carbon by producing lots of nectar (floral or extrafloral) and/or hosting of honeydew-producing insects (Orians & Milewski, 2007). The availability of carbohydrate-rich resources supports the extraordinarily high local ant productivity in Australia's arid regions (Andersen, 2003) and plays an important role in shaping its rainforest ant assemblages (Blüthgen et al., 2004).

The ant diversity and endemicity of Australia are also notable. In most regions of the world, higher ant species richness occurs in rainforest than in arid regions. The opposite is true in Australia (Andersen, 2003; AntWiki, 2010). Local richness of often greater than 100 species per hectare in the arid zone also far exceeds that in similarly arid climates (Andersen, 2007; Andersen, in review). On a continental scale, Australia's approximately 1,600 described ant species (1,576: Economo & Guenard, 2016; 1,622: AntWeb, 2010) is about 10 per cent of the world's described ant species, which is a bit more than expected than if it was only proportional to land mass (6 per cent). However, it harbours 21 per cent of globally described genera (110/518 genera, AntWeb, 2010).

Australia has also suffered from a barrage of tramp ant species arrivals. Several have established and become invasive. All six of the reputedly most invasive ant species (*Linepithema humile*, *Solenopsis invicta*, *S. geminata*, *Anoplolepis gracilipes*, *Pheidole megacephala*, and *Wasmannia auropunctata*) are present in the country and under some degree of management in at least some of their Australian range (Lach & Barker, 2013). Another 24 ant species have established but appear to be less capable of invading outside of human-associated habitats (Majer & Heterick, 2015). Very few, if any, ant incursions are intentional; rather most arrive after stowing away in cargo or passenger vessels.

Once they arrive and establish, invasive ants can have many ecological effects, including on ant-plant interactions. There are many mechanisms through which invasive ants affect ant-plant interactions, and multiple reasons their effects can be different from native ants. I have discussed these in detail in a previous review (Lach, 2003). In brief, invasive ants can affect ant-plant interactions by displacing native ant species and then failing to usurp their role or service or by otherwise changing the nature of the interaction. They can also change the abundance and community composition of herbivores and herbivore natural enemies. Invasive ant species often differ from native ant species in their relative abundance, aggression, and affinity for carbohydrates, and these can all affect interactions with plants (Lach, 2003).

In this chapter, I summarise current knowledge on how invasive ants have affected ant-plant interactions in Australia and compare findings in Australia to those elsewhere. The effects of invasive ants on ant-plant systems globally (Holway et al., 2002; Lach, 2003; Ness & Bronstein, 2004; Lach & Hooper-Bùi, 2010) and in Australia (Lach & Thomas, 2008) have previously been reviewed as part of broader syntheses on the effects of invasive ants. I focus on the advances made since those publications and in particular on seed dispersal, plant protection and trophobiont tending, and ant-plant symbioses.

I focus on interactions involving invasive ants rather than interactions between invasive plants and native ants for a few reasons. Though about 28,000 plant species have been introduced to Australia and 2,700 of these have established in the natural environment (Department of the Environment, 2016), ant-related research on these invasions (e.g., Osunkoya et al., 2011) often investigates how they change ant assemblages, rather than specific interactions. Native ants do harvest and disperse seeds of some weed species, but these behaviours may be of little consequence to the often mass-seed-producing invaders. Moreover, studies that mention interactions between ants and invasive plant seeds are rarely designed to provide specific insights into the evolution or ecology of the interaction itself.

Invasive Ants and Ant-Plant Interactions in Australia

Invasive Ant Distribution

The potential for invasive ants to affect ant-plant interactions in Australia is greatly influenced by the invaders' distributions. Most invasive ant species are found on the coastal fringes of the country and its islands (Figure 15.1). The distribution pattern on the mainland is likely the result of multiple correlated influences. About 85 per cent of the human population lives within 50 km of the coast (Australian Bureau of Statistics, 2004), and all possible points of initial entry (international airports and cargo ports) are in or near coastal cities. Moreover, the arid interior is an inhospitable habitat for most invasive ants of interest. The distribution of some invasive ants on the mainland (*S. invicta*, *A. gracilipes*, *W. auropunctata*) is also influenced by ongoing large-scale management and eradication programmes (Lach & Barker, 2013). Invasions on populated islands (*A. gracilipes* and other tramp ant species on Christmas Island, *L. humile* on Norfolk Island, *P. megacephala* on Lord Howe Island) are likely the result of the importation of goods and benefit from the simplified ant assemblages of the islands, but are also limited by management programs (Lach & Barker, 2013).

At the habitat scale, there has been comparatively little analysis of the distribution limits of invasive ants in Australia. Temperature and humidity certainly play a role (Thomas & Holway, 2005; Hoffmann, 2015; Asfiya et al., 2016). The role of disturbance in the invasion of Australian ecosystems is not well-studied (Hoffmann & Saul, 2010). However, it is clear that invasive ants are capable of establishing in a range of habitats that are of conservation interest and significance, from urban bushland in the southwest to World Heritage rainforest in the northeast.

Seed Dispersal

In Australia, myrmecochory is highest in open dry habitats with Mediterranean-type climates (Berg, 1975; Lengyel, 2010) in the southeast and southwest of the country. As in the other myrmecochory hotspots with Mediterranean-type climates (Europe, South Africa), Australian myrmecochorous species tend to be woody, rather than herbaceous (Berg, 1975). Parts of these regions have been invaded by Argentine ants (*L. humile*) or African big-headed ants (*P. megacephala*). These invaders have lower tolerance for high temperatures and aridity than native ants (Thomas & Holway, 2005; Asfiya et al., 2016). Therefore, overlap at smaller scales is likely dependent on site-specific abiotic conditions.

Myrmecochory was once thought to be a diffuse mutualism in which a large subset of the local ant assemblage was effective at dispersing seeds. More recent research has found that seed dispersal is often accomplished by a few high-quality dispersers that remove >75 per cent of seeds (Gove et al., 2007; Warren & Giladi, 2014). These keystone species are generally subordinate and larger-bodied ants with small transient colonies (Warren & Giladi, 2014, and references therein). Behaviourally

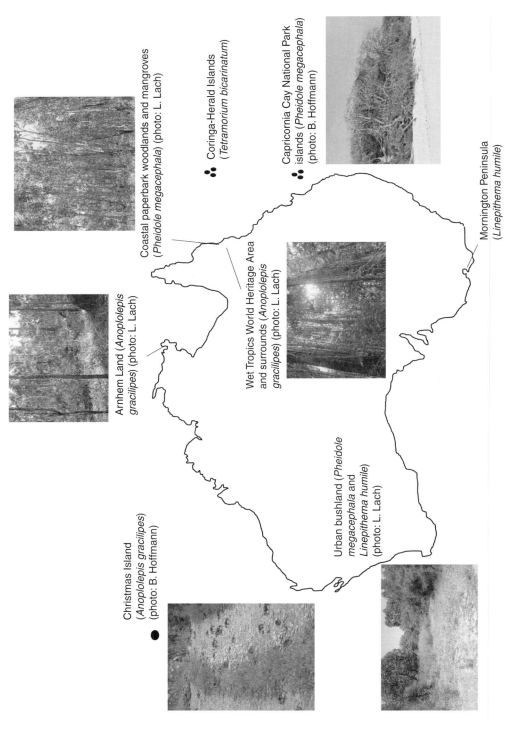

Figure 15.1. Map of mainland Australia and selected islands indicating locations of studies mentioned in this review and the invasive ant species studied.

subordinate species need to remove seeds rapidly to avoid losing them to competitively dominant species (Mesler & Lu, 1983). Larger-bodied ants disperse seeds a greater distance (Chapter 7), are less likely to drop seeds mid-transport, and are more likely to collect seeds and collect a broader range of seed sizes (Ness & Bronstein, 2004; Ness et al., 2004). Seeds will be left behind as colonies move, so species with transient colonies will leave seeds in more locations than more stationary colonies. Key ant dispersers so far identified in Australia, such as *Rhytidoponera metallica* (Westoby et al., 1991b), *R. violacea*, *R. inornata*, and *Melophorus turneri perthensis*, generally conform to these traits (Andersen, 1990; Gove et al., 2007; Lubertazzi et al., 2010; Majer et al., 2011).

Where just a few subordinate ant species are key to seed dispersal, invasive ants need only displace these high-quality dispersers, rather than a large part of the assemblage, to cause significant changes. In Australia, *Rhytidoponera* spp. rarely persist where either Argentine ants or big-headed ants have invaded (Callan & Majer, 2009; Rowles & O'Dowd, 2009a; Majer et al., 2011). However, some *Melophorus* species can persist in all but the densest *P. megacephala* invasions, likely due to its tolerance to high temperatures and consequent temporal niche partitioning with *P. megacephala* (Callan & Majer, 2009).

Despite the potential for more widespread effects, research into the effect of ant invasions on seed dispersal in Australia has been limited, with just one empirical study comparing seed dispersal in sites invaded by the Argentine ant and non-invaded sites. The study, conducted in Mornington Peninsula National Park in Victoria (Figure 15.1), reported no overall collapse of seed dispersal despite displacement of important native seed dispersers, such as *Rhytidoponera victoriae* (Rowles & O'Dowd, 2009b). This contrasts to a global meta-analysis of ten studies (including the Australian study) that found 47 per cent fewer seeds were removed and seedling establishment was 76 per cent lower when Argentine ants were present compared to similar habitats where it was absent (Rodriguez-Cabal et al., 2009). However, in the Australian study, Argentine ants were more likely than ants in non-invaded sites to remove and bury seeds of an invasive plant that had a high reward-to-size ratio. They were also more likely to remove, but often failed to bury, seeds of a large-seeded native plant with a large reward than ants in non-invaded sites, but were less likely to remove the seeds of a small-seeded native plant that had a relatively small reward. Results of an experiment manipulating diaspore and reward size using artificial diaspores also showed the importance of the interaction between diaspore and reward size (Rowles & O'Dowd, 2009b).

In a broad sense, the Australian findings conform to other studies (Christian, 2001; Ness et al., 2004; Warren et al., 2015) that look beyond overall seed removal and conclude that ant invasions are likely to change plant community composition because of differences in how invasive ants interact with seeds from myrmecochorous plants. Invasive ants often numerically dominate their novel habitats and therefore become the most likely to encounter seeds. Invasive ants have variously been found to either consume seeds, remove the elaiosome but fail to move the seeds or bury them, have lower seed collection rates, and/or move seeds shorter distances (reviewed in Ness & Bronstein, 2004). Some of the differences in seed handling have been linked to the smaller body size of invasive ants (Ness et al., 2004 and references therein). Other mechanisms of partner selection between native ants and seeds, such as elaiosome chemistry and synchronicity in seed release and peak keystone ant activity (Warren & Giladi, 2014), have yet to be studied as mechanisms for affecting invasive ant-seed interactions. But if these partner selection mechanisms are highly co-evolved, the seed dispersal mutualism is less likely to be usurped by invasive ant species.

Plant Protection and Trophobiont Tending

Another consequence of the continent's nutrient-poor soils and abundant sunshine is a relative excess of carbon for plants to store or expend (Orians & Milewski, 2007; Morton et al., 2011). Hence, Australia's plants are known to directly or indirectly produce relatively large quantities of exudates, such as nectar (floral and extrafloral) and honeydew, which are known to attract ants, Orians & Milewski, 2007). The effects of carbohydrate availability on ant assemblages have primarily been investigated in tropical rainforests where it has been found that access to carbohydrates drives ant abundance, activity, and the composition and structure of ant assemblages (Davidson et al., 2003; Kaspari et al., 2012). For example, in the lowland tropical rainforest of northeastern Australia, Blüthgen et al. (2004) found strong effects of honeydew and nectar on ant community structure and identified two keystone plant species that affect ant distributions via their hosting of key honeydew producers. A similar bottom-up mechanism was found to drive differences in the proportion of dominant ants in revegetated Acacia and Eucalyptus woodlands in southeastern Australia (Gibb & Cunningham, 2009).

These carbohydrate-rich resources are important in ant invasion dynamics for at least three reasons. First, they attract ants to the plants and drive their interactions with herbivores, herbivore natural enemies (Altfeld & Stiling, 2006; Styrsky & Eubanks, 2007; Lach & Hoffmann, 2011), and pollinators (Lach, 2007; LeVan et al., 2014) through which they then affect plant fitness (Blancafort & Gomez, 2005; Lach et al., 2010), and/or crop yield (Styrsky & Eubanks, 2010). They can also prevent native ants from accessing these resources, which may be a mechanism of displacement (Holway et al., 2002; Wilder et al., 2013). Second, evidence is amassing for the importance of carbohydrates in fuelling ant invasions (Wilder et al., 2011b; Helms, 2013). Invasive ant colonies with access to carbohydrates produce more workers (Kay et al., 2010; Wilder et al., 2011a), are more aggressive (Grover et al., 2007; Savage & Whitney, 2011), and can expand their spatial reach (Rowles & Silverman, 2009), all of which may help them to outcompete native ants and access more resources. Finally, changes in preferences for resources due to their relative availability can affect the attractiveness of toxic baits and the effectiveness of management programmes (Abbott et al., 2014).

Honeydew

One of the best-known examples of a carbohydrate-rich resource fuelling an ant invasion comes from Christmas Island, Australia (Figure 15.1). The yellow crazy ant, *Anoplolepis gracilipes*, which is thought to be from southeast Asia (Wetterer, 2005), was introduced to the island between 1915 and 1934 but did not reach pest levels until 1989 (O'Dowd, 1999) possibly following the arrival of a scale insect (Neumann et al., 2016). Subsequent observations revealed a strong association between outbreaks of the cryptogenic lac scale (*Tachardina aurantiaca*; Kerriidae) and *A. gracilipes* activity and a catastrophic cascade of effects on the rainforest ecosystem (O'Dowd et al., 2003). Toxic baiting of *A. gracilipes* colonies resulted in population collapse of lac scales and soft scales (Coccidae) on two species of host tree (Abbott & Green, 2007). More recently, an experiment in which trees were banded to prevent *A. gracilipes* access to honeydew resulted in a rapid decline in ant population density (Neumann et al., 2016). So tight is the relationship between *A. gracilipes* and the lac scale that biological control of the scale with a parasitoid wasp is being proposed as an indirect way to control the ant. The Christmas Island case is probably the most well-studied and well-documented example of the cascading ecological consequences of an invasive ant-bug association. It has underscored the need to address ant invasions in other environmentally sensitive areas elsewhere in Australia, such as in the Wet Tropics World Heritage Area (Lach & Hoskin, 2015) and other ant-invaded island groups, such as Hawai'i (Loope & Krushelnycky, 2007) and Seychelles (Kaiser-Bunbury et al., 2014).

Another invasive ant-hemipteran relationship once threatened ecosystems on islands off the east coast of Australia but has been contained with biological control agents. The host tree for the association, *Pisonia grandis* (Nyctaginaceae), is a coral cay specialist considered endangered in Australia because the total area of its small dispersed stands is only 190 ha (Greenslade, 2008). Documented population explosions of the scale insect, *Pulvinaria urbicola*, which was first described in Jamaica but is now pan-tropical (Smith, 2004), started in the early 1990s. Scale populations were monitored on two main groups of islands off the east coast, the Coringa-Herald group of islands 400 km east of Cairns, and Capricornia Cays National Park, at the southern end of the Great Barrier Reef Marine Park (Figure 15.1). On islands in both groups, population explosions of the scale began in the early 1990s and led to 80–100 per cent tree death within a few years (Smith, 2004; Greenslade, 2008). Though initially *Pi. grandis* death was attributed to diminished seabird guano inputs, it is now recognised that outbreaks of *Pu. urbicola* are always associated with attendance by an invasive or tramp ant, and the interaction is at least a contributing factor in *Pi. grandis* death (Hoffmann & Kay, 2009; Burwell et al., 2012). Whereas *Tetramorium bicarinatum* is the dominant tending ant on the northern island group, the African big-headed ant *Pheidole megacephala* is the invader associated with *Pu. urbicola* outbreaks in the Capricornia Cays (State of Queensland, 2010; Burwell et al., 2012). Outbreaks of ant-tended *Pulvinaria urbicola* also have been problematic elsewhere in the Indo-Pacific. Toxic baiting of *Pheidole megacephala* on Cousine Island, Seychelles, resulted in a 93 per cent reduction in ant activity and a 100 per cent reduction in *Pu. urbicola* within four months (Gaigher & Samways, 2013). In Australia, both ant suppression and the introduction of multiple natural enemies of the scale insect were needed to achieve declines in *Pu. urbicola* populations (Smith, 2004; State of Queensland, 2010).

Islands are often species-poor ecosystems where addition or removal of a single or small number of species can cause a strong cascade of ecological effects (Chapter 13). In these examples, both of the honeydew-producing insects – *Tachardina auriantiaca* on Christmas Island and *Pulvinaria urbicola* on Coringa-Herald and Capricornia Cay islands – were introduced. Indeed most honeydew-producing insects tended by invasive ants in their adopted range are also non-native (Helms, 2013). With the recognition of the importance of honeydew in many ant invasions, key questions are whether some honeydew producers are exceptionally important or have common traits that make them especially attractive to invasive ants, and whether these interactions are more important or more reported in the simplified ecosystem of islands and agricultural lands (Helms, 2013).

The invasion of yellow crazy ants on mainland Australia has the potential to provide some insights. In northern Queensland, *Anoplolepis gracilipes* has invaded tropical rainforest, residential areas, and sugar cane (Figure 15.1). In the rainforest, *A. gracilipes* stream down trees, their distended gasters clearly indicating that they have consumed a large amount of liquid food (L. Lach, personal observations). The most likely resource is honeydew, though nectar and extrafloral nectar cannot be ruled out without further investigation. The high tree diversity within, and rapidity with which *A. gracilipes* has penetrated this intact rainforest, suggests that if carbohydrates are important they are plentiful and probably come from a variety of sources, including native bugs. But this idea remains to be tested. In nearby sugar cane fields, the native honeydew-producing whitefly, *Neomaskellia bergii*, reaches pest densities only in the presence of *A. gracilipes* (Nader Sallam, Sugar Research Australia, personal commentary; L. Lach, personal observations) (Figure 15.2). Taken together these observations suggest that the origin and traits of the bugs are less important than the traits of the ants, but more data on the composition of honeydew and its availability relative to colony needs would be helpful, as would comparisons to failed invasions.

While elucidating the sources of carbohydrates in order to figure out how their utilisation affects the ecosystem is important for understanding the ecology of ant invasions, understanding the use and availability of these resources is also important in the efforts to control the ants. On Christmas Island, for example, *A. gracilipes* prefers protein-rich lures in the dry season and carbohydrate-rich lures in the wet season. The seasonal shift in preference is thought to reflect the availability of key resources rather than investment in worker or gyne production (Abbott et al., 2014). Honeydew is less available in the wet season when rain washes it from the plants (Abbott et al., 2014). The main toxic bait used for *A. gracilipes* control on Christmas Island is in a proteinaceous matrix (fipronil in fishmeal), and aerial application of

Figure 15.2. Yellow crazy ants (*Anoplolepis gracilipes*) tending whitefly (*Neomaskellia bergii*) on sugarcane near Edmonton, Queensland. (Photo: Mark Bloemberg.) (A black-and-white version of this figure will appear in some formats. For the color version, please refer to the plate section.)

the bait is timed to coincide with the ants' dry-season preference for protein (Lach & Barker, 2013; Abbott et al., 2014). In contrast, in the north Queensland invasion, *A. gracilipes* rarely shows a marked preference for protein, but instead usually shows strong preference for carbohydrate, and the management programme has had to adapt by trialling new bait formulations (Lach, unpublished data).

Extrafloral Nectar

Extrafloral nectar also attracts ants to plants (Chapters 8–10). The prevailing hypothesis is that ants then protect the plants from herbivores (Bentley, 1977; Marazzi et al., 2013; Heil, 2015), an idea supported by several meta-analyses (Chamberlain & Holland, 2009; Rosumek et al., 2009; Trager et al., 2010; Heil, 2015). With 1,070 species, Acacias (sensu lato) comprise about 5 per cent of Australia's vascular flora and are widely distributed across the continent (Australian National Botanic Gardens Australian Flora Statistics, 2009; World Wide Wattle, 2016). Most of these have extrafloral nectaries (EFNs) (Simmons, 2009), although for some of these the nectar may serve as a reward to pollinators rather than to attract ant-guards (Stone et al., 2003). Other plant families known for having species with EFNs (Weber et al., 2015), such as Passifloraceae, Euphorbiaceae, and Malvaceae, are also found in Australia, especially in tropical rainforest. A study in the lowland rainforest of northeast Australia recorded 29 plant species with EFNs in a one hectare plot (Blüthgen & Reifenrath, 2003).

However, few studies in Australia have investigated how invasive ants compare to native ants in their interactions with EFNs and the consequences for the plant. A study comparing invasive *Anoplolepis gracilipes* to native *Oecophylla smaragdina* in Arnhem Land in the Northern Territory (Figure 15.1), found differences between the invasive ant and native ant patrolling behaviour, but no correspondence between patrolling behaviour and herbivory (Lach & Hoffmann, 2011). In this study, patrolling behaviour on saplings in monsoonal savanna was compared between the two ant species by affixing surrogate prey (termites) on *Acacia lamprocarpa* near the most active EFNs at the top of the plant, lower down on the same plant, and in corresponding positions on paired *Eucalyptus tetrodonta*, which offers no EFNs. The invader was more likely than the native ant to attack prey in all locations except those high on *E. tetrodonta*, where the two ant species were equally likely to attack. Moreover, the invasive, but not the native, ant was significantly more likely to attack prey high on *A. lamprocarpa* than high on *E. tetrodonta*, which was interpreted as the invasive ant responding to the availability of EFN on *A. lamprocarpa*. On the basis of patrolling behaviour it was expected that plants in invaded sites would experience lower herbivory than those in non-invaded sites. However, chewing damage by herbivorous insects was greater in invaded sites on old *A. lamprocarpa* phyllodes and on old and new *E. tetrodonta* leaves. Thus the invasive ant was only as effective as the native ant on new *A. lamprocarpa* phyllodes, that is, those with the most active EFNs.

Foliage patrol and effects on plants are not just driven by the availability of extrafloral nectar, however. Ant density and activity levels and the cost (distance) of reaching active EFNs are also likely determinants of the interaction. For example, the availability of extrafloral nectar did not increase *A. gracilipes* (or *O. smaragdina*) herbivore attack on a native vine in tropical rainforest edges (Lach et al., 2016) in northern Queensland. The lack of an EFN effect and the much greater attack rates of *A. gracilipes* compared to *O. smaragdina* in this study were attributed to the relatively high *A. gracilipes* activity levels and the short stature of the vines (Lach et al., 2016).

The context dependency of the consequences of ant invasions for EFN plants revealed in Australian studies is broadly similar to findings globally. It is clear that ant-EFN interactions can be established quickly among non-co-evolved species, but whether the interaction can be considered beneficial to the plant is in part dependent on the comparison of services provided by native ants (Heil, 2015). A 2004 review of invasive ant interactions with EFNs found that in only half (6 of 12) of the studies in the ants' novel range was there strong evidence that the plant benefitted directly or indirectly from the invader (Ness & Bronstein, 2004) but for few of these was there explicit comparison to benefits provided by native ants. In the other six studies, the effect was either neutral, negative, or less convincingly positive (Ness & Bronstein, 2004). Studies of the consequences of invasive ant interactions with EFN plants since then have also revealed a range of consequences for the plant. For example, reduced herbivore abundance on barrel cactus (*Ferocactus viridiscens*: Cactaceae) in *L. humile*-invaded areas did not affect seed mass (Ludka et al., 2015); but *L. humile*-infested flowers on these plants had shorter pollinator visits and reduced seed set relative to those in uninvaded areas (LeVan et al., 2014; Chapter 14). Despite the common traits of invasive ants that affect ant-EFN interactions (Lach, 2003), the effects of invasive ants on EFN plants are not easily generalised or predicted because they are also dependent on other interactors and are best characterised by comparison to the effect of the native ants they have displaced.

Benefits flowing the other direction – from EFN plant to invasive ants – are less studied. Recognition of the role of carbohydrates in fuelling and sustaining ant invasions and influencing behaviour has grown (Grover et al., 2007; Savage & Whitney, 2011; Wilder et al., 2011b), but few studies explicitly investigate the role of EFNs. It would be expected that invasive ants would benefit from tending EFNs; the more intriguing question is whether they benefit *more* than do other ant species. A laboratory experiment in southwest Australia found no differences in survival among workers of invasive *L. humile*, *P. megacephala*, and native *Iridomyrmex chasei* with access to native *Acacia saligna* primed to produce extrafloral nectar, though workers of all three species survived longer than workers with access to *A. saligna* with inactive EFNs and workers with access to only water (Figure 15.3) (Lach et al., 2009). Research in American Samoa found a strong association between *A. gracilipes* abundance and dominance of an EFN plant; other ants at the sites lacked such a correlation (Savage et al., 2009). *Solenopsis invicta* laboratory colonies produce more brood and workers with access to an artificial nectar than without (Wilder et al., 2011a) and exclude native ants from EFNs in field conditions in their introduced range in the United States (Wilder et al., 2011b) but it is unclear that they benefit more than native ants would.

Ant-Plant Symbioses

Myrmecophytes are plants that house ants in exchange for a service, usually protection from herbivores. Africa and the Neotropics are well-known for Acacia (sensu lato) myrmecophytes that are protected by ants (Chapters 10 and 11), and research in these systems has provided a wealth of insights into ant-plant interactions and mutualisms more generally (e.g., Palmer & Brody, 2007; Heil, 2013). In contrast, in Australia, there are no known systems in which plant-ants provide protection from herbivores (Huxley, 1982).

Instead, all known Australian myrmecophytes are epiphytes and are presumed to receive nutrients from their resident ants. Australian myrmecophytes are restricted to the tropical north and fall within three genera: *Myrmecodia, Hydnophytum*

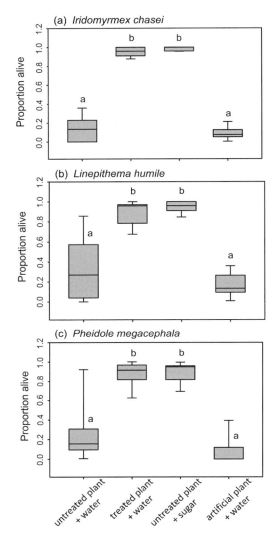

Figure 15.3. The proportion of worker ants alive (out of 25) by treatment for (a) the native ant *Iridomyrmex chasei*, (b) the invasive ant *Linepithema humile*, and (c) the invasive ant *Pheidole megacephala* (n = 10 for each species). Treated plants produced visible extrafloral nectar and untreated plants did not. Different letters above the boxes indicate significant differences between treatments at P < 0.008 for the same species. Modified from Lach et al. (2009) with permission from Springer.

(both Rubiaceae, subtribe Hydnophytinae), and *Dischidia* (Asclepiadaciae) (Huxley, 1982). Epiphytes have no access to soil nutrients and their host plants typically grow in nutrient poor soils (Janzen, 1974). Growth and isotope tracer experiments on *M. cf. tuberosa* in Papua New Guinea revealed ants provide nutrients to their host plants (Huxley, 1978). Herbivore protection by resident ants is thought to be absent across the ~80 species (Chomicki & Renner, 2015) of myrmecophytic Hydnophytinae (Janzen, 1974; Huxley, 1978; Rickson, 1979; Huxley, 1980), which occur predominantly in southeast Asia. These findings have since been replicated in other epiphytic myrmecophytes in the region (reviewed in Mayer et al., 2014), and from this body of work it has been inferred that partner ants provide food for Australian myrmecophytic epiphytes as well.

Both *Hydnophytum* and *Myrmecodia* spp. are primarily associated with the native ants *Philidris cordata* across Australasia (Huxley, 1978). *Myrmecodia beccarii*, an Australian endemic (Figure 15.4), has been reported to host the invasive ant *Pheidole megacephala* (Barrett, 1928; Commons & Waterhouse, 1981; Eastwood & Fraser, 1999). *Pheidole megacephala* is widespread in the region, and an obvious question is whether it interacts with *M. beccarii* in the same way as *Philidris cordata*. A recent pulse-chase experiment (Volp, 2015) in which ^{15}N-enriched glycine in sucrose solution was fed to *Philidris cordata* and *Pheidole megacephala* colonies inhabiting *M. beccarii* plants provides some insights, even with its low sample size (n = 5 plants per ant species). The experiment revealed that though both ant species picked up the ^{15}N label, the δ^{15}N values in leaves at three and six weeks post-ant-feeding were significantly greater than the δ^{15}N values before labelling in plants hosting *Philidris cordata*, but not in plants hosting *Pheidole megacephala*. Thus, though the low sample size indicates the results should be considered cautiously, less nitrogen appears to be taken up by the plant when the invader is hosted than when the native ant is hosted.

Conclusions and Future Directions

The study of invasive ant-plant interactions in Australia has confirmed and extended our knowledge of the ecology of ant-plant interactions and consequences of ant invasions. Since invader traits (e.g., small body size, abundance, aggression, predilection for carbohydrates) play a significant role in which interactions occur, and consequently their outcomes (Lach, 2003), we expect there to be substantial similarity in ant-plant interactions in Australia and elsewhere. However, attributes of the invaded habitat, and the plants and herbivores with which the ants interact, can be very different in Australia, and these differences also play a role in the nature and outcome of the interactions.

Australia is rich in opportunities for scientific discovery and still has many ecological secrets. It has 80 per cent of the land area of the United States, but just 7 per cent of the population the United States has (Brinkhoff, 2015), which means both that there is much more undeveloped land and many fewer people to study its ecology (indeed, Australia has < 1 per cent of the number of tertiary institutions as the United States). The many different invasive and tramp ant species that have established and the numerous urban and peri-urban native bushland reserves mean there are opportunities to study invasive ant-plant interactions in relatively undisturbed habitats. Study of ant-invaded offshore islands has already yielded insights into

Figure 15.4.   *Myrmecodia beccarii* (a) cross section and (b) on a host tree. (Photos: Mark Bloemberg, Lori Lach.)

multitrophic interactions and ant invasions and will continue to be fertile ground for research into dynamics of ant invasions and resources that fuel them.

There are at least two lines of inquiry in which invasive ant-plant research in Australia may be especially fruitful. One is more clearly defining the role of honeydew availability and composition in ant invasions and their management. The introduction of a parasitoid wasp to control *Tachardina auriantiaca*, and thereby indirectly control *Anoplolepis gracilipes*, on Christmas Island will yield insights whether it is successful, as is likely, and provides the ultimate evidence of the link between *A. gracilipes* invasion and honeydew, or whether it is unsuccessful because the relationships and dependencies shifted in some unforeseen way. Even where invasion has not been linked directly to a specific source of carbohydrates, understanding the temporal and spatial dynamics of the availability of these resources may be useful for management because they likely affect ant macronutrient preferences and therefore the take-up of toxic bait. The large-scale management of yellow crazy ants in Arnhem Land and the Wet Tropics are ideal places for such research. Finally, on mainland Australia, many of the typical honeydew-producing groups are largely endemic (e.g., Coccoidea, Eurymelinae, several genera, and species of Aphidae) (Austin et al., 2004), and therefore, there may be important, but as yet unstudied, differences in the quantity and quality of honeydew available in

Australia, with consequences for tending ants. Honeydew composition can alter ant behaviour and outcomes for plant hosts (Pringle et al., 2014) and one hypothesis that remains to be tested is whether poor quality honeydew can prevent invasion (Hoffmann & Kay, 2009).

A second line of inquiry for which Australia may be particularly suited is the evolution and ecology of epiphytic myrmecophytes. Relative to ant-protected myrmecophytes, little is known about how obligate the relationship is for plants that obtain nutrients from ants, why there are so few ant partners, and the mechanisms for partner filtering and nutrient provisioning. At least two species of fungi have been described in *Myrmecodia* in Papua New Guinea, and one of these may be a food source for resident ants, and the other may assist the ants in feeding the plants (Huxley, 1978). *M. beccarii* in Australia also have fungi (Huxley, 1982), and their diversity and role in the ant-plant interaction require investigation. In other myrmecophyte-ant relationships investigated in Africa and South America, ants consume fungi (Blatrix et al., 2012), and fungi mediate the provision of nitrogen to the plants (Defossez et al., 2011; Leroy et al., 2011). Much more remains to be done to understand the nature of these tripartite symbioses. Comparison of native ant functioning in *Myrmecodia beccarii* to that of an ant species that does not share an evolutionary history with the plant or fungi will provide insights into the evolution and ecology of this plant species, as well as to ant-plants more generally.

References

Abbott, K. L. and P. T. Green. (2007). Collapse of an ant-scale mutualism in a rainforest on Christmas Island. *Oikos* 116, 1238–1246.

Abbott, K. L., P. T. Green, and D. J. O'Dowd. (2014). Seasonal shifts in macronutrient preferences in supercolonies of the invasive yellow crazy ant *Anoplolepis gracilipes* (Smith, 1857) (Hymenoptera: Formicidae) on Christmas Island, Indian Ocean. *Austral Entomology* 53, 337–346.

Altfeld, L. and P. Stiling. (2006). Argentine ants strongly affect some but not all common insects on *Baccharis halimifolia*. *Environmental Entomology* 35, 31–36.

Andersen, A. N. (1990). The use of ant communities to evaluate change in Australian terrestrial ecosystems: a review and a recipe. *Proceedings of the Ecological Society of Australia* 16, 347–357.

(2003). Ant biodiversity in arid Australia: productivity, species richness and community organisation. *Records of the South Australian Museum Monograph Series* 7, 79–92.

(2007). Ant diversity in arid Australia: a systematic overview. In *Advances in ant systematics (Hymenoptera: Formicidae): homage to E.O. Wilson–50 years of contributions*, ed. R. R. Snelling, B. L. Fisher, and P. S. Ward. Gainesville, FL: Memories of the American Entomological Institute, pp. 19–51.

AntWeb. (2010) Available from www.antweb.org/taxonomicPage.do?rank=genus&project=australiaants&images=true&genCache=true. Accessed 1 February 2016.

AntWiki. (2010). Australian ant distribution patterns. Available from www.antwiki.org/wiki/Australian_Ant_Distribution_Patterns. Accessed 22 January 2016.

Asfiya, W., P. Yeeles, L. Lach, J. D. Majer, B. Heterick, and R. K. Didham. (2016). Abiotic factors affecting the foraging activity and potential displacement of native ants by the invasive African big-headed ant *Pheidole megacephala* (FABRICIUS, 1793) (Hymenoptera: Formicidae). *Myrmecological News* 22, 43–54.

Austin, A. D., D. K. Yeates, G. Cassis, M. J. Fletcher, J. La Salle, J. F. Lawrence, P. B. McQuillan, L. A. Mound, D. J. Bickel, P. J. Gullan, D. F. Hales, and G. S. Taylor. (2004). Insects 'down under' – diversity, endemism and evolution of the Australian insect fauna: examples from select orders. *Australian Journal of Entomology* 43, 216–234.

Australian Bureau of Statistics. How many people live in Australia's coastal areas? Available from www.abs.gov.au/ausstats/abs@.nsf/Previousproducts/1301.0Feature%20 Article32004. Accessed 15 March 2016.

Australian Floral Estimates. (2009). Available from www.anbg.gov.au/aust-veg/australian-flora-statistics.html. Accessed 1 February 2016.

Barrett, C. (1928). Ant-house plants and their tenants. *Victorian Naturalist*, 133–137.

Bentley, B. L. (1977). Extrafloral nectaries and protection by pugnacious bodyguards. *Annual Review of Ecology and Systematics* 8, 407–427.

Berg, R. (1975). Myrmecochorous plants in Australia and their dispersal by ants. *Australian Journal of Botany* 23, 475–508.

Blancafort, X. and C. Gomez. (2005). Consequences of the Argentine ant, *Linepithema humile* (Mayr), invasion on pollination of *Euphorbia characias* (L.) (Euphorbiaceae). *Acta Oecologica-International Journal of Ecology* 28, 49–55.

Blatrix, R., C. Djieto-Lordon, L. Mondolot, P. La Fisca, H. Voglmayr, and D. McKey. (2012). Plant-ants use symbiotic fungi as a food source: new insight into the nutritional ecology of ant-plant interactions. *Proceedings of the Royal Society B-Biological Sciences* 279,3940–3947.

Blüthgen, N. and K. Reifenrath. (2003). Extrafloral nectaries in an Australian rainforest: structure and distribution. *Australian Journal of Botany* 51, 515–527.

Blüthgen, N., N. E. Stork, and K. Fiedler. (2004). Bottom-up control and co-occurrence in complex communities: honeydew and nectar determine a rainforest ant mosaic. *Oikos* 106, 344–358.

Brinkhoff, Thomas. 2015. City population: World Map. Available from http://world.bymap .org/LandArea.html. Accessed 15 March 2016.

Burwell, C. J., A. Nakamura, A. McDougall, and V. J. Neldner. (2012). Invasive African big-headed ants, *Pheidole megacephala*, on coral cays of the southern Great Barrier Reef: distribution and impacts on other ants. *Journal of Insect Conservation* 16, 777–789.

Callan, S. K. and J. D. Majer. (2009). Impacts of an incursion of African big-headed ants, *Pheidole megacephala* (Fabricius), in urban bushland in Perth, Western Australia. *Pacific Conservation Biology* 15, 102–115.

Chamberlain, S. A. and J. N. Holland. (2009). Quantitative synthesis of context dependency in ant–plant protection mutualisms. *Ecology* 90, 2384–2392.

Chapman, A. D. (2009). Numbers of living species in Australia and the world. Australian Biological Resources Study, Canberra. Available at www.environment.gov.au/ biodiversity/abrs/publications/other/species-numbers/2009/index.html.

Chomicki, G. and S. S. Renner. (2015). Phylogenetics and molecular clocks reveal the repeated evolution of ant-plants after the late Miocene in Africa and the early Miocene in Australasia and the Neotropics. *New Phytologist* 207, 411–424.

Christian, C. E. (2001). Consequences of a biological invasion reveal the importance of mutualism for plant communities. *Nature* 413, 635–639.

Commons, I. F. B. and D. F. Waterhouse. (1981). *Butterflies of Australia*. East Melbourne, Victoria: CSIRO.

Davidson, D. W., S. C. Cook, R. R. Snelling, and T. H. Chua. (2003). Explaining the abundance of ants in lowland tropical rainforest canopies. *Science* 300, 969–972.

Defossez, E., C. Djieto-Lordon, D. McKey, M. A. Selosse, and R. Blatrix. (2011). Plant-ants feed their host plant, but above all a fungal symbiont to recycle nitrogen. *Proceedings of the Royal Society B-Biological Sciences* 278, 1419–1426.

Department of the Environment. 2016. Where do weeds come from? Available from www .environment.gov.au/biodiversity/invasive/weeds/weeds/where/index.html. Accessed 22 January 2016.

Eastwood, R. and A. M. Fraser. (1999). Associations between lycaenid butterflies and ants in Australia. *Australian Journal of Ecology* 24, 503–537.

Economo, E. P. and B. Guenard. (2016). Overall species richness. Available from antmaps. org/? Accessed 1 February 2016.

Environmental Resources Information Network. (2012). *Interim biogeographic regionalization for Australia (IBRA), version 7*. Department of Sustainability, Environment, Water, Population and Communities, Canberra.

Gaigher, R. and M. J. Samways. (2013). Strategic management of an invasive ant-scale mutualism enables recovery of a threatened tropical tree species. *Biotropica* 45, 128–134.

Gibb, H. and S. A. Cunningham. (2009). Does the availability of arboreal honeydew determine the prevalence of ecologically dominant ants in restored habitats? *Insectes Sociaux* 56,405–412.

Gove, A. D., J. D. Majer, and R. R. Dunn. (2007). A keystone ant species promotes seed dispersal in a "diffuse" mutualism. *Oecologia* 153, 687–697.

Greenslade, P. (2008). Climate variability, biological control and an insect pest outbreak on Australia's Coral Sea islets: lessons for invertebrate conservation. *Journal of Insect Conservation* 12, 333–342.

Grover, C. D., A. D. Kay, J. A. Monson, T. C. Marsh, and D. A. Holway. (2007). Linking nutrition and behavioural dominance: carbohydrate scarcity limits aggression and activity in Argentine ants. *Proceedings of the Royal Society B-Biological Sciences* 274,2951–2957.

Heil, M. (2013). Let the best one stay: screening of ant defenders by acacia host plants functions independently of partner choice or host sanctions. *Journal of Ecology* 101, 684–688.

(2015). Extrafloral nectar at the plant-insect interface: a spotlight on chemical ecology, phenotypic plasticity, and food webs. In *Annual Review of Entomology,* Vol. 60, ed. M. R. Berenbaum, pp. 213–232.

Helms, K. R. (2013). Mutualisms between ants (Hymenoptera: Formicidae) and honeydew-producing insects: are they important in ant invasions? *Myrmecological News* 18, 61–71.

Hoffmann, B. D. (2015). Integrating biology into invasive species management is a key principle for eradication success: the case of yellow crazy ant *Anoplolepis gracilipes* in northern Australia. *Bulletin of Entomological Research* 105, 141–151.

Hoffmann, B. D. and A. Kay. (2009). *Pisonia grandis* monocultures limit the spread of an invasive ant-a case of carbohydrate quality? *Biological Invasions* 11, 1403–1410.

Hoffmann, B. D. and W. C. Saul. (2010). Yellow crazy ant (*Anoplolepis gracilipes*) invasions within undisturbed mainland Australian habitats: no support for biotic resistance hypothesis. *Biological Invasions* 12, 3093–3108.

Holway, D. A., L. Lach, A. V. Suarez, N. D. Tsutsui, and T. J. Case. (2002). The causes and consequences of ant invasions. *Annual Review of Ecology and Systematics* 33, 181–233.

Huxley, C. R. (1978). The ant-plants *Myrmecodia* and *Hydnophytum* (Rubiaceae), and the relationships between their morphology, ant occupants, physiology and ecology. *New Phytologist* 80, 231–268.

(1980). Symbiosis between ants and epiphytes. *Biological Reviews of the Cambridge Philosophical Society* 55, 321–340.

(1982). Ant-epiphytes of Australia. In *Ant-plant interactions in Australia,* ed. R. C. Buckley. The Hague: Dr W. Junk Publishers, pp. 63–73.

Janzen, D. H. (1974). Epiphytic myrmecophytes in Sarawak: mutualism through the feeding of plants by ants. *Biotropica* 6, 237–259.

Kaiser-Bunbury, C. N., H. Cuthbert, R. Fox, D. Birch, and N. Bunbury. (2014). Invasion of yellow crazy ant *Anoplolepis gracilipes* in a Seychelles UNESCO palm forest. *NeoBiota* 22, 43–57.

Kaspari, M., D. Donoso, J. A. Lucas, T. Zumbusch, and A. D. Kay. (2012). Using nutritional ecology to predict community structure: a field test in Neotropical ants. *Ecosphere* 3(11),93.

Kay, A. D., T. Zumbusch, J. L. Heinen, T. C. Marsh, and D. A. Holway. (2010). Nutrition and interference competition have interactive effects on the behavior and performance of Argentine ants. *Ecology* 91, 57–64.

Lach, L. (2003). Invasive ants: unwanted partners in ant-plant interactions? *Annals of the Missouri Botanical Garden* 90, 91–108.

(2007). A mutualism with a native membracid facilitates pollinator displacement by Argentine ants. *Ecology* 88, 1994–2004.

Lach, L. and G. Barker. (2013). Assessing the effectiveness of tramp ant projects to reduce impacts on biodiversity. Australian Government Department of Sustainability, Water, Population and Communities, Canberra. Available at www.environment.gov.au/biodi-versity/invasive/publications/tramp-ant-projects.

Lach, L., R. J. Hobbs, and J. D. Majer. (2009). Herbivory-induced extrafloral nectar increases native and invasive ant worker survival. *Population Ecology* 51, 237–243.

Lach, L. and B. D. Hoffmann. (2011). Are invasive ants better plant-defense mutualists? A comparison of foliage patrolling and herbivory in sites with invasive yellow crazy ants and native weaver ants. *Oikos* 120,9–16.

Lach, L. and L. M. Hooper-Bùi. (2010). Consequences of ant invasions. In *Ant ecology,* ed. Lach, L., C. L. Parr, and K. L. Abbott. Oxford: Oxford University Press, pp. 261–286.

Lach, L. and C. Hoskin. (2015). Too much to lose: yellow crazy ants in the Wet Tropics. *Wildlife Australia* 52, 37–41.

Lach, L. and M. L. Thomas. (2008). Invasive ants in Australia: documented and potential ecological consequences. *Australian Journal of Entomology* 47, 275–288.

Lach, L., C. V. Tillberg, and A. V. Suarez. (2010). Contrasting effects of an invasive ant on a native and an invasive plant. *Biological Invasions* 12, 3123–3133.

Lach, L., T. M. Volp, T. A. Greenwood, and A. Rose. (2016). High invasive ant activity drives predation of a native butterfly larva. *Biotropica* 48, 146–149.

Lengyel, S., A. D. Gove, A. M. Latimer, J. D. Majer, and R. R. Dunn. (2010). Convergent evolution of seed dispersal by ants, and phylogeny and biogeography in flowering plants: a global survey. *Perspectives in Plant Ecology Evolution and Systematics* 12, 43–55.

Leroy, C., N. Sejalon-Delmas, A. Jauneau, M. X. Ruiz-Gonzalez, H. Gryta, P. Jargeat, B. Corbara, A. Dejean, and J. Orivel. (2011). Trophic mediation by a fungus in an ant-plant mutualism. *Journal of Ecology* 99, 583–590.

LeVan, K. E., K.-L. J. Hung, K. R. McCann, J. T. Ludka, and D. A. Holway. (2014). Floral visitation by the Argentine ant reduces pollinator visitation and seed set in the coast barrel cactus, *Ferocactus viridescens*. *Oecologia* 174, 163–171.

Loope, L. L. and P. D. Krushelnycky. (2007). Current and potential ant impacts in the Pacific region. *Proceedings of the Hawaiian Entomological Society* 39, 69–73.

Lubertazzi, D., M. A. A. Lubertazzi, N. McCoy, A. D. Gove, J. D. Majer, and R. R. Dunn. (2010). The ecology of a keystone seed disperser, the ant *Rhytidoponera violacea*. *Journal of Insect Science* 10, Article number 158; doi: http://dx.doi.org/10.1673/031.010.14118.

Ludka, J., K. E. Levan, and D. A. Holway. (2015). Infiltration of a facultative ant-plant mutualism by the introduced Argentine ant: effects on mutualist diversity and mutualism benefits. *Ecological Entomology* 40, 437–443.

Majer, J. D., A. D. Gove, S. Sochacki, P. Searle, and C. Portlock. (2011). A comparison of the autecology of two seed-taking ant genera, *Rhytidoponera* and *Melophorus*. *Insectes Sociaux* 58, 115–125.

Majer, J. D. and B. E. Heterick. 2015. Invasive ants on the Australia mainland: the other 24 species. *Anais XXII Simpósio de Mirmecologia: An International Meeting*, Bahia, Ilhéus, Brazil.

Marazzi, B., J. L. Bronstein, and S. Koptur. (2013). The diversity, ecology and evolution of extrafloral nectaries: current perspectives and future challenges. *Annals of Botany* 111, 1243–1250.

Mayer, V. E., M. E. Frederickson, D. McKey, and R. Blatrix. (2014). Current issues in the evolutionary ecology of ant-plant symbioses. *New Phytologist* 202, 749–764.

Mesler, M. R. and K. L. Lu. (1983). Seed dispersal of *Trillium ovatum* (Liliaceae) in second-growth redwood forests. *American Journal of Botany* 70, 1460–1467.

Morton, S. R., D. M. S. Smith, C. R. Dickman, D. L. Dunkerley, M. H. Friedel, R. R. J. McAllister, J. R. W. Reid, D. A. Roshier, M. A. Smith, F. J. Walsh, G. M. Wardle, I. W. Watson, and M. Westoby. (2011). A fresh framework for the ecology of arid Australia. *Journal of Arid Environments* 75, 313–329.

Ness, J. H. and J. L. Bronstein. (2004). The effects of invasive ants on prospective ant mutualists. *Biological Invasions* 6, 445–461.

Ness, J. H., J. L. Bronstein, A. N. Andersen, and J. N. Holland. (2004). Ant body size predicts dispersal distance of ant-adapted seeds: Implications of small-ant invasions. *Ecology* 85, 1244–1250.

Neumann, G., D. J. O'Dowd, P. J. Gullan, and P. T. Green. (2016). Diversity, endemism and origins of scale insects on a tropical oceanic island: Implications for management of an invasive ant. *Journal of Asia-Pacific Entomology* 19, 159–166.

O'Dowd, D. J., P. T. Green, and P. S. Lake. (1999). *Status, Impact, and Recommendations for Research and Management of Exotic Invasive Ants in Christmas Island.* National Park. Center for the Analysis and Management of Biological Invasions, Monash University. (2003). Invasional 'meltdown' on an oceanic island. *Ecology Letters* 6, 812–817.

Orians, G. H. and A. V. Milewski. (2007). Ecology of Australia: the effects of nutrient-poor soils and intense fires. *Biological Reviews* 82, 393–423.

Osunkoya, O. O., C. Polo, and A. N. Andersen. (2011). Invasion impacts on biodiversity: responses of ant communities to infestation by cat's claw creeper vine, *Macfadyena unguis-cati* (Bignoniaceae) in subtropical Australia. *Biological Invasions* 13, 2289–2302.

Palmer, T. M. and A. K. Brody. (2007). Mutualism as reciprocal exploitation: African plant-ants defend foliar but not reproductive structures. *Ecology* 88, 3004–3011.

Pringle, E. G., A. Novo, I. Ableson, R. V. Barbehenn, and R. L. Vannette. (2014). Plant-derived differences in the composition of aphid honeydew and their effects on colonies of aphid-tending ants. *Ecology and Evolution* 4, 4065–4079.

Rickson, F. R. (1979). Absorption of animal tissue breakdown products into a plant stem – feeding of a plant by ants. *American Journal of Botany* 66, 87–90.

Rodriguez-Cabal, M. A., K. L. Stuble, M. A. Nunez, and N. J. Sanders. (2009). Quantitative analysis of the effects of the exotic Argentine ant on seed-dispersal mutualisms. *Biology Letters* 5, 499–502.

Rosumek, F. B., F. A. O. Silveira, F. D. Neves, N. P. D. Barbosa, L. Diniz, Y. Oki, F. Pezzini, G. W. Fernandes, and T. Cornelissen. (2009). Ants on plants: a meta-analysis of the role of ants as plant biotic defenses. *Oecologia* 160, 537–549.

Rowles, A. D. and D. J. O'Dowd. (2009a). Impacts of the invasive Argentine ant on native ants and other invertebrates in coastal scrub in south-eastern Australia. *Austral Ecology* 34, 239–248.

    (2009b). New mutualism for old: indirect disruption and direct facilitation of seed dispersal following Argentine ant invasion. *Oecologia* 158, 709–716.

Rowles, A. D. and J. Silverman. (2009). Carbohydrate supply limits invasion of natural communities by Argentine ants. *Oecologia* 161, 161–171.

Savage, A. M., J. A. Rudgers, and K. D. Whitney. (2009). Elevated dominance of extrafloral nectary-bearing plants is associated with increased abundances of an invasive ant and reduced native ant richness. *Diversity and Distributions* 15, 751–761.

Savage, A. M. and K. D. Whitney. (2011). Trait-mediated indirect interactions in invasions: unique behavioral responses of an invasive ant to plant nectar. *Ecosphere* 2(9),106.

Simmons, M. (2009). Distinctive features of *Acacia*. Available from http://anpsa.org.au/aca-feat.html. Accessed 1 February 2016.

Smith, D. P., D. Haliam, and J. Smith (2004). Biological control of '*Pulvinaria urbicola*' (Cockerell) (Homoptera: Coccidae) in a '*Pisonia grandis*' forest on North East Herald Cay in the Coral Sea [online]. *General and Applied Entomology: The Journal of the Entomological Society of New South Wales* 33, 61–68.

State of Queensland. (2010). Managing scale insect outbreaks in the Capricornia Cays. Queensland Parks and Wildlife Service, State of Queensland, Brisbane. Available at www.nprsr.qld.gov.au/parks/capricornia-cays/pdf/scale-insect.pdf.

Stone, G. N., N. E. Raine, M. Prescott, and P. G. Willmer. (2003). Pollination ecology of acacias (Fabaceae, Mimosoideae). *Australian Systematic Botany* 16, 103–118.

Styrsky, J. D. and M. D. Eubanks. (2007). Ecological consequences of interactions between ants and honeydew-producing insects. *Proceedings of the Royal Society B-Biological Sciences* 274, 151–164.

    (2010). A facultative mutualism between aphids and an invasive ant increases plant reproduction. *Ecological Entomology* 35, 190–199.

Thomas, M. L. and D. A. Holway. (2005). Condition-specific competition between invasive Argentine ants and Australian *Iridomyrmex*. *Journal of Animal Ecology* 74, 532–542.

Trager, M. D., S. Bhotika, J. A. Hostetler, G. V. Andrade, M. A. Rodriguez-Cabal, C. S. McKeon, C. W. Osenberg, and B. M. Bolker. (2010). Benefits for plants in ant-plant protective mutualisms: a meta-analysis. *PLoS ONE* 5,e14308.

Volp, T. M. (2015). Interactions between the epiphytic ant-plant *Myrmecodia beccarii* and its ant inhabitants. Honours thesis. James Cook University, Cairns.

Warren, R. J. and I. Giladi. (2014). Ant-mediated seed dispersal: a few ant species (Hymenoptera: Formicidae) benefit many plants. *Myrmecological News* 20, 129–140.

Warren, R. J., A. McMillan, J. R. King, L. Chick, and M. A. Bradford. (2015). Forest invader replaces predation but not dispersal services by a keystone species. *Biological Invasions* 17, 3153–3162.

Weber, M. G., L. D. Porturas, and K. H. Keeler. (2015). World list of plants with extrafloral nectaries. Available at www.extrafloralnectaries.org. Accessed 3 March 2016.

Westoby, M., K. French, L. Hughes, B. Rice, and L. Rodgerson. (1991a). Why do more plant species use ants for dispersal on infertile compared with fertile soils? *Australian Journal of Ecology* 16, 445–455.

Westoby, M., L. Hughes, and B. L. Rice. (1991b). Seed dispersal by ants; comparing infertile soils with fertile soils. In *Ant-plant interactions,* ed. C. R. Huxley and D. F. Cutler. Oxford: Oxford University Press, pp. 434–447.

Wetterer, J. K. (2005). Worldwide distribution and potential spread of the long-legged ant, *Anoplolepis gracilipes* (Hymenoptera: Formicidae). *Sociobiology* 45, 77–97.

Wilder, S. M., T. R. Barnum, D. A. Holway, A. V. Suarez, and M. D. Eubanks. (2013). Introduced fire ants can exclude native ants from critical mutualist-provided resources. *Oecologia* 172, 197–205.

Wilder, S. M., D. A. Holway, A. V. Suarez, and M. D. Eubanks. (2011a). Macronutrient content of plant-based food affects growth of a carnivorous arthropod. *Ecology* 92, 325–332.

Wilder, S. M., D. A. Holway, A. V. Suarez, E. G. LeBrun, and M. D. Eubanks. (2011b). Intercontinental differences in resource use reveal the importance of mutualisms in fire ant invasions. *Proceedings of the National Academy of Sciences of the United States of America* 108, 20639–20644.

World Wide Wattle. Distribution and phytogeography of *Acacia sens. lat.* Available from http://worldwidewattle.com/infogallery/distribution/. Accessed 15 March 2016.

Part V

Applied Ant Ecology:
Agroecosystems, Ecosystem
Engineering, and Restoration

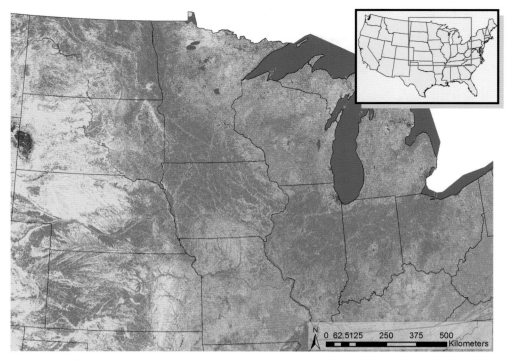

Figure 1.1   Land use and land cover map of the Midwest US region of North America. The major land cover types in this region include cultivated crops (brown), deciduous forest (green), pasture/hay (yellow), grassland (tan), low-intensity developed (pink), and high-intensity developed (red) (Homer et al., 2015, reproduced with permission.)

Figure 2.1.   Arboreal ant nests in tree canopies of New Guinea lowland forests. (a) Nest of *Colobopsis* cf. *macrocephala* in a dead twig – example of internal nest site in host tree. (b) Carton nest of *Polyrhachis luteogaster* – example of external nest site on a leaf of host tree (photographs by P. Klimes). See also Table 2.1.

Figure 3.2. Bird's-nest fern (*Asplenium nidus*) in the high canopy of lowland Dipterocarp rain forest in Malaysian Borneo. The largest ferns reach 200 kg wet weight (Ellwood & Foster 2004) and can support diverse arthropod communities, including multiple colonies of co-existing ants. Inset photograph shows a colony of ant belonging to the genus *Diacamma*, one of many species that excavate nesting cavities in the root mass of these ferns. Main photograph credit Chien C. Lee; inset Tom Fayle.

Figure 7.5.   Seed dispersal and ant-mediated seedling establishment in the Atlantic forest tree, *Clusia criuva*. (a) Open mature fruit-exposing diaspores (each with up to 17 seeds) coated by a red, lipid-rich aril. (b) *Turdus albicollis* (white-necked thrush), the main frugivore and seed disperser in the tree canopy (Photo: Octavio C. Salles). (c) Bird dropping beneath a fruiting tree containing embedded seeds with bits of red aril attached. (d) Worker of *Odontomachus chelifer* transporting a fallen diaspore to its nest, where the aril will be fed to larvae and the viable seeds discarded nearby. (e) Seedlings of *C. criuva* clumped in the vicinity of a nest of *O. chelifer* (tagged "N"). After Passos & Oliveira (2002), photo (d) reproduced with permission from John Wiley and Sons.

Figure 8.1.   (a) Leaf of *Mallotus japonicus* with ants, *Pristomyrmex punctatus*. (b) *P. punctatus* collecting extrafloral nectar of *M. japonicus*. (c) *Pheidole noda* attacking caterpillar of *Parasa lepida* (Lepidoptera: Limacodidae) on leaf of *M. japonicus*. Photos by Akira Yamawo (a, c) and Hiromi Mukai.

Figure 10.3. An African elephant (*Loxodonta africana*) sniffs an *Acacia drepanolobium*, assessing how well-defended the plant is by symbiotic ants. The four acacia ant species produce pungent and different volatile alarm pheromones (Wood et al. 2002) when they detect herbivores, which elephants may use as cues that indicate both the identity of the ant species in residence, as well as the density of the resident colony. Elephants tend to attack *Acacia drepanolobium* with very low densities of acacia ants, or those occupied by *Crematogaster sjostedti*, the least aggressive ant species. Photo: Kathleen Rudolph.

Figure 11.1. The myrmecophytic plant *Leonardoxa africana* subsp. *africana* has hollow internodes in which nests the mutualistic ant *Petalomyrmex phylax*. This symbiosis is endemic to a narrow strip of coastal rainforest in Cameroon and is threatened by anthropogenic pressure, in particular land-use change. Photo credit: Rumsais Blatrix.

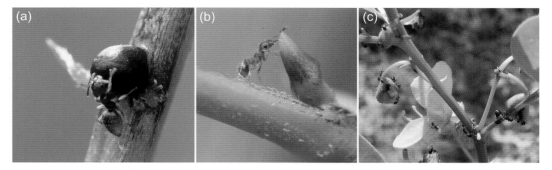

Figure 12.2. *Senna* extrafloral nectaries with ant visitors – left to right: (a) *Senna chapmannii* with *Wasmannia auropunctata*; (b) *Senna ligustrina* with same ant species; (c) *Solenopsis invicta* on *Senna chapmannii*. Photos by Maria Cleopatra Pimienta (a & b) and Ian Jones (c).

Figure 14.1. *Linepithema humile* workers foraging at the extrafloral nectaries of *Ferocactus viridescens*. Plants produce carbohydrate-rich extrafloral nectar from modified spines at the crown on the plant, often in locations associated with the production of reproductive structures. The white depression at top centre indicates that a fruit was previously attached at that location. Photo credit: David A. Holway.

Figure 15.2.   Yellow crazy ants (*Anoplolepis gracilipes*) tending whitefly (*Neomaskellia bergii*) on sugarcane near Edmonton, Queensland. (Photo: Mark Bloemberg.)

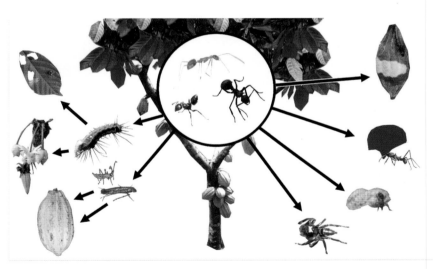

Figure 16.4.   Ecosystem services (left) and disservices (right) mediated by ants in cacao agroforests. Ants protect the plants against insects feeding on the leaves, flowers, and pods. At the same time they can spread pathogens, tend mealybugs and other hemipteran and reduce the numbers of other predators, such as spiders. The strengths of the different interactions depend on the ant species and the pest and disease community, which differ between regions. The species shown are the ants (clockwise starting from the top) *Anoplolepis gracilipes*, *Crematogaster* sp., *Polyrhachis* sp. and, with (clockwise starting from the top right) a cacao pod with black pod disease *Phytophthora palmivora*, a leaf-cutter ant *Atta* sp., mealybugs with *Philidris* cf. *cordata*, a salticid spider, the cacao pod-borer *Conopomorpha cramerella* with symptoms on a cacao pod, a juvenile mirid *Helopeltis sulawesi*, an erebid caterpillar feeding on a cacao flower, and a psychid larvae causing leaf damage. Artwork by Yann Clough.

Figure 17.5 *Crematogaster* ants foraging on coffee beans in a Colombian coffee plantation. Photo by Andrés López.

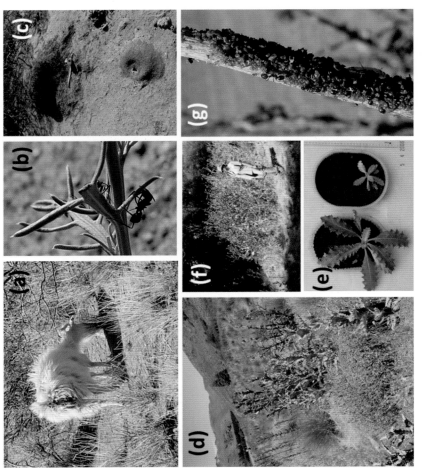

Figure 18.1. Photographs illustrating the consequences of introduced livestock and road maintenance on the role of leaf-cutting ants as ecological engineers in Patagonia, Argentina. (a–c): Influence of exotic livestock on the soil-improver effects of LCA nests in Monte Desert. Livestock reduce plant richness and cover through grazing (a), depleting the diet of the LCA *Acromyrmex lobicornis* (b), with the subsequent reduction of the nutrient content of their external refuse dumps (RD) (c). (d–g): Consequences of road building and maintenance on LCA density, exotic plant species, and associated aphid tending-ants relationships in a Patagonian steppe. Nest of *A. lobicornis* and associated exotic thistles in roadside areas (d). Exotic thistles grow better on external refuse dumps of *A. lobicornis* than on non-nest soils (e), forming huge "exotic plant islands" around ant nests (f). Plants growing on ant refuse dumps also sustain more aphid density, increasing the abundance and activity of aphid-tending ants (g). Photo credits: M. Tadey (a, b); A. G. Farji-Brener (c–g).

16 Services and Disservices of Ant Communities in Tropical Cacao and Coffee Agroforestry Systems

Yann Clough, Stacy Philpott, and Teja Tscharntke

Tropical tree crops such as cacao and coffee are produced around the tropics in diverse, multistrata agroforests as well as monoculture plantations (Box 16.1 and references therein). The smallholders cultivating these systems battle pests and diseases that differ regionally and change over time, but often take a significant part of their yield, and therefore their revenue. In these perennial systems, ants are tremendously diverse and abundant, and affect pests and diseases directly as well as indirectly. Management by farmers of particular ant species to control insect pests has a long history (Offenberg, 2015). It is not until recently that the effects of ants on yields have been quantified. The complex interactions through which ants affect the crop plants, and how their mediation by species- and community-level characteristics, are starting to be better understood. The extent of the impact ants have on yields and revenue justifies the anthropocentric framing of the outcome of these interactions in terms of ecosystem services and disservices. In this chapter we present the current state of knowledge on agroforest ant communities, economically relevant ecological interactions driven by these communities and the way landscape-scale land-use change and climate change can be expected to influence ants and ant effects on insect communities and yields. Finally, we discuss how farmers may adapt their management to support ant-mediated ecosystem services and minimize potential disservices. We refer to Del Toro et al. (2012) and Choate and Drummond (2011) for more broad reviews of the role of ants in agriculture, as providers of biological control and other ecosystem services and disservices.

Taxonomically and Functionally Rich Ant Communities

Ant surveys from cacao and coffee systems from throughout the range of these crops show a very high species richness that in most cases is comparable to that found in undisturbed forests (Table 16.1). Agroforests harbor arboreal and ground-dwelling ants. This includes species that nest in the canopy and trunk of the trees (dead wood, hollow twigs, foliage, sometimes with carton/silk/dirt nesting structures), in the herb layer, in the litter layer, on open ground, in epiphytic and parasitic plants,

Box 16.1

Agroforestry systems combine management of trees and crops. They are multi-strata systems comprising multiple planted species of trees, sometimes with remnant natural forest trees and crop plants. They can be established in a thinned forest plot, as homegardens, or on previously bare/shrubby/tilled land. In their

Figure 16.1.   Little shaded cacao (a), shaded cacao (b), full-sun coffee (c) and shaded coffee (d). Photographs by Yann Clough (a, b) and Stacy Philpott (c, d).

most complex form they strongly resemble natural forests (Figure 16.1b, d), in their simplest they are close to a monoculture with one crop species (Figure 16.1a, c), sometimes a sparse, monospecific shade-tree cover. All intermediate complexity levels are found. For a typology of agroforestry systems see Moguel and Toledo (1999). For cacao (*Theobroma cacao*) and coffee (*Coffea arabica*), but also rubber (*Hevea brasiliensis*), there is a marked worldwide trend away from agroforests to full-sun monocultures documented for all three of these crops (coffee: Philpott, Arendt, Armbrecht et al., 2008; Philpott, Bichier & Rice et al., 2008; Jha et al., 2014; cacao: Clough et al., 2009; Ruf, 2011; Tscharntke et al., 2011; rubber: Ekadinata & Vincent, 2011).

dead wood debris and other plant residues, such as dry cacao pods on the ground or on the tree (Room, 1971; De la Mora et al., 2013; Castaño-Meneses et al., 2015).

The ant communities of forests and agroforests overlap only partly. In Sulawesi, 75 per cent of the ant species recorded in forest plots can also be found in nearby cacao plots (Bos et al., 2007), but overlap can be lower in more intensive, unshaded systems (Asfiya et al., 2015). Ant communities of Indonesian rubber agroforests can readily be distinguished from both forest and rubber monoculture communities, despite the species richness being similar between these systems, as well as oil palm (Rubiana et al., 2015; see Chapter 3). In coffee farms in Colombia, about 25 per cent of ant species found in forests are also found in diverse shade farms, whereas only around 2–6 per cent of species are shared with simplified shade coffee or sun coffee farms and forests (Armbrecht et al., 2005). Agroforests and monoculture plantations, as other disturbed systems, tend to host more generalist, invasive or tramp ant species than undisturbed forest. The presence of some of these species can be associated with locally reduced ant species richness, as shown for *Anoplolepis gracilipes* Smith (Bos et al., 2008) and *Philidris* cf. *cordata* (Wielgoss et al., 2010).

Impact of Ant Communities on the Crop

Most ants are predatory to some extent and can benefit plants by consuming or deterring herbivorous insects. In none of the intensively studied systems, however, can the impact of ants on the crop be summarized as a simple trophic chain linking the ants, herbivores and the crop. Much to the contrary, ants and agroforestry crops are linked by large, complex webs of trophic and non-trophic interactions (Perfecto et al., 2014; Wielgoss et al., 2014, Figures 16.2 and 16.3). Nitrogen stable isotope ratios have been used to differentiate between mainly predatory species and species drawing their resources from plant parts (e.g. elaiosomes) and honeydew. Data from Sulawesi cacao (Gras, 2015) show that ant communities in cacao occupy a wide trophic range (Figure 16.2), similar to that found in forests of Peru and Borneo (Davidson et al., 2003).

Some ants can feed on the crop plants themselves. In Latin America, leaf-cutter ants such as *Atta cephalotes* L., *Atta sexdens sexdens* L. and *Acromyrmex subterraneus brunneus* Forel cut young leaves and flowers (see Chapters 4 and 18), an activity that can retard the development of young cacao trees (Delabie, 1990) potentially causing delayed returns on planting investments. Chewing of bark of buds and shoots by several species was reported from cacao in Southern Bahia, Brazil. The damage this causes has to our knowledge not been quantified economically, but is considered minimal for most species (Delabie, 1990).

For the overwhelming majority of the ant species, however, effects on plants are indirect. The impact of ants occurs via the pests and pathogens that are claiming substantial parts of cacao and coffee yields (Keane & Putter, 1992; Table 16.2), and possibly also via the pollinators of these crops.

Table 16.1 Observed Species Richness of Ant Communities in Tropical Agroforests

Region	Habitat	Shading	Number of morpho-species in tree crop	Number of morpho-species in forest	Method	Extent	Study
Lampung, Sumatra, Indonesia	Coffee		214	171	Multiple		Philpott, Bichier & Rice et al., 2008
Central Sulawesi, Indonesia	Cacao	Low-high	40	40	Fogging	24 trees, different plots	Bos et al., 2007
Ghana	Cacao	unknown	~250		unknown		Bolton, unpublished data; Room et al., 1971
Ghana	Cacao	None	108		Multiple	One farm	Room et al., 1971
Central Cameroon	Cacao	Low-high	38		Hand sampling + baiting on cacao trees	20 farms × 30 trees	Bisseleua et al., 2013
Central Sulawesi, Indonesia	Cacao	Low-high	80		Hand sampling + baiting on cacao trees and ground	16 plots × 10 trees	Rizali et al., 2013b
Chiapas, Mexico	Coffee	Low-High	107	73	Multiple methods	30 coffee sites, 10 forest fragments	De la Mora et al., 2013
Central Sulawesi, Indonesia	Cacao	Low-high	160		Baiting on cacao trees + ground	44 plots × 10 trees	Wielgoss et al., 2010
Jambi, Sumatra, Indonesia	Rubber	High*	42	31–48	Hand sampling and baiting	8 plots per system	Rubiana et al., 2015
Papua New Guinea	Cacao						Room & Smith, 1975
Central Cameroon	Cacao		60			4 plots × 100 trees	Tadu et al., 2014
South Bahia, Brazil	Cacao	unknown	175	130		One hectare	Delabie unpublished; Leston, 1978 cited in Delabie, 1990

Values reported are for the whole study unless stated otherwise. Methods and extent of sampling are reported as an indicator of sampling intensity, which is rarely exhaustive. Direct sampling refers to visual search and capture of foragers.

* Mature rubber trees are often part of the canopy rather than understory trees.

Trophobionts as Mediators of Services and Disservices

While it is true that most ants are predatory (Figure 16.3a), many tend other organisms, most notably Hemiptera, from which they harvest honeydew as a source of carbohydrates, in exchange for some degree of protection from predators (Figure 16.3b). The food providers in such a symbiotic association ('trophobiosis') are termed trophobionts. From the point of view of the plants, trophobionts impose costs, as they divert nutrients and can spread diseases. It is not surprising then that there has been significant debate about the suitability of many trophobiont-tending ant species as biological control agents. In reality, when considering ant-plant-hemiptera interactions, about three-fourth of the studies show net benefits for the plants (Styrsky & Eubanks, 2007), and many ant species important for biological control are also dependent on trophobionts.

The black cacao ant *Dolichoderus thoracicus* Smith found in Southeast Asia is an effective predator or deterrent of mirid pests, the cacao pod borer and mammal pests such as rats (Khoo & Ho, 1992). This translates into substantial beneficial effects on yield and income (24 per cent more on trees with *D. thoracicus* than without ants, Wielgoss et al., 2014). Yet *D. thoracicus*, and the control it provides, is totally dependent on the presence of associated trophobionts, to the extent that efforts to establish the ant in new plantations will fail if trophobionts are not also transferred (Ho & Khoo, 1997).

In Latin America, several trophobiont-tending ant species, including *Azteca* spp., are efficient predators of the coffee berry borer *Hypothenemus hampei* (Ferrari), a small bark beetle that burrows into the coffee fruits (Larsen & Philpott, 2010; Gonthier et al., 2013; Jiménez-Soto et al., 2013; Chapter 17). *Azteca* ants are also negatively related to the incidence of coffee rust *Hemileia vastatrix* Berk & Broome (1869), currently the most serious coffee disease in Latin America. Key to this complex interaction is the trophobiosis between *Azteca sericeasur* Longino (formerly reported as being *Azteca instabilis*) and the coffee green scale *Coccus viridis* (Green). At high densities, the latter is attacked by the white halo fungus *Lecanicillium lecanii* (Zimm.) (Zare & W. Gams, 2001) that also parasitizes the coffee rust (Vandermeer et al., 2009; Jackson et al., 2012; Perfecto et al., 2014). This leads to a negative relationship between the ant and the rust that holds across spatial scales (10 m to 45 ha) (Vandermeer et al., 2009). Not only the presence per se, but also the location at which the trophobionts are concentrated by the ants can mediate the impact of the ants on the crop. Comparing the effect of two dolichoderine ant species on cacao, Wielgoss et al. (2014) found that leaf herbivory was decreased most by the species tending mealybugs on the leaves, while the species tending the same mealybug species on the cacao pods had the highest impact on mirid damage on the pods.

Trophobionts are not always innocuous, however. In West Africa, the Cocoa Swollen Shoot Virus is a very severe disease that kills trees, i.e. with crop losses up to 100 per cent (Ploetz, 2016). To control it, farmers must remove and uproot the

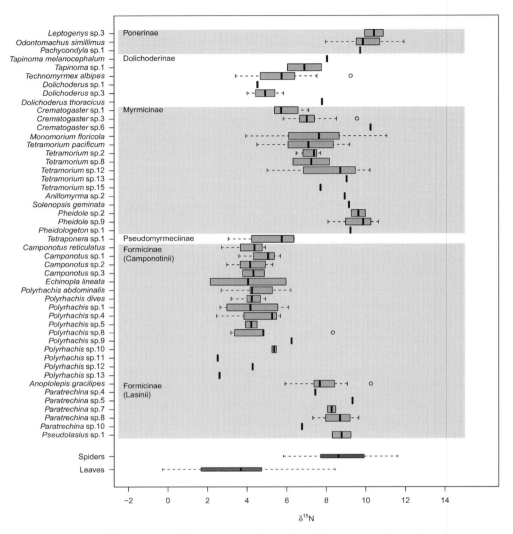

Figure 16.2. Variation in δ^{15}N between different ant species collected from cacao plantations in Sulawesi, Indonesia. Values for cacao leaves and spiders added for comparison. Ants are presented ordered by family (and tribe in the case of Formicidae). The right axis shows number of samples analysed and number of plantations covered by the samples. Reproduced with permission from Gras (2015).

infected trees, endangering the sustainability of the cultivation in parts of the region (Ploetz, 2016). The disease is transmitted by mealybugs (Pseudococcidae) that are tended predominantly by *Pheidole megacephala* Fabricius, a ground-nesting species, and different species of *Crematogaster* that are arboreal and build carton nests and tents (Strickland, 1951). The prospects of being able to manipulate ant assemblages to the detriment of such undesirable species initiated research on ant mosaics, i.e.

Table 16.2 Impacts of Ants on Major Pests and Diseases of Cacao and Coffee

Crop	Region	Pest/Disease	Ant species	Impact	References
Cacao	South-east Asia	Mirids *Helopeltis* spp.	*Dolichoderus thoracicus, Oecophylla smaragdina, Philidris* cf. *cordata*	Reduction	Way & Khoo, 1989; Way & Khoo, 1991; Khoo & Ho, 1992; Wielgoss et al., 2014
	South-east Asia	Cacao pod borer *Conopomorpha cramerella*	*Dolichoderus thoracicus*	Reduction	See & Khoo, 1996;
			Philidris cf. *cordata*	Increase (indirect)	Wielgoss et al., 2014
	South-east Asia	Black pod *Phytophthora palmivora*	*Crematogaster* spp., *Iridomyrmex* spp., *Solenopsis geminata, Philidris* cf. *cordata*	Dissemination	McGregor & Moxon, 1985; Wielgoss et al., 2014
	South-east Asia	Vascular-streak dieback (*VSD*) *Oncobasidium theobroma*		None reported	
	Latin America	Witches broom *Crinipellis* (formerly *Moniliophthora*) *perniciosa*		None reported	
	Latin America	Frosty pod Rot (or Moniliasis) *Moniliophthora roreri*		None reported	
	West Africa	Capsids *Sahlbergella singularis* and *Distantiella theobromae*	*Oecophylla longinoda*	Reduction	Ayenor et al., 2007; Babin et al., unpublished data
	West Africa	Black pod *Phytophthora megakarya* and *P. palmivora*	*Camponotus acvapimensis, Pheidole megacephala, Crematogaster striatula* and others	Dissemination	Evans, 1973; Babacauh, 1982
	West Africa	Cacao swollen-shoot virus (CSSV)	*Crematogaster* spp., *Pheidole* spp.	Tending of virus vector (Coccids)	Strickland, 1951; Hanna et al., 1956
Coffee	Latin America	Coffee rust *Hemileia vastatrix*	*Azteca sericeasur* (formerly referred to as *A. instabilis*)	Reduction, indirect	Vandermeer et al., 2009; Jackson et al., 2012
	Latin America	Coffee berry borer *Hypothenemus hampei*, Ferrari	*A. sericeasur, Pheidole synanthropica, Pseudomyrmex ejectus, P. simplex, Tapinoma* sp., *Wasmannia auropunctata*	Reduction	Armbrecht & Gallego, 2007; Gonthier et al., 2013; Morris et al., 2015
	East Africa	Coffee twig borer, *Xylosandrus compactus*	*Plagiolepis* sp.	Reduction	Egonyu et al., 2015

Figure 16.3. *Philidris* cf. *cordata* workers preying on a caterpillar (a); workers of the same species tending mealybugs under tents (b); the partitioning of harvested cacao pods into healthy (top left, opened pods), with uneven ripening symptoms caused by the cacao pod borer (middle left) and black pod disease (front). Photos by Arno Wielgoss (a) and Yann Clough (b, c).

the spatial patterns created by non-overlapping territories of dominant ant species, which is a major topic in ant community ecology (Leston, 1970; Majer, 1972).

Ants can increase pathogen pressure on the crop not only by tending trophobionts that are vectors, but also by building protective structures over nests and trophobiont aggregations using soil and plant material (Figure 16.3b). Species such as *Philidris* cf. *cordata* in Indonesia, or *Crematogaster striatula*,

Camponotus acvapimensis and *Pheidole megacephala* in West Africa (Evans, 1973), have been associated with increased losses due to black pod disease *Phytophthora palmivora* and *P. megakarya* (Figure 16.3c). The material these species use to build tents, soil particles or plant material from remains of disease-infected cacao pods lying on the ground can contain high loads of disease spores (Evans, 1973). Once brought up into the tree by the ants, these spores act as a source of inoculum for other pods. In the case of *Ph*. cf. *cordata* this mechanism is a major cause for the lower yields observed in cacao plots occupied by this species.

Ants and Pollination

Ants are not frequently considered to be pollinators, and in fact sometimes negatively affect plant reproduction. Ants may deter pollinators from visiting flowers; they can act as nectar thieves (Ghazoul, 2001), flower predators (Galen & Cuba, 2001) and may reduce pollen viability via antibiotic secretions (Beattie et al., 1984; Wagner, 2000). However, in agroforests, positive impacts of ants on the pollination of the crop have been found, despite the ants visiting the flowers without contributing directly to pollination (coffee: Free, 1993; cacao: Clough, personal observation). In an ant/pollinator exclosure study, Philpott et al. (2006) documented that plants to which both ants and flying pollinators had access had higher fruit weight than plants with only pollinators, or neither ants or pollinators, but only in diversified coffee farms with higher pollinator abundance and richness. Although the mechanisms could not yet be identified, ants may influence flying pollinators such that they spend less time on a particular flower and move between plants more, thereby increasing the amount of pollen, or the diversity of pollen deposited on flowers. These high pollen loads may lead to higher pollen tube growth, early fertilization and a longer time period for the maturation of the fruit (Niesenbaum, 1999). Further, flying pollinators are vectors of floral microbes that can affect fruit set and fruit weight – thus some interaction between ants, flying pollinators and microbes may have driven differences in fruit weights (Vannette et al., 2017). In Sulawesi cacao, fruit set was higher on trees occupied by dolichoderine ants than on trees from which ants were excluded, possibly due to higher cross-pollination associated with the more frequent disturbance of the pollinators (Wielgoss et al., 2014).

Net Effects of Ants on Yields and Revenue

Published studies reporting complete exclusions of ants from tropical agroforestry crops in combination with a quantification of the impact on the trees are rare. Using a well-replicated exclusion experiment across multiple sites in cacao agroforests of Sulawesi, Wielgoss et al. (2014) showed that ant exclusion reduced yields by 27 per cent on average. Where ants had access, fruit abortion and leaf herbivory were significantly lower than where ants were excluded, leading to higher yields.

This equals an estimated loss in revenue of 875 US$ ha^{-1} yr^{-1}. The overall result could be confirmed for the same crop in the same region, but at different altitude with differing ant communities, by Gras et al. (2016), with 40 per cent lower yields under ant exclusion on average.

In a correlational study conducted in Ghana, Ayenor et al. (2007) showed significantly higher yields (70 per cent) in cacao plots with colonies of the weaver ant *Oecophylla longinoda* Latreille compared with the control, significantly so in one out of three years.

Philpott et al. (2008b) excluded ants from a subset of coffee trees in the Soconusco region of Chiapas, Mexico, to examine impacts on arthropod communities, including the coffee berry borer and coffee yields. Although the ant communities in control trees partly included *Azteca sericeasur*, shown to be an effective predator of the berry borer in other studies (Gonthier et al., 2013; Morris et al., 2015; see Chapter 17), ant exclusion had no overall effect on arthropod densities or yield. Ant exclusion did have effects on particular arthropod groups, with some orders (Orthoptera) responding positively to ant presence and others (Collembola, Lepidoptera, scale insects) responding negatively to ant presence. Such effects likely resulted in changes in communities on plants with and without ants, but had no impact on overall arthropod numbers.

What to Look For in an Ant Community?

Identifying characteristics of ants and ant communities that are useful for biological control is an important endeavour that could help systematizing and generalizing management recommendations.

Most individual ant species considered useful for biological control are dominant ant species (Way & Khoo, 1992). Dominant ants are numerous in the areas they occupy and can exclude other ant species from their territories. They can be more or less aggressive and may or may not be behaviourally dominant. Wielgoss (2013) found that across cacao trees dominated by different ant species communities, the best predictor for reduction in leaf herbivory in cacao was the ant abundance at tuna baits, corrected for worker size. Using abundance attained by a species could be highly misleading, however, because the effect of ants on plants is much more complex. Wielgoss et al. (2014) introduced two dolichoderine species, *Dolichoderus thoracicus* and *Philidris* cf. *cordata* into cacao plots, comparing them with ant excluded plots and unmanipulated plots with species-rich ant communities. Both introduced dolichoderine species seemed perfectly suited as a biological control agent given their abundance, aggressiveness and willingness to colonize trap-nests (Wielgoss et al., 2014), but it turned out that their effects on the yield could hardly have been more different. In the case of *D. thoracicus*, the yields were similarly high as in the unmanipulated controls. In the case of *Ph.* cf. *cordata*, yields were even lower (34 per cent less than in the unmanipulated plots) than in the ant exclusion, because these ants participate in the spread of *Phytophthora palmivora*, the black pod disease. Furthermore, by decreasing mirid damage, *Ph.* cf. *cordata* has indirect beneficial effects on the most economically important pest, the cacao pod borer *Conopomorpha cramerella* Snellen, whose egg-laying females avoid cacao pods that bear feeding marks by the economically less important mirids (Clough, 2012; Wielgoss et al., 2012, 2014).

The role of ant diversity in the predatory function has been debated (Gove, 2006; Philpott & Armbrecht, 2006). In principle the complementarity of species in temporal and spatial niches and the preference of differently sized predators for differently sized prey should result in a higher predatory pressure exerted in diverse communities, i.e. with more or a higher share of non-dominant species. Using observation at baits and exposure of prey, Philpott et al. (2008a) and Wielgoss (2007) compared different arboreal species in coffee and cacao plantations and show that they differ substantially in their predatory behaviour. This gives support to the complementarity hypothesis. At the same time, their data also show that one or two species tend to account for the largest impact on the prey, at least under somewhat simplified settings (studies of short duration and use of exposed prey). A major obstacle to demonstrating benefits of diversity is the difficulty of manipulating more than one or two ant species in the field in a context where multiple interaction partners, vegetation complexity and variability of environmental conditions across space and time would make a diverse predator assemblage most useful. Wielgoss et al. (2014) showed that ant communities with a higher evenness in abundance among species were associated with higher cacao yields, but this effect cannot be completely separated from that of the identity of the dominant species manipulated in the experiment.

Recently, non-dominant ant species were shown to be critically important in maintaining biocontrol of pests when the dominant ant species is under threat by its own natural enemies. In coffee systems in southern Mexico, *Azteca* ants are responsible for most of the predation on the coffee berry borer, yet other ant species take over the predation of the borer whenever specialist parasitic phorid flies force *Azteca* ants to reduce their activity (Philpott et al., 2012). Phorids are common in Latin American coffee systems, but are also present elsewhere. In Indonesian cacao, the tents built by *Philidris* cf. *cordata* to shield trophobionts and worker pathways could be partly a response to these enemies. While phorids have been observed attacking this or related species in Sulawesi forests (Disney, 1986), they have not yet been observed in the cacao plantations despite many hours of observation.

The similar productivity of cacao trees with unmanipulated, species-rich communities and cacao trees with introduced black cocoa ants suggests it may not be necessary to introduce ants as biological control agents. While dominant ant species can be very important, maintaining diverse communities could be critical to support ant-related ecosystem services over time. The studies reviewed show that an understanding of the species and the key interactions is critical and can be system- and species-specific, but that avoiding facilitating aggressive species that are deleterious to the crop should be a priority.

Figure 16.4.   Ecosystem services (left) and disservices (right) mediated by ants in cacao agroforests. Ants protect the plants against insects feeding on the leaves, flowers and pods. At the same time they can spread pathogens, tend mealybugs and other hemipteran, and reduce the numbers of other predators, such as spiders. The strengths of the different interactions depend on the ant species and the pest and disease community, which differ between regions. The species shown are the ants (clockwise starting from the top) *Anoplolepis gracilipes*, *Crematogaster* sp., *Polyrhachis* sp. and, with (clockwise starting from the top right) a cacao pod with black pod disease *Phytophthora palmivora*, a leaf-cutter ant *Atta* sp., mealybugs with *Philidris* cf. *cordata*, a salticid spider, the cacao pod borer *Conopomorpha cramerella* with symptoms on a cacao pod, a juvenile mirid *Helopeltis sulawesi*, an erebid caterpillar feeding on a cacao flower and a psychid larvae causing leaf damage. Artwork by Yann Clough. (A black-and-white version of this figure will appear in some formats. For the color version, please refer to the plate section.)

Ants and Management of Agroforests

Since the composition of the ant communities drives the balance between services and disservices to the farmer (Figure 16.4), it is important to realize that ant communities are not static over time. From the establishment to the replanting/abandonment of an agroforest, processes such as the growth of the main tree crop, the vegetation succession after forest thinning and crop establishment (Majer, 1972), as well as potential shade tree thinning, pruning (Philpott, 2005) or removal of all trees affect the resources available to ants. As the tree crops grow, the surface of trunk, branches and foliage increases, and cracks, branch stumps and epiphytes appear, creating nesting habitats. The consequence is a change in ant communities and an increase in ant species richness, at least in the initial years (Dejean et al., 2008; Kone et al., 2014; Conceição et al., 2015). Invasive and/or undesirable species such as *Philidris* cf. *cordata* can benefit from nesting resources such as cracks in the bark. The spread of mistletoe in cacao plantations, common in neglected unshaded plantations in West Africa, leads to increases in ant species that can utilize the microhabitats and the mealybugs associated with these parasitic plants. The same ant species can tend the mealybug species that carry Cacao Swollen Shoot Virus, however. The mistletoe thus contributes to the spread of the pathogen (Room, 1972a, 1972b).

Proper pruning of cacao trees to avoid creating cracks in the bark and removal of mistletoes are best practice cultural techniques that can reduce these problems. Overall, knowledge about how cultural techniques affect the impact of ants on pests and diseases is virtually non-existent.

Shading Practices and Ant Communities

The widespread reduction in shade tree cover and the adoption of full-sun management practices (Box 16.1) have a large impact on ant communities. Recent work from Southern Cameroon shows, for instance, that *Oecophylla longinoda* Latreille is associated with sunny areas and *Crematogaster* species with shaded areas, while *Tetramorium aculeatum* (Mayr) occurs irrespective of shade levels (Tadu et al., 2014). All of these species were considered beneficial against mirid pests. In coffee, *Azteca sericeasur*, shown to be effective against a range of coffee pests, is associated with certain shade trees and can be negatively (Philpott, 2005) or positively affected (Jackson et al., 2014) by shade tree removal depending on the extent of the habitat manipulation.

The effects of shade tree removal on ants can also be mediated by the increase in ground vegetation. Areas with low shade often have high grass cover, which can be accompanied by the appearance of ant species such as the fire ant *Solenopsis geminata* that tends root aphids feeding on the grasses (Perfecto & Vandermeer, 1996). This is especially relevant agronomically since *S. geminata* suppresses ant species that predate on the coffee berry borer (Trible & Carroll, 2014).

The joint preference of ants and pests for particular levels of shading has made it difficult to assess the impact of ants under contrasting shade levels using correlational studies (Majer, 1976; Ayenor et al., 2007). To circumvent this issue, Gras et al. (2016) excluded ants from trees across a shade gradient. Ants were found to increase yields most at low to intermediate shade levels, with little effect under high shade levels. Loss of flowers to herbivores was prevented by ants across the shade gradient, but fruit abortion in response to herbivore damage and diseases was prevented by ants only under low to medium shade levels. The causes for this pattern are a combination of higher pest pressure under low to medium shade levels and to the difference in composition in the ant communities across the shade gradient (Gras et al., 2016).

Insecticide Use

The impact of insecticide use on ants in agroforests is not well studied. De la Mora et al. (2013) found no significant effect of insecticides on ant abundance and species richness of ground ants or arboreal ants in coffee in Mexico. In cacao of Sulawesi, effects of insecticides on diversity (abundance was not recorded) could not be detected, but the species composition differed greatly between the insecticide treated and the control plots, with species such as *Oecophylla smaragdina* Fabricius being found only in control plots (Rizali et al., 2013a). In Brazil, the direct targeting of certain ant species (*Azteca* spp.) with insecticides is possibly related to the spread of *Wasmannia auropunctata* (Roger), a species considered to be less beneficial (Delabie, 1990).

Ant Community Manipulation

Besides cultural practices that aim to support desirable ant communities, direct manipulation of ant communities is being practiced in different parts of the world. The tending of the weaver ant *O. smaragdina* as biological control agents in citrus plantations has a long history in Asia (Offenberg, 2015). In Indonesia and Malaysia the black cacao ant *D. thoracicus* and its mealybugs are being managed in cacao plantations (Giesberger, 1983; Hosang et al., 2010, Figure 16.5).

Delabie (1990) mentions that farmers in Bahia, Brazil, introduce *Azteca charifex spiritii* Forel to farm areas where it was previously absent by distributing nest fragments. Offenberg (2015) provides an exhaustive overview of ant introductions to plantations and emphasizes that accompanying actions are often necessary, such as reducing the existing ant species prior to establishment, manipulating the between-tree connections, providing supplemental nest and food resources. In Colombian and Mexican coffee plantations, twig-nesting ant populations are nest-limited and the provision of suitable nesting resources can increase the populations (Armbrecht et al., 2004; Philpott & Foster, 2005). Many of the same species that occupy the artificial nests are known predators of coffee pests; so nests might be used as a way to increase populations of ants that provide pest control services.

A little-investigated means to manipulate ant communities is the provision of artificial sugar sources. The provisioning of sugar feeders can be used to attract ants to those parts of a plantation that are in need of their services (see e.g. Maňák et al., 2013). It can also be used to interfere with the symbiosis between ants and trophobionts. The principle is that ants may neglect and/or even start to prey on trophobionts once saturated with an artificial sugar supply, since they do not need the honeydew produced by hemipterans. This manipulation is probably not investigated in tropical agroforestry but has recently utilized to reduce aphid infestations in apple orchards (Nagy et al., 2013, 2015).

Figure 16.5. Workers of the black cacao ant *Dolichoderus thoracicus* in artificial nest sites made of bamboo and palm leaves, placed in the branches of a cacao tree in Sulawesi, Indonesia. Photo by Yann Clough.

Limits to Predictability

The mosaic of dominant ants can change very rapidly as shown recently in cacao plantations in Cameroon that were exhaustively mapped two years in a row (Tadu et al., 2014). This study shows that while the distribution of ants is linked to the shade, there is substantial variability in the spatial distribution of the ants from year to year. Compared to the large attention given to spatial patterns, as demonstrated by the large literature on ant mosaics, surprisingly little attention has been given to these temporal dynamics. The role of interspecific competition has been emphasized (Greenslade, 1971), but in reality the relative role of management, stochastic changes in resource abundance, predatory pressure (e.g. predation pressure by toads on *A. gracilipes*; Wanger et al., 2011) and interactions between ant colonies is poorly understood. We know that ant communities and the distribution of individual species within agroforests change rapidly over time, but the effects this has on the associated entomofauna, pests, diseases and yields are yet to be investigated.

Effects of Large-Scale Land-Use and Climate Change

Large-scale changes in landscape-scale land-use – such as loss of shade trees, changes in dominant crops and loss of remaining forest areas – are often common where agroforestry crops are grown. At the same time, significant changes in temperature and precipitation are expected during this century for coffee- and cacao-growing regions (Schroth et al., 2015, 2016).

Ants and Landscape Context

In agroforests, ant communities are primarily affected by local factors such as shading practices and other aspects of vegetation complexity and its management, as well as insecticide use. Several studies show, however, that ant communities may depend significantly on the regional land-use context. In a study of landscape and local-scale influences of ant communities in Mexican coffee, De la Mora et al. (2015) showed that the species richness and abundance of twig-nesting and leaf litter ants was correlated with distance to forest, positively and negatively, respectively. Arboreal ants and ants collected from rotten logs were affected only by local factors. In Sulawesi, resampling ant communities surveyed 6–9 years previously showed contrasting community-level changes depending on the surrounding land-use (Rizali et al., 2013a). In a landscape surrounded by forests, ant species richness in cacao was maintained, while in a landscape experiencing large-scale forest encroachment species richness decreased significantly over time. In both regions, ant communities had changed significantly. A marked change was the increase in *Ph.* cf. *cordata*, whose deleterious effects on cacao yields have been detailed earlier. Another is the decrease in *Anoplolepis gracilipes*, whose spread in less-shaded plantations had been highlighted by Bos et al. (2008).

Altitude, Temperature and Precipitation

Altitude and temperature are major drivers of ant species richness and community composition (Bishop et al., 2014), which is important for cacao and coffee agroforests that are often grown across larger altitudinal ranges. In natural forests, intermediate altitudes tend to have the highest abundances and species densities of ants (Samson et al., 1997). At higher altitudes, the decrease in abundance of ants is reflected in decreased predation pressure exerted by ants on caterpillars (Sam et al., 2014). Twig-nesting ant communities in coffee in Chiapas, Mexico, are most abundant and diverse at intermediate altitudes (~900–1100m) (Gillette et al., 2015). In Sulawesi, arboreal and ground ant species richness increased from 400 to 900 m altitude (Wielgoss et al., 2010). Agroforests at lower altitudes were more susceptible to invasive species, with the higher temperatures favouring *Ph.* cf. *cordata* and *A. gracilipes* (Bos et al., 2008; Wielgoss et al., 2010), the former being deleterious to crop productivity. Thus, while the role of ants for predation may increase with temperature at intermediate altitudes, the associated higher pressure by invasive ants can be risky for the crop. Little is known about precipitation, but Rizali et al. (2013b) showed that low precipitation was linked to a decrease in between-cacao tree ant diversity (effects on the crop were not recorded). Future changes in precipitation and temperature patterns are expected to change the ant communities and the associated services and disservices. However, the data currently available does not allow to make any general predictions on the direction or size of the impact, both due to the scarcity of the studies available, but also because the impact is conditional on the joint effect of climate change on ants, pests, diseases and management adaptation by the farmers.

Conclusion and Outlook

Cacao and coffee producers are facing significant challenges, with pests and diseases threatening the sustainability of production in different regions. The work conducted in recent years in cacao and coffee agroforests has demonstrated the importance of ant-mediated interactions for the incidence of pests and diseases and the productivity of agroforests. We are starting to better understand the complex interactions through which ants impact the agroecosystems, but need to better understand how the ants and the interactions are affected by the actions of the farmers. Several other areas, for example, the drivers and consequences of temporal change in ant communities and their distribution in space also commend more attention. The biggest and most important challenge, arguably, will be to help farmers diagnose the ant communities they have on their crops and make ecologically sound management choices that take into account the potential of these communities to provide services, or disservices.

References

Armbrecht, I. and Gallego, M. C. (2007). Testing ant predation on the coffee berry borer in shaded and sun coffee plantations in Colombia. *Entomologia Experimentalis et Applicata*, 124, 261–267.

Armbrecht, I., Perfecto, I. and Vandermeer, J. (2004). Enigmatic biodiversity correlations: ant diversity responds to diverse resources. *Science*, 304, 284–286.

Armbrecht, I., Rivera, L. and Perfecto, I. (2005). Reduced diversity and complexity in the leaf-litter ant assemblage of colombian coffee plantations. *Conservation Biology*, 19, 897–907.

Asfiya, W., Lach, L., Majer, J., Heterick, B. and Didham, R. (2015). Intensive agroforestry practices negatively affect ant (Hymenoptera: Formicidae) diversity and composition. *Asian Myrmecology*, 7, 87–104.

Ayenor, G. K., Van Huis, A., Obeng-Ofori, D., Padi, B. and Roeling, N.G. (2007). Facilitating the use of alternative capsid control methods towards sustainable production of organic cocoa in Ghana. *International Journal of Tropical Insect Science*, 27, 85–94.

Babacauh, K. D. (1982). Role of insect communities and water in the dissemination of *Phytophthora palmivora* (Butl.) Butl. emend. Bras. & Griff. in cacao plantations in the Ivory Coast. *Café Cacao Thé*, 26, 31–36.

Beattie, A. J., Turnbull, C., Knox, R. B. and Williams, E. G. (1984). Ant inhibition of pollen function – a possible reason why ant pollination is rare. *American Journal of Botany*, 71, 421–426.

Bishop, T. R., Robertson, M. P., Rensburg, B. J. and Parr, C. L. (2014). Elevation–diversity patterns through space and time: ant communities of the Maloti-Drakensberg Mountains of southern Africa. *Journal of Biogeography*, 41, 2256–2268.

Bisseleua, H. B. D., Fotio, D., Missoup, A. D. and Vidal, S. (2013). Shade tree diversity, cocoa pest damage, yield compensating inputs and farmers' net returns in West Africa. *PloS one*, 8, e56115.

Bos, M. M., Steffan-Dewenter, I. and Tscharntke, T. (2007). The contribution of cacao agroforests to the conservation of lower canopy ant and beetle diversity in Indonesia. *Biodiversity and Conservation*, 16, 2429–2444.

Bos, M. M., Tylianakis, J. M., Steffan-Dewenter, I. and Tscharntke, T. (2008). The invasive Yellow Crazy Ant and the decline of forest ant diversity in Indonesian cacao agroforests. *Biological Invasions*, 10, 1399–1409.

Castaño-Meneses, G., Mariano, C. S., Rocha, P. et al. (2015). HYMENOPTERA: The ant community and their accompanying arthropods in cacao dry pods: an unexplored diverse habitat. *Dugesiana*, 22, 1.

Choate, B. and Drummond, F. (2011). Ants as biological control agents in agricultural cropping systems. *Terrestrial Arthropod Reviews*, 4, 157–180.

Clough, Y. (2012). A generalized approach to modeling and estimating indirect effects in ecology. *Ecology*, 93, 1809–1815.

Clough, Y., Faust, H. and Tscharntke, T. (2009). Cacao boom and bust: sustainability of agroforests and opportunities for biodiversity conservation. *Conservation Letters*, 2, 197–205.

Conceição, E. S., Delabie, J. H. C., Della Lucia, T. M. C., Costa-Neto, A. D. O. and Majer, J. D. (2015). Structural changes in arboreal ant assemblages (Hymenoptera: Formicidae) in an age sequence of cocoa plantations in the south-east of Bahia, Brazil. *Austral Entomology*, 54, 315–324.

Davidson, D. W., Cook, S. C., Snelling, R. R. and Chua, T. H. (2003). Explaining the abundance of ants in lowland tropical rainforest canopies. *Science*, 300, 969–972.

Dejean, A., Djiéto-Lordon, C., Céréghino, R. and Leponce, M. (2008). Ontogenetic succession and the ant mosaic: an empirical approach using pioneer trees. *Basic and Applied Ecology*, 9, 316–323.

Del Toro, I., Ribbons, R. R. and Pelini, S. L. (2012). The little things that run the world revisited: a review of ant-mediated ecosystem services and disservices (Hymenoptera: Formicidae). *Myrmecological News*, 17, 133–146.

Delabie, J. H. C. (1990). The ant problems of cocoa farms in Brazil. In *Applied Myrmecology: A World Perspective*, ed. R. K. van der Meer, K. Jaffé and A. Cedeño. Boulder, CO: Westview Press, pp. 555–569.

De la Mora, A., García-Ballinas, J. A. and Philpott, S. M. (2015). Effects of local and landscape factors on predatory impacts of ants in coffee landscapes. *Agriculture, Ecosystems, and Environment*, 201, 83–91.

De la Mora, A., Murnen, C. J. and Philpott, S. M. (2013) Local and landscape drivers of ant-communities in Neotropical coffee landscapes. *Biodiversity and Conservation*, 22, 871–888.

Disney, R. H. L. (1986). A new genus and three new species of Phoridae (Diptera) parasitizing ants (Hymenoptera) in Sulawesi. *Journal of Natural History*, 20, 777–787.

Egonyu, J. P., Baguma, J., Ogari, I. et al. (2015). The formicid ant, *Plagiolepis* sp., as a predator of the coffee twig borer, *Xylosandrus compactus*. *Biological Control*, 91, 42–46.

Ekadinata, A. and Vincent, G. (2011). Rubber agroforests in a changing landscape: analysis of land use/cover trajectories in Bungo district, Indonesia. *Forests, Trees and Livelihoods*, 20, 3–14.

Evans, H. C. (1973). Invertebrate vectors of *Phytophthora palmivora*, causing black pod disease of cocoa in Ghana. *Annals of Applied Biology*, 75, 331–345.

Free, J. B. (1993). *Insect Pollination of Crops*. 2nd Enlarged Edition. London: Academic Press.

Galen, C. and Cuba, J. (2001). Down the tube: Pollinators, predators, and the evolution of flower shape in the alpine skypilot. *Polemonium viscosum. Evolution*, 55, 1963–1971.

Ghazoul, J. (2001). Can floral repellents pre-empt potential ant-plant conflicts? *Ecology Letters*, 4, 295–299.

Giesberger, G. (1983). Biological control of the *Helopeltis* pest of cocoa in Java. *Archives of Cocoa Research*, 2, 1900–1950.

Gillette, P. N., Ennis, K. K., Domínguez Martínez, G. and Philpott, S. M. (2016). Change in species richness, abundance, and composition of arboreal twig-nesting ants along an elevational gradient in coffee landscapes. *Biotropica*, 47, 711–722.

Gonthier, D. J., Ennis, K. K., Philpott, S. M., Vandermeer, J. and Perfecto, I. (2013). Ants defend coffee from berry borer. *Biological Control*, 58, 815–820.

Gove, A. D. (2007). Ant biodiversity and the predatory function. (A response to Philpott and Armbrecht, 2006). *Ecological Entomology*, 32, 435.

Gras, P. (2015). Trophic interactions of ants, birds and bats affecting crop yield along shade gradients in tropical agroforestry. PhD thesis, Georg-August University of Göttingen, Germany.

Gras, P., Tscharntke, T., Maas, B. et al. (2016) How ants, birds and bats affect crop yield along shade gradients in tropical cacao agroforestry. *Journal of Applied Ecology.* DOI: 10.1111/1365–2664.12625

Greenslade, P. J. M. (1971). Interspecific competition and frequency changes among ants in Solomon Islands coconut plantations. *Journal of Applied Ecology*, 8, 323–352.

Hanna, A. D., Judenko, E. and Heatherington, W. (1956). The control of *Crematogaster* ants as a means of controlling the mealybugs transmitting the swollen-shoot virus disease of cacao in the Gold Coast. *Bulletin of Entomological Research*, 47, 219–226.

Ho, C. T. and Khoo, K. C. (1997). Partners in biological control of cocoa pests: mutualism between *Dolichoderus thoracicus* (Hymenoptera: Formicidae) and *Cataenococcus hispidus* (Hemiptera: Pseudococcidae). *Bulletin of Entomological Research*, 87, 461–470.

Hosang, M. L. A., Schulze, C. H., Tscharntke, T. and Buchori, D. (2010). The potential of artificial nesting sites for increasing the population density of the black cacao ants. *Indonesian Journal of Agriculture*, 3, 45–50.

Jackson, D., Skillman, J. and Vandermeer, J. (2012). Indirect biological control of the coffee leaf rust, *Hemileia vastatrix*, by the entomogenous fungus *Lecanicillium lecanii* in a complex coffee agroecosystem. *Biological Control*, 61, 89–97.

Jackson, D., Vandermeer, J., Perfecto, I. and Philpott S. M. (2014) Population responses to environmental change in a tropical ant: the interaction of spatial and temporal dynamics. *PLosOne*, 9, e97809.

Jha, S., Bacon, C. M., Philpott, S. M. et al. (2014). Shade coffee: update on a disappearing refuge for biodiversity. *BioScience*, 64, 416–428.

Jiménez-Soto, E., Cruz-Rodríguez, J. A., Vandermeer, J. and Perfecto, I. (2013). *Hypothenemus hampei* (Coleoptera: Curculionidae) and its interactions with *Azteca instabilis* and *Pheidole synanthropica* (Hymenoptera: Formicidae) in a shade coffee agroecosystem. *Environmental Entomology*, 42, 915–924.

Keane, P. J. and Putter, C. A. J. (1992). *Cocoa pest and disease management in Southeast Asia and Australasia*. FAO Plant Production and Protection Paper, 112. Rome: Food & Agriculture Organisation

Khoo, K. C. and Ho, C. T. (1992). The influence of *Dolichoderus thoracicus* (Hymenoptera: Formicidae) on losses due to *Helopeltis theivora* (Heteroptera: Miridae), black pod disease, and mammalian pests in cocoa in Malaysia. *Bulletin of Entomological Research*, 82, 485–491.

Kone, M., Konate, S., Yeo, K., Kouassi, P. K. and Linsenmair, K. E. (2014). Effects of management intensity on ant diversity in cocoa plantation (Oume, centre west Côte d'Ivoire). *Journal of Insect Conservation*, 18, 701–712.

Larsen, A. and Philpott, S. M. (2010). Twig-nesting ants: the hidden predators of the coffee berry borer in Chiapas, Mexico. *Biotropica*, 42, 342–347.

Leston, D. (1970). Entomology of the cocoa farm. *Annual Review of Entomology*, 15, 273–294.

Majer, J. D. (1972). The ant mosaic in Ghana cocoa farms. *Bulletin of Entomological Research*, 62, 151–160.

(1976). The influence of ants and ant manipulation on the cocoa farm fauna. *Journal of Applied Ecology*, 13, 157–175.

Maňák, V., Nordenhem, H., Björklund, N., L. Lenoir and Nordlander, G. (2013). Ants protect conifer seedlings from feeding damage by the pine weevil Hylobius abietis. *Agricultural and Forest Entomology,* 15, 98–105.

McGregor, A. J. and Moxon, J. E. (1985). Potential for biological control of tent building species of ants associated with *Phytophthora palmivora* pod rot of cocoa in Papua New Guinea. *Annals of Applied Biology*, 107, 271–277.

Moguel, P. and Toledo, V. M. (1999). Biodiversity conservation in traditional coffee systems of Mexico. *Conservation Biology*, 13, 11–21.

Morris, J. R., Vandermeer, J. and Perfecto, I. (2015). A keystone ant species provides robust biological control of the coffee berry borer under varying pest densities. *PloS one*, 10, e0142850.

Nagy, C., Cross, J. V. and Markó V. (2013). Sugar feeding of the common black ant, *Lasius niger* (L.), as a possible indirect method for reducing aphid populations on apple by disturbing ant-aphid mutualism. *Biological Control*, 65, 24–36.

Nagy, C., Cross, J. V. and V. Markó. (2015). Can artificial nectaries outcompete aphids in ant-aphid mutualism? Applying artificial sugar sources for ants to support better biological control of rosy apple aphid. *Dysaphis plantaginea* Passerini in apple orchards. *Crop Protection*, 77, 127–138.

Niesenbaum, R. (1999). The effects of pollen load size and donor diversity on pollen performance, selective abortion, and progeny vigor in *Mirabilis jalapa*. *American Journal of Botany*, 86, 261–268.

Offenberg, J. (2015). Ants as tools in sustainable agriculture. *Journal of Applied Ecology*, 52, 1197–1205.

Perfecto, I. and Vandermeer, J. (1996). Microclimatic changes and the indirect loss of ant diversity in a tropical agroecosystem. *Oecologia*, 108, 577–582.

Perfecto, I., Vandermeer, J. and Philpott, S. M. (2014) Complex ecological interactions in the coffee Agroecosystem. *Annual Review of Ecology and Systematics*, 45, 137–158.

Philpott, S. M. (2005). Changes in arboreal ant populations following pruning of coffee shade-trees in Chiapas, Mexico. *Agroforestry Systems*, 64, 219–224.

Philpott, S. M., Arendt, W., Armbrecht, I. et al. (2008) Biodiversity loss in Latin American coffee landscapes: reviewing evidence on ants, birds, and trees. *Conservation Biology*, 22, 1093–1110.

Philpott, S. M. and Armbrecht, I. (2006). Biodiversity in tropical agroforests and the ecological role of ants and ant diversity in predatory function. *Ecological Entomology*, 31, 369–377.

Philpott, S. M., Bichier, P., Rice, R. A. and Greenberg, R. (2008). Biodiversity conservation, yield, and alternative products in coffee agroecosystems in Sumatra, Indonesia. *Biodiversity and Conservation*, 17, 1805–1820.

Philpott, S. M. and Foster, P. F. (2005). Nest-site limitation in coffee agroecosystems: artificial nests maintain diversity of arboreal ants. *Ecological Applications*, 15, 1478–1485.

Philpott, S. M., Greenberg, R., Bichier, P. and Perfecto, I. (2004) Impacts of major predators on tropical agroforest arthropods: comparisons within and across taxa. *Oecologia*, 140, 140–149.

Philpott, S. M., Pardee, G. L. and Gonthier D. (2012). Cryptic biodiversity effects: Importance of functional redundancy revealed through addition of food web complexity. *Ecology*, 93, 992–1001.

Philpott, S. M., Perfecto, I. and Vandermeer, J. (2008a). Behavioral diversity of predatory arboreal ants in coffee agroecosystems. *Environmental Entomology*, 37, 181–191.

Philpott, S. M., Perfecto, I. and Vandermeer, J. (2008b) Effects of predatory ants on lower trophic levels across a gradient of coffee management complexity. *Journal of Animal Ecology*, 77, 505–511.

Philpott, S. M., Uno, S. and Maldonado, J. (2006). The importance of ants and high-shade management to coffee pollination and yield in Chiapas, Mexico. *Biodiversity and Conservation*, 15, 487–501.

Ploetz, R. (2016). The impact of diseases on cacao production: a global overview. In *Cacao Diseases*, ed. B. A. Bailey and L. W. Meinhardt. Switzerland: Springer International Publishing, pp. 33–59.

Rizali, A., Clough, Y., Buchori, D. et al. (2013a). Long-term change of ant community structure in cacao agroforestry landscapes in Indonesia. *Insect Conservation and Diversity*, 6, 328–338.

Rizali, A., Clough, Y., Buchori, D. and Tscharntke, T. (2013b). Dissimilarity of ant communities increases with precipitation, but not reduced land-use intensity, in Indonesian cacao agroforestry. *Diversity*, 5, 26–38.

Room, P. M. (1971). The relative distributions of ant species in Ghana's cocoa farms. *Journal of Animal Ecology*, 40, 735–751.

Room, P.M. (1972a). The constitution and natural history of the fauna of the mistletoe *Tapinanthus bangwensis* (Engl. & K. Krause) growing on cocoa in Ghana. *Journal of Animal Ecology*, 41, 519–535.

Room, P. M. (1972b). The fauna of the mistletoe *Tapinanthus bangwensis* (Engl. & K. Krause) growing on cocoa in Ghana: relationships between fauna and mistletoe. *Journal of Animal Ecology*, 41, 611–621.

Room, P. M. and Smith, E. S. C. (1975). Relative abundance and distribution of insect pests, ants and other components of the cocoa ecosystem in Papua New Guinea. *Journal of Applied Ecology*, 12, 31–46.

Rubiana, R., Rizali, A., Denmead, L. H. et al. (2015). Agricultural land use alters species composition but not species richness of ant communities. *Asian Myrmecology*, 7, 73–85.

Ruf, F. O. (2011). The myth of complex cocoa agroforests: the case of Ghana. *Human Ecology*, 39, 373–388.

Sam, K., Koane, B. and Novotny, V. (2014) Herbivore damage increases avian and ant predation of caterpillars on trees along a complete elevational forest gradient in Papua New Guinea. *Ecography*, 37, 1–8.

Samson, D. A., Rickart, E. A. and Gonzales, P. C. (1997). Ant diversity and abundance along an elevational gradient in the Philippines. *Biotropica*, 29, 349–363.

Schroth, G., Läderach, P., Cuero, D. S. B., Neilson, J. and Bunn, C. (2015). Winner or loser of climate change? A modeling study of current and future climatic suitability of Arabica coffee in Indonesia. *Regional Environmental Change*, 15, 1473–1482.

Schroth, G., Läderach, P., Martinez-Valle, A. I., Bunn, C. and Jassogne, L. (2016). Vulnerability to climate change of cocoa in West Africa: patterns, opportunities and limits to adaptation. *Science of The Total Environment*, 556, 231–241.

See, Y. A. and Khoo, K. C. (1996). Influence of *Dolichoderus thoracicus* (Hymenoptera: Formicidae) on cocoa pod damage by *Conopomorpha cramerella* (Lepidoptera: Gracillariidae) in Malaysia. *Bulletin of Entomological Research*, 86, 467–474.

Strickland, A. H. (1951). The entomology of swollen shoot of cacao. *Bulletin of Entomological Research*, 41, 725–748.

Styrsky, J. D. and Eubanks, M. D. (2007). Ecological consequences of interactions between ants and honeydew-producing insects. *Proceedings of the Royal Society of London B: Biological Sciences*, 274, 151–164.

Tadu, Z., Djiéto-Lordon, C., Youbi, E. M. et al. (2014). Ant mosaics in cocoa agroforestry systems of Southern Cameroon: influence of shade on the occurrence and spatial distribution of dominant ants. *Agroforestry Systems*, 88, 1067–1079.

Trible, W. and Carroll, R. (2014). Manipulating tropical fire ants to reduce the coffee berry borer. *Ecological Entomology*, 39, 603–609.

Tscharntke, T., Clough, Y., Bhagwat, S. A. et al. (2011). Multifunctional shade-tree management in tropical agroforestry landscapes – a review. *Journal of Applied Ecology*, 48, 619–629.

Vandermeer, J., Perfecto, I. and Liere, H. (2009). Evidence for hyperparasitism of coffee rust (*Hemileia vastatrix*) by the entomogenous fungus, *Lecanicillium lecanii*, through a complex ecological web. *Plant Pathology*, 58, 636–641.

Vannette, R. L., Bichier, P. and Philpott, S. M. (2017). The presence of aggressive ants is associated with fewer insect visits to and altered microbe communities in coffee flowers. *Basic and Applied Ecology* (in press). http://doi.org/10.1016/j.baae.2017.02.002.

Wagner, D. (2000). Pollen viability reduction as a potential cost of ant association for *Acacia constricta* (Fabaceae). *American Journal of Botany*, 87, 711–715.

Wanger, T. C., Wielgoss, A. C., Motzke, I. et al. (2011). Endemic predators, invasive prey and native diversity. *Proceedings of the Royal Society of London B: Biological Sciences*, 278, 690–694.

Way, M. J. and Khoo, K. C. (1989). Relationships between *Helopeltis theobromae* damage and ants with special reference to Malaysian cocoa smallholdings. *Journal of Plant Protection in the Tropics*, 6, 1–11.

(1991). Colony dispersion and nesting habits of the ants, *Dolichoderus thoracicus* and *Oecophylla smaragdina* (Hymenoptera: Formicidae), in relation to their success as biological control agents on cocoa. *Bulletin of Entomological Research*, 81, 341–350.

(1992). Role of ants in pest management. *Annual Review of Entomology*, 37, 479–503.

Wielgoss, A. C. (2007). The impacts of ants on pests and diseases of cocoa in Indonesian agroforestry systems. Diploma Thesis, University of Würzburg.

(2013). Services and disservices driven by ant communities in tropical agroforests. PhD Thesis, University of Göttingen.

Wielgoss, A., Clough, Y., Fiala, B., Rumede, A. and Tscharntke, T. (2012). A minor pest reduces yield losses by a major pest: plant-mediated herbivore interactions in Indonesian cacao. *Journal of Applied Ecology*, 49, 465–473.

Wielgoss, A., Tscharntke, T., Buchori, D., Fiala, B. and Clough, Y. (2010). Temperature and a dominant dolichoderine ant species affect ant diversity in Indonesian cacao plantations. *Agriculture, Ecosystems & Environment*, 135, 253–259.

Wielgoss, A., Tscharntke, T., Rumede, A. et al. (2014). Interaction complexity matters: disentangling services and disservices of ant communities driving yield in tropical agroecosystems. *Proceedings of the Royal Society of London B: Biological Sciences*, 281, 2013–2144.

17 Ant-Plant-Herbivore Interactions in Northern Neotropical Agroecosystems

Inge Armbrecht and Ivette Perfecto[*]

Introduction

Ants constitute one of the most ubiquitous and abundant groups of animals in terrestrial ecosystems and play important roles in food production systems. Ants have been evolving and co-evolving along with other components of terrestrial ecosystems over roughly 100 million years (Ward, 2007). Countless examples of adaptations of plants involving ant protection against herbivores have arisen during their long co-evolutionary trajectory. However, in a very short period of time, a single human activity, agriculture, has transformed ecosystems at the local, regional and landscapes levels throughout the world. Although agriculture has existed for more than 10,000 years, the establishment of monocultural plantations during the colonial period, followed by the intensification and spatial expansion of agriculture that occurred after World War II, represents a dramatic alteration of the terrestrial ecosystem and has had stronger impacts on biodiversity than the traditional agriculture that existed for thousands of years before. The relatively sudden human transformation of terrestrial ecosystems through agriculture has affected the long-evolving ant-plant-herbivore interactions found in natural ecosystems.

Although many plants in terrestrial ecosystems have evolved mechanisms to avoid the damage caused by ants, some have evolved mechanisms to take advantage of ants, either to disperse their seeds or to protect themselves against their herbivores. Furthermore, from an evolutionary perspective, it has been argued that ants are better suited to benefit plants than to injure them (Davidson, 2008) because they descend from predatory wasps and are neither well adapted for herbivory nor for digesting cellulose (as termites). Some ants are adapted to feed on plant products that are rich in carbohydrates and easy to digest, but poor in nitrogen. Since ant colonies need both carbon and nitrogen, many ant species that feed on carbon-rich plant products also depend on scavenging and predation to acquire sufficient nitrogen (Davidson, 2008). Ants can interact directly or indirectly with plants and herbivores in many ways, and the ten categories (grouped in three classes) proposed

[*] We are grateful to Valentina Peñaranda Armbrecht for re-drawing and adapting the figures in this document and to two anonymous reviewers who made invaluable comments and corrections.

by Buckley (1982) still appear to be valid: [A] *Predation on plants by ants*: (1) seed harvesting and (2) leaf cutting; [B] *mutualisms*: (3) extrafloral nectaries (EFNs), (4) food bodies and domatia, (5) ant-epiphytes, (6) ant-gardens, (7) seed dispersal and (8) pollination; [C] *indirect interactions*: (9) ant-arthropod-plant systems and (10) soil modification. For agroecosystems several of these categories (e.g. 1, 2, 3, 7, 9 and 10) are very important and some will be discussed further.

The evolutionary history of ant-plant interactions has profound implications for agroecosystems, since the latter provide food, fibre and other products to humans and their animals, and the former can enhance or deter production depending on the context. Unfortunately, ant-plant interactions have been poorly studied in the Neotropics in comparison to other regions of the world (Rico-Gray & Oliveira, 2007).

Neotropical agriculture is especially vulnerable to climate change because high temperatures and changes in the severity and frequency of rainfall regimes may lead to lower production in this region (Altieri & Nicholls, 2013). However, perhaps the most important problem is the expansion of large-scale monocultures now dominating most landscapes in the Neotropics. These large-scale monocultural plantations not only have a very low crop species diversity (only one crop species by definition), but also they frequently have very low genetic diversity (Altieri & Nicholls, 2013). The ecological disruptions associated with the intensification of agriculture cause not only the loss of ant biodiversity, but also the loss of beneficial ant-plant-herbivore interactions. Furthermore, since insects tend to respond both to local and landscape complexity, conservation of biodiversity and conservation/management of ant-plant-herbivore interactions in agriculture need a multi-scale approach (Gonthier et al., 2014).

Four Neotropical Agroecosystems and Their Effects on Ant-Plant-Herbivore Interactions

Sugar Cane

Invasive types of ant-plant interactions might develop better in simplified agroecosystems than in diversified agroecosystems. For example, large sugar cane (*Saccharum officinarum*) monocultures in southwestern Colombia were infested by the symbiotic relationship of the invasive ant *Nylanderia fulva* (formerly *Paratrechina fulva*, Formicidae: Formicinae) associated with *Saccharicoccus sacchari* (Pseudococcidae) and *Pulvinaria* sp. (Coccidae) (Girón et al., 2005). However, in this same region of Colombia, in a study comparing four land uses, organic sugar cane showed the highest diversity of hypogenous ants compared to conventionally managed sugar cane, and the ant species resembled more those of nearby forest than did the ant species of the other land uses (Ramírez et al., 2012). Furthermore, no invasive ant-hemipteran associations have been reported in the more diverse organic sugar cane-producing farms. The presence of weeds (another diversity component) was shown to negatively affect the invasion of *N. fulva* with its hemipteran trophobionts in another sugar cane landscape of Colombia (Hernández et al., 2002). More studies are needed in order to design more biodiversity-friendly ways to incorporate sugar cane in Neotropical farms, which, in turn, might benefit from the complexity of ant-plant-herbivore interactions and avoid invasive pests.

Tropical Pastures

Pastures and other forms of livestock production represent 30 per cent of terrestrial ecosystems globally (FAO, 2006). The dominant management model is that of open grasslands without any woody vegetation. Furthermore, the use of fire is very common to stimulate vigorous re-growth of fresh pasture in treeless grasslands. The territorial extension of these agricultural systems justifies attention to their impact on biodiversity and on potential ecological services from ant-plant interactions.

The biological value for the ants of maintaining isolated trees in cattle pastures is an important focus for management, because many native trees provide extrafloral nectaries and/or hemipteran trophobionts. In a pioneer study, Majer and Delabie (1999) found 77 ant species in samples of isolated trees in cattle pastures and a nearby forest in Brazil. Although there were more exclusive species in the isolated trees (26 in the pasture vs. 14 in the forest), the distance from the forest did not affect the richness in the isolated trees. This suggests that isolated trees promote ant biodiversity in cattle-producing systems; further, restoration through augmentation of vegetation surrounding isolated trees may conserve more ant species (Gove et al., 2009).

Ants could play important roles in cattle grasslands both by removing seeds (Escobar-Ramírez et al., 2012) and by predating on potential pests (Risch & Carroll, 1982). Ant-plant interactions can also be used to evaluate the recovery of forest from pastures. For example, Falcão et al. (2015) found that the diversity of ant-plant interactions (and species richness of both groups) in cattle pastures was lower than in adjacent forests or any of the reforestation sites in Brazil. They also found that these ant-plant relationships were more dependent on just one ant species in the pastures than in the forests, with *Pheidole gertrudae* being responsible for 70 per cent of the interactions with EFN-bearing plants in pastures. Rivera et al. (2013) have shown that removing or augmenting trees in the Andean landscapes of Colombia directly affects the ant diversity and hence the diversity of ant-plant interactions (Figure 17.1).

Ants could play an important role in the efforts to rehabilitate these impoverished agroecosystems (open cattle pastures), through the movement of tree seeds, the creation of microhabitats for tree seed germination and as predators of potential herbivores. It has been proposed that some ant species, such as *Solenopsis geminata,* native of the Neotropics, may play important roles in pasture rehabilitation because it is a keystone predator both of other arthropods and of weed seeds (Risch & Carroll, 1982; Nestel & Dickschen, 1990). Also, in open cattle pastures, its nests may provide appropriate 'microhabitat' for the seed bank of trees, which

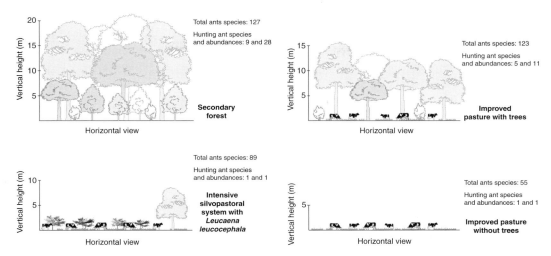

Figure 17.1. Secondary forest (positive control) and three kinds of cattle pastures sampled for ants in Andean cattle pasture landscapes in Western-central Colombia (adapted from Rivera et al., 2013). More ant species and hunting ants were found in cattle pastures with shrubs and trees, than open cattle pastures without trees.

may eventually germinate, since these ants seem to allow seeds in their nests without hurting them (Hurtado et al., 2012). For this reason, this ant species could be important in rehabilitation programmes that rely on the natural succession process to restore forests from degraded pastures. However, the role of *S. geminata* may also be negative since it tends hemipterans and a high infestation may prevent a tree seedling from getting established. This is a problem that could potentially be overcome with good management. For example, it has been hypothesised, and partially supported, that *S. geminata* will prey more on their hemipteran symbionts when a sugary solution is supplied by the farmer (Carabalí et al., 2013). Nonetheless, this species, as well as many other ant species, can have both positive and negative effects in agroecosystems as well as in forest restoration projects (Trible & Carroll, 2014).

Cacao Agroecosystems

Cacao (*Theobroma cacao*) has been of great interest to ant ecologists because the 'ant mosaic' hypothesis was originally proposed based on observations made in this agroecosystem (Majer, 1976; Leston, 1978). The ant mosaic was proposed as a mechanism that generates and maintains biodiversity in tropical forests and agroforests. Leston (1978) stated that, in the tropics, where the vegetation is more or less permanent, there is a limited number of dominant (and abundant) ant species that are stable in temporal and spatial dimensions, and dominate over mutually exclusive territories. Although the ant mosaic is not universally present in all tropical ecosystems (Dejean et al., 2003), it can be useful in certain tropical agroforests. Majer (1993) suggested that the ant mosaic could be managed to enhance predation services by ants in agroecosystems. One management strategy consists of enhancing certain ant-hemipteran interactions (Aldana et al., 2000) to reduce the level of other more damaging herbivores. The practice of introducing both, ants and hemipterans, has been documented for cacao farmers in Malaysia (Way & Khoo, 1992). When the ants that are effective as biological control agents are arboreal and the canopies of the tree crops are not physically connected, farmers can expand the influence area of the ants by providing artificial connectors among the tree crops. This is an ancient practice in China (Huang & Yang, 1987) and was observed (by I. Perfecto) in a cacao plantation in Brazil. Furthermore, it has been suggested that ant-aphid mutualisms in Neotropical cacao plantations could mediate competition among the natural enemies of the hemipterans because ants may deter certain natural enemies (e.g. syrphid flies) more than others (e.g. coccinellid beetles) (Silva & Perfecto, 2013). This means that ecological complexity may result in the coexistence of predators at the plot and the landscape level.

Oil Palm

Oil palm (*Elais guineensis*) cultivation has increased in Neotropical countries over the past decade (Gutiérrez et al., 2011; Chapter 3). As for the cacao agroecosystems, several studies have shown benefits from manipulating the ant mosaic in oil palm plantations (Guzmán et al., 1997; Aldana et al., 2000). Maintaining ant colonies and their hemipteran symbionts might have a positive overall effect when the domesticated plants (as in agroecosystems) have many herbivores, or when some of the herbivores attacked by ants transmit serious diseases. For example, in oil palm plantations in Colombia, a small *Crematogaster* species that nests in the palms is an efficient predator on *Leptopharsa gibbicarina* (Hemiptera: Tingidae), which transmits a pathogenic fungus (Guzmán et al., 1997). This is the only species of the 14 ant species reported for oil palm in Colombia that was found to be an effective predator of *L. gabbicarina*. In the 1990s a pest management programme was developed introducing *Crematogaster* sp. nests every five palms within every five rows (Aldana et al., 2000). The programme resulted in 50 per cent of the palms colonised by the ants and 98 per cent reduction of the bug pest (Guzmán et al., 1997).

Ant-Plant-Herbivore Interactions in Neotropical Coffee Agroecosystems

Coffee Intensification and Ant Biodiversity

Coffee is one of the most emblematic Neotropical agroecosystems regarding biodiversity (Chapter 16). In the 1990s, Perfecto and colleagues (1996) warned about the loss of biodiversity as shaded coffee plantations were converted to unshaded monocultures in Latin America. In the past two decades, many studies using ants as bioindicators have shown that biodiversity of ground-foraging ants decreases along the gradient of increasing intensification of coffee agriculture. This pattern has been consistent throughout the northern Neotropics (Perfecto & Armbrecht, 2003). Three mechanisms have been proposed to explain ant biodiversity loss: microclimatic changes associated with the absence of trees, decrease in food and other resources, and changes in the interactions among species (Perfecto & Snelling, 1995).

Coffee Intensification and Ant-Plant-Herbivore Interactions

The elimination of shaded trees from coffee farms leads to profound changes in the network of interactions among ants, between ants and their trophobionts, and between ants and plants. In gradients of intensification of coffee agriculture in Colombia, Armbrecht et al. (2005) and Urrutia and Armbrecht (2013) found that the number and quality of ant species involved in associations decreased as the system got simplified. In the southwest coffee-growing region in Colombia, Mera-Velasco et al. (2010) found a higher richness of ant-plant-herbivore interactions in shaded than in sun coffee plantations. However, the richness of insect species was higher when ants were excluded from coffee branches, which suggests that ants influence coffee-herbivore interactions.

Dominant arboreal ants may be associated with shade trees in coffee agroforestry system through their interactions with hemipterans and extrafloral nectaries within those trees forming patches where the trees are located. Philpott (2006) found that most numerically dominant ants were patchily distributed in a gradient of intensification of coffee production in Mexico, but only the patches of *Azteca sericeasur* (originally identified as *Azteca instabilis*) were related to the presence of shade trees. This is because *Azteca* forms large colonies that nest in shade trees but forage both on shade trees and the surrounding coffee plants.

Ant-Plant-Herbivore Interactions and Biological Control of Coffee Pests

Several attributes of ants make them potential agents for biological control (Way & Khoo, 1992). Owing to their eusocial behaviour, ants respond to prey density without reaching the satiation point that commonly limits the consumption of individual predators (Risch & Carroll, 1982; Morris et al., 2015). Ants also may stay abundant despite the scarcity of prey because many species are generalist predators and they store food that enables them to stay active despite immediate necessity of protein (Blüthgen & Feldhaar, 2010). In many cases, through their active patrolling activity on plants, ants act as repellent of pests without killing them (Buckley, 1982; Jiménez-Soto et al., 2013). Finally, humans can manipulate ants by augmentation or re-distribution or enhancing contact with the potential prey (Way & Khoo, 1992). In tropical agroecosystems there is an urgent need to understand how biological control and plant protection by ants could be affected by the loss of ant diversity (Philpott & Armbrecht, 2006). Nevertheless, ants can be functionally important both in simplified or complex agroecosystems. Ant diversity is important as a form of 'functional insurance' because, in diverse agroecosystems, such as coffee agroforestry, certain predator species may prove to be functionally significant as new pests arrive to the system (Philpott et al., 2008, 2012).

In coffee agroforests, the management of highly dominant and aggressive arboreal ants through maintaining trees and allowing some infestation of hemipterans may lead to an overall positive effect on the control of the worst insect pest of coffee in Latin America, the coffee berry borer (CBB), *Hypothenemus hampei* (Ferrari) (Coleoptera: Curculionidae: Scolytinae). Perfecto and Vandermeer (2006), in an organic farm of Mexico, found that a mutualistic ant-hemipteran relation between *Azteca sericeasur* and the green coffee scale, *Coccus viridis* (Hemiptera: Coccidae), provided indirect benefit to the coffee plants by deterring the CBB from plants where the hemipterans were being tended. In cases where ants are very effective in controlling the most damaging pest, it may be worth introducing ants with the associated hemipterans to maintain high levels of biological control of the damaging pest.

In a CBB addition/ant exclusion experiment, Gonthier et al. (2013) documented differences among ant species in the protection they provide against CBB infestation. Of the eight species tested, *Tapinoma* sp. had the strongest effect in CBB reduction as compared with the control plants (without ants). This was followed by *Pheidole synanthropica, Azteca sericeasur, Wasmannia auropunctata, Pseudomyrmex ejectus* and *P. simplex*. Surprisingly, *Solenopsis picea* and *Crematogaster* spp. did not show a significant effect on the reduction of this pest in the same study contradicting studies in Colombia (Gallego-Ropero & Armbrecht, 2005; Armbrecht & Gallego, 2007; Posada-Flórez et al., 2009). There are still questions as to whether climate differences could affect the way different predatory ants behave in different ecological contexts.

Larsen and Philpott (2010) discovered that *Pseudomyrmex* twig-nesting ants are important predators in coffee plantations and that predation by these ants on CBB were higher in traditional polyculture as compared to more intensified coffee farms in Mexico. However, generalist open habitat species could also be effective biological control agents in coffee agroecosystems. In Cuba, generalist (or invasive) ants such as *Wasmannia auropunctata, Solenopsis geminata, Pheidole megacephala, Tetramorium bicarinatum* and *Monomorium floricola*, along with *Pseudomyrmex* spp. have shown to be useful for the control of the CBB (Vásquez-Moreno et al., 2009).

An important and frequent ant-plant-herbivore interaction in Neotropical coffee plantations is that of ants associated with the shade tree *Cordia alliodora* (Boraginaceae), which provides domatia or nesting cavities to ant colonies. Experimental evidence from Pringle et al. (2011) indicates that the *Cordia*-ant-hemipteran system is highly effective both in Mexico and Costa Rica. The ants were shown to be more aggressive against herbivores when stimulated by the carbohydrates that hemipteran coccoids supplied. These ant-plant-herbivore interactions may hence benefit coffee growers through biological control of damaging coffee pests by ants.

Ground-foraging ants can also be effective predators in the coffee system. Ant-exclusion experiments in Colombia demonstrated that soil ants reduced the number of CBB by one-third and that this effect was higher in the rainy than in the dry season (Gallego-Ropero & Armbrecht, 2005). Seven ant species, most of them minute ones (e.g. *Solenopsis laeviceps*, *S. decipiens*, *S. picea*, *Tetramorium similimum*), could penetrate into infested beans placed in the ground, which was also corroborated in controlled experiments in laboratory conditions. In 2007, Gallego and Armbrecht used spiral traps to evaluate whether free adults of the CBB (which is the infesting stage of this pest) were killed by ants and found that 30.5 per cent of CBB were removed in shaded coffee farms, while 15.5 per cent were removed in sun coffee farms (Armbrecht & Gallego, 2007). Using living adults of *Drosophila melanogaster* as sentinel prey in Colombia, Henao (2008) found that soil-dwelling ants predated near 100 per cent of the flies within 24 hours and showed that ants can provide biocontrol services both in shaded and sun coffee systems. The same study found that predation was much lower on coffee bushes and on trees than on the soil surface (Henao, 2008). In Mexico, De la Mora et al. (2015) also found lower predation on the coffee stratum than on soil, but in this case, the Vegetational Complexity Index was negatively correlated with prey removal on coffee plants and trees, but not on soil. However, prey removal always increased with ant abundance and richness (De la Mora et al., 2015). In short, from the studies mentioned, it could be derived that ants provide benefits to cultivated plants through biological control of key pests and these benefits can be considered as ecosystem services provided by associated biodiversity. Since, with some exceptions, most studies indicate that these benefits decrease in more intensively managed agroecosystems such as sun coffee plantations, it is possible that the removal of trees and the use of synthetic pesticides associated with the intensification of agriculture constitute an obstacle to ecosystem services provided by complex ant-plant-herbivore interactions in the Neotropics.

Complex Networks of Trait-Mediated-Indirect Interactions and Pest Control

The ecological interactions described never occur out of context. These ecological units themselves interact with other organisms and with other units with potential consequences for pest control. Many of the mutualistic interactions that have been described (with trophobionts and with extrafloral nectaries) involve what is called trait-mediated indirect interactions (TMIIs) (Werner & Peacor, 2003). In these interactions, one species, rather than directly changing the density of another through consumption, induces some morphological, physiological, behavioural or life history changes in another species (Perfecto & Vandermeer, 2015). It has been shown that these changes can have indirect effect on subsequent interactions with other species (Utsumi et al., 2010) forming cascades of TMIIs.

One of the best studies of a complex TMII cascade involves the *Azteca sericeasur* network in coffee farms of Mexico (Perfecto et al., 2014). The *Azteca* ants nest in shade trees but forage on the coffee plants where they tend (take care of) green coffee scales (*C. viridis*), a potential pest in coffee. The ants protect the scale against its natural enemies, including the voracious coccinellid beetle, *Azya orbigera*. The

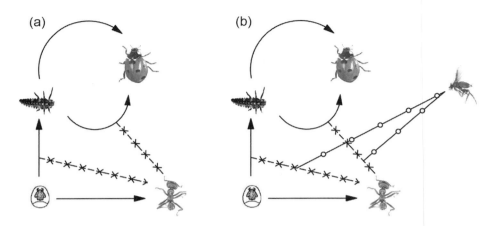

Figure 17.2. Trait-mediated indirect interactions involving the ant *Azteca sericeasur* system in the coffee farms of Mexico. The complex ecological interactions involve two interconnected trait-mediated interaction units (a) and (b), which are represented by the ant (lower right corner of each figure), the beetle (upper right corners), the beetle's larvae (upper left corners) and the scale insects (lower left corners). The parasitic phorid fly is represented only in (b).

adult beetles are constantly harassed by the ants and cannot get close to the scales when the ants are present. However, the beetle larvae are covered with waxy filaments that protect them from the ants, which inadvertently protect the beetle larvae from their own parasitoids (the ants scare away tiny parasitic wasps) (Liere & Perfecto, 2008) (Figure 17.2a). To protect their eggs from ant predation the female *A. orbigera* hides her eggs under the scale insects. But how is it possible for the adult beetle to get close to the scale insects while ants are tending them? The answer emerges from the connection with another TMII. *Pseudacteon lasciniosus* is a phorid fly parasitoid of the *Azteca* ants. The flies have very bad eyesight and need movement to be able to attack the ants (Mathis et al., 2011). As a defensive strategy, when the phorids arrive, the ants stop moving. The ant activity is reduced by about 50 per cent (Philpott et al., 2005) providing the gravid female beetle with a window of opportunity to get close to the scales and hide her eggs under the scale insects (Hsieh et al., 2012) (Figure 17.2b). Surprisingly, the gravid beetles are able to detect the 'phorid alert' pheromones produced by the ants and use this cue to safely approach scale insects (Hsieh et al., 2012). These complex interactions have important consequences for the control of the scale insects in coffee farms since *A. orbigera* is the main biological control of the scale insects. Ironically, the patches where the ants are tending scale insects are refugia for the larvae of the coccinellid beetle, and the adults control scale insects on the rest of the farm (Perfecto et al., 2014).

Not All Ant-Plant Interactions Are Positive for the Coffee Plants

Ant species vary in effectiveness as biological control agents. For example, De la Mora et al. (2008), in Mexico, found that *A. sericeasur* did not reduce the attack of coffee plants by the leaf miner, *Leucoptera coffeella* (Lepidoptera: Lyonetiidae). However, other ant species nesting in twigs on coffee bushes did decrease the attack of this pest, with a significant regression between the number of twig nests and both the percentage of leaves with the miner and the average damage (De la Mora et al., 2008).

The presence of certain ants in agroecosystems can also impact other predators. In complex agroecosystems such as coffee agroforests, the presence of shade trees, which harbour large colonies of ants, may indirectly affect other predatory arthropods and their insect prey. Marin et al. (2015) demonstrated that the abundance of the web spider *Pocobletus* sp. was increased by 100 per cent in the presence of *Azteca sericeasur* and by 40 per cent in the presence of *Pheidole synanthropica*. Ants also interact with other ant species with potential detrimental or positive effects for pest regulation in agroecosystems. Trible and Carroll (2014) hypothesised that the dominant and aggressive fire ant *Solenopsis geminata* negatively interacts with other, more effective predatory ant species with negative consequences for the biological control of the CBB. In an experiment in a coffee plantation in Costa Rica, the predation (disappearance) of CBB was quadrupled (from 6 to 23 per cent) in plots where *S. geminata* was experimentally removed as compared to control plots. On the basis of these results, the authors proposed the suppression of *S. geminata* as management technique for the CBB in Central America. Owing to the evidence of *S. geminata's* positive effect (shown earlier), this species deserves better attention from agroecologists.

Given all the evidence in the literature about the role of ants as predators of insect pests of coffee in the Neotropics, the lack of actual pest management programmes that incorporate ants is surprising. This is partially due to the erroneous perception that ants in general are pests, which is popular among farmers in areas with high abundance of leaf-cutter ants. Furthermore, the associations of ants with hemipterans, some of which are potential pests of crops and in some cases are disease vectors, also hinder the image of ants as beneficial for pest management. Finally, the fact that it is the ant community and its interactions with other species that needs to be managed and not a particular species adds a level of complexity that requires more reliance on ecological knowledge.

Ant-Plant Relations through Extrafloral Nectaries (EFNs) in Agroecosystems

Neotropical agroecosystems have the potential to conserve a very rich array of ant-plant interactions through EFN-bearing species (Chapters 8–10). For example, Aguirre et al. (2013) found 50 EFN-bearing plant species in Los Tuxlas, Mexico, 40 per cent of which were Fabaceae and 85 per cent of all were trees and woody vines. Leguminous trees are very frequently used by farmers in Neotropical agroforests such as coffee and cacao, and may result in high ant abundances and biocontrol services provided by ants associated with the EFNs. Ant interactions with plants through the use of EFN can be generalist. For example, in Mexican forest, species such as *Dolichoderus bispinosus*, *Pseudomyrmex gracilis* and *Camponotus planatus* were found to visit many species of plants with EFNs and were thus considered 'super-generalists' by Aguirre et al. (2013); however, 31 ant species were associated with 29 plant species, showing also a diverse array of ant-plant interactions in this forest.

In coffee plantations throughout the northern Neotropics, the Fabaceae *Inga* spp. are frequently used as shade trees. In two Colombian coffee plantations, ants foraging on EFNs in *Inga* trees were different from ants foraging on leaf litter and soil (Sinisterra et al., 2016). Consistent with the study of Aguirre and others (2013), it can be concluded that ants visiting EFNs were species from the genera *Procryptocerus*, *Crematogaster*, *Camponotus* and *Pseudomyrmex*. *Inga* trees from three species studied in Panama (Bixennmann et al., 2013) produce a constant amount and concentration of nectar per leaf in young leaves, which are better protected by ants. In addition, a secondary metabolite (phenolics) decreases in *Inga* species as the leaves expand, suggesting that there is an additive effect of classes of defences (ants and secondary metabolites) instead of a trade-off between them (Bixennman et al., 2013). In coffee agroecosystems of Colombia, ants using the EFNs of *Inga* trees also forage on coffee plants and may protect the crop in dry times when nectar resources get scarce (Ramírez et al., 2010). An intriguing possibility is whether the trees providing enough energetic rewards for ant colonies will free them from the need of hemipteran trophobionts and whether this carbohydrate resource will result in lower or higher levels of predation activity on the trees or adjacent crops (e.g. Carabalí-Banguero et al., 2013). In a study in Mexico, Livingston et al. (2008) recorded lower numbers of *C. viridis* trophobionts tended on coffee plants underneath *Inga micheliana* (Fabaceae) versus *Alchornea latifolia* (Euphorbiaceae). In this case both of these trees have EFN but ants tended another preferred hemipteran on the *Inga* tree. This suggests that if ants have an alternative and preferred source of carbohydrates on the shade trees, they will reduce interactions with trophobionts on the coffee. The net effect of this shift on coffee production will depend on whether ants tending trophobionts on coffee also protect the plants from other more damaging herbivores, as have been shown for *A. sericeasur* in coffee plantations in Mexico (Perfecto et al., 2014).

Understanding the functional interplay of multiple mutualists associated with the same host plant remains an important challenge for managing ant interactions in agroecosystems. Bottom-up effects have important implications for agroecological management of leguminous crops that benefit from protection by ants, suggesting that the specific conditions of each agroecosystems should be evaluated to balance bottom-up and top-down effects to obtain a higher overall defence of the crop at the lowest cost to the plant.

Seed Movement by Ants in Neotropical Agroecosystems and Habitat Restorations Sites

Seed movement by ants has been intensively and extensively studied in natural ecosystems (Rico-Gray & Oliveira, 2007) and in the context of ecological restoration or rehabilitation. The nature of ant-plant interactions through the movement of seeds may be antagonistic or mutualistic (Chapters 5 and 6). Ants may be antagonistic when they behave as seed predators (generalist omnivores or specialist granivores) or when they retrieve seeds that would otherwise be dispersed by more efficient vertebrate species (Barnett et al., 2015). Nevertheless, these antagonistic interactions may turn out to be beneficial when the seeds escape predation and are able to germinate, when the ants act as secondary seed dispersers (Rico-Gray & Oliveira, 2007), or when ants predate weed seeds.

Seed transportation by ants could be beneficial in the management of Neotropical agroecosystems. In relatively stable agroecosystems such as coffee or cacao agroforests, ants may help in weed management by aiding in the dispersion of 'noble' (or less noxious) weeds, although this idea still needs to be tested. In rehabilitation or restoration projects, seed movement by ants can accelerate secondary succession after land abandonment (Bol & Vroomen, 2008). Ants have been explored as agents for seed movement in cattle pastures in Colombia. Escobar et al. (2007) tested the effect of the removal by ants of three species of seeds, and found that 15 ant species moved 26 per cent of the 1,350 seeds offered in periods of approximately two hours each. Nine, five and four ant species moved the seeds from forest, silvopastoral (cattle pastures that incorporate trees and shrubs) and open cattle grasslands, respectively, and most species were removed by species of *Pheidole*, *Solenopsis* and *Ectatomma ruidum*. Ants removed arillated seeds more often than non-arillated seeds (Escobar et al., 2007). In another study in open cattle pastures, Escobar-Ramírez et al. (2012) studied the role of ants as secondary dispersers of useful dry tropical forest trees for rehabilitation purposes. They found that ants removed 25 per cent of the 1,827 seeds offered in 48 hours (Figure 17.3). Total seed removal did not exceed 40 per cent but levels of removal may be significant in ecological terms. Twenty-one generalist ant species were collected in these pastures using tuna baits, of which *Ectatomma ruidum*, *S. geminata*, *Crematogaster abstinens* and *Pheidole* spp. were found to be important for seed movement.

Furthermore, there was a switch from the ants moving seeds in the day (former list) and after 17:00 h in which seeds were mostly moved by *Cyphomyrmex major* and *Atta cephalotes*. *Ectatomma ruidum,* in particular, seems to be an important functional agent in pastures due to its high potential to move seeds over large distances and its function as an efficient predator (Santamaría et al., 2009). On the contrary, hunting ants belonging to the genera *Odontomachus, Pachycondyla* and *Rhytidoponera* have been shown to be important seed dispersal agents of Neotropical plants in the forests of Mexico (Horvitz & Beattie, 1980; Chapter 7). The former

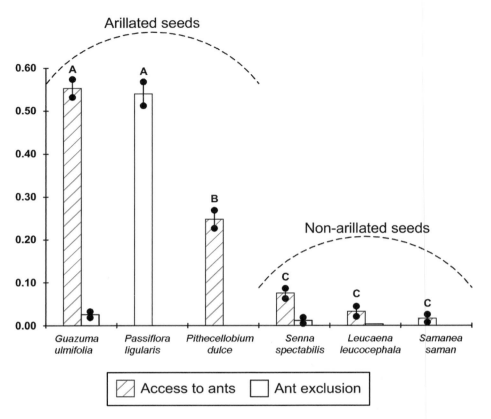

Figure 17.3.   Removal of tropical dry forests tree seeds by ants with rehabilitation purposes after intensive open cattle pasture use (modified from Escobar-Ramírez et al., 2012).

studies suggest that species activity turnover might be important in enhancing seed movement in degraded agroecosystems in the process of regeneration.

A way to induce ants to perform ecological services in agroecosystems is the use of artificial arils to induce seed hauling. The purpose of inducing seed hauling is promoting rehabilitation of native vegetation in degraded cattle pastures and other agroecosystems after intense use. Henao-Gallego et al. (2012) prepared an artificial aril using unflavoured gelatin, oat flakes, sugar and tuna oil and found that generalist ants in open cattle pastures moved 80 per cent of *Senna spectabilis* (Fabaeae) seeds with artificial aril versus 32 per cent of seeds without aril, and the difference was statistically significant (Figure 17.4).

In a tropical montane forest of the Bolivian Andes, Gallegos et al. (2014) found high recruitment of secondarily dispersed seeds of *Clusia trochiformis* (Clusiaceae) in degraded habitats for restoration purposes, because ants apparently enhanced the survival of seeds by reducing predation and increasing germination. *Pheidole, Solenopsis* and *Acromyrmex* ants dispersed seeds and removed the aril after the relocation. Ants moved twice as many seeds with than without arils (the arils were

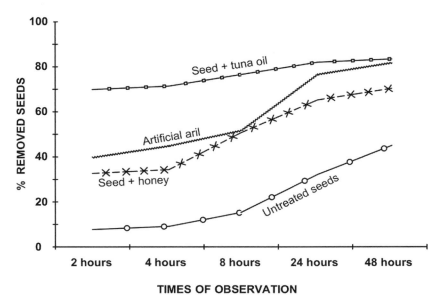

Figure 17.4. The use of artificial arils or attractants to induce generalist ants to seed hauling in degraded agroecosystems may become an important tool for rehabilitation or restoration purposes in Neotropical lands (re-drawn from Henao-Gallego et al. 2012).

removed experimentally by the authors). In this study ants moved 42 per cent of the seeds available in vertebrate exclusion cages. Of these, 20 per cent were carried to their nests and half of other seeds were left on the soil or leaf litter. Interestingly, secondary dispersal positively affected seed germination although germination was higher in forests than in degraded habitats. The authors suggest that in degraded habitats where vertebrate secondary dispersers have been lost, ants might turn out to be keystone ecological agents for restoration or rehabilitation purposes (Gallegos et al., 2014). In sites where human activity completely removed the soil's upper layers, such as coal mining, ants could play important roles both as secondary dispersal agents and in increasing the survival of seeds (Domínguez & Armbrecht, 2011). Griffith (2004) compared post-agricultural succession of tropical forest in Nicaragua and Guatemala along a gradient of agricultural intensification including agroforestry, slash-and-burn agriculture, cattle pastures and intensive monocultures, and found that agroforestry systems experienced faster succession than pastures after abandonment in Guatemala. In pastures and monocultures only pioneer tree species were dispersed, while shade-tolerant species were dispersed in slash-and-burn and agroforestry systems. Ants such as *Atta, Solenopsis* and *Myrmecina* sp. contributed to the movement of seeds, which was higher in agroforestry, slash-and-burn and pasture than in intensive monocultures (Griffith, 2004). We interpret that different plant assemblages throughout the succession process allow different ant assemblages that carry out ecosystem services differently (i.e. dispersing shade or non-shade plant species).

Figure 17.5.   *Crematogaster* ants foraging on coffee beans in a Colombian coffee plantation. Photo by Andrés López. (A black-and-white version of this figure will appear in some formats. For the color version, please refer to the plate section.)

Finally, as the so-called Anthropocene period progresses and vertebrates vanish (Dirzo et al., 2014), many generalist ants will survive the massive extinction and could perform some of the ecological functions that vertebrates once carried out, including seed dispersal. Christianini et al. (2014: 1129) stated, 'As the Anthropocene progresses, invertebrates – notably ants – will be the prospective heirs of those plant-animal interactions (that are lost)'.

Concluding Remarks

Ant-plant-herbivore interactions are major drivers of the functioning of agroecosystems in the Neotropics, with thousands of plant species bearing EFNs, or being involved in complex networks through trophobiont relationships. Millions of years of evolution have shaped these relationships in natural ecosystems in the Neotropics and the rise of agroecosystems is a new phenomenon in Earth's history. All agroecosystems were formerly natural ecosystems, but contrary to the later, agroecosystems are created and controlled by human intelligence/technology and exist because of a human plan. The way ant-plant-herbivore relationships are viewed determines the way agroecosystems are managed, such as it has been shown for coffee agroecosystems (Figure 17.5) and will affect the performance of these systems in terms of human goals.

When agroecosystems are simplified through intensification, ant-plant-herbivore interactions may change and alter the performance of agroeco-systems and their resilience. It is necessary to maintain biodiversity of ants because it maintains these important interactions that also benefit agriculture. However, no blind generalisations can be made, and each type of agroecosystem needs particular attention to understand ant-plant interactions for human ben-efits. Understanding how ant-plant-herbivore interactions affect Neotropical production systems and their ecological mechanisms is necessary in order to design ecologically sound agroecosystems. In general, the loss of diversity of ant-plant-herbivore interactions may negatively affect agricultural production through the loss or weakening of biological control. Practices such as agro-forestry and other diverse farming systems in the Neotropics might allow ant-plant-herbivore interactions to be maintained or reestablished and to recover ecological internal controls within restored habitats and production-based agroecosystems.

Recommendations for Land Managers and for Future Research

In the search of a sustainable, less 'toxic' management of agroecosystems, ant-hemipteran-plant associations are of extreme importance, because pesticides may eliminate natural enemies leaving intact the mutualistic relation. A possibility to manage ant-hemipteran associations when the symbiont ant is also a predator is providing carbohydrate resources to the ant in order to induce it to switch to pre-dation. Introducing trees in tropical agroecosystems is, in general, of benefit for the diversity of ant-plant interactions and complex ecological networks. Providing connectivity among tree canopies that contain dominant predaceous ants is an ancient beneficial practice in tropical agroforestry. However, not always a higher biodiversity automatically provides better biological control, thus, proper manage-ment practices will depend on how profound is the farmer's knowledge of his/her agroecosystem. We recommend land managers and researchers learn the identities of the species present in their agroecosystem, observe their interactions and study their ecology.

Literature cited

Aguirre, A., Coates, R., Cumplido-Barragán, G., Campos-Villanueva, A. and Díaz-Castelazo, C. (2013). Morphological caracterization of extrafloral nectaries and associ-ated ants in tropical vegetation of Los Tuxtlas, Mexico. *Flora*, 208, 147–156.

Aldana, J., Calvache, H. and Aria, D. (2000). Programa comercial de manejo de *Leptopharsa gibbicarina* Froeschner (Hemiptera: Tingidae) con la hormiga *Crematogaster* spp. en una plantación de palma de aceite. *Palmas*, 21, 167–173.

Altieri, M. A. and Nicholls, C. I. (2013). Agroecología y resiliencia al cambio climático: principios y consideraciones metodológicas. *Agroecología*, 8, 7–20.

Armbrecht, I. and Gallego, M. C. (2007). Testing ant predation on the coffee berry borer in shaded and sun coffee plantations in Colombia. *Entomologia Experimentalis et Applicata*, 124, 261–267.

Armbrecht, I., Rivera, L. and Perfecto, I. (2005). Reduced diversity and complexity in the leaf litter ant assemblage of Colombian coffee plantations. *Conservation Biology*, 19, 897–907.

Barnett, A. A., Almeida, T., Andrade, R. et al. (2015). Ants and their plants: *Pseudomyrmex* ants reduce primate, parrot and squirrel predation on *Macrolobium acaciifolium* (Fabaceae) seeds in Amazonian Brazil. *Biological Journal of the Linnean Society*, 114, 260–273.

Bixenmann, R. J., Coley, P. D. and Kursar, T. A. (2013). Developmental changes in direct and indirect defenses in the young leaves of the Neotropical tree genus *Inga* (Fabaceae). *Biotropica*, 45, 175–184.

Blüthgen, N. and Feldhaar, H. (2010). Food and shelter: how resources influence ant ecology. In *Ant ecology*. ed. L. Lach, C. L. Parr and K. L. Abbott. New York: Oxford University Press, pp. 115–136.

Bol, M. and Vroomen, D. (2008). The succession of pasture land towards original cloud forest in the pre-mountain area of Costa Rica. Bachelor thesis research for Tropical Forest. Van Hall Larenstein Institute.

Buckley, R. C. (1982). Ant-plant interactions: a world review. In *Ant-plant interactions in Australia*. ed. R. C. Buckley. The Hague, Australia: Dr W. Junk Publishers, pp. 111–141.

Carabalí-Banguero, D. J., Wyckhuys, K. A. G., Montoya-Lerma, J., Kondo, T. and Lundgren, J. G. (2013). Do additional sugar sources affect the degree of attendance of *Dysmicoccus brevipes* by the fire ant *Solenopsis geminata*? *Entomologia Experimentalis et Applicata*, 148, 65–73.

Christianini, A. V., Oliveira, P. S., Bruna, E. M. and Vasconcelos, H. L. (2014). Fauna in decline: meek shall inherit. *Science*, 345, 1129.

Dáttilo, W. and Dyer, L. (2014). Canopy openness enhances diversity of ant-plant interactions in the Brazilian Amazon rain forest. *Biotropica*, 46, 712–719.

Davidson, D. W. (2008). Ant-plant interactions. In *Encyclopedia of entomology*, ed. J. L. Capinera. Florida: Springer, pp. 166–185.

Davidson, D. W., Cook, S. C., Snelling, R. and Chua, T. H. (2003). Explaining the abundance of ants in lowland tropical rainforest canopies. *Science*, 300, 969–972.

De la Mora, A., García-Ballinas, J. A. and Philpott, S. M. (2015). Local, landscape and diversity drivers of predation services provided by ants in a coffee landscape in Chiapas, Mexico. *Agriculture, Ecosystems and Environment*, 201, 83–91.

De la Mora, A., Livingston, G. and Philpott, S. M. (2008). Arboreal ant abundance and leaf miner damage in coffee agroecosystems in Mexico. *Biotropica*, 40, 742–746.

Dejean, A., Corbara, B., Fernández, F. and Delabie, J. H. C. (2003). Mosaicos de hormigas arbóreas en bosques y plantaciones tropicales. In *Introducción a las hormigas de la región Neotropical*, ed. F. Fernández. Bogotá: Instituto de Investigación de Recursos Biológicos Alexander von Humboldt, pp 149–158.

Del-Claro, K., Rico-Gray, V., Torezan-Silingardi, H. M. et al. (2016). Loss and gains in ant-plant interactions mediated by extrafloral nectar: fidelity, cheats and lies. *Insectes Sociaux*, published online, 15 February 2016. DOI 10.1007/s00040-016-0466-2.

Dirzo, R., Young, H. S., Galetti, M. et al. (2014). Defaunation in the anthropocene. *Science*, 345, 401–406.

Domínguez-Haydar, Y. and Armbrecht, I. (2011). Response of ants and their seed removal in rehabilitation areas and forests at El Cerrejón coal mine in Colombia. *Restoration Ecology*, 19, 178–184.

Escobar, S., Armbrecht, I. and Calle, Z. (2007). Transporte de semillas por hormigas in bosques y agroecosistemas ganaderos de los Andes colombianos. *Agroecología*, 2, 65–84.

Escobar-Ramírez, S., Duque, J. S., Henao, N., Hurtado-Giraldo, A. and Armbrecht, I. (2012). Removal of nomyrmecochorous seeds by ants: role of ants in cattle grasslands. *Psyche*, 2012, Article ID 951029, doi:10.1155/2012/951029, pp. 1–8.

Falcão, J. C. F., Dáttilo, W. and Izzo, T. J. (2015). Efficiency of different planted forests in recovering biodiversity and ecological interactions in Brazilian Amazon. *Forest Ecology and Management*, 339, 105–111.

FAO. (2006). Livestock´s long shadow: environmental bases and options. http://go.nature.com/BFrtHv (retrieved 1 January 2016).

Gallego-Ropero, M. C. and Armbrecht, I. (2005). Depredación por hormigas sobre la broca del café en cafetales cultivados bajo dos niveles de sombra en Colombia. *Revista Manejo Integrado de Plagas y Agroecología (Costa Rica)*, 76, 1–9.

Gallegos, S. C., Hensen, I. and Schleuning, M. (2014). Secondary dispersal by ants promotes forest regeneration after deforestation. *Journal of Ecology*, 102, 659–666.

Girón, K., Lastra, L. A., Gómez, L. A. and Mesa, N. C. (2005). Observaciones acerca de la biología y los enemigos naturales de *Saccaricoccus sacchari* y *Pulvinaria* pos *elongata*, dos homópteros asociados con la hormiga loca en caña de azúcar. *Revista Colombiana de Entomología*, 31, 29–35.

Gonthier, D. J., Ennis, K. K., Farinas, S. et al. (2014). Biodiversity conservation in agriculture requires a multi-scale approach. *Proceedings of the Royal Society B*, 281 20141358. http://dx.doi.org/10.1098/rspb.2014.1358.

Gonthier, D. J., Ennis, K. K., Philpott, S. M., Vandermeer, J. and Perfecto, I. (2013). Ants defend coffee from berry borer colonization. *BioControl*, 58, 815–820.

Gove, A., Majer, J. D. and Rico-Gray, V. 2009. Ant assemblages in isolated trees are more sensitive to species placement than their woodland counterparts. *Basic and Applied Ecology*, 10, 187–195.

Griffith, D. M. (2004). Succession of tropical rain forest along a gradient of agricultural intensification: pattens,mechanisms and implications for conservation. PhD thesis, Ann Arbor, MI: University of Michigan, Department of Ecology and Evolutionary Biology.

Gutiérrez-Vélez, V. H., DeFries, R., Pinedo-Vásquez, M. et al. (2011). High-yield oil palm expansion spares land at the expense of forests in the Peruvian Amazon. *Environmental Research Letters*, 6, 044029. doi:10.1088/1748–9326/6/4/044029.

Guzmán, L., Calvache Guerrero, H., Aldana la Torre, J. and Méndez, A. (1997). Manejo de *Leptopharsa gibbicarina* Froeschner (Hemiptera: Tingidae) con la hormiga *Crematogaster* sp. en una plantación de palma de aceite. *Palmas*, 18, 19–26.

Henao, H. (2008). Análisis de la actividad depredadora por hormigas en cafetales con y sin sombra de árboles de Cauca y Valle. MScTesis. Cali, Colombia: Universidad del Valle, Facultad de Ciencias, Departamento de Biología.

Henao-Gallego, N., Escobar-Ramírez, S., Calle, Z., Montoya-Lerma, J. and Armbrecht, I. (2012). An artificial aril designed to induce seed hauling by ants for ecological rehabilitation purposes. *Restoration Ecology*, 20, 555–560.

Hernández, C. P., Martínez, Y. P., Insuasty, O. et al. (2002). Efecto del control de malezas y la fertilización nitrogenada sobre la población de hormiga loca *Paratrechina fulva* (Hymenoptera: Formicidae). *Revista Colombiana de Entomología*, 28, 83–90.

Horvitz, C. C. and Beattie, A. J. (1980). Ant dispersal of Calathea (Marantaceae) seeds by carnivorous ponerines (Formicidae) in a tropical rain forest. *American Journal of Botany*, 67, 321–326.

Hsieh, H. Y., Liere, H., Soto, E. J. and Perfecto, I. (2012). Cascading trait-mediated interactions induced by ant pheromones. *Ecology and Evolution*, 2, 2181–2191.

Huang, H. T. and Yang, P. (1987). The ancient cultured citrus ant. *BioScience*, 37, 665–671.

Hurtado, A., Escobar, S., Torres, A. M. and Armbrecht, I. (2012). Explorando el papel de la hormiga generalista *Solenopsis geminata* (Formicidae: Myrmicinae) en la germinación de semillas de *Senna spectabilis* (Fabaceae: Caesalpinioideae). *Caldasia*, 34, 127–137.

Jiménez-Soto, E., Cruz-Rodríguez, J. A., Vandermeer, J. and Perfecto, I. (2013). *Hypothenemus hampei* (Coleoptera: Curculionidae) and its interactions with *Azteca instabilis* and *Pheidole synanthropica* (Hymenoptera: Formicidae) in a shade coffee agroecosystem. *Environmental Entomology*, 42, 915–924.

Lange, D. and Del-Claro, K. (2014). Ant-plant interaction in a tropical savanna: may the network structure vary over time and influence on the outcomes of associations? *Plos One*, 9, e105574.

Larsen, A. and Philpott, S. M. (2010). Twig-nesting ants: the hidden predators of the coffee berry borer in Chiapas, Mexico. *Biotropica*, 42, 342–347.

Laurance, W. F., Sayer, J. and Cassman, K. G. (2014). Agricultural expasion and its impact on tropical nature. *Trends in Ecology and Evolution*, 29, 107–116.

Leston, D. (1978). A Neotropical ant mosaic. *Annals of the Entomological Society of America*, 71, 649–653.

Liere, H. and Perfecto, I. (2008). Cheating on a mutualism: indirect benefits of ant attendance to a coccidophagous coccinellid. *Environmental Entomology*, 37, 143–149.

Livingston, G. F., White, A. M. and Kratz, C. J. (2008). Indirect interactions between ant-tended hemipterans, a dominant ant *Azteca instabilis* (Hymenoptera: Formicidae), and shade trees in a tropical agroecosystem. *Environmental Entomology*, 37, 734–740.

Majer, J. D. (1976). The influence of ants and ant manipulation on the cocoa farm fauna. *Journal of Applied Ecology*, 13, 157–175.

  1993. Comparison of the arboreal ant mosaic in Ghana, Brazil, Papua New Guinea, and Australia: Its structure and influence on arthropod diversity. In *Hymenoptera and Biodiversity*, ed. J. LaSalle and I. D. Gauld. Wallingford, UK: CAB International, pp. 115–141.

Majer, J. D. and Delabie, J. H. C. (1999). Impact of tree isolation on arboreal and ground ant communities in cleared pasture in the Atlantic rain forest region of Bahia, Brazil. *Insectes Sociaux*, 46, 281–290.

Marin, L., Jackson, D. and Perfecto, I. (2015). A positive association between ants and spiders and potential mechanisms driving them. *Oikos*, 124, 1078–1088.

Mathis, K. A., Philpott, S. M. and Moreira, R. F. (2011). Parasite lost: chemical and visual cues used by *Pseudacteon* in search of *Azteca instabilis*. *Journal of Insect Behavior*, 24, 186–199.

Mera-Velasco, Y. A., Gallego-Ropero, M. C. and Armbrecht, I. (2010). Asociaciones entre hormigas y otros insectos en follaje de cafetales de sol y sombra, Cauca Colombia. *Revista Colombiana de Entomología*, 36, 116–126.

Morris, J. R., Vandermeer, J. and Perfecto, I. (2015). A keystone ant species provides robust biological control of the coffee berry borer under varying pest densities. *PloS one* 10, e0142850.

Nestel, D. and Dickschen, F. (1990). Foraging kinetics of ground ant communities in different mexican coffee agroecosystems. *Oecologia*, 84, 58–63.

Perfecto, I. and Armbrecht, I. (2003). The coffee agroecosystem in the Neotropics: combining ecological and economic goals. In *Tropical agroecosystems*, ed. J. Vandermeer. Boca Raton, FLA: CRC Press, pp. 159–194

Perfecto, I., Rice, R. A., Greenberg R. and Van der Voort, M. E. (1996). Shade coffee: a disappearing refuge for biodiversity. *Bioscience*, 46, 598–608.

Perfecto, I. and Snelling, R. (1995). Biodiversity and the transformation of a tropical agroecosystem: ants in coffee plantations. *Ecological Applications*, 5, 1084–1097.

Perfecto, I. and Vandermeer, J. (2006). The effect of an ant-hemipteran mutualism on the coffee berry borer (*Hypothenemus hampei*) in southern Mexico. *Agriculture, Ecosystems and Environment*, 117, 218–221.

(2015). *Coffee agroecology,* New York: Earthscan.

Perfecto, I., Vandermeer, J. and Philpott, S. M. (2014). Complex ecological interactions in the coffee agroecosystem. *Annual Review of Ecology, Evolution, and Systematics*, 45, 137–158.

Philpott, S. M. (2006). Ant patchiness: a spatially quantitative test in coffee agroecosystems. *Naturwissenschaften*, 93, 386–392.

Philpott, S. M. and Armbrecht, I. (2006). Biodiversity in tropical agroforests and the ecological role of ants and ant diversity in predatory function. *Ecological Entomology*, 31, 369–377.

Philpott, S. M., Pardee, G. L. and Gonthier, D. J. (2012). Cryptic biodiversity effects: importance of functional redundancy revealed through addition of food web complexity. *Ecology*, 93, 992–1001.

Philpott, S. M., Perfecto, I. and Vandermeer, J. (2008). Behavioral diversity of predatory arboreal ants in coffee agroecosystems. *Environmental Entomology*, 37, 181–191.

Posada-Flórez, F. J., Vélez-Hoyos, M. and Zenner de Polanía, I. (2009). *Hormigas: enemigos naturales de la broca del café,*. Chinchiná, Colombia: Universidad de Ciencias Aplicadas y Ambientales.

Pringle, E. G., Dirzo, R. and Gordon, D. H. (2011). Indirect benefits of symbiotic coccoids for an ant-defended myrmecophytic tree. *Ecology*, 92, 37–46.

Ramírez, M., Chará, J., Pardo-Lorcano, L. C. et al. (2012). Biodiversidad de hormigas hipogeas (Hymenoptera: Formicidae) en agroecosistemas del Cerrito, Valle del Cauca. *Livestock Research for Rural Development*, 241, 1–18.

Ramírez, M., Herrera, J. and Armbrecht, I. (2010). Hormigas que depredan en potreros y cafetales colombianos: ¿bajan de los árboles? *Revista Colombiana de Entomología*, 36, 106–115.

Rico-Gray, V. and Oliveira, P. S. (2007). *The ecology and evolution of ant-plant interactions*, Chicago: University of chicago Press.

Risch, S. J. and Carroll, R. (1982). Effect of a keystone predaceous ant, *Solenopsis geminata* on arthropods in a tropical agroecosystem. *Ecology*, 63, 1979–1983.

Rivera, L. F., Armbrecht, I. and Calle, Z. (2013). Silvopastoral systems and ant diversity conservation in a cattle-dominated landscape of the Colombian Andes. *Agriculture, Ecosystems and Environment*, 181, 188–194.

Santamaría, C., Armbrecht, I. and Lachaud, J. P. (2009). Nest distribution and food preferences of *Ectatomma ruidum* (Hymenoptera: Formicidae) in shaded and open cattle pastures of Colombia. *Sociobiology*, 53, 517–541.

Silva, E. N. and Perfecto, I. (2013). Coexistence of aphid predators in cacao plants: does ant-aphid mutualism play a role? *Sociobiology*, 60, 259–265.

Sinisterra, M. R., Gallego-Ropero, M. C. and Armbrecht, I. (2016). Hormigas asociadas a nectarios extraflorales de árboles de dos especies de *Inga* en cafetales de Cauca, Colombia. *Acta Agronomica*, 65, 9–15.

Trible, W. and Carroll, R. (2014). Manipulating tropical fire ants to reduce the coffee berry borer. *Ecological Entomology*, 39, 603–609.

Urrutia-Escobar, X. and Armbrecht, I. (2013). Effect of two agroecological management strategies on ant (Hymenoptera: Formicidae) diversity on coffee plantations in Southwestern Colombia. *Environmental Entomology*, 42, 194–203.

Utsumi, S., Kishida, O. and Ohgushi, T. (2010). Trait-mediated indirect interactions in ecological communities. *Population ecology*, 52, 457–459.

Vásquez Moreno, L. L., Matienzo Brito, Y., Alfonso Simonetti, J., Moreno Rodríguez, D. and Alvarez Nuñez, A. (2009). Diversidad de especies de hormigas (Hymenoptera: Formicidae) en cafetales afectados por *Hypothenemus hampei* Ferrari (Coleoptera: Curculionidae: Scolytinae), *Fitosanidad*, 13, 163–168.

Ward, P. S. (2007). Phylogeny, classification and species-level taxonomy of ants (Hymenoptera: Formicidae). *Zootaxa*, 1668, 549–563.

Way, M. J. and Khoo, K. C. (1992). Role of ants in pest management. *Annual Review of Entomology*, 37, 479–503.

Werner, E. E. and Peacor, S. D. (2003). A review of trait-mediated indirect interactions in ecological communities. *Ecology*, 84, 1083–1100.

18 Leaf-Cutting Ants in Patagonia: How Human Disturbances Affect Their Role as Ecosystem Engineers on Soil Fertility, Plant Fitness, and Trophic Cascades

Alejandro G. Farji-Brener, Mariana Tadey, and María N. Lescano[*]

Introduction

Human activities affect plant-animal relationships in multiple ways. The conversion of natural habitats into cities, cultivated areas, pastures, and forest plantations may directly reduce the abundance or, even, locally extinguish organisms. In addition, these human-induced changes also indirectly impact biotic components through cascade effects (Laurance et al., 2014). For example, changes in plant assemblages generated by human disturbances produce concomitant changes in a variety of animal communities that use plants for food, nest, and refuge (Sallanbanks et al., 2000; Bestelmeyer & Wiens, 2001; Bieber et al., 2014). To what extent anthropogenic disturbances spread across ecological levels will ultimately depend on the features of affected organisms. In this context, some species, for example, those characterized as ecological engineers, can amplify the effect of human-induced disturbances. Ecosystem engineers were originally defined as organisms that directly or indirectly modulate the availability of resources to other species, by causing physical changes in biotic or abiotic materials (Jones et al., 1994, 1997). Conceptually, ecological engineers incorporate and transform "materials" generating new physical states or "products" that directly affect the performance of specific organisms and indirectly other associated biotic components. For example, several animals change the physical state of trees, creating holes that are then used by other species (Jones et al., 1997) or modify leaves, creating tubular foliage structures influencing other organisms (Fournier et al., 2003). Hence, ecological engineers may be especially likely to trigger cascading effects along trophic and non-trophic chains as consequence of human activities.

[*] We thank Paulo Oliveira and Suzanne Koptur for the invitation to write this chapter. The comments of two anonymous reviewers improved the preliminary versions of this work. We also thank the financial support of the CONICET (grants PIP 5110/11 and 0665/14) and FONCYT (grants PICT 25314 and 1406) to AGFB.

Leaf-cutting ants (hereafter LCA) are one of the most outstanding examples of eco-system engineers because of their extraordinary ability to alter the environment where they inhabit. LCA usually move huge amounts of soil to construct and maintain their nests, cut large quantities of vegetation to cultivate their symbiotic fungus, and pro-duce a great amount of organic waste (Farji-Brener & Illes, 2000; Leal et al., 2004; Chapter 4). This organic waste is the degraded plant material remaining from the fun-gus culture process and is several times richer in organic carbon and nutrients than adjacent soils (Farji-Brener & Werenkraut, 2015). The creation of this novel, nutri-ent-rich substrate produces hot-spots of soil nutrients around the nest area, affecting soil biota, nutrient cycling, and vegetation patterns (Farji-Brener, 2010; Farji-Brener & Werenkraut, 2015). Specifically, when this nutrient-rich organic waste is depos-ited in piles on the soil surface (hereafter, refuse dumps) it can be easily reached and exploited by neighboring plants. Hence, plants growing near refuse dumps often show higher abundance and performance than plants growing on adjacent, non-nest soils (Farji-Brener & Ghermandi, 2008; Farji-Brener et al., 2010). This increased plant per-formance often affects plant-associated organisms, like mutualistic biota, herbivores and their natural enemies (Lescano et al., 2012). Consequently, any human activity modifying LCA density might affect their bottom-up cascading effect. In addition, these human-induced changes are often associated with changes in the composition of species community, affecting the identity of the potential "users" of the "materials" generated by ecological engineers. Since ecosystem engineers as LCA can profoundly affect ecosystem structure and functioning (Farji-Brener & Illes, 2000; Leal et al., 2014), understanding how human activities influence the impact of these organisms is key for conservation and sustainable management of natural habitats.

Although LCA have been proposed as ecosystem engineers, studies of their effects under this conceptual framework have mainly been focused on tropical and subtropical areas, in particular, the effects of the colossal nests of *Atta* spe-cies (Farji-Brener & Illes, 2000; Correa et al., 2010; Leal et al., 2014). A colony of *Atta* may remove up to 40 tons of soil, clearing a surface of 250 m² altering can-opy cover, leaf-litter abundance and soil features. Such physical effects may cascade across multiple levels of biological organization and spatial scales (Correa et al., 2010; Leal et al., 2014). However, the role of LCA as ecosystem engineers is less understood for temperate regions and species from the genus *Acromyrmex,* who built smaller nests than *Atta*. In this chapter we summarize the effects of human-related disturbances on *Acromyrmex lobicornis*, the only LCA species that inhabit temperate regions of Patagonia.

Lonely and Cold: The Role of the Leaf-Cutting ant *Acromyrmex lobicornis* in Arid Patagonia

Acromyrmex lobicornis (Emery) is one of the LCA species with the widest latitudi-nal range, reaching from subtropical areas of southern Brazil and Bolivia (23° S) to northern Patagonia (44° S), where it is the only LCA species (Farji-Brener &

Ruggiero, 1994). In this temperate region, *A. lobicornis* is especially abundant in dry habitats such as steppes and Monte Desert, reaching up to 100 nests/ha (Farji-Brener, 2000; Tadey & Farji-Brener, 2007). Their nests reach depths of 1 m and externally consist of a single mound of twigs, soil and dry plant material of up to 1 m in height and diameter. Organic debris from the fungus culture are removed from the internal fungus garden and dumped on the soil surface near the mound in a few large piles (Figure 18.1c). Preliminary studies indicated that this LCA species plays a relevant role as ecological engineer mainly through the accumulation of external waste piles (Figure 18.2a). Refuse dumps of *A. lobicornis* were several times richer in nutrients and soil biota than adjacent non-nest soils, increasing the rate of soil decomposition and nutrient cycling around nest areas (Farji-Brener & Ghermandi, 2000, 2004, 2008; Tadey & Farji-Brener, 2007; Farji-Brener, 2010; Fernández et al., 2014a). Plants near or growing on refuse dumps benefited from these hot-spots of soil nutrients, frequently presenting higher growth rate, biomass, and fitness than plants on adjacent, non-nest soils (Farji-Brener & Ghermandi, 2000, 2004, 2008; Farji-Brener et al., 2010). These ecological impacts are especially relevant in arid regions of Patagonia, where soils are extremely poor and nutrients are also a key limiting factor for plant establishment, growth, and reproduction (Satti et al., 2003). Consequently, any factor that changes the amount and quality of soil nutrients (i.e., the amount and/or quality of ant refuse dumps) might have important ecological consequences. Here, we describe how two human-associated disturbances, introduced livestock and road maintenance, may influence the role of *A. lobicornis* as soil improvers in the dry areas of Patagonia.

Case 1: How Introduced Livestock Reduces the Quality of LCA as Soil Improvers

In some dry areas of Argentina, the combined effects of insufficient rainfall, poor soils and overgrazing have produced important declines in vegetation and soil fertility (Abril & Bucher, 2001; Tadey, 2006) which, in turn, affect the activity of LCA as foragers and soil-nutrient improvers. The low abundance of palatable plants in this region forces livestock to feed upon a large number of plant species to maintain their nutritional requirements (Golluscio et al., 1998; Tadey, 2006). Therefore, this wider range of consumed species by livestock may decrease the availability of plants for LCA. Previous studies showed that introduced livestock share preferences with LCA for certain plant species, competing with ants for food resources (Robinson & Fowler, 1982; Guillade et al., 2014). This seems to be the case in the Monte Desert, Argentina, where exotic livestock and *A. lobicornis* co-occur.

Monte Desert is a temperate arid region of Patagonia, Argentina (39° S, 68° W), with a mean annual precipitation around 180 mm and mean annual temperature of 15°C (Cabrera, 1953). Vegetation physiognomy corresponds to a xerophytic shrub land dominated by *Larrea cuneifolia* (Cavanilles) and *L. divaricata* (Cavanilles) (Zygophyllaceae) with scarce abundance of grasses and large areas of bare ground (Correa, 1969). In this region, livestock drastically reduced

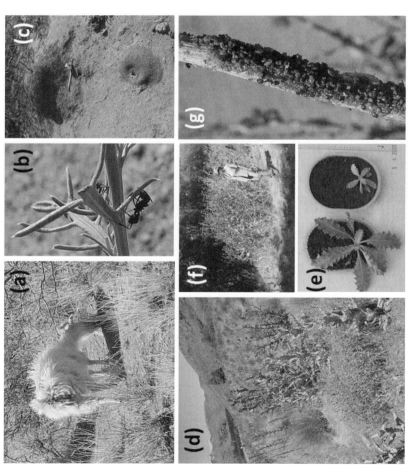

Figure 18.1. Photographs illustrating the consequences of introduced livestock and road maintenance on the role of leaf-cutting ants as ecological engineers in Patagonia, Argentina. (a–c): Influence of exotic livestock on the soil-improver effects of LCA nests in Monte Desert. Livestock reduce plant richness and cover through grazing (a), depleting the diet of the LCA *Acromyrmex lobicornis* (b), with the subsequent reduction of the nutrient content of their external refuse dumps (RD) (c). (d–g): Consequences of road building and maintenance on LCA density, exotic plant species and associated aphid tending-ants relationships in a Patagonian steppe. Nest of *A. lobicornis* and associated exotic thistles in roadside areas (d). Exotic thistles grow better on external refuse dumps of *A. lobicornis* than on non-nest soils (e), forming huge "exotic plant islands" around ant nests (f). Plants growing on ant refuse dumps also sustain more aphid density, increasing the abundance and activity of aphid-tending ants (g). Photo credits: M. Tadey (a, b); A. G. Farji-Brener (c–g). (A black-and-white version of this figure will appear in some formats. For the color version, please refer to the plate section.)

(a)

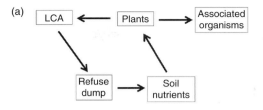

(b)

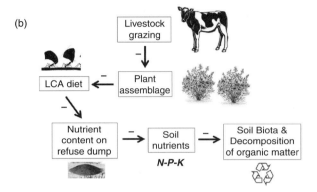

(c)

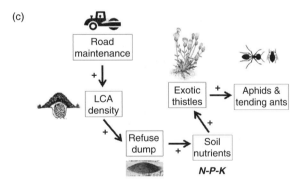

Figure 18.2. The role of leaf-cutting ant *Acromyrmex lobicornis* as ecological engineer in arid Patagonia (a), and the effects of human disturbances on this process (b–c). Plant fragments are harvested by the ants to feed the ant mutualistic fungus. Organic debris from the fungus culture are removed from the internal fungus garden and dumped on the soil surface. This nutrient-rich refuse dump increases the performance of nearby plants affecting plant-associated animals (a). In Monte Deserts, introduced livestock decrease the richness and cover of vegetation, reducing the availability of palatable plants for LCA. Consequently, the organic debris produced by an impoverished ant diet is less nutrient-rich, reducing the positive effect of refuse dumps on soil biota, soil nutrients and decomposition rate of organic matter (b). In arid steppes, the building and maintenance of roads increased ant nest and exotic plant densities. The high density of nutrient-rich refuse dumps are colonized mainly by exotic thistles, which are infested by a huge amount of aphids that attract a large number of aphid-tending ant species (c).

plant cover and diversity (Tadey, 2006; Bertiller & Ares, 2011, Figure 18.1a). Although livestock did not significantly affect the abundance of nests in this region (Tadey & Farji-Brener, 2007), there is evidence suggesting that livestock, through a drastic reduction of plant cover and richness, indirectly affects the ecological role of *A. lobicornis* (Tadey & Farji-Brener, 2007). First, the diet of *A. lobicornis* depends on plant availability, which in turn is affected by the abundance of livestock. With the drastic decrease in plant cover (from 70 to 20 percent) and the increment in grazing intensity, LCA harvested 40 percent less number of plant species possibly due to a decreased chance of encountering palatable plants. Supporting this idea, as grazing intensity increased, the same plant species that declined their cover were those that disappeared from ant diet (Tadey & Farji-Brener, 2007). In addition, the increased proportion of bare soil as a result of decreased vegetation cover led to extreme soil temperatures that limited ant foraging activity. Therefore, a reduction in plant availability due to grazing and the associated increment of bare soil may explain the reduction on the number of plant species harvested by *A. lobicornis*. It is known that the nutrient content on refuse dumps often depends on the number of plant species harvested by the LCA (Wirth et al., 2003; Bucher et al., 2004; Fernández et al., 2014b). Thus, decreasing the availability of the more nutritious plant species by livestock grazing leads to their absence in the LCA diet, and affects the nutrient content of their refuse dumps. Since exotic livestock often prefers to forage on nutrient-rich plant species (Guevara et al., 1996; Bisigato & Bertiller, 1997; Bertiller & Bisigato, 1998), ants might be shifting their foraging towards the remaining less nutritive plant species impoverishing the nutrient content of their refuse dumps. Thereby, the amount of nutrients provided by refuse dumps on soil surface will ultimately depend on the grazing intensity. For example, in the studied area an increment in the grazing intensity of only 0.06 cattle/ha determined a mean net decrease of 80 g/kg of C, 3 g/kg of N and 250 mg/kg of P in the refuse dumps (Tadey & Farji-Brener, 2007). This decrease in nutrient content of refuse dumps might negatively affect plant community. In a common garden experiment under natural conditions, refuse dumps from *A. lobicornis* originated under two extreme grazing intensities (low and high) showed differential effects on plant germination and performance of the most representative plant species of the studied area. Refuse dumps from low-grazed paddocks (i.e., richer in nutrient content) increased the germination rate and seedling vigor more than refuse dumps from the highly grazed paddock (Cerda et al., 2012). Given the high abundance of nests in the study area, we expect that refuse dumps in field conditions bear the same effects than in the performed experiment. To summarize, livestock decreased plant cover depleting leaf-cutting ant diets, which produce refuse dumps with low nutrient contents. Consequently, the introduction of exotic livestock diminishes the known positive effects of refuse dumps on soil biota, nutrient cycling, and plant performance (Figure 18.1a–c, Figure 18.2b).

Case 2: How Roads Amplify the Ecological Engineering Effects of LCA on Plants and Associated Organisms

Roads construction and maintenance are human activities that strongly affect natural environments (reviewed by Spellerberg, 1998; Trombulak & Frissell, 2002). First, the creation of roads in a natural area locally reduces the abundance of the native biota (Fahrig & Rytwinski, 2009). Second, soil disturbances associated with the maintenance of road verges often increase the invasion of exotic plant species (Gelbard & Harrison, 2005). Third, roads increase the rate of fauna mortality by traffic (Spellerberg, 1998). Finally, the existence of roads enhances the establishment of human settlements, exponentially increasing all these effects. Here, we summarize how roads affect the engineering effects of LCA in a Patagonian steppe, with consequences to both exotic and native biota.

The studies were conducted at the east (driest) border of Nahuel Huapi National Park, Patagonia, along the road 237 that is the main paved access to Bariloche city located in the National Park (41° S, 72° W). The vegetation in the study area has a physiognomy of a shrub/herbaceous steppe. The mean annual temperature is 8°C and the mean annual precipitation is approximately 600 mm. Soils are sandy loam andisols from volcanic ashes and aridisols. The dominant plants species are a mixture of native species typical of Patagonian steppes (e.g., *Stipa speciosa Imperata condensata* (Poaceae), *Mulinum spinosum* (Apiaceae), and *Plagiobothrys tinctorius* (Boraginaceae) and exotic weeds characteristics of road verges such us *Bromus tectorum* (Poaceae), *Onopordum acanthium*, *Carduus thoermeri* (Asteraceae) and *Verbascum thapsus* (Scrophulariaceae)) (Correa, 1969). In this natural protected area of Patagonia, the density of exotic thistles such as *Carduus thoermeri* (nodding thistle) and *Onopordum acanthium* (scotch thistle) and nests of the LCA *A. lobicornis* are several orders of magnitude higher near the roads than further in open fields, where exotic thistles and ants are almost absent (Farji-Brener, 1996, 2000; Farji-Brener & Ghermandi, 2008; Figure 18.1d). These human-induced increments of ant nests and exotic plant density trigger a bottom-up cascade effect, positively affecting the abundance of aphids that feed on thistles and the activity of aphid-tending ants (Lescano et al., 2012).

Several studies showed that the occurrence of soil disturbances that reduce native plant cover and the existence of improved soil nutrient patches promote biological invasions (Rickey & Anderson, 2004; Gelbard & Harrison, 2005; Leishman & Thomson, 2005). In Patagonian steppes, nests of *A. lobiornis* are almost only present in road verges at least for two reasons. First, LCA prefer open areas for establishing new nests because incipient colonies appear to require a certain degree of insolation to survive (Jaffe & Vilela, 1989; Vasconcelos, 1990). Second, the most abundant plant species that are in disturbed areas often show low levels of chemical defenses and, therefore, they are more palatable for LCA (Farji-Brener, 2001). This high availability of palatable plant species for LCA on roadsides increases the growth of ant nests and the generation of more incipient colonies (Vieira-Neto &

Vasconcelos, 2010), resulting in a high density of LCA nests. Since each ant nest may generate one or two external refuse dumps available for neighboring plants, their increased abundance near the roads represents hot-spots of soil-nutrients immersed in a "sea" of exotic plants. In the study area, the exotic thistles *Carduus thoermeri* and *Onopordum acanthium* are particularly abundant near road verges and often colonize the refuse dumps of *A. lobicornis*. As discussed earlier, refuse dumps of *A. lobicornis* contain several times higher nutrient levels and a better water retention capacity than adjacent soils (Farji-Brener & Ghermandi, 2000, 2004, 2008). Specifically, refuse dumps in the studied area showed 400 percent more organic matter, 500 percent more Nitrogen, 700 percent more Phosphorous and 600 percent more Potassium than adjacent, non-nest soils. Using stable isotopes, previous studies demonstrated that thistles reached and used nutrients from refuse dumps of *A. lobicornis* (Farji-Brener & Ghermandi, 2008; Lescano et al., 2012). Accordingly, thistles growing on refuse dumps showed up to 10 times more leaf and root biomass and tripled the amount of seeds (Figure 18.1e), comparing with those growing on adjacent soils, producing huge "thistles islands" at the roadsides (Farji-Brener & Ghermandi, 2008, Figure 18.1f).

Contrasting with the co-occurring native plants, in the disturbed areas of Patagonian steppes both thistle species are commonly infested by aphids, which in turn are tended by several ant species such as *Dorymyrmex tener*, *D. wolffuegeli* (Dolichoderinae), *Brachymyrmex patagonica* (Formicinae) and *Solenopsis richteri* (Myrmicinae). Ant-aphid relationships are mutualistic interactions in which aphids provide ants with carbohydrate-rich honeydew, a waste product of their sugar-rich diet of plant sap and, in exchange, ants defend them from other ants, predators and parasitoids (Way, 1963). Thus, the enhanced resource availability generated by *A. lobicornis* affects not only thistles but also the aphids that feed on them and the native tending ants. Since thistles growing on refuse dumps had greater biomass and foliar nutrient content than those growing in non-nest soils, they sustain higher aphid density leading to higher honeydew availability for tending ants, promoting the coexistence of a greater number of aphid-tending ant species and increased tending activity (Farji-Brener et al., 2009; Lescano & Farji-Brener, 2011; Lescano et al., 2012, Figure 18.1g). Overall, the presence of road verges amplifies the bottom-up effects of ant refuse dumps simply by increasing their density, influencing plants, herbivores, and ant-mutualistic interactions (Figure 18.2c). This scenario offers a good opportunity to study both theoretical and applied ecology. First, the previously described system constitutes a good model to understand how bottom-up impacts such as soil disturbances influence biotic interactions, including the mechanisms affecting the establishment and spread of exotic plants. Second, LCA ant nests that sustain higher exotic thistles, which in turn house high aphid density and a rich associated native aphid-tending ant assemblage, represent a challenge in terms of eradication of exotics and conservation of native biota. Exotic plants on refuse dumps generate up to 300 percent more seeds than the same plant species on non-nest soils (Farji-Brener & Ghermandi, 2008), representing a potential threat to neighboring non-invaded communities. Therefore, efforts in the eradication of thistles should be focused on ant refuse dumps, where plants are denser and represent a higher source of propagules. However, these exotic plants also sustain large populations of native ant species through increased aphid colonies. It is known that native ants are important components of ecosystems because they act as seed dispersers, agents of biological control and soil improvers (Del Toro et al., 2012). Therefore, conservationists should consider both – the potential dangerous of invasive exotic plants species and the role of the native ant fauna – to adequately manage exotic plants on ant refuse dumps without causing great damage on the native ant biota.

Concluding Remarks: Toward a Conceptual Framework to Understand How Human Activities Affect the Impacts of Ecosystem Engineers

Leaf-cutting ants, our ecological engineer model, harvest plants ("materials") and transform them into organic waste ("products"), which may favor the performance of neighboring plants and the associated fauna. We identified from our work three hypothetical ways in which human activities may interfere in one or more steps in this cascade process, changing the strength and/or sign of ecological engineering (Figure 18.3). Human activities may affect the role of ecosystem engineers by: (1) disrupting the availability of "materials" for ecological engineers (e.g., reducing the availability of palatable plant species for LCA by the introduction of livestock that compete for forage); (2) directly affecting the density of ecological engineers (e.g., favoring the establishment of LCA through habitat modifications that favor nest establishment and growth), and/or (3) changing the ecological context and thus influencing the type of organisms that may use the "products" of ecosystem engineering (e.g., favoring the abundance of exotic plant species that benefit from refuse dumps). These effects often act synergistically and may affect the role of LCA ants as ecosystem engineers in other environments.

Anthropogenic disturbances may simultaneously affect the density of ecological engineers and alter the abundance of materials that these organisms use. These changes trigger a particular ecological context that affects associated organisms, which may benefit from the ecological engineers. For example, in America, highly fragmented agro-mosaics are already the major scenario across many previously forested tropical lands (Tabarelli et al., 2004). In these anthropogenic landscapes, some leaf-cutting ant species are among the most successful organisms, increasing colony density up to 20 times (Wirth et al., 2007; Meyer et al., 2009). As previously discussed, part of this enhanced abundance is because human-induced changes often increase the availability of nesting sites and palatable plant species for LCA (Vasconcelos, 1990; Farji-Brener, 2001). In other words, forest edges and the conversion of pristine forests into agricultural lands and forest plantations promote the proliferation of pioneer plant species for LCA, which support higher colony density. Since exotic and pioneer plant species are common in these disturbed habitats, the ecological engineering effects of LCA as soil enrichers and seed dispersers may

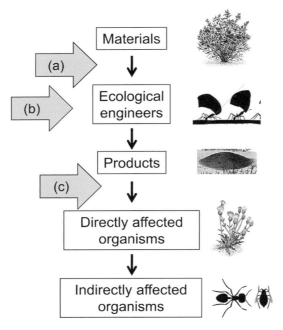

Figure 18.3. Diagram illustrating the potential effects of human disturbances (horizontal thickened arrows, a–c) on the role of ecological engineers. Through the processing of "materials", ecological engineers generate "products" that directly affect the performance of certain organisms, which in turn influence other associated biotic components. Human activities may affect the role of ecological engineers by: (a) disrupting the ingress of "materials"; (b) directly affecting the density of ecological engineers and (c) affecting the identity of organisms that benefit from ecological engineering and changing the way in which the "products" of ecological engineers affect other organisms. Illustrations show the example of the cascade effect discussed in this chapter: leaf-cutting ants, ant refuse dumps, plants, aphids, and aphid-tending ants.

particularly favor this type of vegetation, enhancing environmental homogenization (Leal et al., 2014).

Identifying the mechanisms through which humans affect the role of ecosystem engineers is relevant to understand and mitigate the impact of human activities on ecosystems. According to the proposed conceptual framework, humans may modify at least three key factors associated with the activities of ecosystems engineers: (1) the density of ecological engineers; (2) the availability of "materials" used by these organisms, and (3) the characteristics of the species that use their "products" (Figure 18.3). We showed that in Patagonia, as in other ecosystems (see Leal et al., 2014), the role of LCA as ecological engineers is context-dependent and may be modulated by anthropogenic activities. Specifically, we described how the introduction of exotic livestock and the maintenance of road verges affect the role of LCA as soil improvers, with concomitant effects on associated flora and fauna. Reducing stocking rates to mitigate the competition with ants for palatable forage may help to maintain the role of ant-nests as hot spots of nutrient-rich soils in this poor-soil environment. Finally, reducing the disturbances associated with the maintenance of road verges may help to control the proliferation of exotic weeds that take advantage of refuse dumps, without excessively affecting the native aphid-tending ants.

References

Abril, A., & Bucher, E. H. (2001). Overgrazing and soil carbon dynamics in the western Chaco of Argentina. *Applied Soil Ecology*, 16, 243–249.

Ares, J., Beeskow, A., Bertiller, M., Rostagno, C., Irisarri, M., Anchorena, J., Defossé, G., & Meroni, C. (1990). Structural and dynamics characteristics of overgrazed lands of Northern Patagonia, Argentina. In Bremeyer, A. (ed.). *Managed grasslands: regional studies*. Amsterdam: Elsevier Science Publishers, pp. 149–175.

Bennett, A. F. (1991). Roads, roadsides and wildlife conservation: a review. In Saunders, D. A. & Hobbs, R. J. (eds.). *Nature conservation 2: the role of corridors*. Chipping Norton, Australia: Surrey Beatty, pp. 99–117.

Bertiller, M. B., & Ares, J. O. (2011). Does sheep selectivity along grazing paths negatively affect biological crusts and soil seed banks in arid shrublands? A case study in the Patagonian Monte, Argentina. *Journal of Environmental Management*, 92, 2091–2096.

Bertiller, M. B., & Bisigato, A. (1998). Vegetation dynamics under grazing disturbance. The state-and-transition model for the Patagonian steppes. *Ecología Austral*, 8, 191–199.

Bestelmeyer, B., & Wiens, J. (2001) Ant biodiversity in semiarid landscape mosaics: the consequence of grazing vs. natural heterogeneity. *Ecological Applications*, 11, 1123–1140.

Bieber, A. G., Silva, P. S. D., Sendoya S. F., & P. S. Oliveira (2014). Assessing the impact of deforestation of the Atlantic Rainforest on ant-fruit interactions: a field experiment using synthetic fruits. *PlosOne*, 9, e90369.

Bisigato, A., & Bertiller, M. (1997). Grazing effects on patchy dryland vegetation in northern Patagonia. *Journal of Arid Environments.*, 36, 639–653.

Bucher, E. H., Marchesini, V., & Abril, A. (2004). Herbivory by leaf-cutting ants: nutrient balance between harvested and refuse material. *Biotropica*, 36, 327–332.

Cerdá, N., Tadey, M., Farji-Brener, A. G., & Navarro, M. (2012). Effects of leaf-cutting ant refuse on native plant performance under two levels of grazing intensity in the Monte Desert of Argentina. *Applied Vegetation Science*, 15, 479–487.

Correa, M. N. (1969). *Flora patagónica*. Buenos Aires, Argentina: INTA-Buenos Aires.

Corrêa, M. M., Silva, P. S. D., Wirth, R., Tabarelli, M., & Leal, I. R. (2010). How leaf-cutting ants impact forests: drastic nest effects on light environment and plant assemblages. *Oecologia*, 162, 103–115.

Del Toro, I., Ribbons, R. R., & Pelini, S. L. (2012). The little things that run the world revisited: a review of ant-mediated ecosystem services and disservices (Hymenoptera: Formicidae). *Myrmecological News*, 17, 133–146.

Fahrig, L., & T. Rytwinski (2009). Effects of roads on animal abundance: an empirical review and synthesis. *Ecology and Society*, 14, 21.

Farji-Brener, A. G. (1996). Posibles vías de expansión de la hormiga cortadora de hojas *Acromyrmex lobicornis* hacia la Patagonia. *Ecología Austral*, 6, 144–150.

(2000). Leaf-cutting ant nests in temperate environments: mounds, mound damages and mortality rates in *Acromyrmex lobicornis*. *Studies of Neotropical Fauna and Environment*, 35, 131–138.

(2001). Why are leaf-cutting ants more common in early secondary forests than in old-growth tropical forests? An evaluation of the palatable forage hypothesis. *Oikos*, 92, 169–177.

(2010). Leaf-cutting ant nests and soil biota abundance in a semi-arid steppe of northwestern Patagonia. *Sociobiology,* 56, 549–557.

Farji-Brener, A. G., & Ghermandi, L. (2000). The influence of nests of leaf-cutting ants on plant species diversity in road verges of northern Patagonia. *Journal of Vegetation Science*, 11, 453–460.

(2004). Seedling recruitment in the semi-arid Patagonian steppe: facilitative effects of refuse dumps of leaf-cutting ants. *Journal of Vegetation Science,* 15, 823–830.

(2008). Leaf-cutting ant nests near roads increase fitness of exotic plant species in natural protected areas. *Proceedings of the Royal Society – Series B*, 275, 1431–1440.

Farji-Brener, A. G., Gianoli, E., & Molina-Montenegro, M. (2009). Small-scale disturbances spread along trophic chains: leaf-cutting ant nests, plants, aphids and tending ants. *Ecological Research*, 24, 139–145.

Farji-Brener, A. G., & Illes A. E. (2000) Do leaf-cutting ant nests make "bottom-up" gaps in neotropical rain forests?: a critical review of the evidence. *Ecology Letters*, 3, 219–227.

Farji-Brener, A. G., Lescano, N., & Ghermandi, L. (2010). Ecological engineering by a native leaf-cutting ant increases the performance of exotic plant species. *Oecologia*, 163, 163–169.

Farji-Brener, A. G., & Ruggiero, A. (1994). Leaf-cutting ants (*Atta* and *Acromyrmex*) inhabiting Argentina: patterns in species richness and geographical range sizes. *Journal of Biogeography*, 21, 391–399.

Farji-Brener, A. G., & Werenkraut, V. (2015). A meta-analysis of leaf-cutting ant nest effects on soil fertility and plant performance. *Ecological Entomology*, 40, 150–158.

Fernández, A., Farji-Brener, A. G., & Satti, P. (2014a). Factores que influyen sobre la actividad microbiana en basureros de hormigas cortadoras de hojas. *Ecología Austral*, 24,103–110.

Fernández, A., Farji-Brener, A. G., & Satti, P. (2014b). Moisture enhances the positive effect of leaf-cutting ant refuse dumps on soil biota activity. *Austral Ecology,* 39, 198–203.

Fournier, V., Rosenheim, J., Laney L., & Johnson, M. (2003). Herbivorous mites as ecological engineers: indirect effects on arthropods inhabiting papaya foliage. *Oecologia*, 135, 442–450.

Gelbard, J., & Harrison, S. (2005). Invasibility of roadless grasslands: an experimental study of yellow starthistle. *Ecological Applications*, 15, 1570–1580.

Golluscio, R. A., Deregibus, V. A., & Paruelo, J. M. (1998). Sustainability and range management in the Patagonian steppes. *Ecología Austral*, 8, 265–284.

Guevara J. C., Stasi, C. R., & Estevez, O. R. (1996). Seasonal specific selectivity by cattle on rangeland in the Monte desert of Mendoza, Argentina. *Journal of Arid Environments*, 34, 125–132.

Guillade, A. C., & Folgarait, P. J. (2014). Competition between grass-cutting *Atta vollenweideri* ants (Hymenoptera: Formicidae) and domestic cattle (*Artiodactyla: Bovidae*) in Argentine rangelands. *Agricultural and Forest Entomology*, 17, 113–119.

Hölldobler, B., & Wilson, E. O. (2011). *The leafcutter ants: civilization by instinct*. London: W. W. Norton and Company, Inc.

Jaffe, K., & Vilela, E. (1989). On nest densities of the leaf-cutting ant Atta cephalotes in tropical primary forest. *Biotropica*, 48, 234–236.

Jones, C., Lawton J., & Shachar M. (1994) Organisms as ecosystem engineers. *Oikos, 69*, 373–386.

(1997). Positive and negative effects of organisms as physical ecosystem engineers. *Ecology*, 78, 839–841.

Laurance, W. F., Sayer, J., & Cassman, K. G. (2014). Agricultural expansion and its impacts on tropical nature. *Trends in Ecology & Evolution*, 29, 107–116.

Leal, I. R., Wirth, R., & Tabarelli, M. (2014). The multiple impacts of leaf-cutting ants and their novel ecological role in human-modified neotropical forests. *Biotropica*, 46, 516–528.

Leishman, M. R., & Thomson, V. P. (2005). Experimental evidence for the effects of additional water, nutrients and physical disturbance on invasive plants in low fertility Hawkesbury Sandstone soils, Sydney, Australia. *Journal of Ecology*, 93, 38–49.

Lescano, M. N., & Farji-Brener, A. G. (2011). Exotic thistles increase native ant abundance through the maintenance of enhanced aphid populations. *Ecological Research*, 26, 827–834.

Lescano, M. N., Farji-Brener, A. G., & Gianoli, E. (2015). Outcomes of competitive interactions after a natural increment of resources: the assemblage of aphid-tending ants in northern Patagonia. *Insect Sociaux*, 62, 199–205.

Lescano, M. N., Farji-Brener, A. G., Gianoli, E., & Carlo, T. (2012). Bottom-up effects may not reach the top: the influence of ant-aphid interactions on the spread of soil disturbances through trophic chains. *Proceedings of the Royal Society – Series B*, 279, 3779–3787.

Markl, U., Schleuning, M., Forget, P., Jordano, P., Lambert, J., Traveset, A., Wight, J., & Bohning-Gaese, K. (2012). Meta-analysis of the effects of human disturbance on seed dispersal by animals. *Conservation Biology*, 26, 1072–1081.

Meyer, S. T., Leal, I. R., & Wirth, R. (2009). Persisting hyper-abundance of keystone herbivores (Atta spp.) at the edge of an old Brazilian Atlantic Forest fragment. *Biotropica*, 41, 711–716.

Montoya-Lerma, J., Giraldo-Echeverri, C., Armbrecht, I., Farji-Brener, A. G., & Calle, Z. (2012). Leaf-cutting ants revisited: towards rational management and control. *International Journal of Pest Management*, 58, 225–247.

Raymond, A., Moranz, R. A., Debinski, D. M., Winkler, L., Trager, J., Mc Granahan, D., Engle, D., & Miller, J. (2013). Effects of grassland management practices on ant functional groups in central North America. *Journal of Insect Conservation, 17*, 699–713.

Rickey, M. A., & Anderson, R. C. (2004). Effects of nitrogen addition on the invasive grass Phragmites australis and a native competitor Spartina pectinata. *Journal of Applied Ecology*, 41, 888–896.

Robinson, S. W., & Fowler, H. G. (1982). Foraging and pest potential of Paraguayan grass-cutting ants (*Atta* and *Acromyrmex*) to the cattle industry. *Zeitschrift fuer angewandte Entomologie*, 93, 42–54.

Sallanbanks, R., Arnett, E. B., & Marzluff, J. M. (2000). An evaluation of research on the effects of timber harvest on bird populations. *Wildlife Society Bulletin*, 28, 1144–1155.

Satti, P., Mazzarino, M. J., Gobbi, M., Funes, F., Roselli, L., & Fernandez, H. (2003). Soil N dynamics in relation to leaf litter quality and soil fertility in north-western Patagonian forests. *Journal of Ecology*, 91, 173–181.

Spellerberg, I. F. (1998). Ecological effects of roads and traffic: a literature review. *Global Ecology and Biogeography*, 7, 317–333.

Tabarelli, M., Silva, J. & Gascon, C. (2004). Forest fragmentation, synergisms and the impoverishment of neotropical forests. *Biodiversity and Conservation.*, 13, 1419–1425.

Tadey, M. (2006). Grazing without grasses: effects of introduced livestock on plant community of an arid ecosystem in northern Patagonia. *Applied Vegetation Science, 9,* 109–116.

Tadey, M., & Farji-Brener, A. G. (2007). Indirect effects of exotic grazers: livestock decreases the nutrient content of refuse dumps of leaf-cutting ants through vegetation impoverishment. *Journal of Applied Ecology*, 44, 1209–1218.

Thompson, I. A., Baker, J. A., & Ter-Mikaelian, M. (2003). A review of the long-term effects of post-harvest silviculture on vertebrate wildlife, and predictive models, with an emphasis on boreal forest in Ontario, Canada. *Forest Ecology and Management,* 177, 441–469.

Trombulak, S. C., & Frissell, C. A. (2002). Review of ecological effects of roads on terrestrial and aquatic communities. *Conservation Biology*, 14, 18–30.

Vasconcelos, H. L. (1990). Habitat selection by the queens of the leaf-cutting ant Atta sexdens L. in Brazil. *Journal of Tropical Ecology*, 6, 249–252.

Vieira-Neto, E. H. M., & Vasconcelos, H. L. (2010). Developmental changes in factors limiting colony survival and growth of the leaf-cutter ant *Atta laevigata*. *Ecography*, 33, 538–544.

Way, M. J. (1963). Mutualism between ants and honeydew producing Homoptera. *Annual Review of Entomology*, 8, 307–344.

Wirth, R., Herz, H., Rye, I., Beyschlag, W., & Hölldobler, B. (2003). *Herbivory of leaf-cutting ants*. Berlin: Springer.

Wirth, R., Meyer, S. T., Almeida, W. R., Araújo, M. V., Jr., Barbosa, V. S., & Leal, I. R. (2007). Increasing densities of leaf-cutting ants (*Atta* spp.) with proximity to the edge in a Brazilian Atlantic forest. *Journal of Tropical Ecology,* 23, 501–505.

Part VI

Perspectives

19 The Study of Interspecific Interactions in Habitats under Anthropogenic Disturbance: Importance and Applications

Martin Heil and Marcia González-Teuber*

Introduction

Ant-plant mutualisms are a common element in all ecosystems in which ants occur and reach particularly high dominance and diversity in tropical lowland forests (Chapter 2; Davidson & McKey, 1993; Bronstein, 1998; Heil & McKey, 2003). Although this geographic distribution should make these mutualisms prime candidates to suffer from deforestation and climate change, ant-plant mutualisms might be under the few winners of anthropogenic global change. For example, increased water stress – a change predicted for many ecosystems worldwide – can strengthen an obligate ant-plant mutualism (Pringle et al., 2013). Tabarelli and colleagues show in this book that leaf-cutter ants clearly benefit from anthropogenic changes in tropical and sub-tropical ecosystems (Chapter 4). As discussed by Farji-Brener and colleagues, the mere increase in roads increases the density of nests of these light-demanding ants (Chapter 18). Oliveira and colleagues show in Chapter 7 that the decline of vertebrate frugivores in disturbed sites can benefit this type of ant-plant interactions. Even when they are no true winners, ant-plant mutualisms appear to suffer surprisingly little negative effects from habitat fragmentation, deforestation and the invasion by novel species (Chapter 3). In this chapter, we discuss the specific characteristics of ant-plant mutualisms that are likely to explain (1) the somehow counterintuitive observation that many of these mutualisms increase in frequency in disturbed areas, particularly in the tropics, (2) which ones of these mutualisms can easily be established *de novo* among non-co-evolved species (i.e., invasive ants and local plants or vice versa), and (3) how these phenomena can be used in scientific research as well as the biological control of pest insects.

* Marcia González-Teuber gratefully thanks Fondecyt (no. 11130039) and the Max Planck Society for their financial support during the development of this work. Martin Heil acknowledges support from CONACyT de México (grant numbers 212715 and 251102), and both authors apologise to all authors whose works could not be cited due to length limitations.

In general, ant-plant mutualisms can occur as facultative or obligate mutualisms. Facultative mutualisms are usually established among extrafloral nectar (EFN) secreting plants and ants that visit these plants to obtain a food reward that is rich in calories and – in most cases – also in essential amino acids (González-Teuber & Heil, 2009; Heil, 2015). Extrafloral nectar represents an induced defensive trait in the vast majority of species that have been investigated so far (Heil, 2011). Plants respond to herbivory with an activation of the octadecanoid signalling pathway in which jasmonic acid (JA) represents the central and mobile hormone, and one of the multiple JA-responsive plant defensive traits is EFN (Heil, 2004, 2008, 2015; Heil et al., 2001; Heil, Greiner & Meimberg, 2004). Representing a food source that is not constitutively available and for which ants have to compete with many other arthropods (Rudgers & Gardener, 2004; Heil, 2015), ant-plant mutualisms that depend exclusively on EFN are necessarily facultative in nature. EFN-secreting plants are taxonomically extremely widespread (Marazzi, Bronstein & Koptur, 2013; Weber & Keeler, 2013) and represent a very common element in multiple ecosystems. Particularly in certain tropical ecosystems, EFN secretion can characterise up to 30 per cent of all plant species (Keeler, 1979; Keeler, 1980; Oliveira & Brandão, 1991; Schupp & Feener, 1991; Fiala & Linsenmair, 1995). The observation that EFN secretion seemingly has evolved independently in multiple orders of plants (Weber & Keeler, 2013) already indicates that in general, EFN secretion is a successful strategy. In spite of some studies that could not find evidence for the protection hypothesis (Freitas et al., 2000), hundreds of studies reported that EFN secretion contributes significantly to the defence of plants against herbivores (Bentley, 1976, 1977; Keeler, 1980; Heil & McKey, 2003). In fact, several recent meta-analyses demonstrated unambiguously that ant-plant interactions that are mediated by EFN in general benefit the plant via the reduction of herbivory (Chamberlain & Holland, 2009; Rosumek et al., 2009; Romero & Koricheva, 2011; Koricheva & Romero, 2012; Trager et al., 2010).

Obligate mutualisms formed among 'plant-ants' and 'ant-plants' (so-called myrmecophytes) are taxonomically less frequent than EFN-mediated facultative ant-plant mutualisms and seem to be restricted to lowland tropical ecosystems (Davidson & McKey, 1993; Heil & McKey, 2003). The most characteristic element of this category of ant-plant mutualisms is the provisioning of nesting space by the plant in the form of hollow structures, so-called domatia (Fiala & Maschwitz, 1992a). Domatia can be formed by leaf pouches, hollow thorns, bulbs and even hollow leaf stalks (Huxley, 1978; Yu & Pierce, 1998; Blüthgen & Wesenberg, 2001; Brouat et al., 2001; Moog et al., 2002; Kato et al., 2004; Webber et al., 2007; Ferreira et al., 2011; González-Teuber & Heil, 2015; Chomicki et al., 2016). However, by far the most common type of domatia appears to be formed by hollow stems and twigs. These shoot domatia might even possess pre-formed entrance holes or open spontaneously to facilitate their colonisation by specific species of ants (Brouat et al., 2001; Federle et al., 2001; Moog et al., 2002; Webber et al., 2007). In the nutritive ant-plant mutualisms, which are frequently formed by epiphytes with limited access to soil nutrients (Huxley, 1978), nesting space is usually the only resource that plants provide to ants. However, a recent study reported an additional ant food reward in the form of a post-anthetic secretion of 'concealed' floral nectar of epiphytic Rubiaceae on Fiji Islands (Chomicki et al., 2016). By contrast, host plants in the defensive mutualisms – which form the topic of this chapter – in general provide a food reward as well: EFN, or cellular food bodies (FBs), or both (Rickson, 1971; Rickson & Risch, 1984; Fiala & Maschwitz, 1991, 1992b; Folgarait & Davidson, 1995; Fischer et al., 2002; Heil et al., 1997; Heil, Baumann & Krüger, 2004; Heil et al., 2009; Palmer et al., 2008). In these interactions, ants significantly contribute to the defence of their host plant (Janzen, 1974; Bronstein, 1998; Heil & McKey, 2003).

Taxa that engage in ant-plant mutualisms are frequently diverse (Fiala & Linsenmair, 1995; Fiala et al., 1999; Ward, 1999; Blattner et al., 2001; Palmer et al., 2003; Quek et al., 2004; Marazzi, Conti & Sanderson, 2013; Weber & Keeler, 2013; González-Teuber & Heil, 2015). For example, the EFN-secreting clade within *Senna* comprised 80 per cent of all species of this genus (Marazzi & Sanderson, 2010). Particularly high numbers of species in EFN-secreting taxa have also been observed for the Fabaceae and the Rosaceae (Weber & Keeler, 2013). Thus, it has been suggested that EFN secretion might represent a key innovation that enables rapid speciation (Marazzi & Sanderson, 2010) and trait-dependent diversification models indeed confirmed that EFN-secreting plant taxa show a tendency towards increased rates of diversification (Weber & Agrawal, 2014). Although generalisations are difficult to draw, it also appears that ant-plants are enriched in pioneer species (Bentley, 1976; Longino, 1989; Ferguson et al., 1995; Piovia-Scott, 2011; Leal et al., 2015). Many ant-plants are also tolerant to fire (Janzen, 1967; Alves-Silva and Del-Claro, 2013; Fagundes et al., 2015) and, finally, ant-plant mutualisms are frequently reported to involve invasive species (Lach, 2003, 2010; Ness & Bronstein, 2004; Savage & Whitney, 201; Ness et al., 2013; Carrillo et al., 2014; Leal et al., 2015). What makes ant-plants such particularly successful pioneers, and why are facultative ant-plant mutualisms so easily established, both in ecological and evolutionary timescales? Here, we discuss how a dependency of EFN secretion on a reduced set of universal plant genes can explain the easy evolution of this trait and that ecological fitting sensu Janzen (1985) can explain the easy de novo assembly of ant-plant mutualism among non-co-evolved species. Finally, we suggest that these general patterns point towards an underused potential of ant-plant mutualisms in biological pest control and that the mutualisms among invasive ants and native plants and vice versa bear an underused scientific potential for the understanding of mutualisms.

Extrafloral Nectar Is a Rapidly Evolving Mediator of Ant-Plant Mutualisms

Nectar is the most common mediator of mutualistic interactions among plants and animals (Heil, 2011). Extrafloral nectaries have been reported from an extremely diverse array of plant species (Marazzi, Bronstein & Koptur, 2013). Even ferns secrete EFN (Darwin, 1876; Schremmer, 1969; Tempel, 1983; Koptur et al., 1998; White & Turner, 2012) and can gain an enhanced anti-herbivore defence (Jones & Paine, 2012; Koptur et al., 2013), although several earlier studies failed to find positive evidence for the protectionist hypothesis for this ancient group of plants (Darwin, 1876; Schremmer, 1969; Tempel, 1983). On the basis of the phylogenetic evidence available to date, it seems most likely that EFN secretion, after having been evolved already in the ferns, was then lost and regained multiple times (Weber & Keeler, 2013). Moreover, extrafloral nectaries are morphologically extremely diverse and their traits can differ quite strongly within a family (Melo et al., 2010; Aguirre et al., 2013) or genus (Escalante-Perez et al., 2012; Marazzi, Conti & Sanderson, 2013), and even the response of EFN secretion herbivore-inflicted damage might differ among invasive and native populations of a plant species (Wang et al., 2013). Thus, the presence, the morphology and the secretion mechanism of extrafloral nectaries are evolutionarily highly flexible and at least the morphologically simple nectaries can be gained and lost quickly. Intrinsically, this evolutionary flexibility means that the functioning of a defensive, EFN-based ant-plant mutualism does not require co-evolution among the interacting species.

Why is the presence and functioning of extrafloral nectaries so flexible? Here, we suggest that the rapid evolution of these nectaries can be explained by the assumption that a few conserved metabolic genes such as cell-wall invertase (CWIN), the sucrose transporter, SWEET9 and sucrose phosphate synthase (Ruhlmann et al., 2010; Heil, 2011; Escalante-Perez & Heil, 2012; Lin et al., 2014; Millán-Cañongo et al., 2014) are centrally involved in the secretion of nectar. Even the inducibility of EFN can be explained by the conserved JA-responsiveness of cell wall invertases in general (Millán-Cañongo et al., 2014) and, thus, does not require any specific evolutionary changes. If we propose a simple mechanism that controls the coordinated expression of these genes in the nectary tissue, or simply in the parenchyma between phloem and the surface, mutations in a single gene could cause the phenotypic appearance and disappearance of functioning extrafloral nectaries (Heil, 2015). This hypothesis is strengthened by the observation on 'gestaltless' nectaries in *Brassica juncea* (Mathur et al., 2013) and of an EFN-like wound secretion on the leaves of *Solanum dulcamara* (Lortzing et al., 2016), hence, by the occurrence of defensive EFN-like secretions in single species within genera that possess now known species with extrafloral nectaries (Heil, 2016).

The high phylogenetic instability of this trait would then be caused by the context-dependency of the resulting defensive effect, which might exert strong selective pressure either on its maintenance or its rapid loss, depending on the average conditions in the environment in which the plant is growing. Furthermore, the easy appearance of EFN secretion and its frequent evolutionary fixation also indicate that EFN secretion is likely to provide plants and ants with an immediate benefit, even when the interacting species have no co-evolutionary history. In this context, Klimes shows in this book that even tree-nesting ants are usually not specialised to particular tree species (Chapter 2), meaning that individuals of a plant species that just is about to evolve EFN secretion are likely to already carry at least some ant species. Evidently, this easy assembly among non-specialised species makes EFN-mediated mutualism particularly prone to involve invasive species as well on the side of both, plants and ants.

Ant-Plant Mutualisms Are Frequently Formed among Invasive and Native Partners

Invasive species are a hot topic in ecological research (Chapters 12–15). Several researchers investigated EFN in the 'invasive' context and considered, for example, the consequences for the EFN-producing plants when invasive ants replace native ants (Lach & Hoffmann, 2011; Savage & Rudgers, 2013) or when invasive, EFN-bearing plants are visited by native or invasive ants (Lach et al., 2010) and the putative importance of EFN in the success of invasive ants (Lach et al., 2009) or plants (Wang et al., 2013). Many invasive ant species visit extrafloral nectaries and enhance the predation pressure on herbivores, which themselves also can be native or invasive (Ness and Bronstein, 2004). Thus, invasive ants commonly establish mutualisms with native plants (Ness & Bronstein, 2004; Savage and Whitney, 2011; Bleil et al., 2011). However, they might be less efficient defenders than some co-evolved native ants (Ness & Bronstein, 2004; Lach & Hoffmann, 2011) and their presence can cause ecological costs, at least in comparison to the interaction with the native species. For example, the invasive big-headed ant (*Pheidole megacephala*) replaced the symbiotic, defending *Crematogaster* ants from whistling thorn acacia (*Acacia drepanolobium*) trees, a situation which strongly enhanced predation by elephants (Riginos et al., 2015). Similarly, the release of the invasive plant from its natural enemies can reduce the protective effect of ants in spite of a damage-induced EFN secretion (Ness et al., 2013) and then cause lower or zero net defensive effects of EFN secretion.

In summary, the defensive effects of ant-plant mutualisms that involve at least one invasive partner are highly context-dependent. On the one hand, the invasive *Technomyrmex albipes* on Ile aux Aigrettes in the Indian Ocean was the most common and abundant ant visitor to the native shrub *Scaevola taccada* plant. However, the ants frequently tended sap-sucking hemipterans on this plant, and the experimental exclusion of ants increased the plant's growth and fruit production (Lach et al., 2010). On the other hand, this ant also visited the invasive tree, *Leucaena leucocephala*, and successfully deterred the primary herbivore of *L. leucocephala*, the Leucaena psyllid (*Heteropsylla cubana*). In this case, experimentally excluding ants from the plant resulted in decreased growth and seed production (Lach et al., 2010). Similarly, *Anoplolepis gracilipes*, a highly invasive ant species on the Samoan Archipelago, quickly responded to increased availability of EFN-secreting plants and discovered, attacked and removed more prey than other ant species (Savage & Whitney, 2011). On the side of the plant, higher rates of EFN secretion have been reported for plants of invasive populations as compared to native populations of tallow tree (*Triadica sebifera*) (Wang et al., 2013). In these cases, it is tempting to speculate that invasive ants are the better defenders and that EFN can contribute to the invasiveness of a plant. The latter scenario is also presented by Blütghen and colleagues, who show in this book that plants that are introduced to formerly ant-free habitats (i.e., islands) possess traits that protect their flowers from nectar-robbing ants and thus might be defended better from invasive ants than local species that never experienced social ants in their evolutionary history (Chapter 13). Thus, ant-plant mutualisms, like other mutualisms (Traveset & Richardson, 2014), can act as facilitators of species invasion.

The multiple outcomes of the interactions of invasive, EFN-secreting plants species and invasive ant species with native species indicate that these interactions represent a still underexplored, gigantic natural experiment that allows to study from its very beginnings the 'Living together in novel habitats' (Chapter 3) and *de-novo* establishment of EFN-mediated ant-plant mutualisms. Making better use of this experiment will help to understand an import aspect that is likely to have contributed – and still contribute – to the tremendous importance of ant-plant mutualisms in the evolution of the involved species and the functioning of ecosystems: ant-plant mutualisms can be easily established among non-co-evolved species.

Many Ant-Plants Are Pioneer Species

As reviewed by Yamawo in Chapter 8, ant-plant protection mutualisms are found in many pioneer plant species. For example, *Macaranga* in South-East Asia (Fiala & Maschwitz, 1990; Fiala et al., 1999), the 'swollen-thorn' acacias in Africa and Central America (Janzen, 1967; Janzen, 1974; Young et al., 1997; Goheen & Palmer, 2010) and *Cecropia* in Central and South America (Putz & Holbrook, 1988; Ferguson et al., 1995; Folgarait & Davidson, 1995) are genera that are dominated by pioneer species. *Macaranga* spp. in particular are successful on deforested and eroded areas and in heavily disturbed ecosystems such as road-sides, pastures and sub-urban areas, and the ant-Cecropias and ant-Acacias dominate in similarly disturbed ecosystems. Surprisingly, very few studies have aimed at understanding the causal link between using ants as a biotic defence mechanism and the capacity of a plant species to colonise disturbed areas.

As mentioned, several of the obligate myrmecophytes are tolerant to fire (Janzen, 1967; Alves-Silva & Del-Claro, 2013; Fagundes et al., 2015) and, in fact, it has been suggested that the pruning of the surrounding vegetation that is performed by several genera of plant-ants (Janzen, 1969; Yumoto & Maruhashi, 1999; Federle et al., 2002) helps the host plant to survive fire, simply because it is spatially separated from the rest of the burning vegetation (Janzen, 1967). However, a high frequency in fire successional ecosystems was also reported for plants that only engage in EFN-mediated facultative interactions with ants (Koptur et al., 2010). A part of the explanation might be provided by the observation that individuals of the shrub *Banisteriopsis campestris* (Malpighiaceae) that were resprouting after fire had higher sugar concentration in their EFN and were characterised by higher ant abundance than control plants that remained undamaged by fire (Alves-Silva & Del-Claro, 2013). These effects might be important, because food availability was an important determinant for the recovery of ant diversity after fire disturbance (Fagundes et al., 2015).

A second, non-exclusive explanation might be that ants in general represent a very effective defence mechanism. In the obligate myrmecophytes such as *Acacia*, *Cecropia* or *Macaranga*, ants frequently take over a large part of the defence of their host plant against herbivores (Janzen, 1966, 1969, 1972, 1974; Schupp, 1986; Rocha & Bergallo, 1992; Bronstein, 1998; Bizerril & Vieira, 2002) and can contribute to its protection from pathogens (Heil et al., 1999; González-Teuber et al., 2014). Ants that colonise a myrmecophyte protect their host even against mammals and therefore help, for example, the swollen-thorn acacias to establish on pastures and in other disturbed ecosystems with high herbivore pressure from cattle, sheep, donkeys and even elephants (Brown, 1960; Stapley, 1998; Palmer et al., 2008). As Palmer and Young show in Chapter 10, the protective effects of plant-ants clearly have large-scale effects that make them a major determinant of ecosystem structure. Similarly, a study that simulated the damage to buttonwood mangrove (*Conocarpus erectus*) caused by hurricanes revealed that damaged plants provided more EFN to ant mutualists and that ant-exclusion increased leaf damage (Piovia-Scott, 2011). Pioneers are frequently building large stands that are dominated by a single species. Owing to their high apparency (Feeny, 1976) and the easy movements of herbivores and pathogens from one plant to the next, natural as well as agronomic monocultures are particularly prone to infestations and infections by specialist plant enemies (Barrett & Heil, 2012). Thus, it is tempting to speculate that the high efficiency of ants as defenders against taxonomically and functionally diverse plant enemies including herbivores, pathogens and competing plants might represent an important factor that helps the involved plants to act as pioneers and establish large stands on eroded areas and in other heavily disturbed areas.

Ecological Fitting Facilitates the Assembly of Ant-Plant Mutualisms

As we have discussed in the previous sections, ant-plant mutualisms, particularly those that are mediated by EFN, can quickly be established among non-co-evolved species. In most cases, the resulting interaction has at least some protective effect on the plant and, thus, can be considered mutualistic. Why are these mutualisms so successful and versatile? We argue that ant-plant mutualisms are classical 'by-product mutualisms': each partner trades a reward or service that is cheap for it to offer to obtain a valuable reward or service.

First, EFN secretion is a cheap investment and does not appear to be subject to the classical growth-versus-defence trade-off (Herms & Mattson, 1992). In fact, high EFN secretion rates have been observed on regrowing plants, for example, after pruning or burning (Piovia-Scott, 2011; Alves-Silva et al., 2013). Since the uploading of sucrose from the phloem represents a limiting step in concurrent EFN secretion (Millán-Cañongo et al., 2014), we can expect a generally positive correlation of growth rate with EFN secretion. For the obligate mutualism of *Azteca* ants and *Cordia alliodora* trees, water stress was reported to increase plant investment via phloem-feeding scale insects and, consequently, ant-mediated defence (Pringle et al., 2013). Thus, maintaining ants might be a defence mechanism that can easily be maintained when other defences become too costly.

Second, predators and parasitoids usually suffer from stronger energetic limitations than herbivores. For example, increased EFN secretion rates have been related to higher survival rates of ant workers (Lach et al., 2009) and can increase their activity and aggressiveness, even in facultative ant-plant interactions (Sobrinho et al., 2002; Ness, 2006; Ness et al., 2009). In addition, the general nutritive demands concerning energy and amino acids of adult insect predators and parasitoids and their restriction to liquid food are very widespread traits, a situation that makes it easy for plants to evolve a generalised food source that meets the demands of the most potential defenders.

Third, herbivores evolve to cope with plant-derived secondary compounds rather than the predatory and defensive traits of other animals. For most animals, the defensive chemistry of ants in particular seems almost impossible to cope with, and more so for species that have a co-evolutionary history with plants rather than with ants (Heil, 2015). The other way round, even the protease inhibitors (PIs) that convert *Acacia* FBs into 'exclusive rewards', indigestible for herbivores, are likely not a result of any specific (co-) evolution: legume PIs have low inhibitory activity on typical ant proteases, likely because ants do not exert any selective pressure on plant PIs as anti-herbivore defence (Orona-Tamayo et al., 2013).

Finally, as Klimes shows in Chapter 2, tree-nesting ants are usually not specialised to particular tree species, meaning that individuals of a species that just is about to evolve EFN secretion are likely to already carry at least some ant species. As pointed out before, these ants are equally likely to be limited by energy and should readily visit the new food source, gaining a benefit which – since all ants to some degree defend their area – would immediately feed back to the plant. In summary, EFN is more likely to attract predatory arthropods and parasitoids rather than herbivores, potential consumers of EFN are likely to be present when this trait newly emerges, and several, if not most, of these species will exert at least some predation pressure on herbivores. In summary, 'ecological fitting' sensu Janzen (1985) is likely to be a central reason for the wide taxonomic range of EFN-secreting plants and the general ecological success of this trait.

Ant-Plant Interactions and Biocontrol

As we have summarised, many ant-plants are pioneers, and the provisioning of EFN or nesting space allows for an easy establishment of ant-plant mutualisms among native and invasive species. Since EFN-mediated defence does not require co-evolution between plants and predators, and since the defence is active against a wide range of pests, ant-plant interactions usually will benefit the plant to some degree without requiring any specific adaptations on the side of the ants. All crops and the majority of horticultural and orchard species are pioneers as well and in most cases, this characteristic already applies to their wild ancestors (e.g., cereals, beans and vegetables). Moreover, most crops are cultivated outside their centre of domestication, which means that in the end, they represent invasive species. Therefore, it does not seem surprising when Clough and colleagues report in this book that ants can serve as biocontrol agents that benefit the yield of multiple crops, as well as horticultural or orchard species (Chapter 16; Jones et al., 2016). For example, the presence of extrafloral nectaries in peach (*Prunus persicae*) was related to lower herbivory rates and higher productivity (Mathews et al., 2007, 2009). Providing ants with artificial nesting space enhanced the predation pressure on pests of coffee (*Coffea arabica*) (Philpott & Foster, 2005), and ant exclusion reduced cacao yield by 50 per cent (Gras et al., 2016).

Why are EFN-mediated tritrophic interactions not an integrative part of common biocontrol programmes? As exemplified elsewhere (Stenberg et al., 2015), a major problem might be that breeding has impaired the beneficial interactions of crops with the third trophic level. For example, nectariless lines have been deliberately bred in cotton because the attraction of insects to EFN was considered detrimental (Beach et al., 1985). However, the earlier-discussed positive correlation of growth rate with EFN secretion indicates the potential to return this traits at least to those of our crops whose ancestors secreted EFN. The previously cited studies on invasive species demonstrated that the defensive outcomes of newly established ant-plant interactions are highly context-dependent, and such context dependency represents a relevant limitation for the reliable use of tritrophic interactions in agriculture. However, EFN is cheap to produce, is naturally produced by many cultivated species and provides direct benefits to multiple beneficial insects. Planting field margins with EFN-bearing species might be another promising avenue to explore (Olson & Wäckers, 2007; Géneau et al., 2012). Similarly, the active release of beneficial arthropods in combination with planting EFN-producing crops, or intercropping with EFN-secreting species, provides interesting perspectives. In this scenario, EFN would keep the biocontrol agents at stable population levels within the agricultural field even during pest-free periods.

Conclusions

In this chapter, we summarised evidence for the somehow provocative statement that protective ant-plant interactions are among the winners of anthropogenic changes in ecosystem structures and local as well as global climatic conditions. We also propose that invasive species of ants and plants can easily engage in these mutualisms in the novel ecosystems. Although global change and the establishment of novel species in specific areas are generally tagged as negative events, we further argue that these observations indicate an underused potential of ant-plant mutualisms for scientific research and the biological control of pests. Scientifically speaking, EFN secretion, particularly, represents a trait that shows an extremely high ecological and evolutionary plasticity, which means that it can emerge easily in evolutionary terms and then can become fixed, or lost. The study of the establishment of EFN-mediated mutualisms and their net fitness effects for native and invasive species will allow a better understanding of the evolution of this very common plant defensive trait. Moreover, the hypothesis that defensive ant-plant interactions are by-product mutualisms that are subject to significant ecological fitting also gives hope for an application of these interactions in biological pest control. We suggest that future studies on defensive ant-plant interactions should move away from biased research aimed at finding negative effects of invasive species and global change and should focus more on the other, positive side of the coin. Interactions among native and invasive plants and ants represent a gigantic and underused natural experiment which can open novel research avenues towards the understanding of the evolution of ant-plant interactions and their better use in biological pest control.

References

Aguirre, A., Coates, R., Cumplido-Barragán, G., Campos-Villanueva, A. and Díaz-Castelazo, C. (2013) Morphological characterization of extrafloral nectaries and associated ants in tropical vegetation of Los Tuxtlas, Mexico. *Flora*, 208, 147–156.

Alves-Silva, E., Baronio, G. J., Torezan-Silingardi, H. M. and Del-Claro, K. (2013) Foraging behavior of *Brachygastra lecheguana* (Hymenoptera: Vespidae) on *Banisteriopsis malifolia* (Malpighiaceae): extrafloral nectar consumption and herbivore predation in a tending ant system. *Entomological Science*, 16, 162–169.

Alves-Silva, E. and Del-Claro, K. (2013) Effect of post-fire resprouting on leaf fluctuating asymmetry, extrafloral nectar quality, and ant-plant-herbivore interactions. *Naturwissenschaften*, 100, 525–532.

Barrett, L. G. and Heil, M. (2012) Unifying concepts and mechanisms in the specificity of plant-enemy interactions. *Trends in Plant Science*, 17, 282–292.

Beach, R. M., Todd, J. W. and Baker, S. H. (1985) Nectaried and nectariless cotton cultivars as nectar sources for the adult soybean looper. *Journal of Entomological Science*, 20, 233–236.

Bentley, B. L. (1977) Extrafloral nectaries and protection by pugnacious bodyguards. *Annual Review of Ecology and Systematics*, 8, 407–427.

(1976) Plants bearing extrafloral nectaries and the associated ant community: interhabitat differences in the reduction of herbivore damage. *Ecology*, 57, 815–820.

Bizerril, M. X. A. and Vieira, E. M. (2002) *Azteca* ants as antiherbivore agents of *Tococa formicaria* (Melastomataceae) in Brazilian Cerrado. *Studies on Neotropical Fauna and Environment*, 37, 145–149.

Blattner, F. R., Weising, K., Bänfer, G., Maschwitz, U. and Fiala, B. (2001) Molecular analysis of phylogenetic relationships among myrmecophytic *Macaranga* species (Euphorbiaceae). *Molecular Phylogenetics and Evolution*, 19, 331–344.

Bleil, R., Bluethgen, N. and Junker, R. R. (2011) Ant-plant mutualism in Hawai'i? Invasive ants reduce flower parasitism but also exploit floral nectar of the endemic shrub *Vaccinium reticulatum* (Ericaceae). *Pacific Science*, 65, 291–300.

Blüthgen, N. and Wesenberg, J. (2001) Ants induce domatia in a rain forest tree (*Vochysia vismiaefolia*). *Biotropica*, 33, 637–642.

Bronstein, J. L. (1998) The contribution of ant-plant protection studies to our understanding of mutualism. *Biotropica*, 30, 150–161.

Brouat, C., Garcia, N., Andary, C. and McKey, D. (2001) Plant lock and ant key: pairwise coevolution of an exclusion filter in an ant-plant mutualism. *Proceedings of the Royal Society of London, Series B*, 268, 2131–2141.

Brown, W. L. J. (1960) Ants, acacias, and browsing mammals. *Ecology*, 41, 587–592.

Carrillo, J., McDermott, D. and Siemann, E. (2014) Loss of specificity: native but not invasive populations of Triadica sebifera vary in tolerance to different herbivores. *Oecologia*, 174, 863–871.

Chamberlain, S. A. and Holland, J. N. (2009) Quantitative synthesis of context dependency in ant-plant protection mutualisms. *Ecology*, 90, 2384–2392.

Chomicki, G., M. Staedler, Y., Schönenberger, J. and Renner, S. S. (2016) Partner choice through concealed floral sugar rewards evolved with the specialization of ant-plant mutualisms. *New Phytologist*, 211, 1358–1370.

Darwin, F. (1876) On the glandular bodies on *Acacia sphaerocephala* and *Cecropia peltata* serving as food for ants. With an appendix on the nectar-glands of the common brake fern, *Pteris Aquilina*. *Botanical Journal of the Linnean Society of London*, 15, 398–409.

Davidson, D. W. and McKey, D. (1993) Ant-plant symbioses: stalking the Chuyachaqui. *Trends in Ecology and Evolution*, 8, 326–332.

Escalante-Perez, M. and Heil, M. (2012) The production and protection of Nectars. In U. Lüttge and W. Beyschlag, eds. *Progress in Botany, Vol.* 74. Heidelberg: Springer, pp. 239–261.

Escalante-Perez, M., Jaborsky, M., Lautner, S., Fromm, J., Muller, T. et al. (2012) Poplar extrafloral nectaries: two types, two strategies of indirect defenses against herbivores. *Plant Physiology*, 159, 1176–1191.

Fagundes, R., Anjos, D. V., Carvalho, R. and Del-Claro, K. (2015) Availability of food and nesting-sites as regulatory mechanisms for the recovery of ant diversity after fire disturbance. *Sociobiology*, 62, 1–9.

Federle, W., Fiala, B., Zizka, G. and Maschwitz, U. (2001) Incident daylight as orientation cue for hole-boring ants: prostomata in *Macaranga* ant-plants. *Insectes Sociaux*, 48, 165–177.

Federle, W., Maschwitz, U. and Holldobler, B. (2002) Pruning of host plant neighbours as defence against enemy ant invasions: *Crematogaster* ant partners of *Macaranga* protected by 'wax barriers' prune less than their congeners. *Oecologia*, 132, 264–270.

Feeny, P. (1976) Plant apparency and chemical defense. *Recent Advances in Phytochemistry*, 10, 1–40.

Ferguson, B. G., Boucher, D. H. and Maribel Pizzi, C. R. (1995) Recruitment and decay of a pulse of *Cecropia* in Nicaraguan rain forest damaged by hurricane Joan: relation to mutualism with *Azteca* ants. *Biotropica*, 27, 455–460.

Ferreira, J. A. M., Cunha, D. F. S., Pallini, A., Sabelis, M. W. and Janssen, A. (2011) Leaf domatia reduce intraguild predation among predatory mites. *Ecological Entomology*, 36, 435–441.

Fiala, B., Jakob, A., Maschwitz, U. and Linsenmair, K. E. (1999) Diversity, evolutionary specialisation and geographic distribution of a mutualistic ant-plant complex: *Macaranga* and *Crematogaster* in South East Asia. *Biological Journal of the Linnean Society*, 66, 305–331.

Fiala, B. and Linsenmair, K. E. (1995) Distribution and abundance of plants with extrafloral nectaries in the woody flora of a lowland primary forest in Malaysia. *Biodiversity and Conservation*, 4, 165–182.

Fiala, B. and Maschwitz, U. (1992a) Domatia as most important adaptions in the evolution of myrmecophytes in the paleotropical tree genus *Macaranga* (Euphorbiaceae). *Plant Systematics and Evolution*, 180, 53–64.

(1991) Extrafloral nectaries in the genus *Macaranga* (Euphorbiaceae) in Malaysia: comparative studies of their possible significance as predispositions for myrmecophytism. *Biological Journal of the Linnean Society*, 44, 287–305.

(1992b) Food bodies and their significance for obligate ant-association in the tree genus *Macaranga* (Euphorbiaceae). *Botanical Journal of the Linnean Society*, 110, 61–75.

(1990) Studies on the south east asian ant-plant association *Crematogaster borneensis/Macaranga*: adaptations of the ant partner. *Insectes Sociaux*, 37, 212–231.

Fischer, R. C., Richter, A., Wanek, W. and Mayer, V. (2002) Plants feed ants: food bodies of myrmecophytic *Piper* and their significance for the interaction with *Pheidole bicornis* ants. *Oecologia*, 133, 186–192.

Folgarait, P. J. and Davidson, D. W. (1995) Myrmecophytic *Cecropia*: antiherbivore defenses under different nutrient treatments. *Oecologia*, 104, 189–206.

Freitas, L., Galetto, L., Bernardello, G. and Paoli, A. A. S. (2000) Ant exclusion and reproduction of *Croton sarcopetalus* (Euphorbiaceae). *Flora*, 195, 398–402.

Géneau, C. E., Wäckers, F. L., Luka, H., Daniel, C. and Balmer, O. (2012) Selective flowers to enhance biological control of cabbage pests by parasitoids. *Basic and Applied Ecology*, 13, 85–93.

Goheen, J. R. and Palmer, T. M. (2010) Defensive plant-ants stabilize megaherbivore-driven landscape change in an African savanna. *Current Biology*, 20, 1766–1772.

González-Teuber, M. and Heil, M. (2015) Comparative anatomy and physiology of myrmecophytes: ecological and evolutionary perspectives. *Research and Reports in Biodiversity Studies*, 4, 21–32.

(2009) Nectar chemistry is tailored for both attraction of mutualists and protection from exploiters. *Plant Signaling and Behavior*, 4, 809–813.

González-Teuber, M., Kaltenpoth, M. and Boland, W. (2014) Mutualistic ants as an indirect defence against leaf pathogens. *New Phytologist*, 202, 640–650.

Gras, P., Tscharntke, T., Maas, B., Tjoa, A., Hafsah, A. et al. (2016) How ants, birds and bats affect crop yield along shade gradients in tropical cacao agroforestry. *Journal of Applied Ecology*, 53, 953–963.

Heil, M. (2015) Extrafloral nectar at the plant-insect interface: a spotlight on chemical ecology, phenotypic plasticity, and food webs. *Annual Review of Entomology*, 60, 213–232.

(2008) Indirect defence via tritrophic interactions. *New Phytologist*, 178, 41–61.

(2004) Induction of two indirect defences benefits Lima bean (*Phaseolus lunatus*, Fabaceae) in nature. *Journal of Ecology*, 92, 527–536.

(2011) Nectar: generation, regulation and ecological functions. *Trends in Plant Science*, 16, 191–200.

(2016) Nightshade wound secretion: the world's simplest extrafloral nectar? *Trends in Plant Science*, 21, 637–638.

Heil, M., Baumann, B., Krüger, R. and Linsenmair, K. E. (2004) Main nutrient compounds in food bodies of Mexican *Acacia* ant-plants. *Chemoecology*, 14, 45–52.

Heil, M., Fiala, B., Boller, T. and Linsenmair, K. E. (1999) Reduced chitinase activities in ant plants of the genus *Macaranga*. *Naturwissenschaften*, 86, 146–149.

Heil, M., Fiala, B., Linsenmair, K. E., Zotz, G., Menke, P. et al. (1997) Food body production in *Macaranga triloba* (Euphorbiaceae): a plant investment in anti-herbivore defence via mutualistic ant partners. *Journal of Ecology*, 85, 847–861.

Heil, M., González-Teuber, M., Clement, L. W., Kautz, S., Verhaagh, M. et al. (2009) Divergent investment strategies of *Acacia* myrmecophytes and the coexistence of mutualists and exploiters. *Proceedings of the National Academy of Science USA*, 106, 18091–18096.

Heil, M., Greiner, S., Meimberg, H., Krüger, R., Noyer, J.-L. et al. (2004b) Evolutionary change from induced to constitutive expression of an indirect plant resistance. *Nature*, 430, 205–208.

Heil, M., Koch, T., Hilpert, A., Fiala, B., Boland, W. et al. (2001) Extrafloral nectar production of the ant-associated plant, *Macaranga tanarius*, is an induced, indirect, defensive response elicited by jasmonic acid. *Proceedings of the National Academy of Sciences of the USA*, 98, 1083–1088.

Heil, M. and McKey, D. (2003) Protective ant-plant interactions as model systems in ecological and evolutionary research. *Annual Review of Ecology, Evolution, and Systematics*, 34, 425–453.

Herms, D. A. and Mattson, W. J. (1992) The dilemma of plants: to grow or to defend. *The Quarterly Review of Biology*, 67, 283–335.

Huxley, C. R. (1978) The ant-plants *Myrmecodia* and *Hydnophytum* (Rubiaceae), and relationships between their morphology, ant occupants, physiology and ecology. *New Phytologist*, 80, 231–268.

Janzen, D. H. (1969) Allelopathy by myrmecophytes: the ant *Azteca* as an allelopathic agent of Cecropia. *Ecology*, 50, 147–153.

　(1966) Coevolution of mutualism between ants and acacias in Central America. *Evolution*, 20, 249–275.

　(1967) Fire, vegetation structure, and the ant x *Acacia interaction* in Central America. *Ecology*, 48, 26–35.

　(1985) On ecological fitting. *Oikos*, 45, 308–310.

　(1972) Protection of *Barteria* (Passifloraceae) by *Pachysima* ants (Pseudomyrmecinae) in a Nigerian rain forest. *Ecology*, 53, 885–892.

　(1974) *Swollen-Thorn Acacias of Central America: Smithsonian Contributions to Botany*, Vol. 13. Washington, DC: Smithsonian Institution Press.

Jones, I. M., Koptur, S. and von Wettberg, E. (2016) The use of extrafloral nectar in pest management: overcoming context dependence. *Journal of Applied Ecology*, 54, 489–499.

Jones, M. E. and Paine, T. D. (2012) Ants impact sawfly oviposition on bracken fern in southern California. *Arthropod-Plant Interactions*, 6, 283–287.

Kato, H., Yamane, S. and Phengklai, C. (2004) Ant-colonized domatia on fruits of *Mucuna interrupta* (Leguminosae). *Journal of Plant Research*, 117, 319–321.

Keeler, K. H. (1979) Distribution of plants with extrafloral nectaries and ants at two different elevations in Jamaica. *Biotropica*, 11, 152–154.

　(1980) Distribution of plants with extrafloral nectaries in temperate communities. *The American Midland Naturalist*, 104, 274–279.

Koptur, S., Palacios-Rios, M., Diaz-Castelazo, C., Mackay, W. P. and Rico-Gray, V. (2013) Nectar secretion on fern fronds associated with lower levels of herbivore damage: field experiments with a widespread epiphyte of Mexican cloud forest remnants. *Annals of Botany*, 111, 1277–1283.

Koptur, S., Rico-Gray, V. and Palacios-Rios, M. (1998) Ant protection of the nectaried fern *Polypodium plebeium* in Central Mexico. *American Journal of Botany*, 85, 736–739.

Koptur, S., William, P. and Olive, Z. (2010) Ants and plants with extrafloral nectaries in fire successional habitats on Andros (Bahamas). *Florida Entomologist*, 93, 89–99.

Koricheva, J. and Romero, G. Q. (2012) You get what you pay for: reward-specific trade-offs among direct and ant-mediated defences in plants. *Biology Letters*, 8, 628–630.

Lach, L. (2003) Invasive ants: unwanted partners in ant-plant interactions? *Annals of the Missouri Botanical Garden*, 90, 91–108.

Lach, L., Hobbs, R. J. and Majer, J. D. (2009) Herbivory-induced extrafloral nectar increases native and invasive ant worker survival. *Population Ecology*, 51, 237–243.

Lach, L. and Hoffmann, B. D. (2011) Are invasive ants better plant-defense mutualists? A comparison of foliage patrolling and herbivory in sites with invasive yellow crazy ants and native weaver ants. *Oikos*, 120, 9–16.

Lach, L., Tillberg, C. V. and Suarez, A. V. (2010) Contrasting effects of an invasive ant on a native and an invasive plant. *Biological Invasions*, 12, 3123–3133.

Leal, L. C., Andersen, A. N. and Leal, I. R. (2015) Disturbance winners or losers? Plants bearing extrafloral nectaries in Brazilian Caatinga. *Biotropica*, 47, 468–474.

Lin, I. W., Sosso, D., Chen, L.-Q., Gase, K., Kim, S.-G. et al. (2014) Nectar secretion requires sucrose phosphate synthases and the sugar transporter SWEET9. *Nature*, 508, 546–549.

Longino, J. T. (1989) Geographic variation and community structure in an an-plant mutualism: *Azteca* and *Cecropia* in Costa Rica. *Biotropica*, 21, 126–132.

Lortzing, T., Calf, O. W., Böhlke, M., Schwachtje, J., Kopka, J. et al. (2016) Extrafloral nectar secretion from wounds of *Solanum dulcamara*. *Nature Plants*, 2, Article number: 16056, doi:10.1038/nplants.2016.56.

Marazzi, B., Bronstein, J. and Koptur, S. (2013) The diversity, ecology and evolution of extrafloral nectaries: current perspectives and future challenges. *Annals of Botany*, 111, 1243–1250.

Marazzi, B., Conti, E., Sanderson, M. J., McMahon, M. M. and Bronstein, J. L. (2013) Diversity and evolution of a trait mediating ant-plant interactions: insights from extrafloral nectaries in *Senna* (Leguminosae). *Annals of Botany*, 111, 1263–1275.

Marazzi, B. and Sanderson, M. J. (2010) Large-scale patterns of diverstfication in the widspread legume genus *Senna* and evolutionary role of extrafloral nectaries. *Evolution*, 64, 3570–3592.

Mathews, C. R., Bottrell, D. G. and Brown, M. W. (2009) Extrafloral nectaries alter arthropod community structure and mediate peach (*Prunus persica*) plant defense. *Ecological Applications*, 19, 722–730.

Mathews, C. R., Brown, M. W. and Bottrell, D. G. (2007) Leaf extrafloral nectaries enhance biological control of a key economic pest, *Grapholita molesta* (Lepidoptera: Tortricidae), in peach (Rosales: Rosaceae). *Environmental Entomology*, 36, 383–389.

Mathur, V., Wagenaar, R., Caissard, J.-C., Reddy, A. S., Vet, L. E. M. et al. (2013) A novel indirect defence in Brassicaceae: Structure and function of extrafloral nectaries in *Brassica juncea*. *Plant, Cell & Environment*, 36, 528–541.

Melo, Y., Machado, S. R. and Alves, M. (2010) Anatomy of extrafloral nectaries in Fabaceae from dry-seasonal forest in Brazil. *Botanical Journal of the Linnean Society*, 163, 87–98.

Millán-Cañongo, C., Orona-Tamayo, D. and Heil, M. (2014) Phloem sugar flux and jasmonic acid-responsive cell wall invertase control extrafloral nectar secretion in *Ricinus communis*. *Journal of Chemical Ecology*, 40, 760–769.

Moog, J., Feldhaar, H. and Maschwitz, U. (2002) On the caulinary domatia of the SE-Asian ant-plant *Zanthoxylum myriacanthum* Wall. ex Hook. f. (Rutaceae) their influence on branch statics, and the protection against herbivory. *Sociobiology*, 40, 547–574.

Ness, J. H. (2006) A mutualism's indirect costs: the most aggressive plant bodyguards also deter pollinators. *Oikos*, 113, 506–514.

Ness, J. H. and Bronstein, J. L. (2004) The effects of invasive ants on prospective ant mutualists. *Biological Invasions*, 6, 445–461.

Ness, J. H., Morales, M. A., Kenison, E., Leduc, E., Leipzig-Scott, P. et al. (2013) Reciprocally beneficial interactions between introduced plants and ants are induced by the presence of a third introduced species. *Oikos*, 122, 695–704.

Ness, J. H., Morris, W. F. and Bronstein, J. L. (2009) For ant-protected plants, the best defense is a hungry offense. *Ecology*, 90, 2823–2831.

Oliveira, P. S. and Brandão, C. R. F. (1991) The ant community associated with extrafloral nectaries in the Brazilian cerrados. In C. R. Huxley and D. F. Cutler, eds. *Ant-plant interactions*. Oxford: Oxford University Press, pp. 198–212.

Olson, D. M. and Wäckers, F. L. (2007) Management of field margins to maximize multiple ecological services. *Journal of Applied Ecology*, 44, 13–21.

Orona-Tamayo, D., Wielsch, N., Blanco-Labra, A., Svatos, A., Faría-Rodríguez, R. et al. (2013) Exclusive rewards in mutualisms: ant proteases and plant protease inhibitors create a lock-key system to protect *Acacia* food bodies from exploitation. *Molecular Ecology*, 22, 4087–4100.

Palmer, T. M., Stanton, M. L. and Young, T. P. (2003) Competition and coexistence: Exploring mechanisms that restrict and maintain diversity within mutualist guilds. *American Naturalist*, 162, S63–S79.

Palmer, T. M., Stanton, M. L., Young, T. P., Goheen, J. R., Pringle, R. M. et al. (2008) Breakdown of an ant-plant mutualism follows the loss of large herbivores from an African savanna. *Science*, 319, 192–195.

Philpott, S. M. and Foster, P. F. (2005) Nest-site limitation in coffee agroecosystems: Artificial nests maintain diversity of arboreal ants. *Ecological Applications*, 15, 1478–1485.

Piovia-Scott, J. (2011) The effect of disturbance on an ant-plant mutualism. *Oecologia*, 166, 411–420.

Pringle, E. G., Akçay, E., Raab, T. K., Dirzo, R. and Gordon, D. M. (2013) Water stress strengthens mutualism among ants, trees, and scale insects. *PLoS Biol*, 11, e1001705.

Putz, F. E. and Holbrook, N. M. (1988) Further observations on the dissolution of mutualism between *Cecropia* and its ants: the Malaysian case. *Oikos*, 53, 121–125.

Quek, S. P., Davies, S. J., Itino, T. and Pierce, N. E. (2004) Codiversification in an ant-plant mutualism: Stem texture and the evolution of host use in *Crematogaster* (Formicidae: Myrmicinae) inhabitants of *Macaranga* (Euphorbiaceae). *Evolution*, 58, 554–570.

Rickson, F. R. (1971) Glycogen plastids in Müllerian body cells of *Cecropia peltata* – a higher green plant. *Science*, 173, 344–347.

Rickson, F. R. and Risch, S. J. (1984) Anatomical and ultrastructural aspects of the ant-food cell of *Piper cenocladum* C.DC (Piperaceae). *American Journal of Botany*, 71, 1268–1274.

Riginos, C., Karande, M. A., Rubenstein, D. I. and Palmer, T. M. (2015) Disruption of a protective ant-plant mutualism by an invasive ant increases elephant damage to savanna trees. *Ecology*, 96, 654–661.

Rocha, C. F. D. and Bergallo, H. G. (1992) Bigger ant colonies reduce herbivory and herbivore residence time on leaves of an ant-plant: *Azteca muelleri* vs. *Coelomera ruficornis* on *Cecropia pachystachia*. *Oecologia*, 91, 249–252.

Romero, G. Q. and Koricheva, J. (2011) Contrasting cascade effects of carnivores on plant fitness: a meta-analysis. *Journal of Animal Ecology*, 80, 696–704.

Rosumek, F. B., Silveira, F. A. O., Neves, F. D., Barbosa, N. P. D., Diniz, L. et al. (2009) Ants on plants: a meta-analysis of the role of ants as plant biotic defenses. *Oecologia*, 160, 537–549.

Rudgers, J. and Gardener, M. C. (2004) Extrafloral nectar as a resource mediating multispecies interactions. *Ecology*, 85, 1495–1502.

Ruhlmann, J. M., Kram, B. W. and Carter, C. J. (2010) CELL WALL INVERTASE 4 is required for nectar production in *Arabidopsis*. *Journal of Experimental Botany*, 61, 395–404.

Savage, A. M. and Rudgers, J. A. (2013) Non-additive benefit or cost? Disentangling the indirect effects that occur when plants bearing extrafloral nectaries and honeydew-producing insects share exotic ant mutualists. *Annals of Botany*, 111, 1295–1307.

Savage, A. M. and Whitney, K. D. (2011) Trait-mediated indirect interactions in invasions: unique behavioral responses of an invasive ant to plant nectar. *Ecosphere*, 2, Article 106.

Schremmer, F. (1969) Extranuptiale Nektarien. Beobachtungen an *Salix eleagnos* Scop. und *Pteridium aquilinum* (L.) Kuhn. *Plant Systematics and Evolution (Österr. Bot. Z.)*, 117, 205–222.

Schupp, E. W. (1986) *Atzeca* protection of *Cecropia*: ant occupation benefits juvenile trees. *Oecologia*, 70, 379–385.

Schupp, E. W. and Feener, D. H. (1991) Phylogeny, life form, and habitat dependence of ant-defended plants in a Panamanian forest. In C. R. Huxley and D. F. Cutler, eds. *Ant-plant interactions*. Oxford: Oxford University Press, pp. 175–197.

Sobrinho, T. G., Schoereder, J. H., Rodrigues, L. L. and Collevatti, R. G. (2002) Ant visitation (Hymenoptera: Formicidae) to extrafloral nectaries increases seed set and seed viability in the tropical weed *Triumfetta semitriloba*. *Sociobiology*, 39, 353–368.

Stapley, L. (1998) The interaction of thorns and symbiotic ants as an effective defence mechanism of swollen-thorn acacias. *Oecologia*, 115, 401–405.

Stenberg, J. A., Heil, M., Åhman, I. and Björkman, C. (2015) Optimizing crops for biocontrol of pests and disease. *Trends in Plant Science*, 20, 698–712.

Tempel, A. S. (1983) Bracken fern (*Pteridium aquilinum*) and nectar-feeding ants: a nonmutualistic interaction. *Ecology*, 64, 1411–1422.

Trager, M. D., Bhotika, S., Hostetler, J. A., Andrade, G. V., Rodriguez-Cabal, M. A. et al. (2010) Benefits for plants in ant-plant protective mutualisms: a meta-analysis. *Plos One*, 5, e14308.

Traveset, A. and Richardson, D. M. (2014) Mutualistic interactions and biological invasions. *Annual Review of Ecology, Evolution, and Systematics*, 45, 89–113.

Wang, Y., Carrillo, J., Siemann, E., Wheeler, G. S., Zhu, L. et al. (2013) Specificity of extrafloral nectar induction by herbivores differs among native and invasive populations of tallow tree. *Annals of Botany*, 112, 751–756.

Ward, P. S. (1999) Systematics, biogeography and host plant associations of the *Pseudomyrmex viduus* group (Hymenoptera: Formicidae), *Triplaris-* and *Tachigali*-inhabiting ants. *Zoological Journal of the Linnean Society*, 126, 451–540.

Webber, B. L., Moog, J., Curtis, A. S. O. and Woodrow, I. E. (2007) The diversity of ant-plant interactions in the rainforest understorey tree, *Ryparosa* (Achariaceae): food bodies, domatia, prostomata and hemipteran trophobionts. *Botanical Journal of the Linnean Society*, 154, 353–371.

Weber, M. G. and Agrawal, A. (2014) Defense mutualisms enhance plant diversification. *Proceedings of the National Academy of Science USA*, 111, 16442–16447.

Weber, M. G. and Keeler, K. H. (2013) The phylogenetic distribution of extrafloral nectaries in plants. *Annals of Botany*, 111, 1251–1261.

White, R. A. and Turner, M. D. (2012) The anatomy and occurrence of foliar nectaries in *Cyathea* (Cyatheaceae). *American Fern Journal*, 102, 91–113.

Young, T. P., Stubblefield, C. H. and Isbell, L. A. (1997) Ants on swollen-thorn acacias: species coexistence in a simple system. *Oecologia*, 109, 98–107.

Yu, D. W. and Pierce, N. E. (1998) A castration parasite of an ant-plant mutualism. *Proceedings of the Royal Society of London Series B-Biological Sciences*, 265, 375–382.

Yumoto, T. and Maruhashi, T. (1999) Pruning behavior and intercolony competition of *Tetraponera* (Pachysima) *aethiops* (Pseudomyrmecinae, Hymenoptera) in *Barteria fistulosa* in a tropical forest, Democratic Republic of Congo. *Ecological Research*, 14, 393–404.

20 Why Study Ant-Plant Interactions?

Andrew J. Beattie

The question has to be asked, but first consider this: The human population is projected to reach 10–11 billion by the end of this century (Emmott, 2013; United Nations, 2015), a number surely too great for the safety and comfort of everyone (Wilson, 2016). Further, there is a growing conviction that global civilisation itself faces collapse because of species extinctions, soil, water and air pollution, pandemics, global warming, land degradation, poor choice of technologies and overconsumption – all driven by increasing numbers of human beings exerting burgeoning demands on the environment (Ehrlich & Ehrlich, 2012; Cribb, 2016). So now the question has to be asked: given the parlous state of Earth and the precarious plight of humanity, is there really a case for the continued study of ant-plant interactions?

Before this question might be answered, the ecosystems in which ant-plant interactions occur have to be placed into a realistic context: Humanity has modified over 50 per cent of Earth's surface and a mere 15 per cent is dedicated to nature conservation in some form (World Bank, 2014). Overall, 75 per cent of ice-free land can no longer be considered wild (Hooke & Martin-Duque, 2012) and there is good evidence that biodiversity is currently undergoing a massive extinction event (Barnosky et al., 2011; Wilson 2016). With respect to the areas formally designated as 'wilderness', about 30 million km² remain intact but, as there was an approximately 10 per cent loss since the early 1990s, it is clear that wilderness is under continuing threat worldwide (Watson et al., 2016). Further, the human footprint increasingly impinges on areas of high biodiversity (Venter et al., 2016). These grim figures mean that 'native' ant-plant interactions, i.e. those where evolution by natural selection is the dominant driver of their future, occupy dwindling areas under increasing threat. By contrast, the ant-plant interactions in a significant majority of terrestrial areas are more likely to be subject to 'unnatural' or artificial selection driven by a wide range of intrusive and disruptive human activities.

In the context of continuing declines in wilderness areas and global increases in the human disruption of terrestrial ecosystems, what is the case for the study of ant-plant interactions? Figure 20.1 (modified from Rastogi, 2011) is

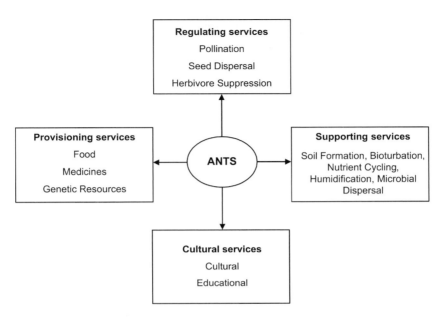

Figure 20.1. Ecosystem services provided by ants. Modified from Rastogi (2011), with permission from the author.

an 'ant-thropocentric' view of the world that places ants in the context of the four great groups of ecosystem services identified by the Millennium Ecosystem Assessment (MEA, 2005). The goal of the MEA was to establish the connection between ecosystem function and human well-being which, astonishing as it may seem, was a concept largely unknown to many national governments of the time, or at least widely ignored by them. This chapter is written in the context of the MEA as it embraces both the fundamental and applied aspects of ant-plant interactions, although the dividing line between these two kinds of research is increasingly irrelevant.

The chances are high that every plant in the world, together with the soil in which it grows, has been visited by at least one ant, and probably many more, during its lifetime. This translates into a web of interactions from the perspective of their importance to humanity, as summarised in Figure 20.1. Ants have both indirect and direct effects on plants, and ant-plant interactions fall within the 'Supporting Services' and 'Regulating Services' categories although there is much overlap. Pollination, seed dispersal and herbivore suppression are critical ecosystem processes influencing the structure, stability and function of natural communities and therefore important to the social and political processes that actuate conservation outcomes (MEA, 2005; Kremen et al., 2007). Among the most ubiquitous of their indirect effects are those on the physical and chemical properties of soils (Folgarait, 1998). Changes in soil properties resulting from the activities of one to many ant species can be both extensive and profound, leading to major changes in vegetation

(Cammeraat et al., 2002; Culver & Beattie, 1983; see Chapters 4 and 18 for leaf-cutter effects). Seed dispersal, biological control (herbivore suppression) and pollination by ants may involve suites or guilds of species (Beattie & Culver, 1981), and even when the interaction is highly specific, a range of abiotic and biotic interactions may affect the supply of specialist species (Chapter 11).

The enormous strength of the research outlined in this book, and the ant-plant literature in general, is that it has created a very large literature that functions as a substantial multi-national, multi-ecosystem database of species interactions between a group of the most common animals on Earth and the dominant vegetation. This is of irreplaceable value and provides the rationale for continuing research into ant-plant interactions in at least five categories:

1.   Evolutionary Biology and Ecological Theory

Existing ant-plant databases provide the material for understanding deep evolutionary and ecological processes. One of the most striking examples is the study of myrmecochory that has led to profound insights into evolutionary biology, providing a textbook example of convergent evolution, showing similar adaptations in a wide range of flowering plant families to ant services, and that these ant-seed interactions have been a significant factor in driving evolutionary diversification within the Angiosperms (Lengyel et al., 2009a, 2009b). This example illustrates how ant-plant research contributes to another panel in Figure 20.1, 'Cultural Services', particularly under the rubric of education, as they provide clear, evidence-based examples of evolution which unfortunately remain a critical weapon in the defence of science even today (Fitzroy & Papyrakis, 2016; Rosenau, 2016). Important insights into the ecology of species interactions have been documented in many chapters in this book as well as in a very large literature elegantly summarised in Rico-Gray and Oliveira (2007) and Lach et al. (2010). Ant-plant interactions involving extrafloral nectaries and herbivore suppression are so frequent and diverse that they make ideal models for the investigation of a variety of ecological and evolutionary processes, not least the structuring of food webs (Davidson, 1997; Heil & McKey, 2003).

2.   Measuring Environmental Degradation and Restoration Success

Changes in the diversity, distribution and abundance of particular animal taxa such as birds, butterflies and mammals have been widely used to document the impacts of human activities on the environment. In this context, one of the chief roles of ants has been as monitors of environmental degradation and restoration (Lach et al., 2010; Andersen et al., 2002) and measuring human impacts on ecosystems as varied as soils (Lobry de Bruyn, 1999), grassland remnants (Dauber et al., 2006) and forests (Ribas et al., 2012). However, this book shows that environmental monitoring can achieve higher levels of subtlety and information content through the use of species interactions. As background to this proposition, recall that Bronstein et al. (2004) proposed that all species are involved in at least one mutualistic interaction and Price (1980) suggested that parasitism is the most common species interaction on Earth. Eleven thousand species of flowering plants are ant dispersed (Lengyel et al., 2009b) and disruptions to them can cause ecosystem change (Bond, 1994). While the ubiquity of species interactions and their importance to food web structure suggest their potential for environmental monitoring, a body of litera-ture such as that for ant-plant interactions is effectively the database that makes it possible. Keirs et al. (2010) tracked three indicators of human disturbance: Shifts from mutualism to antagonism, switches to novel partners and the abandonment of mutualism in a database of 179 interactions, including some myrmecochorous and extrafloral nectary systems. Human impacts mostly degraded the interactions, although some showed no effects and others appeared to be enhanced. Their results enabled them to suggest the characteristics of mutualisms most resilient to disturb-ance. Overall, species interactions are particularly sensitive to human disturbances, especially global warming, often demonstrating more far-reaching effects than previously detected (see next section; Tylianakis et al., 2008; Mayer et al., 2014; Valiente-Banuet et al., 2015). The utility of ant-plant interactions as environmental indicators relies heavily on the existing literature/databases because they provide the critical baseline of information on natural variability, without which it is diffi-cult to distinguish between anthropogenic disturbance and the range of variation when unaffected by human activities (Underwood, 1991; Tylianakis et al., 2008).

3.   Understanding Coextinction

One of the most important insights into the human disturbance of ant-plant inter-actions described widely in this book is the disruption and modification of the diver-sity and abundance of the interacting species. Continuing research in this area will create the opportunity to predict: (1) The characteristics such as nestedness and modularity of species interactions networks that form the 'architecture' of ecosys-tems and their vulnerability to human disturbance and (2) the application of this knowledge to the development of ants as biological control agents (Heil & McKey, 2003; Bascompte & Jordano, 2007; Valiente-Banuet, 2015, Schleuning et al., 2015; Chapter 19).

While there is general agreement that habitat loss, global warming and invasive species are among the most important drivers of species extinctions (Sodhi et al., 2009), this book has already made it clear that the processes involved are often highly complex. Thus, because of interaction networks, extinctions are associ-ated with coextinctions, sometimes in cascading sequences (e.g. Dunn et al., 2009; Asian et al., 2013). The process of coextinction is itself complex, for example, 'widow' species – those bereft of mutualistic partners – may persist by abandoning the former interaction (Keirs et al., 2010) or by acquiring new partners (Rowles

& O'Dowd, 2009), or the nature of the interaction may change dramatically, for example, from mutualism to antagonism (Palmer et al., 2008). Chapter 19 documents the comparative resilience of many ant-plant interactions to human disturbance through the acquisition of novel partners but, as Rowles and O'Dowd (2008), Valiente-Banuet et al. (2015) and Tylianakis et al. (2008) have warned, while the short-term circumstances may favour such novel interactions, the long-term prognosis is unknown. Novel, multi-faceted, competitive and multi-trophic interactions may result in undesirable outcomes such as local extinction, altered genetic structure, declines in population density, changes in distribution, changes in reproductive output and perverse services provided to invasive species (Chapter 11; Bond, 1994; Rowles & O'Dowd, 2008; Trager et al. 2010; Asian et al., 2013). Phenological changes resulting from global warming can significantly degrade species interactions (Thackerey et al., 2016). Much remains to be discovered in this field (Dunn, 2009).

4.   Biological Control

This topic is thoroughly discussed in Chapter 19 which emphasises the conditional nature of the outcomes of ant-crop interactions. Conditionality results both from the effects of ant and plant variables on the outcomes of the interactions nested within the networks of species in which crops are grown. A variety of drivers make this research more imperative, including the decreasing efficacy of pesticides, the increasing alarm over pesticide pollution, the growing importance of smallholder operations that can afford neither the financial nor the health costs of pesticides, growing demands for organic produce and the desire of growers to adopt more holistic approaches to production. The successful biological control of crop plants by ants is in a very real sense the ultimate test of our understanding of ant-plant interactions and the communities in which they occur. McKey and Blatrix (Chapter 11) point out that ant-plant mutualisms often involve many species with different life histories, each of which responds to a complex of abiotic and biotic factors. In addition, Gordon (1999) showed that different colonies of the same ant species may vary greatly. The interaction rate between species such as ants and the herbivores on which they prey is the local summation or aggregate of all the different levels of variation in the system – functional facilitation (e.g. Klein et al., 2008). Interaction rates also require the assessment of both quantitative and qualitative effects. This is the kind of complex issue that ant-plant researchers developing biological control will have to address. Rudgers (2004) and Rudgers and Strauss (2004), working with cotton EFNs, provided insights into experimentation that helps untangle the contributions of species-species interaction nested within community-level dynamics. This kind of research underpins trait-based analyses in which the fate of individual mutualistic systems (Rutter & Rausher, 2004) or that of entire species networks are explored (Schleuning et al., 2015). Proneness to extinction is a function of multiple traits that mediate responses to the environment which, in the case of species involved in mutualisms, includes matching response traits of other species. This is a rich area for future research (Aizen et al., 2012).

5. Conservation Action

The authors of the preceding chapters point the way forward, emphasising that we should all be conservationists. But do ant-plant interactions offer any special opportunities? I think they do: This may be controversial but it is something worth considering – that some ant-plant interactions should be promoted as 'charismatic' to decision makers and to the general public. Much of the world is well aware of the way that Sir David Attenborough, through careful imagery and explanation, can turn a little-known animal or plant into an object of wonder, respect and even concern. Such a process might be straight forward for ant-plant interactions such as mutualisms as Google Images and other comparable websites demonstrate. The stunning imagery that already exists, together with imaginative but scientifically accurate story-telling, emphasising human parallels of mutualistic behaviour perhaps reflecting the main currencies of exchange between ant-plant partners (food and protection, Mayer et al. 2014) might be a platform worth trying to energise conservation action. Scientists are often unaware that phenomena which they know and understand intimately are unknown even to locals. For example, many residents of northern Australia are delighted to learn the story of ant plants such as *Myrmecodia*. This approach thus may be most effective on local scales where personal experiences are possible.

As we have seen in this book, the abundance, distribution and participating species in ant-plant interactions reflect the impacts of human activities. Turning this around, we find that ant-plant interactions are useful in monitoring the effects of human disturbance such as forest fragmentation, selective logging, road-building, deforestation, mining exploration, climate change and many others. Longitudinal studies of interactions such as these would surely be a robust measure of sustainability in a variety of ecosystems. Claims such as these could be profitably researched because new indicators for environmental monitoring are still in demand (Laurence et al. 2012a, 2012b). The availability of 'charismatic' interactions, especially on a local scale, would be a bonus for attracting attention and funding. One small example was the study of myrmecochory in *Sanguinaria* (Pudlo et al., 1980) that attracted attention from regional newspapers because the plant species was an attractive harbinger of spring and its elaiosome was visually impressive.

I think these suggestions are important because so much of our research is academic, which is fine, but rarely, if ever, has a direct bearing on real world conservation (Laurance et al., 2012a) and has little effect on legislation (Guber, 2003; Johns, 2009). This is a contentious area where some argue passionately that we do not need more data but we need more action (Knight et al., 2010) while others point out that the continuation of data collection is essential in many circumstances (Stuart et al., 2010). The global paucity of conservation areas and outlook for ant-plant interactions suggests that a significant part of ant-plant research in the future should be dedicated to conservation and to the detection and alleviation of the forces that threaten them. One problem is that academics may not ask conservation practitioners what kind of questions they need answering and may be averse to the kinds of crisis situations in which they might be useful (Laurence, 2012a). As Johns (2009) averred, a lot of research should be applied outside academia and it would be a major step forward to examine our own abilities, preferences and potential to determine whether or not we enter the time-critical and highly complex arenas where conservation, economics and politics collide.

References

Aizen, M. A., Sabatino, M. & Tylianakis, J. M. (2012). Specialization and rarity predict non-random loss of interactions from mutualistic networks. *Science*, 335, 1486–1489.

Andersen, A. N., Hoffmann, B. D., Muller, W. J. & Griffiths, A. D. (2002). Using ants as bioindicators in land management: simplifying assessment of ant community responses. *Journal of Applied Ecology*, 39, 8–17.

Asian, C. E., Zavaleta, E. S., Tershy, B. & Croll, D. (2013). Mutualism disruption threatens global plant diversity: a systematic review. *PLoS ONE*, 8, e66993.

Barnosky, A. D., Matzke, N., Tomiya, S. et al. (2011). Has the Earth's sixth mass extinction already arrived? *Nature*, 471, 51–57.

Bascompte, J. & Jordano, P. (2007). Plant-animal mutualistic networks; the architecture of biodiversity. *Annual Review of Ecology, Evolution and Systematics*, 38, 567–593.

Beattie, A. J. & Culver, D. C. (1981). The guild of myrmecochores in the herbaceous flora of West Virginia forests. West Virginia forests. *Ecology*, 62, 107–119.

Bond, W. J. (1994). Do mutualisms matter? Assessing the impact of pollinator and disperser disruption on plant extinction. *Philosophical Transactions of the Royal Society B*, 344, 83–90.

Bronstein, J. L., Dieckmann, U. & Ferriere, R. (2004). Coevolutionary dynamics and the conservation of mutualisms. In *Evolutionary Conservation Biology*, ed. Fierriere, R., Dieckmann, U. and Couvet, D. Cambridge: Cambridge University Press, pp. 305–326.

Cammeraat, L. H., Willott, S. J., Compton, S. G. & Incoll, L. D. (2002). The effect of ants' nests on the physical, chemical and hydrological properties of a rangelend soil in semi-arid Spain. *Geoderma*, 105, 1–20.

Cribb, J. (2016). *Surviving the 21st Century*. New York: Springer International.

Culver, D. C. & Beattie, A. J. (1983). Effects of ant mounds on soil chemistry and vegetation patterns in Colorado. *Ecology*, 64, 485–492.

Dauber,J., Bengtsson, J. & Lenoir, L. (2006). Evaluating effects of habitat loss and land-sue continuity on ant species richness in seminatural grassland remnants. *Conservation Biology*, 20, 1150–1160.

Davidson, D. W. (1997). The role of resource imbalances in the evolutionary ecology of tropical arboreal ants. *Biological Journal of the Linnean Society*, 61, 153–181.

Dunn, R. R. (2009). Coextinction: anecdotes, models and speculation. In *Holocene Extinctions*, Turvey, S. T. Oxford Scholarship Online, doi: 10.1093/acprof.oso/9780199535095.003.0008.

Dunn, R. R., Harris, N. C., Colwell, R. K., Koh, L. P. & Sodhi, N. S. (2009). The sixth mass coextinction: are most endangered species parasites and mutualists? *Proceedings of the Royal Society Series B*, 276, 3037–3045.

Ehrlich, P. R. & Ehrlich, A. H. (2012). Can a collapse of global civilization be avoided? *Proceedings of the Royal Society Series B*, 280, 20122845.

Emmott, S. (2013). *10 Billion*. London: Penguin Books.

Fitzroy, F. R. & Papyrakis, E. (2016). *An Introduction to Climate Change Economics and Policy, 2nd ed.* London: Routledge.

Folgarait, P. J. (1998). Ant biodiversity and its relationships to ecosystem functioning: a review. *Biodiversity & Conservation*, 7, 1221–1244.

Gordon, D. (1999). *Ants at Work*. New York: Free Press.

Guber, D. (2003). *The Grassroots of a Green Revolution*. Cambridge, MA: MIT Press.

Heil, M. & McKey, D. (2003). Protective ant-plant interactions as model systems in ecological and evolutionary research. *Annual Review of Ecology. Evolution. and Systematics*, 34, 425–453.

Hooke, R. L. & Martin-Duque, J. F. (2012). Land transformation by humans: a review. *Geological Society of America (GSA)*, 22, 4–10.

Johns, D. (2009). *New Conservation Politics*. Chichester: John Wiley & Sons.

Kiers, E. T., Palmer, T. M., Ives, A. R., Bruno, J. F. & Bronstein, J. L. (2010). Mutualisms in a changing world. *Ecology Letters*, 13, 1459–1474.

Klein, A. M., Cunningham, S. A., Bos, M. & Steffan-Dewenter, I. (2008). Advances in pollination ecology from tropical plantation crops. *Ecology*, 89, 935–943.

Knight, A. T., Bode, M., Fuller, R. A. et al. (2010). Barometer of life: more action, not more data. *Science*, 329, 141.

Kremen, C., Williams, N. M., Aizen, M. A., Gemmill-Herren, B. et al. (2007). Pollination and other ecosystem services by mobile organisms: a conceptual framework of land-use change. *Ecology Letters*, 10, 299–314.

Lach. L., Parr, C. & Abbott, K. (2010). *Ant Ecology*. Oxford: Oxford University Press.

Laurence, W. F., Koster, H., Grooten, M. et al. (2012a). Making conservation research more relevant for conservation practitioners. *Biological Conservation*, 153, 164–168.

Laurence, W. F., Useche, D., Rendeiro, J. et al. (2012b). Averting biodiversity collapse in tropical forest protected areas. *Nature*, 489, 290–294.

Lengyel, S., Gove, A. D., Latimer, A. M., Majer, J. D. & Dunn, R. R. (2009a). Convergent evolution of seed dispersal by ants, and phylogeny and biogeography in flowering plants: a global survey. *Perspectives in Plant Ecology, Evolution and Systematics*, 12, 43–55.

(2009b). Ants sow the seeds of global diversification in flowering plants. *PLoS ONE*, 4, 1–6.

Lobry de Bruyn, L. A. (1999). Ants as bioindicators of soil function in rural environments. *Agriculture Ecosystems & Environment*, 74, 425–441.

Mayer, V. E., Frederickson, M. E., MvKey, D. & Blatrix, R. (2014). Current issues in the evolutionary ecology of ant-plant symbioses. *New Phytologist*, 202, 749–764.

Millennium Ecosystem Assessment (2005). *Ecosystems and Human Well-Being: A Framework for Assessment*. Washington, DC: Island Press.

Palmer, T. M., Stanton, M. L., Young, T. P., Goheen, J. R., Pringle, R. M. & Karban, R. 2008 Breakdown of an ant-plant mutualism follows loss of large herbivores from an African savanna. *Science* 319, 192–195.

Price, P. W. (1980). *Evolutionary Biology of Parasites*. Princeton: Princeton University Press.

Pudlo, R. J., Beattie, A. J. & Culver, D. C. (1980). Population consequences of changes in an ant-seed mutualism in *Sanguinaria Canadensis*. *Oecologia*, 46, 32–37.

Rastogi, N. (2011). Provisioning services from ants: food and pharamceuticals. *Asian Myrmecology*, 4, 103–120.

Ribas, C. R., Campos, R. B. F., Schmidt, F. A. & Solar, R. R. C. (2012). Ants as indicators in Brazil: a review with suggestions to improve the use of ants in environmental monitoring programs. *Psyche*, 636749, doi:10.1155/2012/636749

Rico-Gray, V. & Oliveira, P. S. (2007). *The Ecology and Evolution of Ant-Plant Interactions.* Chicago: University of Chicago Press.

Rosenau, J. (2016). That sinking feeling. *New Scientist*, 231, 18–19.

Rowles, A. D. & O'Dowd, D. J. (2009). New mutualism for old: indirect disruption and direct facilitation of seed dispersal following Argentine ant invasion. *Oecologia*, 158, 709–716.

Rudgers, J. A. (2004). Enemies of herbivores can shape plant traits: selection in facultative ant-plant mutualism. *Ecology*, 85, 192–205.

   & Strauss, S. Y. (2004). A selection mosaic in the facultative mutualism between ants and wild cotton. *Proceedings of the Royal Society B*, 271, 2481–4288.

Rutter, M. T. & Rausher, M. D. (2004). Natural selection of extrafloral nectar production on *Chamaecrista fasciculate*: the costs and benefits of a mutualism trait. *Evolution*, 58, 2657–2668.

Schleuning, M., Frund, J. & Garcia, D. (2015). Predicting ecosystem functions from biodiversity and mutualistic networks: an extension of trait-based concepts to plant-animal interactions. *Ecography*, 38, 380–392.

Sodhi, N. S., Brook, B. W. & Bradshaw, C. J. A. (2009). Causes and Consequences of Species Extinctions. In *Princeton Guide to Ecology*, ed. Levin, S.A. Princeton: Princeton University Press, pp. 514–520.

Stuart, S. N., Wilson, E. O., McNeely, J. A., Mittermeier, R. A. & Rodriguez, J. P. (2010). Barometer of life: response. *Science*, 329, 141–142.

Thackerey, S. J., Henrys, P. A., Hemmeing, D. et al. (2016). Phenological sensitivity to climate across taxa and trophic levels. *Nature*, 535, 241–245.

Trager, M. D., Bhotika, S., Hostetler, J. A. et al. (2010). Benefits for plants in ant-plant protective mutualisms: a meta-analysis. *PLoS ONE*, 5, e14308.

Tylianakis, J. M., Didham, R. K., Bascompte, J. & Wardle, D. A. (2008). Global change and species interactions in terrestrial ecosystems. *Ecology Letters*, 11, 1351–1363.

Underwood, A. J. (1991). Beyond BACI: experimental designs for detecting human environmental impacts on temporal variations in natural populations. *Marine & Freshwater Research*, 42, 569–587.

United Nations (2015). https://esa.un.org/unpd/wpp/publications/files/key_findings_wpp_2015.pdf.

Valiente-Banuet, A., Aizen, M. A., Alcantara, J. M. et al. (2015) Beyond species loss: the extinction of ecological interactions in a changing world. *Functional Ecology*, 29, 299–307.

Venter, O., Sanderson, E. W., Magrach, A. et al. (2106). Sixteen years of change in the global terrestrial human footprint and implications for biodiversity conservation. *Nature Communications* 7, 12558, doi:10.1038/ncomms12558.

Watson, J. E. M., Shanahan, D. F., Di Marco, M. et al. (2016). Catastrophic declines in wilderness areas undermine global environment targets. *Current Biology*, 26, 1–6.

Wilson, E. O. (2016). *Half Earth: Our Planet's Fight for Life.* New York: Liveright Publishing Corporation.

Wilson, E. O. & Holldobler, B. (2005). The rise of the ants: a phylogenetic and ecological explanation. *PNAS*, 102, 7411–7414.

World Bank (2014). http://data.worldbank.org/indicator/ER.LND.PTLD.ZS.

Index